Luminescence Nanomaterials and Applications

Luminescence Nanomaterials and Applications

Editors

Wei Chen
Derong Cao

MDPI • Basel • Beijing • Wuhan • Barcelona • Belgrade • Manchester • Tokyo • Cluj • Tianjin

Editors
Wei Chen
The University of Texas at Arlington
USA

Derong Cao
South China University of Technology
China

Editorial Office
MDPI
St. Alban-Anlage 66
4052 Basel, Switzerland

This is a reprint of articles from the Special Issue published online in the open access journal *Nanomaterials* (ISSN 2079-4991) (available at: https://www.mdpi.com/journal/nanomaterials/special_issues/luminescent_nano_appl).

For citation purposes, cite each article independently as indicated on the article page online and as indicated below:

LastName, A.A.; LastName, B.B.; LastName, C.C. Article Title. *Journal Name* **Year**, *Volume Number*, Page Range.

ISBN 978-3-0365-7212-3 (Hbk)
ISBN 978-3-0365-7213-0 (PDF)

© 2023 by the authors. Articles in this book are Open Access and distributed under the Creative Commons Attribution (CC BY) license, which allows users to download, copy and build upon published articles, as long as the author and publisher are properly credited, which ensures maximum dissemination and a wider impact of our publications.

The book as a whole is distributed by MDPI under the terms and conditions of the Creative Commons license CC BY-NC-ND.

Contents

About the Editors vii

Wei Chen and Derong Cao
Luminescence Nanomaterials and Applications
Reprinted from: *Nanomaterials* 2023, 13, 1047, doi:10.3390/nano13061047 1

Cunjin Gao, Pengrui Zheng, Quanxiao Liu, Shuang Han, Dongli Li, Shiyong Luo, et al.
Recent Advances of Upconversion Nanomaterials in the Biological Field
Reprinted from: *Nanomaterials* 2021, 11, 2474, doi:10.3390/nano11102474 5

Mingkai Wang, Hanlin Wei, Shuai Wang, Chuanyu Hu and Qianqian Su
Dye Sensitization for Ultraviolet Upconversion Enhancement
Reprinted from: *Nanomaterials* 2021, 11, 3114, doi:10.3390/nano11113114 25

Zhihong Sun, Aaqib Khurshid, Muhammad Sohail, Weidong Qiu, Derong Cao and Shi-Jian Su
Encapsulation of Dyes in Luminescent Metal-Organic Frameworks for White Light Emitting Diodes
Reprinted from: *Nanomaterials* 2021, 11, 2761, doi:10.3390/nano11102761 37

Dangli Gao, Peng Wang, Feng Gao, William Nguyen and Wei Chen
Tuning Multicolor Emission of Manganese-Activated Gallogermanate Nanophosphors by Regulating Mn Ions Occupying Sites for Multiple Anti-Counterfeiting Application
Reprinted from: *Nanomaterials* 2022, 12, 2029, doi:10.3390/nano12122029 53

Huiyong Wang, Hongmei Yu, Ayman AL-Zubi, Xiuhui Zhu, Guochao Nie, Shaoyan Wang, et al.
Self-Matrix N-Doped Room Temperature Phosphorescent Carbon Dots Triggered by Visible and Ultraviolet Light Dual Modes
Reprinted from: *Nanomaterials* 2022, 12, 2210, doi:10.3390/nano12132210 67

Zhou Ding, Yue He, Hongtao Rao, Le Zhang, William Nguyen, Jingjing Wang, et al.
Novel Fluorescent Probe Based on Rare-Earth Doped Upconversion Nanomaterials and Its Applications in Early Cancer Detection
Reprinted from: *Nanomaterials* 2022, 12, 1787, doi:10.3390/nano12111787 81

Fanghui Ma, Qing Zhou, Minghui Yang, Jianglin Zhang and Xiang Chen
Microwave-Assisted Synthesis of Sulfur Quantum Dots for Detection of Alkaline Phosphatase Activity
Reprinted from: *Nanomaterials* 2022, 12, 2787, doi:10.3390/nano12162787 99

Hammam Abdurabu Thabit, Norlaili A. Kabir, Abd Khamim Ismail, Shoroog Alraddadi, Abdullah Bafaqeer and Muneer Aziz Saleh
Development of Ag-Doped ZnO Thin Films and Thermoluminescence (TLD) Characteristics for Radiation Technology
Reprinted from: *Nanomaterials* 2022, 12, 3068, doi:10.3390/nano12173068 109

Ya-Nan Hao, Cong-Cong Qu, Yang Shu, Jian-Hua Wang and Wei Chen
Construction of Novel Nanocomposites (Cu-MOF/GOD@HA) for Chemodynamic Therapy
Reprinted from: *Nanomaterials* 2021, 11, 1843, doi:10.3390/nano11071843 131

Weiwei Zhang, Zhao Kuang, Ping Song, Wanzhen Li, Lin Gui, Chuchu Tang, et al.
Synthesis of a Two-Dimensional Molybdenum Disulfide Nanosheet and Ultrasensitive Trapping of *Staphylococcus Aureus* for Enhanced Photothermal and Antibacterial Wound-Healing Therapy
Reprinted from: *Nanomaterials* **2022**, *12*, 1865, doi:10.3390/nano12111865 143

Oleg A. Yeshchenko, Nataliya V. Kutsevol, Anastasiya V. Tomchuk, Pavlo S. Khort, Pavlo A. Virych, Vasyl A. Chumachenko, et al.
Thermoresponsive Zinc TetraPhenylPorphyrin Photosensitizer/Dextran Graft Poly(N-IsoPropylAcrylAmide) Copolymer/Au Nanoparticles Hybrid Nanosystem: Potential for Photodynamic Therapy Applications
Reprinted from: *Nanomaterials* **2022**, *12*, 2655, doi:10.3390/nano12152655 161

Radek Ostruszka, Giorgio Zoppellaro, Ondřej Tomanec, Dominik Pinkas, Vlada Filimonenko and Karolína Šišková
Evidence of Au(II) and Au(0) States in Bovine Serum Albumin-Au Nanoclusters Revealed by CW-EPR/LEPR and Peculiarities in HR-TEM/STEM Imaging
Reprinted from: *Nanomaterials* **2022**, *12*, 1425, doi:10.3390/nano12091425 181

Haibin Li, Xiang Luo, Ziwen Long, Guoyou Huang and Ligang Zhu
Plasmonic Ag Nanoparticle-Loaded n-p $Bi_2O_2CO_3$/α-Bi_2O_3 Heterojunction Microtubes with Enhanced Visible-Light-Driven Photocatalytic Activity
Reprinted from: *Nanomaterials* **2022**, *12*, 1608, doi:10.3390/nano12091608 199

Jin Huang, Zhen Chu, Christina Xing, Wenting Li, Zhongxin Liu and Wei Chen
Luminescence Reduced Graphene Oxide Based Photothermal Purification of Seawater for Drinkable Purpose
Reprinted from: *Nanomaterials* **2022**, *12*, 1622, doi:10.3390/nano12101622 215

Zhian Xu, Liang Xiao, Xuetao Fan, Dongtao Lin, Liting Ma, Guochao Nie, et al.
Spray-Assisted Interfacial Polymerization to Form $Cu^{II/I}$@CMC-PANI Film: An Efficient Dip Catalyst for A^3 Reaction
Reprinted from: *Nanomaterials* **2022**, *12*, 1641, doi:10.3390/nano12101641 231

About the Editors

Wei Chen

Dr. Wei Chen is a professor in the department of physics. He has been engaged in cutting-edge nanotechnology research for many years, and is an internationally renowned expert in nanomedicine and cancer nanotechnology. At present, he has published more than 330 articles in internationally renowned scientific journals, directed the preparation of a monograph (three volumes), and is in the process of publishing two books. His papers have been cited more than 14,900 times, with an H index of 62, and the highest single paper of these has been cited 729 times. He has obtained 22 US patents, and his scientific research has received widespread attention, as Chen has been interviewed and reported on by the US TV program CBS. He proposed the concept of "Nanoparticle Self-lighting Photodynamic Therapy" to treat deep cancer, and invented the fourth-generation photosensitizer, copper cysteamine. This new kind of photosensitizer can generate reactive oxygen species in ultraviolet light, X-rays, microwaves, and ultrasound for the treatment of cancer and infectious diseases. In 2021, Professor Chen was elected as a Fellow of the Royal Society of Chemistry. In 2022, the International Association of Advanced Materials awarded Professor Wei Chen the Distinguished Scientist Award to recognize his outstanding contributions to nanotechnology. In the same year, he was elected as a fellow for the National Academy of Inventors, and in 2023 he was elected as a member of the University Academy of Distinguished Scholars.

Derong Cao

Derong Cao, Ph.D., is a professor in organic chemistry in the School of Chemistry and Chemical Engineering at the South China University of Technology, Guangzhou, China (since 2005). His research focuses on studying the synthesis and applications of organic functional compounds/materials, in particular in the syntheses of pillararenes, photovoltaic materials, and luminescent materials and devices. He received a master's degree from Lanzhou University under the supervision of Prof. Zhixing Su (1986) and obtained a Ph.D. in organic chemistry from the University of Mainz/Lanzhou University under the supervision of Prof. Herbert Meier and Prof. Yulin Li. His previous academic positions were at Lanzhou University as a teacher (1986-1994); at the Shanghai Institute of Organic Chemistry, Chinese Academy of Sciences, as a postdoctoral fellow (1997-1999); and at Guangzhou Institute of Chemistry, Chinese Academy of Sciences, as a professor (2000-2005). He has published more than 300 papers.

Editorial

Luminescence Nanomaterials and Applications

Wei Chen [1,*] and Derong Cao [2,*]

1 Department of Physics, The University of Texas at Arlington, Arlington, TX 76019-0059, USA
2 Department of Chemistry, South China University of Technology, Guangzhou 510641, China
* Correspondence: weichen@uta.edu (W.C.); drcao@scut.edu.cn (D.C.)

1. Introduction

We are pleased to introduce to you this Special Issue of *Nanomaterials* on 'Luminescence Nanomaterials and Applications'. Luminescence is a phenomenon that we experience daily in our work and lives. Emerging nanotechnology and quantum dots offer a new class of materials to make our lives better. For example, the applications of luminescent nanoparticles for medical research are taking advantage of their high quantum yield, multi-colors, high photostability, large surface-to-volume ratio, surface functionality, and small size. For solid-state displays, they can provide more colors by simply adjusting the size of the particles. In this Special Issue, we discuss nanomaterials with quantum size confinement, photoluminescence, upconversion, thermoluminescence, and long persistence, as well as their potential applications in cell labeling, imaging, detection, and sensing. This Special Issue also covers the synthesis of luminescence nanomaterials, applications for in vitro and in vivo imaging, detection based on fluorescence resonance energy, and the applications of luminescent nanoparticles for photodynamic activation and solid-state displays, as well as new materials and structures, such as perovskite quantum dots, and novel phenomena, such as aggregation-induced emissions. In total, we have 15 papers (2 reviews and 13 research articles) for this Special Issue, which are summarized below:

2. Upconversion Luminescence Nanomaterials

Upconversion nanomaterials can emit high-energy lights when excited with two or more low-energy photons. They can produce ultraviolet (UV)-visible or near-infrared (NIR) light upon excitation with NIR light, depending on size or dopants, owing to their unique properties, such as good optical stability, narrow emission band, large anti-Stokes spectral shift, high levels of light penetration in biological tissues, long luminescence lifetime, and high signal-to-noise ratio. The review paper by Dr. Jigang Wang and his collaborators systematically introduced the physical mechanism of upconversion luminescence nanomaterials and their potential applications in bioimaging, detection, photodynamic therapy, and therapeutics [1].

Upconversion nanocrystals converting near-infrared light into high-energy UV emissions may provide many exciting opportunities for drug release, photocatalysis, photodynamic therapy, and solid-state lasing. However, a key challenge is their low conversion efficiency. For that, Dr. Chuanyu Hu's team [2] proposed and developed dye-sensitized and heterogeneous core–shell lanthanide nanostructures for ultraviolet upconversion improvement. They systematically investigated the main factors on ultraviolet upconversion emission. Interestingly, they found a method for a largely promoted multiphoton upconversion, which provides more opportunities for applications in biomedicine, photo-catalysis, and environmental science.

3. Luminescence for Solid-State Lighting, Displays, and Anti-Counterfeiting

White-light-emitting diodes show great promise for replacing traditional lighting devices because of their high efficiency, low energy consumption, and long lifetime. Metal

Citation: Chen, W.; Cao, D. Luminescence Nanomaterials and Applications. *Nanomaterials* 2023, 13, 1047. https://doi.org/10.3390/nano13061047

Received: 24 February 2023
Accepted: 7 March 2023
Published: 14 March 2023

Copyright: © 2023 by the authors. Licensee MDPI, Basel, Switzerland. This article is an open access article distributed under the terms and conditions of the Creative Commons Attribution (CC BY) license (https://creativecommons.org/licenses/by/4.0/).

organic frameworks (MOFs) are good materials for white-light emissions. The encapsulation of organic dyes is a simple way to obtain luminescent MOFs. In a review, Dr. Derong Cao and his collaboration team summarized the recent research on the design and construction of dye-encapsulated MOFs phosphors and their potential applications [3].

Dr. Dangli Gao and her collaboration team reported the optical properties of $Zn_3Ga_2GeO_8$:Mn phosphors that could be modified by different preparation methods, including a hydrothermal method and solid-state diffusion combined with a non-equivalent ion-doping strategy [4]. Consequently, Mn-doped $Zn_3Ga_2GeO_8$ phosphors prepared by a hydrothermal method showed an enhanced red emission at 701 nm and a green persistent luminescence at 540 nm, while the phosphors prepared by solid-state diffusion in combination with hetero-valent doping only exhibited an enhancement in the single-band red emission. Furthermore, the substitution of hetero-valent dopant ion Li^+ into different sites can change the emission colors. These fantastic phenomena were discussed in detail in the paper [4].

The study of room-temperature phosphorescent carbon quantum dots is important for various applications. Dr. Hongmei Yu et al. [5] successfully fabricated matrix-free carbonized polymer dots (CPDs) that can produce green room-temperature phosphorescence under dual-mode visible- and ultraviolet-light excitations. Hydrogen bonding can provide a space protection and stably excite the triplet state. This self-matrix structure effectively avoids the non-radiative transition by blocking the intramolecular motion of CPDs. The long lasting room-temperature phosphorescence is good for applications in anti-counterfeiting.

4. Particle Based Sensing Technology

Early cancer detection is important, and plenty of sensors are being explored. Dr. Liu and her collaboration team reported novel lanthanide-upconverted $NaYF_4$:Yb,Tm fluorescence probes, which can detect cancer-related specific miRNAs in very low concentrations [6]. The detection is based on emissions at 345, 646, and 802 nm upon excitation at 980 nm. The two common proteins, miRNA-155 and miRNA-150, were captured by the designed fluorescent probes. The probes can effectively distinguish miRNA-155 from partial- and complete-base mismatched miRNA-155, which is critical for early cancer detection.

Sulfur quantum dots (SQDs) are considered potential green nanomaterials because they have no heavy metals. Dr. Yang [7] and his collaboration team prepared SQDs by a microwave-assisted method using sulfur powder as a precursor. SQDs show the highest emission at 470 nm when excited at 380 nm and have a good sensitivity and selectivity in alkaline phosphatase detection.

Radiation detection and dosimetry are old questions but pose new challenges for security and safety. As such, Dr. Abd Khamim Ismail et al. [8] examined the thermoluminescence dosimetry behaviors of Ag-ZnO films. The dose–response revealed high linearity up to 4 Gy. The proposed sensitivity was 1.8 times higher than the TLD 100 dosimeters.

5. Particle-Based Therapeutics

Chemodynamic therapy (CDT) has received extensive research attention in recent years. However, the efficiency of CDT is influenced by H_2O_2 limitations in the tumor. In this issue, Dr. Yang Shu and her collaborators [9] described a novel core–shell nanostructure, namely a Cu-metal organic framework (Cu-MOF)/glucose oxidase (GOD)@hyaluronic acid (HA) (Cu-MOF/GOD@HA), for the purpose of improving CDT efficacy by increasing H_2O_2 concentration and cancer cell targeting. The CDT enhancement as a result of GOD and HA effects in Cu-MOF/GOD@HA was confirmed for both in vitro cell and in vivo animal studies.

Photothermal therapy has been widely tested in treating bacterial infections [10,11]. Weiwei Zhang et al. [12] tested the growth inhibition of Staphylococcus aureus by using a very low concentration of vancomycin and applying photothermal therapy with MoS_2.

MoS$_2$-Van-FITC with near-infrared irradiation significantly inhibited S. aureus growth, reaching an inhibition rate of 94.5%, indicating its possible use as a wound-healing agent.

Dr. Oleg A. Yeshchenko et al. [13] presented the thermoresponsive Zinc-TetraPhenyl Porphyrin-based hybrid nanosystem. The shrinking of D-g-PNIPAM macromolecules during a thermally induced phase transition leads to the release of both ZnTPP molecules and Au NPs from the ZnTPP/D-g-PNIPAM/AuNPs macromolecule. The three-fold enhancement of singlet oxygen production with surface plasmon resonance is critical for clinic applications.

6. New Materials and Structures

Bovine-serum-albumin-embedded Au nanoclusters are thoroughly investigated by Radek Ostruszka [14] using continuous-wave electron paramagnetic resonance, light-induced EPR, etc. In addition to the presence of Au(0) and Au(I) oxidation states in BSA-AuNCs, a significant amount of Au(II) was detected, which may come from a disproportionation event occurring within NCs: 2Au(I) − Au(II) + Au(0).

Haibin Li et al. [15] reported on n-p Bi$_2$O$_2$CO$_3$/α-Bi$_2$O$_3$ heterojunction microtubes and studied their photocatalytic activities under visible-light irradiation. The results indicated that Bi$_2$O$_2$CO$_3$/α-Bi$_2$O$_3$ with a Bi$_2$O$_2$CO$_3$ mass fraction of 6.1% exhibited higher photocatalytic activity than α-Bi$_2$O$_3$.

Obtaining drinking water from seawater has always been a long-term goal. Here, Dr. Zhongxin Liu et al. [16] reported on the use of graphene-loaded nonwoven fabric membranes coated with graphene oxide for seawater purification. The photothermal membrane is expected to be suitable for regional water purification and seawater desalination due to its high light absorption, strong heating effect, and its evaporation rate, which is about five times higher than that of non-woven fabric.

In this issue Dr. Yiqun Li et al. [17] synthesized a novel carboxymethylcellulose–polyaniline-film-supported copper catalyst (CuII/I@CMC-PANI) and used it as a dip catalyst for aldehyde–alkyne–amine coupling reactions with a high yield of 97%. They found that CuII/I@CMC-PANI, as a good dip catalyst, is very useful in organic synthesis due to its easy fabrication, convenient deployment, superior catalytic activity, and high reusability.

We would like to thank the *Nanomaterials* Editorial Office for the opportunity to edit this Special Issue, as well as all the authors for their valuable contributions and reviewers for their valuable comments. This Special Issue would not have been possible without them. We hope that this Special Issue can offer some valuable information and guidance for future research directions.

Conflicts of Interest: The authors declare no conflict of interest.

References

1. Gao, C.; Zheng, P.; Liu, Q.; Han, S.; Wang, J.; Li, D.; Luo, S.; Temple, H.; Xing, C.; Wei, Y.; et al. Recent Advances of Upconversion Nanomaterials in the Biological Field. *Nanomaterials* **2021**, *11*, 2474. [CrossRef] [PubMed]
2. Wang, M.; Wei, H.; Wang, S.; Hu, C.; Su, Q. Dye Sensitization for Ultraviolet Upconversion Enhancement. *Nanomaterials* **2021**, *11*, 3114. [CrossRef] [PubMed]
3. Sun, Z.; Khurshid, A.; Sohail, M.; Qiu, W.; Cao, D.; Su, S.-J. Encapsulation of Dyes in Luminescent Metal-Organic Frameworks for White Light Emitting Diodes. *Nanomaterials* **2021**, *11*, 2761. [CrossRef] [PubMed]
4. Gao, D.; Wang, P.; Gao, F.; Nguyen, W.; Chen, W. Tuning Multicolor Emission of Manganese-Activated Gallogermanate Nanophosphors by Regulating Mn Ions Occupying Sites for Multiple Anti-Counterfeiting Application. *Nanomaterials* **2022**, *12*, 2029. [CrossRef] [PubMed]
5. Wang, H.; Yu, H.; AL-Zubi, A.; Zhu, X.; Nie, G.; Wang, S.; Chen, W. Self-Matrix N-Doped Room Temperature Phosphorescent Carbon Dots Triggered by Visible and Ultraviolet Light Dual Modes. *Nanomaterials* **2022**, *12*, 2210. [CrossRef] [PubMed]
6. Ding, Z.; He, Y.; Rao, H.; Zhang, L.; Nguyen, W.; Wang, J.; Wu, Y.; Han, C.; Xing, C.; Yan, C.; et al. Novel Fluorescent Probe Based on Rare-Earth Doped Upconversion Nanomaterials and Its Applications in Early Cancer Detection. *Nanomaterials* **2022**, *12*, 1787. [CrossRef] [PubMed]
7. Ma, F.; Zhou, Q.; Yang, M.; Zhang, J.; Chen, X. Microwave-Assisted Synthesis of Sulfur Quantum Dots for Detection of Alkaline Phosphatase Activity. *Nanomaterials* **2022**, *12*, 2787. [CrossRef] [PubMed]

8. Thabit, H.A.; Kabir, N.A.; Ismail, A.K.; Alraddadi, S.; Bafaqeer, A.; Saleh, M.A. Development of Ag-Doped ZnO Thin Films and Thermoluminescence (TLD) Characteristics for Radiation Technology. *Nanomaterials* **2022**, *12*, 3068. [CrossRef] [PubMed]
9. Hao, Y.-N.; Qu, C.-C.; Shu, Y.; Wang, J.-H.; Chen, W. Construction of Novel Nanocomposites (Cu-MOF/GOD@HA) for Chemodynamic Therapy. *Nanomaterials* **2021**, *11*, 1843. [CrossRef] [PubMed]
10. Zhen, X.; Chudal, L.; Pandey, N.K.; Phan, J.; Ran, X.; Amador, E.; Huang, X.; Johnson, O.; Ran, Y.; Chen, W.; et al. A Powerful Combination of Copper-Cysteamine Nanoparticles with Potassium Iodide for Bacterial Destruction. *Mater. Sci. Eng. C* **2020**, *110*, 110659. [CrossRef] [PubMed]
11. Zhen, X.; Pandey, N.K.; Amador, E.; Hu, W.; Liu, B.; Nong, W.; Chen, W.; Huang, L. Potassium Iodide Enhances the Anti-Hepatocellular Carcinoma Effect of Copper-Cysteamine Nanoparticle Mediated Photodynamic Therapy on Cancer Treatment. *Mater. Today Phys.* **2022**, *27*, 100838. [CrossRef]
12. Zhang, W.; Kuang, Z.; Song, P.; Li, W.; Gui, L.; Tang, C.; Tao, Y.; Ge, F.; Zhu, L. Synthesis of a Two-Dimensional Molybdenum Disulfide Nanosheet and Ultrasensitive Trapping of Staphylococcus Aureus for Enhanced Photothermal and Antibacterial Wound-Healing Therapy. *Nanomaterials* **2022**, *12*, 1865. [CrossRef] [PubMed]
13. Yeshchenko, O.A.; Kutsevol, N.V.; Tomchuk, A.V.; Khort, P.S.; Virych, P.A.; Chumachenko, V.A.; Kuziv, Y.I.; Marinin, A.I.; Cheng, L.; Nie, G. Thermoresponsive Zinc TetraPhenylPorphyrin Photosensitizer/Dextran Graft Poly(N-IsoPropylAcrylAmide) Copolymer/Au Nanoparticles Hybrid Nanosystem: Potential for Photodynamic Therapy Applications. *Nanomaterials* **2022**, *12*, 2655. [CrossRef] [PubMed]
14. Ostruszka, R.; Zoppellaro, G.; Tomanec, O.; Pinkas, D.; Filimonenko, V.; Šišková, K. Evidence of Au(II) and Au(0) States in Bovine Serum Albumin-Au Nanoclusters Revealed by CW-EPR/LEPR and Peculiarities in HR-TEM/STEM Imaging. *Nanomaterials* **2022**, *12*, 1425. [CrossRef] [PubMed]
15. Li, H.; Luo, X.; Long, Z.; Huang, G.; Zhu, L. Plasmonic Ag Nanoparticle-Loaded n-p $Bi_2O_2CO_3$/-Bi_2O_3 Heterojunction Microtubes with Enhanced Visible-Light-Driven Photocatalytic Activity. *Nanomaterials* **2022**, *12*, 1608. [CrossRef] [PubMed]
16. Huang, J.; Chu, Z.; Xing, C.; Li, W.; Liu, Z.; Chen, W. Luminescence Reduced Graphene Oxide Based Photothermal Purification of Seawater for Drinkable Purpose. *Nanomaterials* **2022**, *12*, 1622. [CrossRef] [PubMed]
17. Xu, Z.; Xiao, L.; Fan, X.; Lin, D.; Ma, L.; Nie, G.; Li, Y. Spray-Assisted Interfacial Polymerization to Form CuII/I@CMC-PANI Film: An Efficient Dip Catalyst for A3 Reaction. *Nanomaterials* **2022**, *12*, 1641. [CrossRef]

Disclaimer/Publisher's Note: The statements, opinions and data contained in all publications are solely those of the individual author(s) and contributor(s) and not of MDPI and/or the editor(s). MDPI and/or the editor(s) disclaim responsibility for any injury to people or property resulting from any ideas, methods, instructions or products referred to in the content.

Review

Recent Advances of Upconversion Nanomaterials in the Biological Field

Cunjin Gao [1], Pengrui Zheng [1], Quanxiao Liu [1], Shuang Han [1], Dongli Li [1], Shiyong Luo [1], Hunter Temple [2], Christina Xing [2], Jigang Wang [1,*], Yanling Wei [3,*], Tao Jiang [4,*] and Wei Chen [2,5,*]

[1] Beijing Key Laboratory of Printing and Packaging Materials and Technology, Beijing Institute of Graphic Communication, Beijing 102600, China; gcj2019015220@163.com (C.G.); zhengpengrui9999@126.com (P.Z.); drllqx@163.com (Q.L.); 13552797385@163.com (S.H.); lidongli@bigc.edu.cn (D.L.); luoshiyong@bigc.edu.cn (S.L.)
[2] Department of Physics, The University of Texas at Arlington, Arlington, TX 76019-0059, USA; hunter.temple@uta.edu (H.T.); Christina.Xing@uta.edu (C.X.)
[3] Faculty of Applied Sciences, Jilin Engineering Normal University, Changchun 130052, China
[4] CAS Center for Excellence in Nanoscience, Beijing Key Laboratory of Micro-Nano Energy and Sensor, Beijing Institute of Nanoenergy and Nanosystems, Chinese Academy of Sciences, Beijing 101400, China
[5] Medical Technology Research Centre, Chelmsford Campus, Anglia Ruskin University, Chelmsford CM1 1SQ, UK
* Correspondence: jigangwang@bigc.edu.cn (J.W.); weiyanling@jlenu.edu.cn (Y.W.); jiangtao@binn.cas.cn (T.J.); weichen@uta.edu (W.C.)

Abstract: Rare Earth Upconversion nanoparticles (UCNPs) are a type of material that emits high-energy photons by absorbing two or more low-energy photons caused by the anti-stokes process. It can emit ultraviolet (UV) visible light or near-infrared (NIR) luminescence upon NIR light excitation. Due to its excellent physical and chemical properties, including exceptional optical stability, narrow emission band, enormous Anti-Stokes spectral shift, high light penetration in biological tissues, long luminescent lifetime, and a high signal-to-noise ratio, it shows a prodigious application potential for bio-imaging and photodynamic therapy. This paper will briefly introduce the physical mechanism of upconversion luminescence (UCL) and focus on their research progress and achievements in bio-imaging, bio-detection, and photodynamic therapy.

Keywords: UCNPs; UCL; RE-doped; bio-imaging; PDT; bio-detection

1. Introduction

Luminescent materials are functional materials that absorb incident energy and subsequently emit incident energy in photons. Most luminescent materials, including organic dyes and quantum dots (QDs) [1], follow Stokes Law. They absorb high-energy photons and emit low-energy photons (energy reduction), known as downconversion materials [2–7]. On the contrary, UCNPs relate to the conversion of long-wavelength NIR light with lower energy into ultraviolet or visible light with higher energy. This process is also known as the anti-stokes process, which is so-called the UCL phenomenon.

When almost all the trivalent rare earth ions (RE^{3+}) are excited externally, the rare earth elements of 15 lanthanides, yttrium, and scandium can move between different energy levels, because they have a unique 4f electron configuration energy level, such as shown in Figure 1. In addition, due to the shielding of the secondary $5s^25p^6$ shell, the external environment has little effect on rare earth ions [8–11]. Under near infrared (NIR) laser irradiation, typical lanthanide-doped upconversion nanoparticles exhibit anti-Stokes shifted visible light and ultraviolet light emission, while the autofluorescence background is minimal, and the light scattering of biological tissues is greatly reduced. The reduction in light scattering caused by this near-infrared excitation will make the penetration depth of biological tissues far greater than the penetration depth under ultraviolet or visible light excitation, which makes UCNP have great potential in biological applications [12–21], including biological tissues imaging [22–24], biological detection [25–30], photodynamic therapy, and other biomedical fields [31–36].

In this review, we briefly describe the luminescence mechanism and synthetic methods of UCNPs. However, the surface of UCNPs obtained through conventional surface modification methods lacks the active groups that are binding to biomolecules, which limits their application in the biological field to a certain extent. Here, we introduce several surface modification methods and systematically discuss the latest progress of UCNPS in biological imaging and photodynamic therapy.

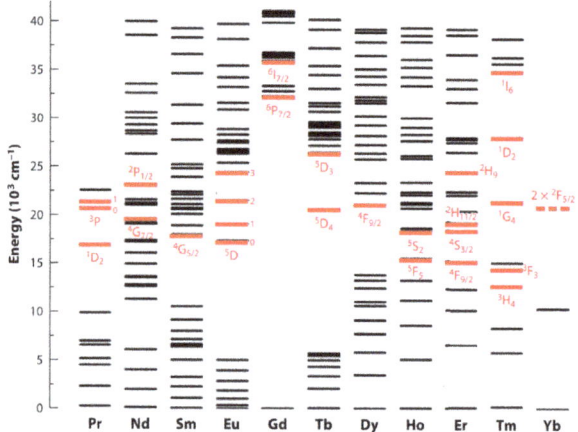

Figure 1. Energy-level diagrams of rare-earth ions. Typical upconversion emissive excited states are highlighted by a red bold line. Reprinted with permission from ref. [11]. Copyright 2015 Annual Reviews.

2. Mechanism of Upconversion

In general, UCNPs are composed of inorganic crystal matrix and rare-earth ions, which do not constitute the light-emitting energy level. The main function is to provide the lattice structure of rare-earth ions to ensure they have appropriate luminescent conditions. According to current research, there are five upconversion mechanisms: excited-state absorption (ESA), energy transfer upconversion (ETU), cooperative sensitization upconversion (CSU), cross-relaxation (CR), and photon avalanche (PA) [37,38].

2.1. Excited State Absorption (ESA)

Excited state absorption (ESA) occurs in the form of continuous absorption of pump photons by a single ion through the trapezoidal structure of a simple multistage system. As shown in Figure 2a, it is realized by a three-level system that continuously absorbs two photons. This mechanism is due to the equal separation degree of E1 and E2, E2 and E3, and the storage capacity of intermediate E2. When an ion is excited from the ground state to the E2 level, because the lifetime of the E2 level is very long, the other pump photon is likely to absorb another photon before it decays to the ground state, so as to promote the ion from the E2 level to the higher E3 level, resulting in the upconversion emission of the E3 level. In order to meet the above process and achieve efficient ESA, the energy states of lanthanide elements need to be arranged in a trapezoid, and only a few lanthanide ions, such as Er^{3+}, Ho^{3+}, Tm^{3+} and Nd^{3+} have such energy level structure [39]. In addition, it is found that the output wavelength of commercial diode lasers (about 975 nm and/or 808 nm) can well match the excitation wavelength of these energy level structures.

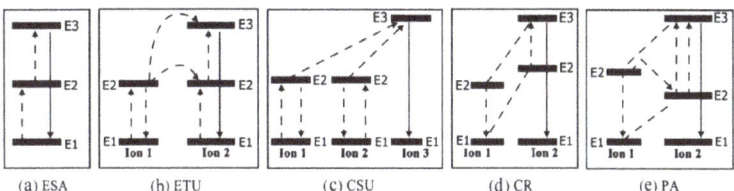

Figure 2. Upconversion mechanisms of RE-doped UCNPs: (**a**) ESA, (**b**) ETU, (**c**) CSU, (**d**) CR and (**e**) PA.

2.2. Energy Transfer Upconversion (ETU)

Unlike ESA, Energy transfer upconversion (ETU) involves two identical or different ions. In this photophysical process, ion 1 transitions from the ground state to E2 by absorbing excitation light; then the absorbed photon energy is transferred to the ground state E1 and excited state E2 of ion 2, so that they are excited again, while ion 1 relaxes back to ground state E1. In this process, the quantum yield of upconversion largely depends on the average distance between ion 1 and ion 2, and is mainly determined by the concentration of doped rare earth ions. The ETU process is the most important photophysical process in upconversion luminescence, because so far, the use of upconversion nanoparticles for therapeutic diagnostics and other applications is through the sensitizer Yb^{3+}, which has a strong absorption of the excitation light at 975 nm, thus making activators (Er^{3+}, etc.) produce more efficient fluorescence emission [40–45].

Moreover, the scattering and absorption of biological tissues at 975 nm are relatively small, and no optical interference occurs. Here, Yb^{3+} has a very good effect as a sensitizer because it has a sufficiently large absorption cross-section in the near-infrared region of about 975 nm. In addition, since Yb^{3+} has only two energy levels, its optimal concentration can be maintained at a high level (20–100% for fluorinated nanoparticles) without causing harmful cross-relaxation. To date, most research has focused on the development of Yb^{3+} sensitized upconversion nanoparticles pumped at about 975 nm. Using lanthanide ions themselves as sensitizers, high-efficiency ETU can also be observed in single lanthanide doping systems, for example, long-wavelength 1490 nm excitation of Er^{3+}-doped $LiYF_4$ [46]; or doping under 1200 nm excitation $NaGdF_4$ nanoparticles doped with Ho^{3+} [47]. The use of other sensitizers can be used to quench and enhance the luminous intensity of certain emission bands. For example, Nd^{3+}, Ce^{3+}, and Ho^{3+} are used as sensitizers to enhance the blue emission band of Tm^{3+}, the red emission band of Ho^{3+}, and the near-infrared emission band of Tm^{3+}, respectively [48–51].

2.3. Cooperative Sensitization Upconversion (CSU)

Cooperative sensitization upconversion (CSU) in Figure 2c is a photophysical process of the interaction of three rare earth ions (two types). Ion 1 and ion 3 usually belong to the same sensitizer, such as Yb^{3+}. After being excited by light, ion 1 and ion 3 transition to an excited state. However, ion 1 or ion 3 alone cannot excite ion 2, because their excited state energy levels are quite different, so ion 1 and ion 3 need to be co-excited to produce a virtual excited state energy level that can be compared with the excited state energy level of ion 2. The ion 2 absorbs the energy of cooperative sensitization and emits a higher energy photon. The photophysical process of CSU is often uncommon, because the energy transfer efficiency is low, and the para-virtual pair energy levels in the transfer process are involved. These energy levels must be described by quantum mechanics in higher disturbances. Nevertheless, limiting the excitation to compensate for the low efficiency provides a possibility to achieve high-resolution imaging, which is not possible with other upconversion mechanisms. At present, the CSU mechanism of Yb^{3+}/Tb^{3+} [52], Yb^{3+}/Eu^{3+} [53], Yb^{3+}/Pr^{3+} [54] ion pairs has been reported.

2.4. Cross-Relaxation (CR)

The CR process involves two identical or different ions. In the CR process, ion 1 transfers part of the energy of the E2 energy level to ion 2, causing ion 2 to transfer to a higher excited state, while ion 1 returns to a lower energy level through a non-radiative relaxation process (Figure 2d). Although CR is related to the concentration quenching effect, it can be used to adjust the emission spectra of UCNPs. The related studies currently reported include Y_2O_3: Yb^{3+}/Er^{3+} [55], $GdPO_4$:Sm^{3+} [56], KYF_4:Tb^{3+}, Yb^{3+} [57], etc.

2.5. Photon Avalanche (PA)

PA includes ESA process and CR process. Photon avalanche (PA) in Figure 2e is a process of upconversion above a certain excitation power threshold. Once excited, the upconversion fluorescence luminescence intensity will increase by orders of magnitude. In addition, the excited state energy levels of rare earth ions are also required to have a relatively high lifetime. In this process, the CR process mentioned above is also required. Ion 2 transitions from E2 back to E1 and releases energy at the same time. Ion 1 absorbs the energy and transitions from E1 to E2, and then transfers the energy to the E1 energy level of ion 2. At this time, the E1 energy level of ion 2 will absorb photons through the ESA process, which increases the population of the E1 energy level exponentially, leading to the PA process. This photophysical process is often uncommon because of the higher excitation power density required. Er^{3+}/Yb^{3+} co-doped $NaBi(WO_4)_2$ phosphor produces strong green upconversion luminescence through the photon avalanche process [58]. Under excitation at 980 nm, the seven-photon PA upconversion (UC) behavior of Er^{3+} ions and four-photon NIR emission [59].

3. Synthesis Strategy and Surface Modification of UCNPs

The luminescence properties of upconversion nanomaterials are closely related to their preparation methods. Different preparation methods will affect the size, morphology, and corresponding microstructure of luminescent materials, making their application directions more diversified. Among them, thermal decomposition, hydrothermal decomposition, and co-precipitation are the three most commonly used methods, and their advantages and limitations are shown in Table 1. Other synthesis methods, including sol-gel method and combustion method, are also discussed for comparison [60–63].

Table 1. Advantages and disadvantages of upconversion nanoparticles synthesized by different methods and examples.

Method	Advantages	Disadvantages	Examples
Thermal Decomposition	Large product volume; small size distribution	The equipment is expensive; the precursor is sensitive to air; toxic by-products	ReF_3 (Re = Y,La) [64] $NaLuF_4$ [65] $NaYbF_4$ [65] MF_2 (M = Ca,Sr,Ba) [66] $LiYF_4$ [67] $NaGdF_4$ [68] $BaREF_5$ (RE = Y,Gd) [69] KY_3F_{10} [70] RE_2O_3 (RE = Y,La,Gd) [71] REOF (RE = Y,La,Gd) [72]
Hydrothermal Decomposition	Inexpensive precursors; no need for post-processing; precise size and shape control	Need an autoclave; the reaction process is unobservable and uncontrollable	REF_3 (RE = Y,La,Ce,Gd) [73] MF_2 (M = Ca,Sr,Ba) [74] $NaYF_4$ [75] $NaLaF_4$ [76] $NaLuF_4$ [77] $KMnF_4$ [78] $BaGdF_5$ [79] RE_2O_3 (RE = Y,Gd,Er) [80] LaOF [81] $REPO_4$ (RE = Ga,Yb,Lu) [82]

Table 1. Cont.

Method	Advantages	Disadvantages	Examples
Co-precipitation	Fast synthesis speed; inexpensive equipment and safe precursors	Need post-processing	NaYF$_4$ [83] NaGdF$_4$ [84] NaTbF$_4$ [85] NaLuF$_4$ [86] KGdF$_4$ [87] CaF$_2$ [88] LaF$_3$ [89] NaScF$_4$ [90]
Sol-gel Method	Inexpensive precursors; small product size	The precursor preparation process is complicated; product is easy to reunite	BaIn$_2$O$_4$:Yb^{3+}/Tm^{3+}/RE^{3+} (RE = Er^{3+}, Ho^{3+}) [91] NaPbLa (MoO$_4$)$_3$: Er^{3+}/Yb^{3+} [92] La$_4$Ti$_3$O$_{12}$ [93] Gd$_2$O$_3$: Er^{3+}/Yb^{3+}/Bi^{3+} [94] CaTi$_4$O$_9$: Er^{3+}/Yb^{3+} [95]
Combustion Method	Fast synthesis speed; energy saving; controllable product quantity	Expensive equipment; high temperature; the particle size of the material is large and easy to agglomerate	Ba$_5$ (PO$_4$)$_3$OH: Er^{3+}/Yb^{3+} [96] Na$_3$Y (PO$_4$)$_2$: Er^{3+}/Yb^{3+} [97] BaLaAlO$_4$:Er^{3+}/Yb^{3+} [98] ZrO$_2$: Ho^{3+}/Yb^{3+} [99] LaO$_3$: Er^{3+}/Tm^{3+} [100]

3.1. Thermal Decomposition Method

The thermal decomposition method is based on the high-temperature decomposition of organometallic precursors (such as metal trifluoroacetate) in high-boiling organic solvents (such as 1-octadecene). Surfactants are long-chain hydrocarbons and functional groups, such as -COOH, -NH$_2$ or -PO$_3$H (such as oleic acid, oleyl amine, trioctyl phosphine and trioctyl phosphine oxide), which are used as ligands, thus preventing the aggregation of nanoparticles. This method was originally developed by Professor Yan and others for the synthesis of LaF$_3$ nanoparticles [64], and later extended to the synthesis of high-quality NaYF$_4$ upconversion nanoparticles [65]. Capobianco et al. used trifluoroacetate precursors to synthesize Yb, Er and Yb, Tm co-doped NaYF$_4$ nanoparticles [101,102]. The asymmetric molecular structure of octadecene (with a high boiling point of 315 °C) is used as a solvent, and oleic acid is used as a passivation ligand. Based on the separation of nucleation and growth of nanocrystals, an improved method has been developed [100], that is, the precursor is slowly added, followed by heating, to synthesize highly monodispersed nanoparticles. Using the same strategy, Murray and colleagues prepared hexagonal NaYF$_4$:Yb,Er nanoparticle bodies, and precisely controlled their morphology and size [103]. Lim et al. The NaGdF$_4$:Er^{3+}/Yb^{3+} colloidal particles with an average particle size of 32 nm were successfully synthesized by the thermal decomposition method. The prepared phosphor shows bright green upconversion luminescence under a 976 nm semiconductor laser. Phosphor particles increase the scattering of optical coherence tomography (OCT) scanning radiation, thereby observing higher image contrast [104]. So far, this method has been widely used to synthesize a series of upconversion fluoride and oxyfluoride nanoparticles (Table 1). Although this method can produce high-quality upconversion nanoparticles, it also has some disadvantages, including expensive materials, air-sensitive precursors, and the production of toxic byproducts (such as hydrogen fluoride). Recently, the Pu team reported a safe and environmentally friendly method to replace octadecene (ODE) with paraffin liquid as a high-boiling non-coordinating solvent [105]. This method is biologically cheaper and sustainable.

3.2. Hydrothermal Method

Hydrothermal synthesis is another cheap method to obtain high-quality nanocrystals. This process is usually carried out at high pressure and high temperature above the boiling point or even the critical point. A special container called autoclave is used. The main disadvantages are the opacity of the reactor and the lack of in-situ reaction process and mechanism research. On the other hand, the advantages of high crystallinity, good dispersion, and no post-treatment make it one of the most popular synthesis technologies of lanthanum-doped nanoparticles. Organic ligands/surfactants such as oleic acid [106], ethylenediamine tetraacetic acid [107,108], cetyltrimethylammonium bromide [107], and polyethyleneimine [109] are added together with precursors to achieve synchronous control of size, morphology, crystalline phase, and surface properties. For example, Zhao and colleagues reported the hydrothermal process of oleic acid mediated upconversion $NaYF_4$ nanocrystal synthesis, which has different shapes and morphology, including nanotubes, nanorods, and flower patterned nano disks [110]. Liu et al., used Ga^{3+} doping method to control the crystal phase [111]. They found that after on adding Ga^{3+} (accurately controlling the concentration), the required reaction temperature and time were greatly reduced, and the ultra-small $NaYF_4$ upconversion nanoparticles underwent a rapid cubic to hexagonal phase transition. Li et al., demonstrated a synthesis strategy of multiphase, interface controlled monodisperse nanoparticles based on Liquid-Solid-Solid transfer and separation mechanism [112]. With some improvements, this method is also used for the preparation of upconversion nanoparticles with different fluorine and oxyfluorine contents (Table 1).

3.3. Co-Precipitation Method

Among the various methods of preparing nanocrystals, the co-precipitation method is the most promising technology, providing a convenient, safe, and economical method for preparing ultra-small and monodisperse upconversion nanoparticles, without the need for expensive equipment and toxic chemicals. These nanoparticles usually need post-treatment (calcination or annealing) to improve the crystallinity of the material. This method was first used by Veggel et al. for the synthesis of lanthanide ions (Eu, Er, Nd, and Ho) doped LaF_3 downconversion nanoparticles [113]. Yi et al., subsequently demonstrated the application of this method in the preparation of upconversion nanoparticles. They used water-soluble precursors and octadecyl dithiophosphoric acid restriction ligands to prepare ultra-small (5 nm) monodisperse nanoparticles [89]. Guo et al. synthesized monodisperse $NaYF_4$:Yb,Er upconversion nanoparticles of different sizes (37–166 nm) by adjusting the molar ratio of ethylenediaminetetraacetic acid to total lanthanides [114]. They also found that annealing these nanoparticles at a temperature of 400–600 °C can achieve a great increase in luminous intensity (up to 40 times). Recently, Huang et al.'s team prepared Sc^{3+}-doped upconversion nanoparticles by lanthanide ion precipitation in the presence of oleic acid and octadecene [90]. They found that the crystalline phase transition depends on the volume ratio of oleic acid to octadecene, through the intermediate monoclinic/hexagonal coexisting phase (oleic acid:octadecene = 3:9), from the pure monoclinic phase Na_3ScF_6 (Oleic acid: octadecene = 3:17) to pure hexagonal phase $NaScF_4$ (oleic acid: octadecene = 3:7). Due to the small radius of Sc^{3+}, the emission of Na_xScF_{3+x}:Yb,Er is quite different from that of $NaYF_4$:Yb,Er, which can extend the application range of upconversion luminescent nanoparticles from optical communication to disease diagnosis.

3.4. Sol–Gel Method

The sol–gel method uses metal organics or inorganic salts as the matrix, generates gelatin through the process of hydrolysis and polycondensation, and then is dried or sintered to obtain UCNP. It has been successfully applied to the preparation of thin film coatings and glass materials for upconversion luminescent materials [115,116], but usually requires high-temperature post-processing to improve its crystallinity in order to obtain better luminescence effects. Prasad and colleagues first prepared erbium-doped ZrO_2 nanoparticles using sol-latex-gel technology [115]. However, the size of UCNP is difficult to control, and the agglomeration after high temperature calcination brings some difficulties to the surface modification of the material and limits its biomedical applications [116–121].

3.5. Combustion Method

Different from the solvothermal method that uses lower temperature and longer reaction time, the combustion method provides a time-saving (a few minutes) method for the preparation of rare earth doped nanoparticles, and the reaction temperature is generally 500–3000 °C. This process takes place in the form of a combustion wave under a controlled explosion. The combustion method is based on a highly exothermic reaction. A variety of oxides and oxysulfide-containing single nucleotide chain nucleotides were synthesized by this method. For example, Zhang et al., used the glycine-nitrate process to synthesize monoclinic phase $Gd_2O_3:Er^{3+}$ upconversion nanoparticles [122]. Dissolve the Gd_2O_3 and Er_2O_3 upconversion nanoparticles in diluted nitric acid, evaporate and heat. Rapid self-sufficient combustion will produce fluffy powder, which is heated to 600 °C for 1 h to remove nitrates and organic residues. The combustion method has the advantages of saving time and energy, but at the same time there is inevitably the phenomenon of aggregation of synthetic materials.

It is also worth mentioning that flame synthesis is another fast method for preparing upconversion nanoparticles. Ju et al., synthesized Y_2O_3:Yb,Er (or Tm,Ho) nanoparticles with a gas-phase flame one-step method, with an average size of less than 30 nm [123]. The results show that temperature has a strong influence on particle size, morphology, and photoluminescence intensity.

4. Surface Modification of UCNPs

The surface of UCNP obtained by various methods usually contains water-transporting organic ligands (amine oleate, octadecene, oleic acid, etc.). This makes UCNPs difficult to dissolve in water, which affects their biomedical applications to a certain extent. Therefore, it is necessary to modify the surface of UCNPs. So far, various surface modification methods have been reported, including silica coating, ligand exchange, ligand oxidation, ligand attraction, and layer-by-layer assembly [124–133], as shown in Figure 3.

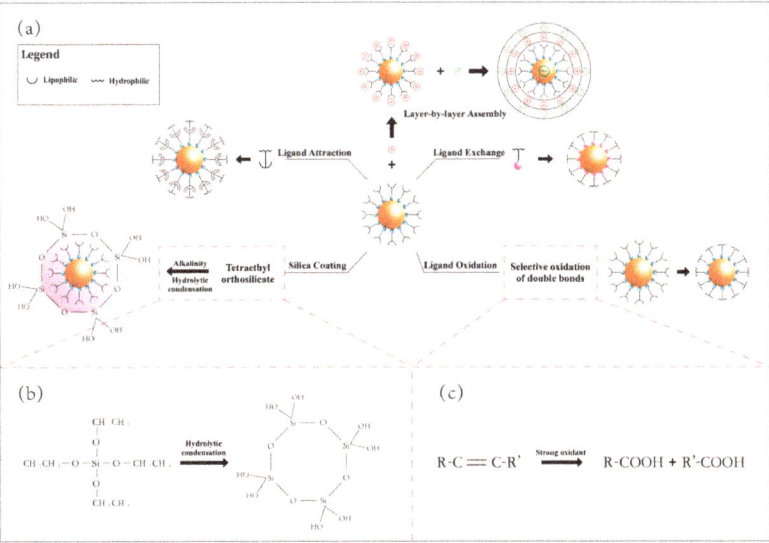

Figure 3. (**a**) Schematic flow diagram of different surface modification methods (silica coating, ligand exchange, ligand oxidation, ligand attraction, and layer–by–layer assembly). (**b**) Chemical reaction formula for preparing silica coating with tetraethyl orthosilicate. (**c**) Basic reaction formula of ligand oxidation process.

4.1. Silica Coating

Surface silanization (or coated silica coating) is an inorganic surface treatment strategy that can make nanoparticles have water solubility and biocompatibility. It is known that silica is highly stable, biocompatible, and optically transparent. When used as a coating material, the method of surface silanization can flexibly provide abundant functional groups (such as -COOH, -NH$_2$, -SH, etc.) to meet the various needs of binding with biomolecules. Wang's group reported that the surface of UCNP is coated with polyhedral oligomeric silsesquioxane (POSS) to make the particles highly hydrophobic. They also described the preparation of liquid marbles based on optically and magnetically active dual-functional UCNPs, and their use as microreactors for the study of photodynamic therapy of cancer cells [134]. Veggel et al., reported the synthesis of SiO$_2$-coated LaF$_3$:Yb^{3+}/Er^{3+} UCNPs using the Stöber method, and the thickness of the silica shell was controlled below 15 nm [135].

4.2. Ligand Exchange

Ligand exchange is a surface modification method that replaces the original hydrophobic ligands with some hydrophilic ligands without significantly affecting the chemical and optical properties of UCNP itself. Chow's group prepared oil-soluble upconversion luminescent nanocrystals by pyrolysis, which are wrapped by oleylamine molecules. Then they exchanged ligands with dicarboxylic acid polyethylene glycol polymers and oleylamine molecules on the surface. Hydrophilic dicarboxylic acid PEG molecules can not only convert nanocrystals into water-soluble form, but also further couple the surface carboxyl groups with biomolecules [136]. Murray and colleagues reported the use of nitrosotetrafluoroborate (NOBF$_4$) to replace the OA and OM ligands attached to the surface of nanoparticles, so that the nanoparticles can exist stably in a variety of polar media for a long time without producing aggregation or precipitation [137].

4.3. Ligand Oxidation

Ligand oxidation is another effective method to obtain water-soluble UCNP based on the selective oxidation of surface carbon-carbon double bonds (R−CH = CH−R′). This method requires the presence of unsaturated bonds in the original ligand. For example, the OA ligand on the surface of UCNPs can be oxidized to azelaic acid (HOOC(CH$_2$)$_7$ COOH), thereby making UCNP a hydrophilic material. Li's group uses m-chloroperoxybenzoic acid as an epoxidizing reagent to oxidize the oleic acid ligand with carbon double bonds on the surface to a ternary epoxy compound, and then interact with organic molecules containing active functional groups (such as MPEG oh) to carry out a ring-opening reaction. Hydrophilic molecules are grafted onto oleic acid molecules to form the water-soluble conversion of luminescent nanoparticles [133]. Yan et al. used a clean and easily available strong oxidant ozone to oxidize the OA on the surface of UCNPs to azelaaldehyde or azelaic acid through ozone decomposition under certain conditions [138]. However, the ligand oxidation method has disadvantages such as long time and low efficiency, which is not conducive to actual production.

4.4. Ligand Attraction

The amphiphilic substance with both hydrophilic and hydrophobic functional groups acts on the surface of the nanoparticles. The hydrophobic functional groups in the substance are adsorbed by the oleic acid ligands on the surface of UCNPs through hydrophobic interaction, while the hydrophilic functional groups act as surface modification. Yi et al., first synthesized NaYF$_4$:Yb/Er@NaYF$_4$ upconversion luminescent nanoparticles with surface oleylamine modified by thermal decomposition method, using polyacrylic acid embedded with octylamine and isopropylamine as modifiers, using the hydrophobicity of polymer molecules. The hydrophobic interaction between the octyl and isopropyl groups of the UCNPs and the hydrophobic hydrocarbon groups on the surface of the UCNPs makes the polymer coat the surface of the UCNPs. The carboxyl groups in the polymer molecules make the hydrophobic UCNPs hydrophilic, realizing the resistance to UCNPs.

4.5. Layer-by-Layer Assembly

Layer-by-layer assembly involves the electrostatic adsorption of oppositely charged anions or cations on the UCNP surface. Electrostatic attraction is one of the strongest and most stable interactions known in nature. Specific layer-by-layer modification usually contributes to the specific biological applications of UCNP. The advantage of this method is that it can prepare coated colloids of different shapes and sizes, with uniform layers of different compositions and controllable thickness [139]. Most importantly, this method can control the surface potential, size, and incorporated functional group ligands of UCNP, which is very important for cell internalization and biological targeting. Polyacrylic acid (PAA) coated with Yi groups on the surface of upconversion luminescent nanocrystals by the layer assembly method. PAA contains 25% octylamino groups and 40% isopropylamine groups, which can transfer oil-soluble upconversion luminescent nanocrystals to the water phase and further couple with biomolecules [140].

In addition to the improvement of biocompatibility, surface modification can also improve the optical properties of UCNPs, laying a foundation for the further application of UCNPs in bioimaging and other fields.

5. Biological Applications of UCNPs

Compared with traditional luminescent materials (including organic dyes and quantum dots), UCNPs have the advantages of high chemical stability, good optical stability, and narrow band gap emission. In addition, under the excitation of near-infrared light, it has strong biological tissue penetration, no damage to biological tissue, high signal-to-noise ratio, and has been widely used in the biological field. This paper also focuses on the application of up conversion luminescent materials in biological imaging, biological detection, and photodynamic therapy [141,142].

For highly infiltrative cancers, including glioblastoma multiform (GBM), it cannot be resected completely. Routinely, the conventional direct introduction to UCNPs suspension into the tissue will cause a series of biocompatibility problems. Thus, implantation of optical fiber has a heavy burden on the patient's body and a high risk of infection. Daniel et al. [143] reported the fabrication of an optical-guided UCNPs implant. Comparatively, the highly biocompatible UCNPs implant is placed into the brain via craniotomy and sealed. Then, the NIR light source passes through the healed scalp and points in the implant to stimulate the UCNPs implant, which will emit visible light to target the photosensitive metabolite protoporphyrin-IX (PpIX) in brain tumors. The flexible light guide with FEP coating approved by the FDA can also maintain NIR to visible light spectrum transduction when the implant is bent to 90°, as shown in Figure 4a. The tumor size of PDT-treated mice was significantly smaller than untreated mice, as shown in Figure 4b. Implant-based UCNPs also allow them to recover from the tissue when they are no longer needed, which cannot be achieved by direct injection of UCNPs suspension into the tissue. A wide range of emission spectrum can be engineered into UCNPs so that the implant can activate multiple drugs simultaneously without cross-interference.

Wang et al. [144] reported a method that can change intracellular pH UCNP@ZIF + TPP + PA nanometer material. The intracellular pH is usually weakly alkaline, which is not conducive to the application of acid-responsible nanomaterials. The proposed new nanomaterials (UCNP@ZIF + TPP + PA) were designed with $NaYF_4$: Yb: Tm (UCNPs) inner core coated by the framework of porous imidazolium zeolite (ZIF-8) for co-loading of photoacid (PA) and acid-responsive porphyrin (TPP). With 980 nm laser irradiation, the emission light of UCNPs activated PA to release H^+. In the new acidic environment, TPP was protonated to increase its water solubility and reduce its aggregation. Simultaneously, transformed UV Vis also activates protonated TPP, which produces more singlet oxygen ($^1O_2$) to kill cancer cells, to enhance the therapeutic effect of photodynamic therapy (PDT); the synthesis route and therapeutic principle are shown in Figure 5.

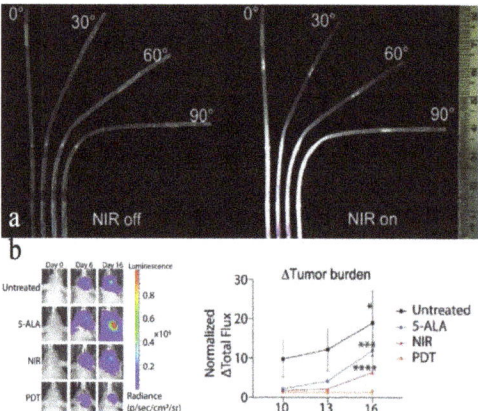

Figure 4. (**a**) UCNPs implant when not excited with NIR and emission intensities visualization at different angles of UCNPs implant bending, excited with 1583 mW cm^{-2} of NIR. (**b**) IVIS imaging indicates that PDT mouse tumors were regressing, as compared to other control groups. The normalized change of tumor burden in all experiment groups over time (n = 5 mice/group. * p = 0.0327, *** p = 0.0002, **** p < 0.0001. Two-way analysis of variance (ANOVA) with Bonferroni's multiple comparison test), reprinted with permission from ref. [143]. Copyright 2020 John Wiley and Sons.

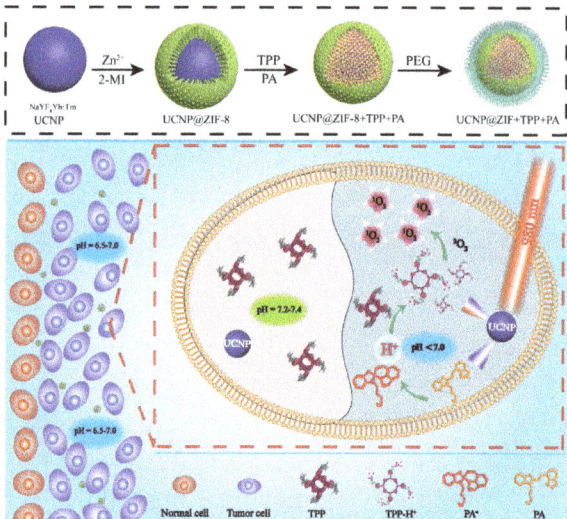

Figure 5. Synthetic route and anticancer mechanism of UCNP@ZIF + TPP + PA. Responding to weak acid tumor microenvironment, UCNP@ZIF + TPP + PA released TPP. When acid-responsive TPP entered the weak alkaline cell, it aggregated again. With 980 nm light irradiation, UCNP emitted UV-Vis light. Photoacid absorbed UV-Vis light and changed its structure to produce H+, which restructured the intracellular pH value. In the new weak pH, the aggregation of TPP was decreased by its protonation. Meanwhile, protonated TPP was activated by the transformed UV-Vis light and produced more $^1O_2$ for enhancing PDT, reprinted with permission from ref. [144]. Copyright 2014 Royal Society of Chemistry.

One of the key challenges in the process of PDT therapy is the accurate killing of cancer cells without destroying normal cells to achieve the desired therapeutic effect. Li et al. [145] designed a type of upconversion nanoprobes (mUCNPs) for intracellular cathepsin B (CAB) reactive PDT, composed of multi-shell upconversion nanoparticlesNaYF$_4$: Gd@NaYF$_4$:Er, Yb@NaYF$_4$: Nd, Yb), with the function of in situ self-tuning therapy effect prediction, as shown in Figure 6.

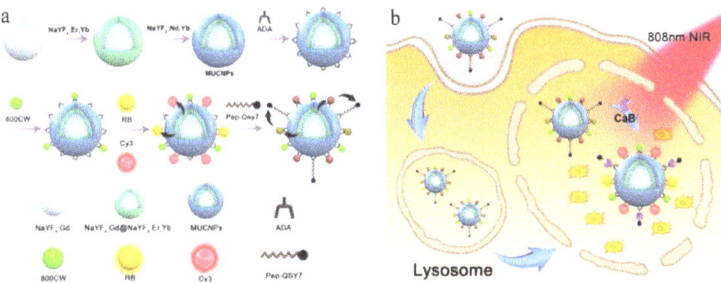

Figure 6. Schematic illustrations of (**a**) synthesis of the upconversion nanoprobe and (**b**) Intracellular CaB-Activated PDT with CaB imaging for Therapeutic Effect Prediction, reprinted with permission from ref. [145]. Copyright 2020 American Chemical Society.

Similarly, the Zhang group [146] designed an amplifier with multiple upconversion luminescence, composed of photo-caged DNA nano-combs and upconversion nanoparticles (UCNPs) sensitized with IRDye® 800CW, to realize the near-infrared light switch cascade reaction triggered by specific microRNA and accurate photodynamic therapy for early cancer. Under 808 nm light irradiation, the generated ultraviolet light cuts off the "photozipper" to induce the cascade hybridization reaction of the microRNA response. This activates the photosensitizer connected to different hairpins to produce reactive oxygen species (ROS) under the blue light emitted simultaneously, to carry out effective PDT, as shown in Figure 7. The amplifier showed desirable serum stability, excellent controllability of reactive oxygen species generation, high specificity for target cancer, and sensitivity to specific microRNA expression. In vivo and in vitro experiments showed strong inhibition on cell proliferation, strong ability to induce apoptosis of tumor cells, and distinct inhibition of tumor growth.

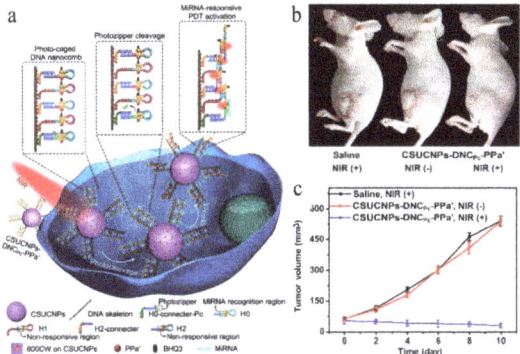

Figure 7. (**a**) Schematic illustration of NIR photo-switched miRNA amplifier for precise PDT. (**b**) Representative images at day 10 and (**c**) tumor volumes of early-stage breast cancer-bearing mice, treated with saline, CSUCNPs-DNC'Pc-PPa' and CSUCNPs-DNCPc-PPa' before and after 808 nm light irradiation at 1 W/cm^2. Error bars indicate means ± SD (n = 5), reprinted with permission from ref. [146]. Copyright 2020 John Wiley and Sons.

Lin et al. [147] designed a spindle-like UCNPs nanoprobe, coated with a layer of gold nanoparticles to enhance upconversion luminescence (UCL), as shown in Figure 8. The results of biocompatibility, blood routine, bioimaging, and anti-cancer tests showed that it was easier for the spindle-like nanoprobes to enter biological tissues. In addition, the combination of SPS@Au and ZnPc (SPSZ) is a potential candidate for synergistic immune photodynamic therapy (PDT), with enhanced UCL effect and excellent biocompatibility.

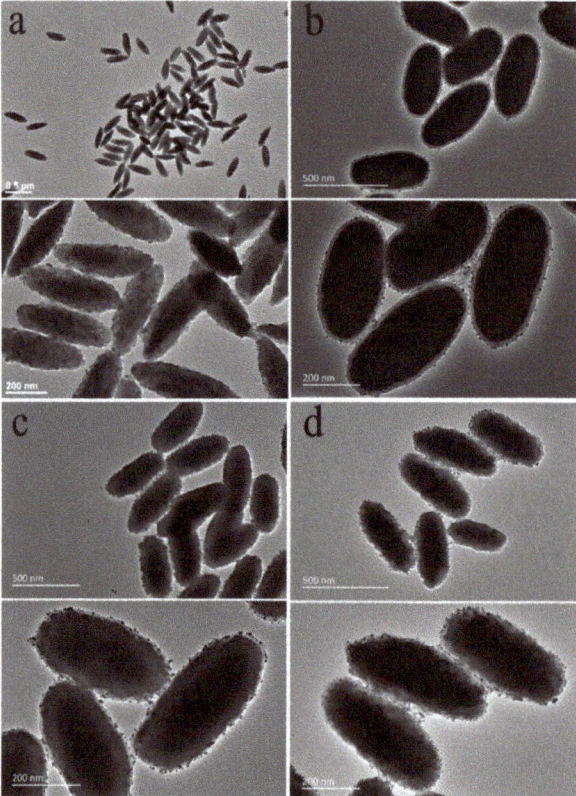

Figure 8. TEM image of (**a**) spindle precursor and (**b**) SPS@Au (LBL1) and (**c**) SPS@Au (LBL2) and (**d**) SPS@Au (LBL3). Reprinted with permission from ref. [147]. Copyright 2020 American Chemistry Society.

The Li group [148] designed core-satellite metal-organic through electrostatic self-assembly framework@UCNP superstructures, composed of a single metal-organic framework (MOF) NP as the core, and Nd^{3+}-sensitized UCNPs as the satellites. In vitro and in vivo experiments show that the double photosensitizer superstructure has a three-mode (magnetic resonance/UCL/ fluorescence) imaging function and excellent anti-tumor effect under the excitation of 808 nm near-infrared light, avoiding overheating caused by laser irradiation. After being exposed to an 808 nm laser for 5 min, the temperature of the irradiated area was lower than 42 °C, and without damage to mice. However, under the same conditions, a 980 nm laser can heat the irradiated area to above 50 °C and severely burn the skin. These findings indicate that 808 nm excitation has a much weaker tissue thermal effect and is more suitable for biological applications.

Sun et al. [149] designed and prepared one kind of lanthanide (Ln^{3+})-doped upconversion nanocomposites with multi-functions, which can not only provide temperature feedback in PTT process, but also play the photodynamic therapy (PDT) function for the synergistic effect of tumor therapy. Based on NaYF$_4$:Yb, Er upconversion nanoparticles (UCNPs), mesoporous SiO$_2$ was modified on the surface combined with photosensitizer Chlorin e6 (Ce6) molecules, which could be excited by red emission of Er^{3+} under the 980 nm laser. Cit-CuS NPs were further linked on the surface of the composite as a photothermal conversion agent, therefore, the temperature of the PTT site can be monitored by recording the ratio of I$_{525}$/I$_{545}$ of green emissions, especially within the physiological range, as shown in Figure 9. Based on the guidance obtained from spectral experiments, they further investigated the dual-modal therapy effect both in vitro and in vivo, respectively, and acquired decent results.

In addition, these rare earth doped nanoparticles also have strong scintillation luminescence that can be used for X-ray induced photodynamic therapy, which is a very hot area, as this new therapy can be used for deep as well as skin cancer treatment [150–163].

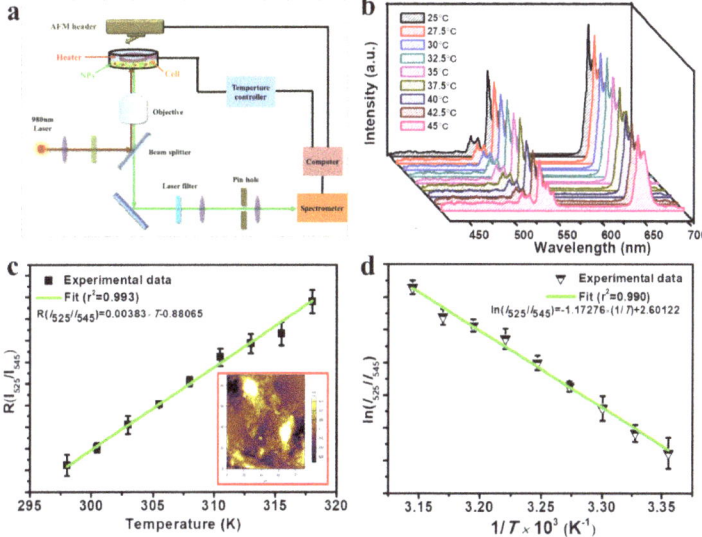

Figure 9. (a) Schematic diagram of the detection of the temperature and emission spectrum of UCNPs-Ce6@mSiO$_2$-CuS incubated with cells in physiological range; (b) UCL emission spectrum of UCNPs-Ce6@mSiO$_2$-CuS incubated with cells at different temperatures by external heating. The peaks were normalized at 525 nm; (c) FIR of the green UC emissions for the ^{2}H$_{11/2}$/^{4}S$_{3/2}$ → ^{4}I$_{15/2}$ transitions relative to the temperature of UCNPs-Ce6@mSiO$_2$-CuS incubated with cells. The inset picture is the AFM image of the cell after spectral detection; (d) A plot of ln(I$_{525}$/I$_{545}$) versus 1/T to calibrate the thermometric scale for UCNPs-Ce6@mSiO$_2$-CuS incubated with cells. Reprinted with permission from ref. [149]. Copyright 2019 Elsevier.

6. Summary and Perspective

Over the last decades, UCNPs have made remarkable advances in the treatment of critical diseases, greatly promoting the application of modern precision medicine in the life system with its enhanced therapeutic effect, high space-time controllability, deep tissue penetration, and minimal invasion. However, despite the remarkable achievements, there are still some challenges of UCNPs. (1) The stability of luminous efficiency after surface modification: in the surface modification, the oil-soluble molecules will be modified to improve biocompatibility. Nevertheless, the dispersing ability of UCNPs in oil and water is different. After surface modification, it is easy to cause fluorescence quenching and reduce the upconversion efficiency. (2) The biological toxicity of UCNPs: many studies have shown that the reasonable optimization of chemical composition, particle size distribution, and surface modification can significantly improve the biocompatibility of UCNPs, which can be used in biomedical applications. However, there are no tests to evaluate its long-term toxicity, including potential immune response and mutagenic effect. (3) Technical gaps in clinical trials: so far, UCNPs-based phototherapy has not been applied to human beings, mainly due to biosafety or therapeutic effect. There is still a long way to go from laboratory animals to human-level technical standard updates. In summary, UCNPs offer a tremendous opportunity to practice precision medicine. We expect that stable surface modification, low toxicity, and clinical trials will make UCNPs more competitive in the biological field.

Author Contributions: Conceptualization, methodology, Q.L.; validation, formal analysis. and; investigation, resources, visualization; writing—original draft preparation, C.G.; data curation, P.Z. and S.H.; supervision, writing—review and editing, J.W., C.X., T.J., H.T. and W.C.; project administration and instruction, J.W.; funding acquisition, Y.W., D.L. and S.L. All authors have read and agreed to the published version of the manuscript.

Funding: This research was funded by the National Science Foundation of China No. 11904125, the Open Research Subject of Key Laboratory of Dielectric and Electrolyte Functional Material Hebei Province No. HKDE201902, BIGC Project No. 22150121034/026, No. 22150121003/050, No. Eb202105, and JENU No. XZD201802. W. C. would like to thank Solgro Inc. and UT Arlington distinguished award.

Institutional Review Board Statement: Not applicable.

Informed Consent Statement: Not applicable.

Data Availability Statement: The study did not report any data.

Conflicts of Interest: The authors declare no conflict of interest.

References

1. Ouyang, J.; Ripmeester, J.A.; Wu, X.; Kingston, D.; Yu, K.; Joly, A.G.; Chen, W. Upconversion luminescence of colloidal CdS and ZnCdS semiconductor quantum dots. *J. Phys. Chem. C* **2007**, *111*, 16261–16266. [CrossRef]
2. Porter, J.F., Jr. Fluorescence excitation by the absorption of two consecutive photons. *Phys. Rev. Lett.* **1961**, *7*, 414. [CrossRef]
3. Yao, J.; Huang, C.; Liu, C.; Yang, M. Upconversion luminescence nanomaterials: A versatile platform for imaging, sensing, and therapy. *Talanta* **2020**, *208*, 120157. [CrossRef] [PubMed]
4. Ansari, A.A.; Parchur, A.K.; Thorat, N.D.; Chen, G. New advances in pre-clinical diagnostic imaging perspectives of functionalized upconversion nanoparticle-based nanomedicine. *Coord. Chem. Rev.* **2021**, *440*, 213971. [CrossRef]
5. Xin, N.; Wei, D.; Zhu, Y.; Yang, M.; Ramakrishna, S.; Lee, O.; Luo, H.; Fan, H. Upconversion nanomaterials: A platform for biosensing, theranostic and photoregulation. *Mater. Today Chem.* **2020**, *17*, 100329. [CrossRef]
6. Wang, J.; Sheng, T.; Zhu, X.; Li, Q.; Wu, Y.; Zhang, J.; Liu, J.; Zhang, Y. Spectral engineering of lanthanide-doped upconversion nanoparticles and their biosensing applications. *Mater. Chem. Front.* **2021**, *5*, 1743–1770. [CrossRef]
7. Peltomaa, R.; Benito-Peña, E.; Gorris, H.H.; Moreno-Bondi, M.C. Biosensing based on upconversion nanoparticles for food quality and safety applications. *Analyst* **2021**, *146*, 13–32. [CrossRef]
8. Freeman, A.J.; Watson, R.E. Theoretical investigation of some magnetic and spectroscopic properties of rare-earth ions. *Phys. Rev.* **1962**, *127*, 2058. [CrossRef]
9. Yan, C.; Jia, J.; Liao, C.; Wu, S.; Xu, G. Rare earth separation in China. *Tsinghua Sci. Technol.* **2006**, *11*, 241–247. [CrossRef]
10. Bünzli, J.C.G. Benefiting from the unique properties of lanthanide ions. *Acc. Chem. Res.* **2006**, *39*, 53–61. [CrossRef]
11. Sun, L.D.; Dong, H.; Zhang, P.Z.; Yan, C.H. Upconversion of rare earth nanomaterials. *Annu. Rev. Phys. Chem.* **2015**, *66*, 619–642. [CrossRef]
12. Tian, B.; Bravo, A.F.; Najafiaghdam, H.; Torquato, N.A.; Altoe, M.V.P.; Teitelboim, A.; Tajon, C.A.; Tian, Y.; Borys, N.J.; Barnard, E.S.; et al. Low irradiance multiphoton imaging with alloyed lanthanide nanocrystals. *Nat. Commun.* **2018**, *9*, 1–8. [CrossRef] [PubMed]
13. Zhan, Q.; Liu, H.; Wang, B.; Wu, Q.; Pu, R.; Zhou, C.; Huang, B.; Peng, X.; Agren, H.; He, S. Achieving high-efficiency emission depletion nanoscopy by employing cross relaxation in upconversion nanoparticles. *Nat. Commun.* **2017**, *8*, 1–11. [CrossRef] [PubMed]
14. Qiu, Z.; Shu, J.; Tang, D. Near-infrared-to-ultraviolet light-mediated photoelectrochemical aptasensing platform for cancer biomarker based on core–shell $NaYF_4$: Yb, Tm@ TiO_2 upconversion microrods. *Anal. Chem.* **2018**, *90*, 1021–1028. [CrossRef] [PubMed]

15. Zhang, K.; Song, S.; Huang, S.; Yang, L.; Min, Q.; Wu, X.; Lu, F.; Zhu, J.J. Lighting Up MicroRNA in Living Cells by the Disassembly of Lock-Like DNA-Programmed UCNPs-AuNPs through the Target Cycling Amplification Strategy. *Small* **2018**, *14*, 1802292. [CrossRef]
16. Yao, C.; Wang, P.; Li, X.; Hu, X.; Hou, J.; Wang, L.; Zhang, F. Near-infrared-triggered azobenzene-liposome/upconversion nanoparticle hybrid vesicles for remotely controlled drug delivery to overcome cancer multidrug resistance. *Adv. Mater.* **2016**, *28*, 9341–9348. [CrossRef]
17. Chen, X.; Tang, Y.; Liu, A.; Zhu, Y.; Gao, D.; Yang, Y.; Sun, J.; Fan, H.; Zhang, X. NIR-to-red upconversion nanoparticles with minimized heating effect for synchronous multidrug resistance tumor imaging and therapy. *ACS Appl. Mater. Interfaces* **2018**, *10*, 14378–14388. [CrossRef]
18. Chen, X.; Sun, J.; Zhao, H.; Yang, K.; Zhu, Y.; Luo, H.; Yu, K.; Fan, K.; Zhang, X. Theranostic system based on NaY(Mn)F$_4$: Yb/Er upconversion nanoparticles with multi-drug resistance reversing ability. *J. Mater. Chem. B* **2018**, *6*, 3586–3599. [CrossRef]
19. Yao, J.; Yang, M.; Duan, Y. Chemistry, biology, and medicine of fluorescent nanomaterials and related systems: New insights into biosensing, bioimaging, genomics, diagnostics, and therapy. *Chem. Rev.* **2014**, *114*, 6130–6178. [CrossRef]
20. Chen, W.; Joly, A.G.; McCready, D.E. Upconversion luminescence from CdSe nanoparticles. *J. Chem. Phys.* **2005**, *122*, 224708. [CrossRef]
21. Joly, A.G.; Chen, W.; McCready, D.E.; Malm, J.O.; Bovin, J.O. Upconversion luminescence of CdTe nanoparticles. *Phys. Rev. B* **2005**, *71*, 165304. [CrossRef]
22. Shen, J.; Sun, L.D.; Yan, C.H. Luminescent rare earth nanomaterials for bioprobe applications. *Dalton Trans.* **2008**, 5687–5697. [CrossRef]
23. Wang, F.; Banerjee, D.; Liu, Y.; Chen, X.; Liu, X. Upconversion nanoparticles in biological labeling, imaging, and therapy. *Analyst* **2010**, *135*, 1839–1854. [CrossRef] [PubMed]
24. Zhou, J.; Liu, Z.; Li, F. Upconversion nanophosphors for small-animal imaging. *Chem. Soc. Rev.* **2012**, *41*, 1323–1349. [CrossRef] [PubMed]
25. Hampl, J.; Hall, M.; Mufti, N.A.; Yung-mae, M.Y.; MacQueen, D.B.; Wright, W.H.; Cooper, D.E. Upconverting phosphor reporters in immunochromatographic assays. *Anal. Biochem.* **2001**, *288*, 176–187. [CrossRef] [PubMed]
26. Van De Rijke, F.; Zijlmans, H.; Li, S.; Vail, T.; Raap, A.K.; Niedbala, R.S.; Tanke, H.J. Up-converting phosphor reporters for nucleic acid microarrays. *Nat. Biotechnol.* **2001**, *19*, 273–276. [CrossRef]
27. Zhang, P.; Rogelj, S.; Nguyen, K.; Wheeler, D. Design of a highly sensitive and specific nucleotide sensor based on photon upconverting particles. *J. Am. Chem. Soc.* **2006**, *128*, 12410–12411. [CrossRef]
28. Chatterjee, D.K.; Gnanasammandhan, M.K.; Zhang, Y. Small upconverting fluorescent nanoparticles for biomedical applications. *Small* **2010**, *6*, 2781–2795. [CrossRef]
29. Cheng, L.; Wang, C.; Liu, Z. Upconversion nanoparticles and their composite nanostructures for biomedical imaging and cancer therapy. *Nanoscale* **2013**, *5*, 23–37. [CrossRef]
30. Gu, Z.; Yan, L.; Tian, G.; Li, S.; Chai, Z.; Zhao, Y. Recent advances in design and fabrication of upconversion nanoparticles and their safe theranostic applications. *Adv. Mater.* **2013**, *25*, 3758–3779. [CrossRef]
31. Wilson, B.C.; Patterson, M.S. The physics, biophysics and technology of photodynamic therapy. *Phys. Med. Biol.* **2008**, *53*, R61. [CrossRef]
32. Li, S.; Cui, S.; Yin, D.; Zhu, Q.; Ma, Y.; Qian, Z.; Gu, Y. Dual antibacterial activities of a chitosan-modified upconversion photodynamic therapy system against drug-resistant bacteria in deep tissue. *Nanoscale* **2017**, *9*, 3912–3924. [CrossRef] [PubMed]
33. Lucky, S.S.; Idris, N.M.; Huang, K.; Kim, J.; Li, Z.; Thong, P.S.P.; Xu, R.; Soo, K.C.; Zhang, Y. In vivo biocompatibility, biodistribution and therapeutic efficiency of titania coated upconversion nanoparticles for photodynamic therapy of solid oral cancers. *Theranostics* **2016**, *6*, 1844–1865. [CrossRef] [PubMed]
34. Wang, D.; Liu, B.; Quan, Z.; Li, C.; Hou, Z.; Xing, B.; Lin, J. New advances on the marrying of UCNPs and photothermal agents for imaging-guided diagnosis and the therapy of tumors. *J. Mater. Chem. B* **2017**, *5*, 2209–2230. [CrossRef] [PubMed]
35. Chen, W.; Joly, A.G.; Malm, J.O.; Bovin, J.O. Upconversion luminescence of Eu^{3+} and Mn^{2+} in ZnS: Mn^{2+}, Eu^{3+} codoped nanoparticles. *J. Appl. Phys.* **2004**, *95*, 667–672. [CrossRef]
36. Chen, W.; Joly, A.G.; Zhang, J.Z. Up-conversion luminescence of Mn^{2+} in ZnS:Mn^{2+} nanoparticles. *Phys. Rev. B* **2001**, *64*, 041202. [CrossRef]
37. Bruschini, C.; Homulle, H.; Antolovic, I.M.; Burri, S.; Charbon, E. Single-photon avalanche diode imagers in biophotonics: Review and outlook. *Light Sci. Appl.* **2019**, *8*, 1–28. [CrossRef] [PubMed]
38. Li, Z.; Ding, X.; Cong, H.; Wang, S.; Yu, B.; Shen, Y. Recent advances on inorganic lanthanide-doped NIR-II fluorescence nanoprobes for bioapplication. *J. Lumin.* **2020**, *228*, 117627. [CrossRef]
39. Auzel, F. Upconversion and anti-stokes processes with f and d ions in solids. *Chem. Rev.* **2004**, *104*, 139–174. [CrossRef]
40. Sun, Y.; Chen, Y.; Tian, L.; Yu, Y.; Kong, X.; Zhao, J.; Zhang, H. Controlled synthesis and morphology dependent upconversion luminescence of NaYF$_4$: Yb, Er nanocrystals. *Nanotechnology* **2007**, *18*, 275609. [CrossRef]
41. Wang, L.; Li, Y. Green upconversion nanocrystals for DNA detection. *Chem. Commun.* **2006**, *131*, 2557–2559. [CrossRef] [PubMed]
42. Wang, L.; Li, Y. Na(Y$_{1.5}$Na$_{0.5}$)F$_6$ single-crystal nanorods as multicolor luminescent materials. *Nano Lett.* **2006**, *6*, 1645–1649. [CrossRef] [PubMed]

43. De la Rosa, E.; Salas, P.; Desirena, H.; Angeles, C.; Rodriguez, R.A. Strong green upconversion emission in ZrO_2: Yb^{3+}-Ho^{3+} nanocrystals. *Appl. Phys. Lett.* **2005**, *87*, 241912. [CrossRef]
44. Wang, G.; Peng, Q.; Li, Y. Upconversion luminescence of monodisperse CaF_2: Yb^{3+}/Er^{3+} nanocrystals. *J. Am. Chem. Soc.* **2009**, *131*, 14200–14201. [CrossRef] [PubMed]
45. Wang, F.; Liu, X. Upconversion multicolor fine-tuning: Visible to near-infrared emission from lanthanide-doped $NaYF_4$ nanoparticles. *J. Am. Chem. Soc.* **2008**, *130*, 5642–5643. [CrossRef]
46. Chen, G.; Ohulchanskyy, T.Y.; Kachynski, A.; Ågren, H.; Prasad, P.N. Intense visible and near-infrared upconversion photoluminescence in colloidal $LiYF_4$: Er^{3+} nanocrystals under excitation at 1490 nm. *ACS Nano* **2011**, *5*, 4981–4986. [CrossRef]
47. Chen, D.; Lei, L.; Yang, A.; Wang, Z.; Wang, Y. Ultra-broadband near-infrared excitable upconversion core/shell nanocrystals. *Chem. Commun.* **2012**, *48*, 5898–5900. [CrossRef]
48. Chen, G.; Ohulchanskyy, T.Y.; Liu, S.; Law, W.C.; Wu, F.; Swihart, M.T.; Ågren, H.; Prasad, P.N. Core/shell $NaGdF_4$: Nd^{3+}/$NaGdF_4$ nanocrystals with efficient near-infrared to near-infrared downconversion photoluminescence for bioimaging applications. *ACS Nano* **2012**, *6*, 2969–2977. [CrossRef]
49. Rui-Rong, W.; Guo, J.; Zhi-Heng, F.; Wei, W.; Xiang-Fu, M.; Zhi-Yong, X.; Fan, Z. Broadband time-resolved elliptical crystal spectrometer for X-ray spectroscopic measurements in laser-produced plasmas. *Chin. Phys. B* **2014**, *23*, 113201.
50. Rakov, N.; Maciel, G.S.; Sundheimer, M.L.; de S. Menezes, L.; Gomes, A.S.L.; Messaddeq, Y.; Cassanjes, F.C.; Poirier, G.; Ribeiro, S.J.L. Blue upconversion enhancement by a factor of 200 in Tm^{3+}-doped tellurite glass by codoping with Nd^{3+} ions. *J. Appl. Phys.* **2002**, *92*, 6337–6339. [CrossRef]
51. Wang, L.; Qin, W.; Liu, Z.; Zhao, D.; Qin, G.; Di, W.; He, C. Improved 800 nm emission of Tm^{3+} sensitized by Yb^{3+} and Ho^{3+} in β-$NaYF_4$ nanocrystals under 980 nm excitation. *Opt. Express* **2012**, *20*, 7602–7607. [CrossRef]
52. Liang, H.; Chen, G.; Li, L.; Liu, Y.; Qin, F.; Zhang, Z. Upconversion luminescence in Yb^{3+}/Tb^{3+}-codoped monodisperse $NaYF_4$ nanocrystals. *Opt. Commun.* **2009**, *282*, 3028–3031. [CrossRef]
53. Dwivedi, Y.; Thakur, S.N.; Rai, S.B. Study of frequency upconversion in Yb^{3+}/Eu^{3+} by cooperative energy transfer in oxyfluoroborate glass matrix. *Appl. Phys. B* **2007**, *89*, 45–51. [CrossRef]
54. Uvarova, T.V.; Kiiko, V.V. Up-conversion multiwave (White) luminescence in the visible spectral range under excitation by IR laser diodes in the active BaY_2F_8: Yb^{3+}, Pr^{3+} medium. *Opt. Spectrosc.* **2011**, *111*, 273–276.
55. Xu, B.; Song, C.; Huang, R.; Song, J.; Lin, Z.; Song, J.; Liu, J. Luminescence properties related to energy transfer process and cross relaxation process of Y_2O_3: Yb^{3+}/Er^{3+} thin films doped with K^+ ion. *Opt. Mater.* **2021**, *118*, 111290. [CrossRef]
56. Ouertani, G.; Ferhi, M.; Horchani-Naifer, K.; Ferid, M. Effect of Sm^{3+} concentration and excitation wavelength on spectroscopic properties of $GdPO_4$: Sm^{3+} phosphor. *J. Alloys Compd.* **2021**, *885*, 161178. [CrossRef]
57. Zheng, B.; Hong, J.; Chen, B.; Chen, Y.; Lin, R.; Huang, C.; Zhang, C.; Wang, J.; Lin, L.; Zheng, Z. Quantum cutting properties in KYF_4: Tb^{3+}, Yb^{3+} phosphors: Judd-Ofelt analysis, rate equation models and dynamic processes. *Results Phys.* **2021**, *28*, 104595. [CrossRef]
58. Kumar, A.; Bahadur, A. Intense green upconversion emission by photon avalanche process from Er^{3+}/Yb^{3+} co-doped $NaBi(WO_4)_2$ phosphor. *J. Alloys Compd.* **2021**, *857*, 158196. [CrossRef]
59. Liu, T.; Song, Y.; Wang, S.; Li, Y.; Yin, Z.; Qiu, J.; Yang, Z.; Song, Z. Two distinct simultaneous NIR looping behaviours of Er^{3+} singly doped BiOBr: The underlying nature of the Er^{3+} ion photon avalanche emission induced by a layered structure. *J. Alloys Compd.* **2019**, *779*, 440–449. [CrossRef]
60. Wang, F.; Liu, X. Recent advances in the chemistry of lanthanide-doped upconversion nanocrystals. *Chem. Soc. Rev.* **2009**, *38*, 976–989. [CrossRef]
61. Chen, G.; Qiu, H.; Prasad, P.N.; Chen, X. Upconversion nanoparticles: Design, nanochemistry, and applications in theranostics. *Chem. Rev.* **2014**, *114*, 5161–5214. [CrossRef] [PubMed]
62. Liu, Y.; Tu, D.; Zhu, H.; Chen, X. Lanthanide-doped luminescent nanoprobes: Controlled synthesis, optical spectroscopy, and bioapplications. *Chem. Soc. Rev.* **2013**, *42*, 6924–6958. [CrossRef] [PubMed]
63. Feng, W.; Han, C.; Li, F. Upconversion-nanophosphor-based functional nanocomposites. *Adv. Mater.* **2013**, *25*, 5287–5303. [CrossRef]
64. Zhang, Y.W.; Sun, X.; Si, R.; You, L.P.; Yan, C.H. Single-crystalline and monodisperse LaF_3 triangular nanoplates from a single-source precursor. *J. Am. Chem. Soc.* **2005**, *127*, 3260–3261. [CrossRef] [PubMed]
65. Mai, H.X.; Zhang, Y.W.; Si, R.; Yan, Z.G.; Sun, L.D.; You, L.P.; Yan, C.H. High-quality sodium rare-earth fluoride nanocrystals: Controlled synthesis and optical properties. *J. Am. Chem. Soc.* **2006**, *128*, 6426–6436. [CrossRef]
66. Du, Y.P.; Sun, X.; Zhang, Y.W.; Yan, Z.G.; Sun, L.D.; Yan, C.H. Uniform alkaline earth fluoride nanocrystals with diverse shapes grown from thermolysis of metal trifluoroacetates in hot surfactant solutions. *Cryst. Growth Des.* **2009**, *9*, 2013–2019. [CrossRef]
67. Mahalingam, V.; Vetrone, F.; Naccache, R.; Speghini, A.; Capobianco, J.A. Colloidal Tm^{3+}/Yb^{3+}-doped $LiYF_4$ nanocrystals: Multiple luminescence spanning the UV to NIR regions via low-energy excitation. *Adv. Mater.* **2009**, *21*, 4025–4028. [CrossRef]
68. Vetrone, F.; Naccache, R.; Mahalingam, V.; Morgan, C.G.; Capobianco, J.A. The active-core/active-shell approach: A strategy to enhance the upconversion luminescence in lanthanide-doped nanoparticles. *Adv. Funct. Mater.* **2009**, *19*, 2924–2929. [CrossRef]
69. Yang, D.; Li, C.; Li, G.; Shang, M.; Kang, X.; Lin, J. Colloidal synthesis and remarkable enhancement of the upconversion luminescence of $BaGdF_5$: Yb^{3+}/Er^{3+} nanoparticles by active-shell modification. *J. Mater. Chem.* **2011**, *21*, 5923–5927. [CrossRef]

70. Mahalingam, V.; Vetrone, F.; Naccache, R.; Speghini, A.; Capobianco, J.A. Structural and optical investigation of colloidal Ln^{3+}/Yb^{3+} co-doped KY$_3$F$_{10}$ nanocrystals. *J. Mater. Chem.* **2009**, *19*, 3149–3152. [CrossRef]
71. Wang, H.; Nakamura, H.; Uehara, M.; Yamaguchi, Y.; Miyazaki, M.; Maeda, H. Highly luminescent CdSe/ZnS nanocrystals synthesized using a single-molecular ZnS source in a microfluidic reactor. *Adv. Funct. Mater.* **2005**, *15*, 603–608. [CrossRef]
72. Sun, X.; Zhang, Y.W.; Du, Y.P.; Yan, Z.G.; Si, R.; You, L.P.; Yan, C.H. From trifluoroacetate complex precursors to monodisperse rare-earth fluoride and oxyfluoride nanocrystals with diverse shapes through controlled fluorination in solution phase. *Chem. A Eur. J.* **2007**, *13*, 2320–2332. [CrossRef]
73. Chen, D.; Huang, P.; Yu, Y.; Huang, F.; Yang, A.; Wang, Y. Dopant-induced phase transition: A new strategy of synthesizing hexagonal upconversion NaYF$_4$ at low temperature. *Chem. Commun.* **2011**, *47*, 5801–5803. [CrossRef] [PubMed]
74. Chen, D.; Yu, Y.; Huang, F.; Huang, P.; Yang, A.; Wang, Y. Modifying the size and shape of monodisperse bifunctional alkaline-earth fluoride nanocrystals through lanthanide doping. *J. Am. Chem. Soc.* **2010**, *132*, 9976–9978. [CrossRef] [PubMed]
75. Wang, F.; Chatterjee, D.K.; Li, Z.; Zhang, Y.; Fan, X.; Wang, M. Synthesis of polyethylenimine/NaYF$_4$ nanoparticles with upconversion fluorescence. *Nanotechnology* **2006**, *17*, 5786. [CrossRef]
76. Wang, L.; Li, P.; Li, Y. Down-and up-conversion luminescent nanorods. *Adv. Mater.* **2007**, *19*, 3304–3307. [CrossRef]
77. Zeng, S.; Xiao, J.; Yang, Q.; Hao, J. Bi-functional NaLuF$_4$: Gd^{3+}/Yb^{3+}/Tm^{3+} nanocrystals: Structure controlled synthesis, near-infrared upconversion emission and tunable magnetic properties. *J. Mater. Chem.* **2012**, *22*, 9870–9874. [CrossRef]
78. Zeng, J.H.; Xie, T.; Li, Z.H.; Li, Y. Monodispersed nanocrystalline fluoroperovskite up-conversion phosphors. *Cryst. Growth Des.* **2007**, *7*, 2774–2777. [CrossRef]
79. Zeng, S.; Tsang, M.K.; Chan, C.F.; Wong, K.L.; Hao, J. PEG modified BaGdF$_5$: Yb/Er nanoprobes for multi-modal upconversion fluorescent, in vivo X-ray computed tomography and biomagnetic imaging. *Biomaterials* **2012**, *33*, 9232–9238. [CrossRef]
80. Vetrone, F.; Boyer, J.C.; Capobianco, J.A.; Speghini, A.; Bettinelli, M. Significance of Yb^{3+} concentration on the upconversion mechanisms in codoped Y$_2$O$_3$: Er^{3+}, Yb^{3+} nanocrystals. *J. Appl. Phys.* **2004**, *96*, 661–667. [CrossRef]
81. Shang, M.; Li, G.; Kang, X.; Yang, D.; Geng, D.; Peng, C.; Cheng, Z.; Lian, H.; Lin, J. LaOF: Eu^{3+} nanocrystals: Hydrothermal synthesis, white and color-tuning emission properties. *Dalton Trans.* **2012**, *41*, 5571–5580. [CrossRef] [PubMed]
82. Heer, S.; Lehmann, O.; Haase, M.; Guedel, H.U. Blue, green, and red upconversion emission from lanthanide-doped LuPO$_4$ and YbPO$_4$ nanocrystals in a transparent colloidal solution. *Angew. Chem. Int. Ed.* **2003**, *42*, 3179–3182. [CrossRef]
83. Schäfer, H.; Ptacek, P.; Eickmeier, H.; Haase, M. Synthesis of Hexagonal Yb^{3+}, Er^{3+}-Doped NaYF$_4$ Nanocrystals at Low Temperature. *Adv. Funct. Mater.* **2009**, *19*, 3091–3097. [CrossRef]
84. Ptacek, P.; Schäfer, H.; Kömpe, K.; Haase, M. Crystal Phase Control of Luminescing α-NaGdF$_4$: Eu^{3+} and β-NaGdF$_4$: Eu^{3+} Nanocrystals. *Adv. Funct. Mater.* **2007**, *17*, 3843–3848. [CrossRef]
85. Gai, S.; Yang, G.; Li, X.; Li, C.; Dai, Y.; He, F.; Yang, P. Facile synthesis and up-conversion properties of monodisperse rare earth fluoride nanocrystals. *Dalton Trans.* **2012**, *41*, 11716–11724. [CrossRef]
86. Yang, T.; Sun, Y.; Liu, Q.; Feng, W.; Yang, P.; Li, F. Cubic sub-20 nm NaLuF$_4$-based upconversion nanophosphors for high-contrast bioimaging in different animal species. *Biomaterials* **2012**, *33*, 3733–3742. [CrossRef] [PubMed]
87. Wong, H.T.; Vetrone, F.; Naccache, R.; Chan, H.L.W.; Hao, J.; Capobianco, J.A. Water dispersible ultra-small multifunctional KGdF$_4$: Tm^{3+}, Yb^{3+} nanoparticles with near-infrared to near-infrared upconversion. *J. Mater. Chem.* **2011**, *21*, 16589–16596. [CrossRef]
88. Dai, Y.; Zhang, C.; Cheng, Z.; Li, C.; Kang, X.; Yang, D.; Lin, J. pH-responsive drug delivery system based on luminescent CaF$_2$: Ce^{3+}/Tb^{3+}-poly (acrylic acid) hybrid microspheres. *Biomaterials* **2012**, *33*, 2583–2592. [CrossRef] [PubMed]
89. Yi, G.S.; Chow, G.M. Colloidal LaF$_3$: Yb, Er, LaF$_3$: Yb, Ho and LaF$_3$: Yb, Tm nanocrystals with multicolor upconversion fluorescence. *J. Mater. Chem.* **2005**, *15*, 4460–4464. [CrossRef]
90. Teng, X.; Zhu, Y.; Wei, W.; Wang, S.; Huang, J.; Naccache, R.; Hu, W.; Tok, A.; Han, Y.; Zhang, Q.; et al. Lanthanide-doped Na$_x$ScF$_{3+x}$ nanocrystals: Crystal structure evolution and multicolor tuning. *J. Am. Chem. Soc.* **2012**, *134*, 8340–8343. [CrossRef]
91. Liu, H.; Liu, M.; Wang, K.; Wang, B.; Jian, X.; Bai, G.; Zhang, Y. Efficient upconversion emission and high-sensitivity thermometry of BaIn$_2$O$_4$: Yb^{3+}/Tm^{3+}/RE^{3+} (RE = Er^{3+}, Ho^{3+}) phosphor. *Dalton Trans.* **2021**, *50*, 12107–12117. [CrossRef] [PubMed]
92. Lim, C.S.; Aleksandrovsky, A.S.; Atuchin, V.V.; Molokeev, M.S.; Oreshonkov, A.S. Microwave sol-gel synthesis, microstructural and spectroscopic properties of scheelite-type ternary molybdate upconversion phosphor NaPbLa (MoO$_4$)$_3$: Er^{3+}/Yb^{3+}. *J. Alloys Compd.* **2020**, *826*, 152095. [CrossRef]
93. Wang, G.; Jia, G.; Wang, J.; Kong, H.; Lu, Y.; Zhang, C. Novel rare earth activator ions-doped perovskite-type La$_4$Ti$_3$O$_{12}$ phosphors: Facile synthesis, structure, multicolor emissions, and potential applications. *J. Alloys Compd.* **2021**, *877*, 160217. [CrossRef]
94. Zhang, Y.; Jia, H.; He, X.; Zheng, Y.; Bai, R.; Liu, H. Investigation of enhancing upconversion and temperature sensing performance of Er^{3+}, Yb^{3+}, Bi^{3+} codoped Gd$_2$O$_3$ phosphor. *J. Lumin.* **2021**, *236*, 118111. [CrossRef]
95. Singh, P.; Jain, N.; Tiwari, A.K.; Shukla, S.; Baranwal, V.; Singh, J.; Pandey, A.C. Near-infrared light-mediated Er^{3+} and Yb^{3+} co-doped CaTi$_4$O$_9$ for optical temperature sensing behavior. *J. Lumin.* **2021**, *233*, 117737. [CrossRef]
96. Mokoena, P.P.; Oluwole, D.O.; Nyokong, T.; Swart, H.C.; Ntwaeaborwa, O.M. Enhanced upconversion emission of Er^{3+}-Yb^{3+} co-doped Ba$_5$(PO$_4$)$_3$OH powder phosphor for application in photodynamic therapy. *Sens. Actuators A Phys.* **2021**, *331*, 113014. [CrossRef]
97. Khajuria, P.; Bedyal, A.K.; Manhas, M.; Swart, H.C.; Durani, F.; Kumar, V. Spectral, surface and thermometric investigations of upconverting Er^{3+}/Yb^{3+} co-doped Na$_3$Y(PO$_4$)$_2$ phosphor. *J. Alloys Compd.* **2021**, *877*, 160327. [CrossRef]

98. Oliva, J.; Chávez, D.; González-Galván, A.; Viesca-Villanueva, E.; Díaz-Torres, L.A.; Fraga, J.; García, C.R. Tunable green/yellow upconversion emission and enhancement of the NIR luminescence in BaLaAlO$_4$: Er^{3+} phosphors by codoping with Yb^{3+} ions. *Optik* **2021**, *241*, 167011. [CrossRef]
99. Li, M.; Xie, T.; Tu, X.; Xu, J.; Lei, R.; Zhao, S.; Xu, S. Effects of doping concentration and excitation density on optical thermometric behaviors in Ho^{3+}/Yb^{3+} co-doped ZrO$_2$ upconversion nanocrystals. *Opt. Mater.* **2019**, *97*, 109478. [CrossRef]
100. Siai, A.; Ajili, L.; Horchani-Naifer, K.; Ferid, M. Tm^{3+} Modifying Er^{3+} Red Emission and Dielectric Properties of Tm^{3+}-Doped LaErO$_3$ Perovskite. *J. Electron. Mater.* **2020**, *49*, 3096–3105. [CrossRef]
101. Boyer, J.C.; Vetrone, F.; Cuccia, L.A.; Capobianco, J.A. Synthesis of colloidal upconverting NaYF$_4$ nanocrystals doped with Er^{3+}, Yb^{3+} and Tm^{3+}, Yb^{3+} via thermal decomposition of lanthanide trifluoroacetate precursors. *J. Am. Chem. Soc.* **2006**, *128*, 7444–7445. [CrossRef]
102. Boyer, J.C.; Cuccia, L.A.; Capobianco, J.A. Synthesis of colloidal upconverting NaYF$_4$: Er^{3+}/Yb^{3+} and Tm^{3+}/Yb^{3+} monodisperse nanocrystals. *Nano Lett.* **2007**, *7*, 847–852. [CrossRef]
103. Ye, X.; Collins, J.E.; Kang, Y.; Chen, J.; Chen, D.T.; Yodh, A.G.; Murray, C.B. Morphologically controlled synthesis of colloidal upconversion nanophosphors and their shape-directed self-assembly. *Proc. Natl. Acad. Sci. USA* **2010**, *107*, 22430–22435. [CrossRef]
104. Maurya, S.K.; da Silva, J.E.; Mohan, M.; Poddar, R.; Kumar, K. Assessment of colloidal NaGdF$_4$: Er^{3+}/Yb^{3+} upconversion phosphor as contrast enhancer for optical coherence tomography. *J. Alloys Compd.* **2021**, *865*, 158737. [CrossRef]
105. Pu, Y.; Lin, L.; Wang, D.; Wang, J.X.; Qian, J.; Chen, J.F. Green synthesis of highly dispersed ytterbium and thulium co-doped sodium yttrium fluoride microphosphors for in situ light upconversion from near-infrared to blue in animals. *J. Colloid Interface Sci.* **2018**, *511*, 243–250. [CrossRef] [PubMed]
106. Kang, X.J.; Dai, Y.L.; Ma, P.A.; Yang, D.M.; Li, C.X.; Hou, Z.Y.; Cheng, Z.Y.; Lin, J. Poly (acrylic acid)-modified Fe$_3$O$_4$ microspheres for magnetic-targeted and ph-triggered anticancer drug delivery. *Chem. A Eur. J.* **2012**, *18*, 15676–15682. [CrossRef] [PubMed]
107. Zeng, J.H.; Su, J.; Li, Z.H.; Yan, R.X.; Li, Y.D. Synthesis and upconversion luminescence of hexagonal-phase NaYF$_4$: Yb, Er^{3+} phosphors of controlled size and morphology. *Adv. Mater.* **2005**, *17*, 2119–2123.
108. Qiu, H.; Chen, G.; Sun, L.; Hao, S.; Han, G.; Yang, C. Ethylenediaminetetraacetic acid (EDTA)-controlled synthesis of multicolor lanthanide doped BaYF$_5$ upconversion nanocrystals. *J. Mater. Chem.* **2011**, *21*, 17202–17208. [CrossRef]
109. Yang, X.; Xiao, Q.; Niu, C.; Jin, N.; Ouyang, J.; Xiao, X.; He, D. Multifunctional core–shell upconversion nanoparticles for targeted tumor cells induced by near-infrared light. *J. Mater. Chem. B* **2013**, *1*, 2757–2763. [CrossRef] [PubMed]
110. Zhang, F.; Wan, Y.; Yu, T.; Zhang, F.; Shi, Y.; Xie, S.; Li, Y.; Xu, L.; Tu, B.; Zhao, D. Uniform nanostructured arrays of sodium rare-earth fluorides for highly efficient multicolor upconversion luminescence. *Angew. Chem.* **2007**, *119*, 8122–8125. [CrossRef]
111. Wang, F.; Han, Y.; Lim, C.S.; Lu, Y.; Wang, J.; Xu, J.; Chen, H.; Zhang, C.; Hong, M.; Liu, X. Simultaneous phase and size control of upconversion nanocrystals through lanthanide doping. *Nature* **2010**, *463*, 1061–1065. [CrossRef]
112. Wang, X.; Zhuang, J.; Peng, Q.; Li, Y. A general strategy for nanocrystal synthesis. *Nature* **2005**, *437*, 121–124. [CrossRef]
113. Stouwdam, J.W.; van Veggel, F.C.J.M. Near-infrared emission of redispersible Er^{3+}, Nd^{3+}, and Ho^{3+} doped LaF$_3$ nanoparticles. *Nano Lett.* **2002**, *2*, 733–737. [CrossRef]
114. Yi, G.; Lu, H.; Zhao, S.; Ge, Y.; Yang, W.; Chen, D.; Guo, L.H. Synthesis, characterization, and biological application of size-controlled nanocrystalline NaYF$_4$: Yb, Er infrared-to-visible up-conversion phosphors. *Nano Lett.* **2004**, *4*, 2191–2196. [CrossRef]
115. Patra, A.; Friend, C.S.; Kapoor, R.; Prasad, P.N. Upconversion in Er^{3+}: ZrO$_2$ nanocrystals. *J. Phys. Chem. B* **2002**, *106*, 1909–1912. [CrossRef]
116. Venkatramu, V.; Falcomer, D.; Speghini, A.; Bettinelli, M.; Jayasankar, C.K. Synthesis and luminescence properties of Er^{3+}-doped Lu$_3$Ga$_5$O$_{12}$ nanocrystals. *J. Lumin.* **2008**, *128*, 811–813. [CrossRef]
117. Shang, H.; Zhang, X.; Xu, J.; Han, Y. Effects of preparation methods on the activity of CuO/CeO$_2$ catalysts for CO oxidation. *Front. Chem. Sci. Eng.* **2017**, *11*, 603–612. [CrossRef]
118. Park, H.; Yoo, G.Y.; Kim, M.S.; Kim, K.; Lee, C.; Park, S.; Kim, W. Thin film fabrication of upconversion lanthanide-doped NaYF$_4$ by a sol-gel method and soft lithographical nanopatterning. *J. Alloys Compd.* **2017**, *728*, 927–935. [CrossRef]
119. Liang, Z.; Wang, X.; Zhu, W.; Zhang, P.; Yang, Y.; Sun, C.; Zhang, J.; Wang, X.; Xu, Z.; Zhao, Y.; et al. Upconversion nanocrystals mediated lateral-flow nanoplatform for in vitro detection. *ACS Appl. Mater. Interfaces* **2017**, *9*, 3497–3504. [CrossRef] [PubMed]
120. Mahalingam, V.; Mangiarini, F.; Vetrone, F.; Venkatramu, V.; Bettinelli, M.; Speghini, A.; Capobianco, J.A. Bright white upconversion emission from Tm^{3+}/Yb^{3+}/Er^{3+}-doped Lu$_3$Ga$_5$O$_{12}$ nanocrystals. *J. Phys. Chem. C* **2008**, *112*, 17745–17749. [CrossRef]
121. Wen, T.; Zhou, Y.; Guo, Y.; Zhao, C.; Yang, B.; Wang, Y. Color-tunable and single-band red upconversion luminescence from rare-earth doped Vernier phase ytterbium oxyfluoride nanoparticles. *J. Mater. Chem. C* **2016**, *4*, 684–690. [CrossRef]
122. Xu, L.; Yu, Y.; Li, X.; Somesfalean, G.; Zhang, Y.; Gao, H.; Zhang, Z. Synthesis and upconversion properties of monoclinic Gd$_2$O$_3$: Er^{3+} nanocrystals. *Opt. Mater.* **2008**, *30*, 1284–1288. [CrossRef]
123. Qin, X.; Yokomori, T.; Ju, Y. Flame synthesis and characterization of rare-earth (Er^{3+}, Ho^{3+}, and Tm^{3+}) doped upconversion nanophosphors. *Appl. Phys. Lett.* **2007**, *90*, 073104. [CrossRef]
124. Liu, J.; Bu, W.; Pan, L.; Shi, J. NIR-triggered anticancer drug delivery by upconverting nanoparticles with integrated azoben-zene-modified mesoporous silica. *Angew. Chem.* **2013**, *125*, 4471–4475. [CrossRef]
125. Liu, Y.; Hou, W.; Sun, H.; Cui, C.; Zhang, L.; Jiang, Y.; Wu, Y.; Wang, Y.; Li, J.; Sumerlin, B.S.; et al. Thiol-ene click chemistry: A biocompatible way for orthogonal bioconjugation of colloidal nanoparticles. *Chem. Sci.* **2017**, *8*, 6182–6187. [CrossRef] [PubMed]

126. Kong, W.; Sun, T.; Chen, B.; Chen, X.; Ai, F.; Zhu, X.; Li, M.; Zhang, W.; Zhu, G.; Wang, F. A general strategy for ligand exchange on upconversion nanoparticles. *Inorg. Chem.* **2017**, *56*, 872–877. [CrossRef] [PubMed]
127. Bao, G.; Wen, S.; Lin, G.; Yuan, J.; Lin, J.; Wong, K.L.; Bünzli, J.C.G.; Jin, D. Learning from lanthanide complexes: The development of dye-lanthanide nanoparticles and their biomedical applications. *Coord. Chem. Rev.* **2020**, *429*, 213642. [CrossRef]
128. Das, G.K.; Stark, D.T.; Kennedy, I.M. Potential toxicity of up-converting nanoparticles encapsulated with a bilayer formed by ligand attraction. *Langmuir* **2014**, *30*, 8167–8176. [CrossRef] [PubMed]
129. Li, X.; Shen, D.; Yang, J.; Yao, C.; Che, R.; Zhang, F.; Zhao, D. Successive layer-by-layer strategy for multi-shell epitaxial growth: Shell thickness and doping position dependence in upconverting optical properties. *Chem. Mater.* **2013**, *25*, 106–112. [CrossRef]
130. Shen, J.; Li, K.; Cheng, L.; Liu, Z.; Lee, S.T.; Liu, J. Specific detection and simultaneously localized photothermal treatment of cancer cells using layer-by-layer assembled multifunctional nanoparticles. *ACS Appl. Mater. Interfaces* **2014**, *6*, 6443–6452. [CrossRef]
131. Escudero, A.; Carrillo-Carrión, C.; Zyuzin, M.V.; Parak, W.J. Luminescent rare-earth-based nanoparticles: A summarized overview of their synthesis, functionalization, and applications. *Top. Curr. Chem.* **2016**, *374*, 1–15. [CrossRef]
132. Lai, W.F.; Rogach, A.L.; Wong, W.T. Molecular design of upconversion nanoparticles for gene delivery. *Chem. Sci.* **2017**, *8*, 7339–7358. [CrossRef]
133. Xu, C.T.; Zhan, Q.; Liu, H.; Somesfalean, G.; Qian, J.; He, S.; Andersson-Engels, S. Upconverting nanoparticles for pre-clinical diffuse optical imaging, microscopy and sensing: Current trends and future challenges. *Laser Photonics Rev.* **2013**, *7*, 663–697. [CrossRef]
134. Chen, Z.; Chen, H.; Hu, H.; Yu, M.; Li, F.; Zhang, Q.; Zhou, Z.; Yi, T.; Huang, C. Versatile synthesis strategy for carboxylic acid-functionalized upconverting nanophosphors as biological labels. *J. Am. Chem. Soc.* **2008**, *130*, 3023–3029. [CrossRef] [PubMed]
135. Sivakumar, S.; Diamente, P.R.; van Veggel, F.C.J.M. Silica-coated Ln^{3+}-doped LaF_3 nanoparticles as robust down-and upconverting biolabels. *Chem. Eur. J.* **2006**, *12*, 5878–5884. [CrossRef]
136. Yi, G.S.; Chow, G.M. Synthesis of hexagonal-phase $NaYF_4$: Yb, Er and $NaYF_4$: Yb, Tm nanocrystals with efficient up-conversion fluorescence. *Adv. Funct. Mater.* **2006**, *16*, 2324–2329. [CrossRef]
137. Dong, A.; Ye, X.; Chen, J.; Kang, Y.; Gordon, T.; Kikkawa, J.M.; Murray, C.B. A generalized ligand-exchange strategy enabling sequential surface functionalization of colloidal nanocrystals. *J. Am. Chem. Soc.* **2011**, *133*, 998–1006. [CrossRef] [PubMed]
138. Zhou, H.P.; Xu, C.H.; Sun, W.; Yan, C.H. Clean and flexible modification strategy for carboxyl/aldehyde-functionalized upconversion nanoparticles and their optical applications. *Adv. Funct. Mater.* **2009**, *19*, 3892–3900. [CrossRef]
139. Bao, Y.; Luu, Q.A.N.; Lin, C.; Schloss, J.M.; May, P.S.; Jiang, C. Layer-by-layer assembly of freestanding thin films with homogeneously distributed upconversion nanocrystals. *J. Mater. Chem.* **2010**, *20*, 8356–8361. [CrossRef]
140. Yi, G.S.; Chow, G.M. Water-soluble $NaYF_4$: Yb, Er (Tm)/$NaYF_4$/polymer core/shell/shell nanoparticles with significant enhancement of upconversion fluorescence. *Chem. Mater.* **2007**, *19*, 341–343. [CrossRef]
141. Sharipov, M.; Tawfik, S.M.; Gerelkhuu, Z.; Huy, B.T.; Lee, Y.I. Phospholipase A2-responsive phosphate micelle-loaded UCNPs for bioimaging of prostate cancer cells. *Sci. Rep.* **2017**, *7*, 1–9.
142. Wang, D.; Zhu, L.; Pu, Y.; Wang, J.X.; Chen, J.F.; Dai, L. Transferrin-coated magnetic upconversion nanoparticles for efficient photodynamic therapy with near-infrared irradiation and luminescence bioimaging. *Nanoscale* **2017**, *9*, 11214–11221. [CrossRef] [PubMed]
143. Teh, D.B.L.; Bansal, A.; Chai, C.; Toh, T.B.; Tucker, R.A.J.; Gammad, G.G.L.; Yeo, Y.; Lei, Z.; Zheng, X.; Yang, F.; et al. A Flexi-PEGDA Upconversion Implant for Wireless Brain Photodynamic Therapy. *Adv. Mater.* **2020**, *32*, 2001459. [CrossRef]
144. Wang, C.; Zhao, P.; Yang, G.; Chen, X.; Jiang, Y.; Jiang, X.; Wu, Y.; Liu, Y.; Zhang, W.; Bu, W. Reconstructing the intracellular pH microenvironment for enhancing photodynamic therapy. *Mater. Horiz.* **2020**, *7*, 1180–1185. [CrossRef]
145. Li, Y.; Zhang, X.; Zhang, Y.; Zhang, Y.; He, Y.; Liu, Y.; Ju, H. Activatable photodynamic therapy with therapeutic effect prediction based on a self-correction upconversion nanoprobe. *ACS Appl. Mater. Interfaces* **2020**, *12*, 19313–19323. [CrossRef]
146. Zhang, Y.; Chen, W.; Zhang, Y.; Zhang, X.; Liu, Y.; Ju, H. A Near-Infrared Photo-Switched MicroRNA Amplifier for Precise Photodynamic Therapy of Early-Stage Cancers. *Angew. Chem.* **2020**, *132*, 21638–21643. [CrossRef]
147. Lin, B.; Liu, J.; Wang, Y.; Yang, F.; Huang, L.; Lv, R. Enhanced upconversion luminescence-guided synergistic antitumor therapy based on photody-namic therapy and immune checkpoint blockade. *Chem. Mater.* **2020**, *32*, 4627–4640. [CrossRef]
148. Li, Z.; Qiao, X.; He, G.; Sun, X.; Feng, D.; Hu, L.; Xu, H.; Xu, H.B.; Ma, S.; Tian, J. Core-satellite metal-organic framework@ upconversion nanoparticle superstructures via electrostatic self-assembly for efficient photodynamic theranostics. *Nano Res.* **2020**, *13*, 3377–3386. [CrossRef]
149. Sun, W.; Sun, J.; Dong, B.; Huang, G.; Zhang, L.; Zhou, W.; Lv, J.; Zhang, X.; Liu, M.; Xu, L.; et al. Noninvasive temperature monitoring for dual-modal tumor therapy based on lanthanide-doped up-conversion nanocomposites. *Biomaterials* **2019**, *201*, 42–52. [CrossRef] [PubMed]
150. Liu, Z.; Xiong, L.; Ouyang, G.; Ma, L.; Sahi, S.; Wang, K.; Lin, L.; Huang, H.; Miao, X.; Chen, W.; et al. The Investigation of Copper Cysteamine Nanoparticles as a new type of radiosensitizers for Colorectal Carcinoma. *Sci. Rep.* **2017**, *7*, 9290. [CrossRef] [PubMed]
151. Chen, W.; Zhang, J. Using Nanoparticles to Enable Simultaneous Radiation and Photodynamic Therapies for Cancer Treatment. *J. Nanosci. Nanotechnol.* **2006**, *6*, 1159–1166. [CrossRef]

152. Chen, W. Nanoparticle Self-Lighting Photodynamic Therapy for Cancer Treatment. *J. Biomed. Nanotechnol.* **2008**, *4*, 369–376. [CrossRef]
153. Ma, L.; Chen, W.; Schatte, G.; Wang, W.; Joly, A.G.; Huang, Y.; Sammynaiken, R.; Hossu, M. A new Cu–cysteamine complex: Structure and optical properties. *J. Mater. Chem. C* **2014**, *2*, 4239–4246. [CrossRef]
154. Ma, L.; Zou, X.; Chen, W. A New X-ray Induced Nanoparticle Photosensitizers for Cancer Treatment. *J. Biomed. Nanotechnol.* **2014**, *10*, 1501–1508. [CrossRef]
155. Shrestha, S.; Wu, J.; Sah, B.; Vanasse, A.; Cooper, L.N.; Ma, L.; Li, G.; Zheng, H.; Chen, W.; Antosh, M.P. X-ray induced photodynamic therapy with copper-cysteamine nanoparticles in mice tumors. *Proc. Natl. Acad. Sci. USA* **2019**, *116*, 16823–16828. [CrossRef] [PubMed]
156. Shi, L.; Liu, P.; Wu, J.; Ma, L.; Zheng, H.; Antosh, M.P.; Zhang, H.; Wang, B.; Chen, W.; Wang, X. The effectiveness and safety of copper-cysteamine nanoparticle mediated X-PDT for cutaneous squamous cell carcinoma and melanom. *Nanomedicine* **2019**, *14*, 2027–2043. [CrossRef]
157. Zhang, Q.; Guo, X.; Cheng, Y.; Chudal, L.; Pandey, N.K.; Zhang, J.; Ma, L.; Xi, Q.; Yang, G.; Chen, Y.; et al. Use of copper-cysteamine nanoparticles to simultaneously enable radiotherapy oxidative therapy and immunotherapy for melanoma treatment. *Signal Transduct. Target. Ther.* **2020**, *5*, 58. [CrossRef]
158. Sah, B.; Wu, J.; Vanasse, A.; Pandey, N.K.; Chudal, L.; Huang, Z.; Song, W.; Yu, H.; Ma, L.; Chen, W.; et al. Effects of Nanoparticle Size and Radiation Energy on Copper-Cysteamine Nanoparticles for X-ray Induced Photodynamic Therapy. *Nanomaterials* **2020**, *10*, 1087. [CrossRef]
159. Chen, X.; Liu, J.; Li, Y.; Pandey, N.K.; Chen, T.; Wang, L.; Amador, E.H.; Chen, W.; Liu, F.; Xiao, E.; et al. Study of copper-cysteamine based X-ray induced photodynamic therapy and its effects on cancer cell proliferation and migration in a clinical mimic setting. *Bioact. Mater.* **2021**, *7*, 504–514. [CrossRef]
160. Wang, Y.; Alkhaldi, N.D.; Pandey, N.K.; Chudal, L.; Wang, L.Y.; Lin, L.W.; Zhang, M.B.; Yong, Y.X.; Amador, E.H.; Huda, M.N.; et al. A new type of cuprous-cysteamine sensitizers: Synthesis, optical properties and potential applications. *Mater. Today Phys.* **2021**, *19*, 100435. [CrossRef]
161. Liu, F.; Chen, W.; Wang, S.P.; Joly, A.G. Investigation of Water-Soluble X-ray Luminescence Nanoparticles For Photodynamic Activation. *Appl. Phys. Lett.* **2008**, *92*, 43901. [CrossRef]
162. Liu, Z.; Xiong, L.; Ouyang, G.; Ma, L.; Sahi, S.; Wang, K.; Lin, L.; Huang, H.; Miao, X.; Chen, W.; et al. Investigation of copper-cysteamine nanoparticles as a new photosensitizer for an-ti-hepatocellular carcinoma. *Cancer Biol. Ther.* **2019**, *20*, 812–825.
163. Pandey, N.K.; Chudal, L.; Phan, J.; Lin, L.; Johnson, O.; Xing, M.; Liu, J.P.; Li, H.; Huang, X.; Shu, Y.; et al. A facile method for synthesis of copper-cysteamine nanoparticles and study of ROS production for cancer treatment. *J. Mater. Chem. B* **2019**, *7*, 6630–6642. [CrossRef] [PubMed]

Article
Dye Sensitization for Ultraviolet Upconversion Enhancement

Mingkai Wang [1], Hanlin Wei [1], Shuai Wang [1], Chuanyu Hu [2,3,4,*] and Qianqian Su [1,*]

[1] Institute of Nanochemistry and Nanobiology, Shanghai University, Shanghai 200444, China; wmk19961108@163.com (M.W.); hanliwei@cityu.edu.hk (H.W.); m18937353231@163.com (S.W.)
[2] Department of Stomatology, Tongji Hospital, Tongji Medical College, Huazhong University of Science and Technology, Wuhan 430030, China
[3] School of Stomatology, Tongji Medical College, Huazhong University of Science and Technology, Wuhan 430030, China
[4] Hubei Province Key Laboratory of Oral and Maxillofacial Development and Regeneration, Wuhan 430022, China
* Correspondence: chuanyuhu@hust.edu.cn (C.H.); chmsqq@shu.edu.cn (Q.S.)

Abstract: Upconversion nanocrystals that converted near-infrared radiation into emission in the ultraviolet spectral region offer many exciting opportunities for drug release, photocatalysis, photodynamic therapy, and solid-state lasing. However, a key challenge is the development of lanthanide-doped nanocrystals with efficient ultraviolet emission, due to low conversion efficiency. Here, we develop a dye-sensitized, heterogeneous core–multishelled lanthanide nanoparticle for ultraviolet upconversion enhancement. We systematically study the main influencing factors on ultraviolet upconversion emission, including dye concentration, excitation wavelength, and dye-sensitizer distance. Interestingly, our experimental results demonstrate a largely promoted multiphoton upconversion. The underlying mechanism and detailed energy transfer pathway are illustrated. These findings offer insights into future developments of highly ultraviolet-emissive nanohybrids and provide more opportunities for applications in photo-catalysis, biomedicine, and environmental science.

Keywords: lanthanide nanoparticles; ultraviolet upconversion; dye sensitization; heterogeneous nanoparticles; energy transfer; luminescence enhancement

1. Introduction

Lanthanide-doped upconversion nanoparticles can absorb near-infrared (NIR) laser light and emit visible and ultraviolet light, with potential applications in bioimaging [1–5], biotherapy [6–12], and so on. In particular, the applications of these nanoparticles in optogenetic [13,14], photothermal [15,16], and photodynamic [17–19] therapy could be achieved via ultraviolet (UV) light emission under NIR excitation. Although UV light can be obtained by Nd^{3+}- and Yb^{3+}-sensitized upconversion [17,18,20,21], it is challenging to realize the high luminescence intensity needed to satisfy the minimum requirement of biological applications. This obstacle can be addressed in several ways: by controlling dopant composition [22], nanoparticle phase and size [23], excitation beam pulse width [24], and nanoparticle core–shell design [21,25–29]. Very recently, our group has made significant progress in overcoming the difficulty using an upconverted excitation lock-in (UCEL) strategy [30].

Hybrid systems are composed of inorganic nanoparticles and an organic dye, which can significantly strengthen the absorbance and expand the absorbance spectra of inorganic nanoparticles [31], leading to enhancement of their emission intensities. It has been demonstrated that NIR dye can effectively enhance the upconversion emission of lanthanide-doped nanoparticles [14,32–34]. However, previous studies have mainly focused on the analysis of visible upconversion emission. Little effort has been made to develop a hybrid nanoparticle with enhanced UV luminescence.

In this study, we developed IR-806-loaded upconversion nanoparticles (Gd-CS$_Y$S$_2$S$_3$@IR-806) with enhanced upconversion emission in the UV spectral region. The key factors that influence upconversion emission were studied, including dye concentration, excitation wavelength, and distance between the dye and the sensitizer Nd^{3+} (Scheme 1). We also demonstrated the dominant effect of ligand loading on multiphoton upconversion. In addition, the upconversion mechanism and the energy transfer pathway in Gd-CS$_Y$S$_2$S$_3$@IR-806 hybrid nanoparticles were carefully studied. This study provides new insights into the mechanistic understanding of UV upconversion luminescence in hybrid nanoparticles and enables new opportunities for these nanomaterials in a broad range of applications.

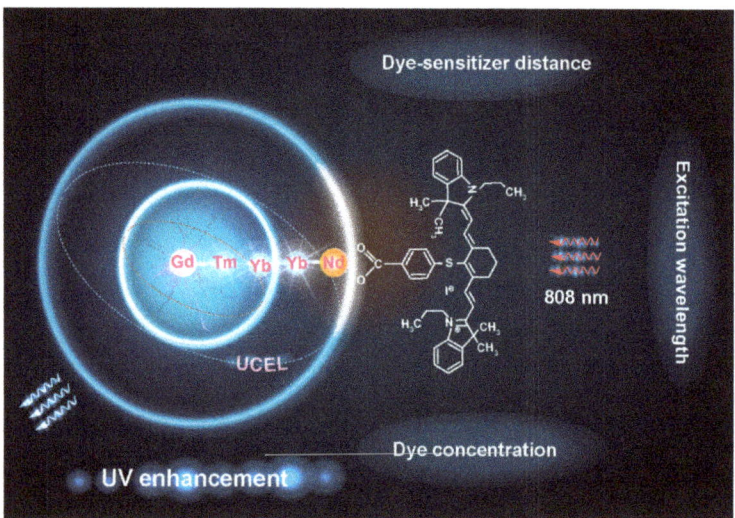

Scheme 1. Schematic illustration of the key factors that influence UV enhancement in IR-806-loaded upconversion nanoparticles, including dye concentration, excitation wavelength, and dye–sensitizer distance.

2. Materials and Methods

2.1. Materials

Gd(CH$_3$CO$_2$)$_3$·xH$_2$O (99.9%), Nd(CH$_3$CO$_2$)$_3$·xH$_2$O (99.9%), Y(CH$_3$CO$_2$)$_3$·xH$_2$O (99.9%), Yb(CH$_3$CO$_2$)$_3$·xH$_2$O (99.9%), Tm(CH$_3$CO$_2$)$_3$·xH$_2$O (99.9%), NaOH (>98%), NH$_4$F (>98%), chloroform(99.9%), oleic acid (OA, 90%), and 1-octadecene (ODE, 90%) were all purchased from Sigma-Aldrich (Shanghai, China). IR-806 is supported by Dr. Sanyang Han from the University of Cambridge. IR-780 iodide, 4-mercaptobenzoic acid, and N,N-dimethylformamide (DMF, anhydrous, 99.8%), as raw materials of IR-806, were obtained from Sigma-Aldrich (London, UK). Dichloromethane (DCM, AR) and diethyl ether (AR), as solvents for the synthesis of IR-806, were obtained from Lab-Scan (London, United Kingdom). Chloroform (AR), cyclohexane (AR), and ethanol (AR) were purchased from Sinopharm Chemical Reagent Co., Ltd. (Shanghai, China). All chemicals and reagents were used as received without further purification unless otherwise noted.

2.2. Characterization

Luminescence emission measurements were obtained with an FS5 (Edinburgh, UK) conjugated with 808 nm (CNI, MDL-III-808-2.5W, China), 793 nm (CNI, MDL-III-793-2.0W, China), and 980 nm (CNI, MDLIII-980-2.0W, China) diode lasers at room temperature. The decay curves were recorded by a lifetime spectrometer (FS5, Edinburgh, UK), in conjunction with pulsed 808 nm, 793 nm, and 980 nm diode lasers and a picosecond pulsed light emitting diode (EPLED-270). Low-resolution transmission electron microscopy (TEM) measurements were carried out on an HT7700 field emission transmission electron microscope operated at an acceleration voltage of 120 kV. The energy-dispersive X-ray (EDX) spectrum was obtained with an HT7700 field emission transmission electron microscope equipped with an Oxford Instruments system. High-resolution TEM images were obtained using an FEI Talos F200S transmission electron microscope operated at an acceleration voltage of 200 kV. HAADF-STEM and elemental mapping images were obtained using an FEI Talos F200X transmission electron microscope. Powder X-ray diffraction (XRD) analysis was performed on a Rigaku D/Max-2200 system equipped with a rotating anode and a Cu Kα radiation source (λ = 0.15418 nm). The excitation power density was measured using a TS5 laser power densitometer (Changchun New Industries Optoelectronics Technology, China). UV–vis absorption spectra were obtained using a PerkinElmer LAMBDA 750 ultraviolet–visible–near-infrared spectrometer and a Hitachi U-3010 spectrophotometer. All spectra were recorded under identical experimental conditions unless otherwise noted. Key experiments were repeated three times, and all other experiments were repeated twice.

2.3. Method

2.3.1. Synthesis of $NaGdF_4$:49%Yb,1%Tm Core Nanocrystals

$NaGdF_4$ doped with 49 mol % of Yb and 1 mol % of Tm ($NaGdF_4$:49%Yb,1%Tm) was synthesized via a modified literature procedure [28,35,36]. A water solution of $Gd(CH_3CO_2)_3$ (0.067 g; 0.2 mmol), $Yb(CH_3CO_2)_3$ (0.069 g; 0.196 mmol), and $Tm(CH_3CO_2)_3$ (0.001 g; 0.004 mmol) was combined with OA (5 mL) and ODE (5 mL) in a 50 mL two-neck round-bottom flask. The mixture was heated to 150 °C and maintained at this temperature for 1.5 h to form the lanthanide–oleate precursor. After cooling to 50 °C, a methanol solution consisting of NH_4F (0.05 g; 1.36 mmol) and NaOH (0.04 g; 1 mmol) was added to the mixture and stirred for 30 min. The solution was heated to 100 °C for 20 min in vacuo to remove methanol. The resulting solution was quickly heated to 300 °C and maintained at this temperature for 1.5 h with nitrogen before cooling to room temperature. The obtained nanocrystals were precipitated by centrifugation at 8000 rpm for 5 min and then washed with cyclohexane and ethanol three times. The core nanoparticles were dispersed in cyclohexane (4 mL) for further shell coating.

2.3.2. Synthesis of $NaGdF_4$:49%Yb,1%Tm@$NaYF_4$:20%Yb Core–Shell Nanocrystals

The synthesis procedure for core–shell nanoparticles was similar to that in our previous paper [36]. We use the obtained $NaGdF_4$:49%Yb,1%Tm nanocrystals as seeds for subsequent shell coating. $NaYF_4$ with 20 mol % of Yb ($NaYF_4$:20%Yb) precursor was prepared via the same procedure as mentioned above, except that different amounts of OA (3 mL) and ODE (7 mL) were used. After cooling to 80 °C, the cyclohexane solution of $NaGdF_4$:Yb/Tm nanoparticles (4 mL) was added and kept at 80 °C for 30 min to remove cyclohexane. Then, a methanol solution of NH_4F (0.05 g; 1.36 mmol) and NaOH (0.04 g; 1 mmol) was added to the mixture and stirred at 50 °C for 30 min. Subsequently, the mixture was heated to 100 °C for 20 min in vacuo to remove methanol. The solution was then heated to 300 °C for 1.5 h under a nitrogen atmosphere. After cooling to room temperature, the core–shell nanoparticles were collected and washed using the same post-treatment approach as for core nanocrystals. $NaGdF_4$@$NaGdF_4$:49%Yb,1%Tm and $NaYF_4$@$NaGdF_4$:49%Yb,1%Tm were synthesized using a similar method to core–shell nanocrystals except for the use of $NaGdF_4$ and $NaYF_4$ as seeds.

2.3.3. Synthesis of $NaGdF_4$:49%Yb,1%Tm@$NaYF_4$:20%Yb@$NaGdF_4$:50%Nd,10%Yb and $NaGdF_4$:49%Yb,1%Tm@$NaYF_4$:20%Yb@$NaGdF_4$:50%Nd,10%Yb@$NaGdF_4$ Core–Multishell Nanocrystals

The following multishelled core–shell nanoparticles were prepared using a procedure similar to the $NaGdF_4$:49%Yb,1%Tm@$NaYF_4$:20%Yb core–shell nanoparticles: $NaGdF_4$@$NaGdF_4$:49%Yb,1%Tm@ $NaYF_4$:20%Yb; $NaYF_4$@$NaGdF_4$:49%Yb,1%Tm@$NaYF_4$: 20%Yb; $NaGdF_4$@$NaGdF_4$:49%Yb, 1%Tm@$NaYF_4$:20%Yb@$NaGdF_4$:50%Nd,10%Yb; $NaYF_4$ @$NaGdF_4$: 49%Yb,1%Tm@$NaYF_4$:20%Yb@$NaGdF_4$:50%Nd,10%Yb;$NaGdF_4$@$NaGdF_4$:49%Yb,1%Tm@

NaYF$_4$:20%Yb@NaGdF$_4$:50%Nd,10%Yb@NaGdF$_4$; NaYF$_4$@NaGdF$_4$:49%Yb,1% Tm@NaYF$_4$: 20%Yb@ NaGdF$_4$:50%Nd,10%Yb@NaGdF$_4$.

2.3.4. Preparation of Dye-Sensitized Upconversion Nanoparticles

The synthesis of IR-806 followed a well-established method [32]. Then, the IR-806 was dissolved in CHCl$_3$ (0.01 mg/mL). The as-prepared core–multishell nanocrystals were centrifuged and dissolved in CHCl$_3$ to a final concentration of 0.375 mg/mL. The samples were prepared by adding different amounts of IR-806 to Gd-CS$_Y$S$_2$S$_3$ CHCl$_3$ solution (4 mL) and stirring for 2 h at a speed of 700 rpm at room temperature before UV–vis–NIR absorption and standard fluorescence measurements. All samples were prepared and measured in a dark environment.

3. Results

3.1. Synthesis of Core–Multishell Upconversion Nanoparticles

We previously designed a heterogeneous core–multishell nanoparticle with enhanced UV upconversion emission, involving six- and five-photon upconversion processes [30]. The optimum doping concentration and nanoparticle design were determined according to our previous reports [36]. From our previous photoluminescence results, the optimized nanostructure was determined to be NaGdF$_4$:49%Yb/1%Tm@NaGdF$_4$:20%Yb@ NaGdF$_4$:10%Yb/50%Nd@NaGdF$_4$. Recently, we found that when the NaGdF$_4$:20%Yb was replaced with NaYF$_4$:20%Yb, UV emission was significantly enhanced due to the effective suppression of energy consumption induced by interior energy traps. Herein, we chose this heterogeneous nanostructure as an experimental model to further enhance upconversion emission in the UV range. We first synthesized the core–multishell NaGdF$_4$:49%Yb,1%Tm@NaYF$_4$:20%Yb@NaGdF$_4$:10%Yb,50%Nd@NaGdF$_4$ (Gd-CS$_Y$S$_2$S$_3$) nanoparticles using a layer epitaxial growth method. Transmission electron microscopy (TEM) images showed that the nanoparticles had a uniform size of about 28 nm and the thickness of each layer was ~2 nm (Figure S1). The as-prepared nanoparticles were identified as the hexagonal phase by X-ray powder diffraction (XRD, JCPDS file number 27-0699, Figure S2). In addition, the constitution of the heterogeneous core–multishell nanostructures was confirmed by high-angle annular dark-field scanning transmission electron microscopy (HAADF-STEM), elemental mapping images and energy dispersive X-ray (EDX) spectra (Figure 1b,c, Figures S3 and S4), where brighter regions correspond to heavier elements (Gd, Yb, and Nd) and lighter regions correspond to lighter elements (Y).

3.2. Remarkable UV Enhancement

To enhance upconversion emission in the UV range, we chose a near-infrared (NIR) fluorescent dye (IR-806) to sensitize upconversion nanoparticles, due to its intense absorption in the NIR range [33]. As shown in Figure 2a, the fluorescence spectrum of IR-806 has considerable overlap with the absorbance of Gd-CS$_Y$S$_2$S$_3$ nanoparticles with Nd^{3+} (^{4}F$_{3/2}$→^{4}I$_{9/2}$) sensitizer, ensuring an effective energy transfer from IR-806 to the nanoparticles. We then utilized a modified Hummelen's method to load the IR-806 onto the surface of Gd-CS$_Y$S$_2$S$_3$ nanoparticles [32]. In addition, free IR-806 has an absorption band at 1708 cm^{-1} in the FTIR spectrum, corresponding to the stretching mode of –COOH. The absorption band at 1708 cm^{-1} disappeared when IR-806 was bound on to the surface of the nanoparticles. Nevertheless, absorption bands at 1560 and 1450 cm^{-1} were observed on Gd-CS$_Y$S$_2$S$_3$@IR-806, corresponding to the antisymmetric and symmetric vibration modes of the –COO$^-$ group. This indicates that the IR-806 carboxylic acid group was bound onto the surface of Gd-CS$_Y$S$_2$S$_3$, since the carboxylic region changed [33]. The successful preparation of IR-806-loaded Gd-CS$_Y$S$_2$S$_3$ nanoparticles was demonstrated by Fourier-transform infrared spectroscopy (FTIR) analysis (Figure 2b), which is consistent with the previous report [32]. In addition, we also compared the absorption spectra of Gd-CS$_Y$S$_2$S$_3$, IR-806, and Gd-CS$_Y$S$_2$S$_3$@IR-806. As shown in Figure 2c, Gd-CS$_Y$S$_2$S$_3$@IR-806 nanoparticles showed an intense absorbance band peaking at ~800 nm, further proving the successful loading of IR-806. Consequently, we observed more than 70-fold enhancements in Tm^{3+} emission over the whole wavelength range from 240–700 nm by Gd-CS$_Y$S$_2$S$_3$@IR-806 compared with Gd-CS$_Y$S$_2$S$_3$ nanoparticles, owing to the fact that the absorption cross section of Gd-CS$_Y$S$_2$S$_3$ was significantly enhanced after IR-806 loading. Furthermore, we also observed more than 600-fold, 300-fold, 150-fold, and 30-fold enhancements in UVC (240–280 nm), UVB (280–320 nm), UVA (320–400 nm), and visible (400–700 nm) regions, respectively (Figure 2e). Similarly, we synthesized the NaGdF$_4$:18%Yb,2%Er@NaYF$_4$:20%Yb@NaGdF$_4$:10%Yb, 50%Nd@NaGdF$_4$ nanoparticles with IR-806 loading. The emission intensity in the UV spectral region increased by more than 60 times, while the intensity in the visible region increased by only 30 times (Figure S5). Taken together, these results demonstrated that the overall enhancements were dominated by increased emission in the UV spectral regions, which is consistent with the dominant effect of ligand coordination on multiphoton upconversion [37]. Notably, the enhancement factors in the UV spectral region are remarkably larger than those in the visible region, offering enticing prospects for NIR light-mediated UV upconversion nanoparticles.

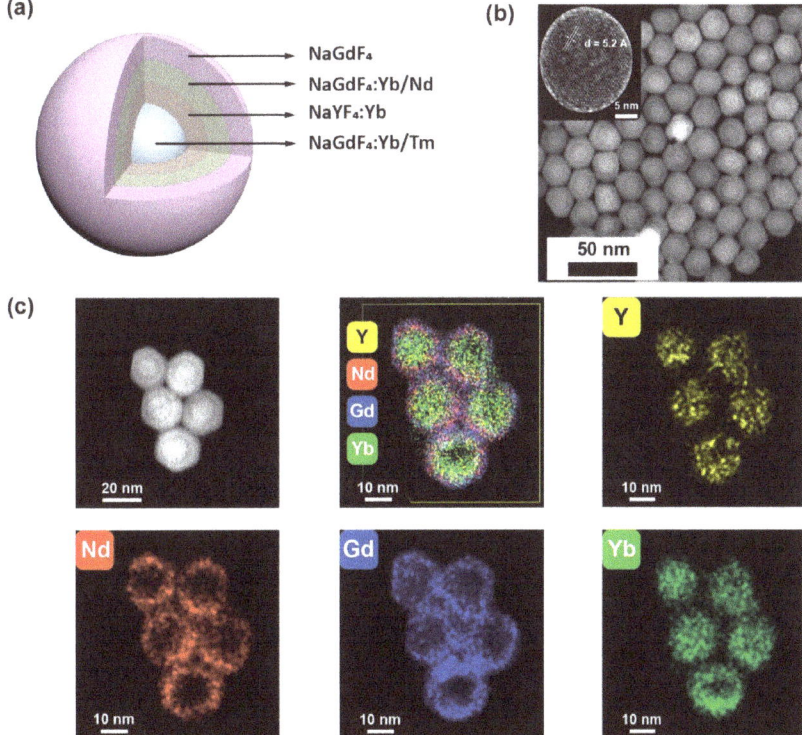

Figure 1. Schematic illustration and characterization of Gd-CS$_Y$S$_2$S$_3$ heterogeneous nanoparticles. (**a**) Diagrammatic representation of Gd-CS$_Y$S$_2$S$_3$ nanostructure. (**b**) High-angle annular dark-field scanning transmission electron microscopy (HAADF-STEM) image of Gd-CS$_Y$S$_2$S$_3$ nanoparticles. Inset: high-resolution TEM of as-prepared Gd-CS$_Y$S$_2$S$_3$ nanoparticle. (**c**) HAADF-STEM image and elemental mapping image of Gd-CS$_Y$S$_2$S$_3$ nanoparticles, revealing the spatial distribution of the Y, Nd, Gd, and Yb elements in the heterogeneous nanoparticles.

3.3. Optimum Weight Ratio between IR-806 and Nanoparticles

We determined the optimum weight ratio of Gd-CS$_Y$S$_2$S$_3$:IR-806 by setting a series of weight gradients from 120:1 to 180:1 (m$_{NPs}$:m$_{IR-806}$). As shown in Figure 3a, the optimum weight ratio was determined to be 160:1. The optimized number of dye molecules on the surface of Gd-CS$_Y$S$_2$S$_3$ nanoparticles was calculated to be 395 [32]. Note that the absorbance of Gd-CS$_Y$S$_2$S$_3$@IR-806 increased as IR-806 increased. However, when the weight ratio of Gd-CS$_Y$S$_2$S$_3$: IR-806 was smaller than 160:1, the emission intensity decreased due to fluorescence quenching caused by dye self-quenching. Due to the critical role of the Nd^{3+} sensitizers in mediating energy transfer from the dye to the upconversion nanoparticles, we verified that the optimum doping concentration of Nd^{3+} was 50 mol% (Figure S6). We then quantified the energy transfer efficiency of IR-806 to Gd-CS$_Y$S$_2$S$_3$ by measuring the lifetime of the IR-806 in a pair of Gd-CS$_Y$S$_2$S$_3$ samples with and without Nd^{3+} nanoparticles. Due to energy trapping by Nd^{3+}, the lifetime is shortened from 1.20 ns to 1.13 ns for Gd-CS$_Y$S$_2$S$_3$@IR-806. However, the lifetime of IR-806 was essentially unchanged after loading on Gd-CS$_{Y90\%Y,10\%Yb}$S$_3$@IR-806, due to the absence of Nd^{3+} dopants. The energy transfer efficiency was calculated to be 5.8% according to the following equation [38]:

$$E = 1 - \frac{\tau_{DA}}{\tau_D} \quad (1)$$

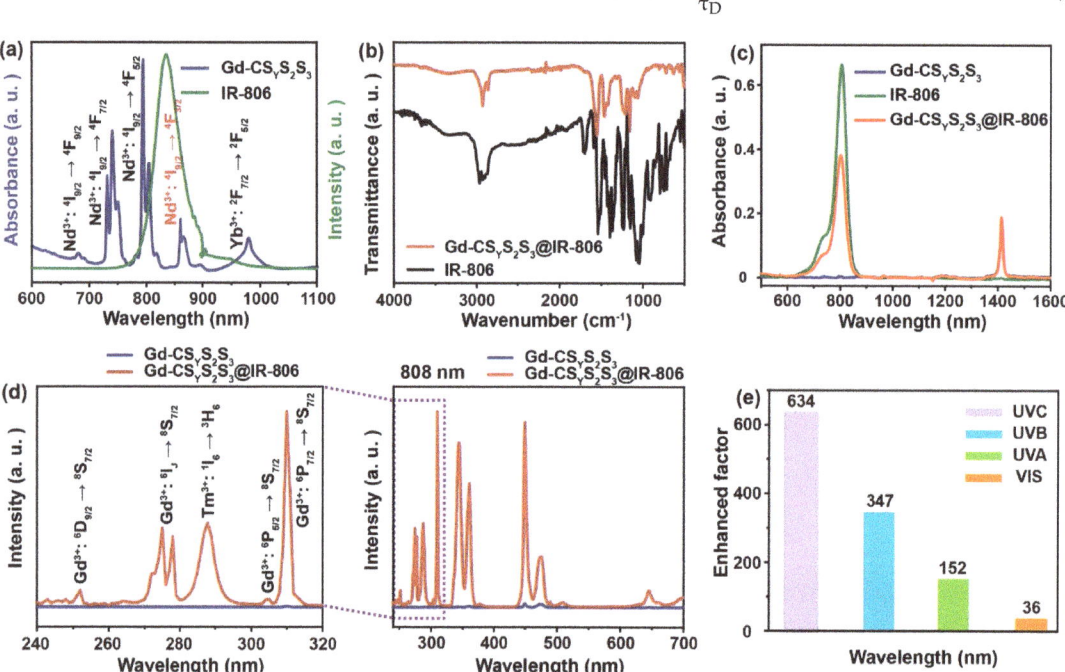

Figure 2. Preparation and characterization of Gd-CS$_Y$S$_2$S$_3$@IR-806. (**a**) IR-806 emission spectrum and Gd-CS$_Y$S$_2$S$_3$ nanoparticles absorption spectrum. (**b**) FTIR of Gd-CS$_Y$S$_2$S$_3$@IR-806 and IR-806. (**c**) The absorption spectra of Gd-CS$_Y$S$_2$S$_3$, IR-806, and Gd-CS$_Y$S$_2$S$_3$@IR-806. (**d**) Emission spectra of Gd-CS$_Y$S$_2$S$_3$ with and without IR-806 loading under 808 nm CW diode laser at a power density of 10 W/cm^2. (**e**) The enhancement factors of upconversion emission were obtained by comparing the results for samples with and without IR-806 loading. The emission intensities were calculated by integrating the spectral intensities in the UVC (240–280 nm), UVB (280–320 nm), UVA (320–400 nm), and visible (400–650 nm) ranges.

3.4. The Effect of Excitation Wavelength on UV Upconversion Emission

To investigate the enhancement effect on upconversion emission under 793, 808, and 980 nm excitation, we measured two series of Gd-CS$_Y$S$_2$S$_3$ nanoparticles with different amounts of IR-806 loading. As shown in Figure 3d, the emission intensities of Gd-CS$_Y$S$_2$S$_3$ were slightly improved after IR-806 loading under 793 nm excitation. In contrast, their emission intensities decreased under 980 nm excitation (Figure 3e). These results can be ascribed to poor matching between the excitation wavelengths (793 nm and 980 nm) and the absorption of IR-806. We then normalized the luminescence spectra of Gd-CS$_Y$S$_2$S$_3$

nanoparticles under three different excitation wavelengths. We found that the ratio was unchanged for UVC, UVB, UVA, and visible spectral regions under 793 nm and 980 nm excitation. In contrast, the normalized intensity of the UVC spectral region clearly increased (Figure S7), indicating effective energy transfer from IR-806 to the nanoparticles under 808 nm excitation.

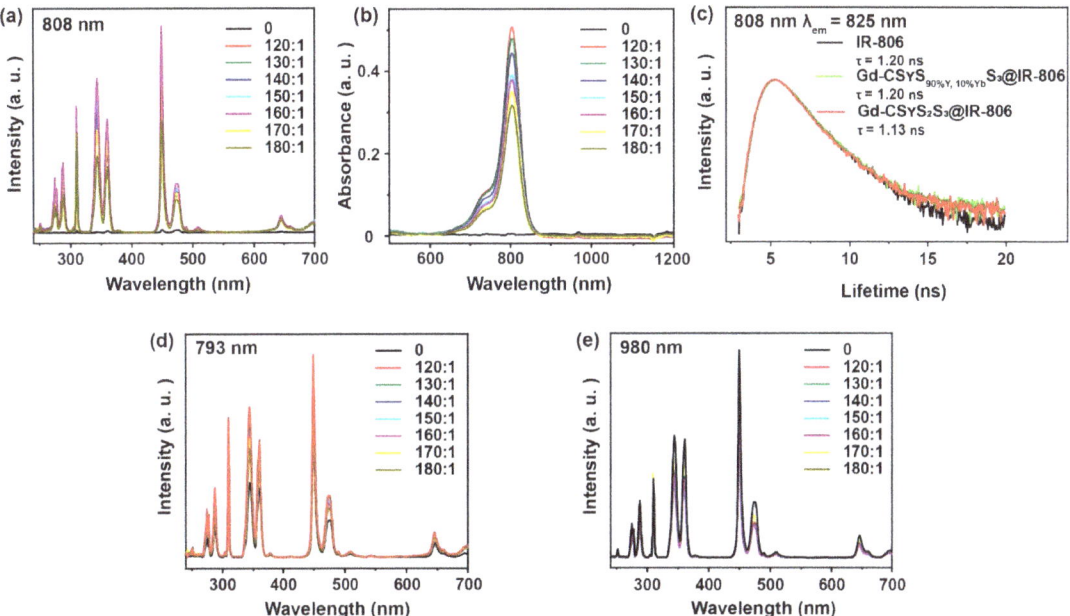

Figure 3. Optimizing the weight ratio of Gd-CS$_Y$S$_2$S$_3$ to IR-806 and calculating the energy transfer efficiency. (**a**) The emission spectrum of Gd-CS$_Y$S$_2$S$_3$ (4 mL in CHCl$_3$, 0.375 mg/mL) after adding various masses of IR-806 dye under 808 nm excitation. (**b**) The absorption spectrum of Gd-CS$_Y$S$_2$S$_3$ (4 mL in CHCl$_3$, 0.375 mg/mL) with various masses IR-806 dye. (**c**) The decay curves of Gd-CS$_Y$S$_2$S$_3$, Gd-CS$_Y$S$_{2(90\%,\,10\%Yb)}$S$_3$@IR-806, and Gd-CS$_Y$S$_2$S$_3$@IR-806. (**d,e**) The emission spectra of Gd-CS$_Y$S$_2$S$_3$ (4 mL in CHCl$_3$, 0.375 mg/mL) after adding various masses of IR-806 dye under 793 nm and 980 nm excitation, respectively.

3.5. The Effect of IR-806 Sensitizer Distance on UV Upconversion

To study the effect of the distance between Nd^{3+} and IR-806 on UV upconversion emission, we synthesized a pair of nanoparticles: Gd-CS$_Y$S$_2$S$_3$ and Gd-CS$_Y$S$_2$ (without the third shell protection) shown in Figure 4a. Comparing the emission intensities of Gd-CS$_Y$S$_2$S$_3$ and Gd-CS$_Y$S$_2$ with and without IR-806, the emission intensities of the Gd-CS$_Y$S$_2$ nanoparticles without shell protection increased by more than 230 times overall, while only 70-fold enhancement was observed in Gd-CS$_Y$S$_2$S$_3$, which has 2 nm thickness shell protection. Furthermore, UV and visible emission intensities increased more than 500-fold and 130-fold, respectively, for the nanoparticles without shell protection (Figure 4b). Notably, the transfer efficiency decreased as 1/R^6 [39]. Therefore, the enhancement factor decreased as the distance between the dye and the sensitizer increased.

Similarly, we synthesized two pairs of nanoparticles: NaGdF$_4$@ NaGdF$_4$:49%Yb,1%Tm @NaYF$_4$:20%Yb@NaGdF$_4$:10%Yb,50%Nd@NaGdF$_4$ (Gd-CS$_1$S$_Y$S$_3$S$_4$) vs. NaGdF$_4$@NaGdF$_4$: 49%Yb,1%Tm@NaYF$_4$:20%Yb@NaGdF$_4$:10%Yb,50%Nd (Gd-CS$_1$S$_Y$S$_3$) and NaYF$_4$@NaGdF$_4$: 49%Yb,1%Tm@NaYF$_4$:20%Yb@NaGdF$_4$:10%Yb,50%Nd@ NaGdF$_4$ (Y-CS$_1$S$_Y$S$_3$S$_4$) vs. NaYF$_4$ @NaGdF$_4$:49%Yb,1%Tm@NaYF$_4$:20%Yb@NaGdF$_4$:10%Yb, 50%Nd (Y-CS$_1$S$_Y$S$_3$) (Figure S8). The core–multishell structures are illustrated in Figure S9. To study the effect of different structures on emission enhancement, NaGdF$_4$ and NaYF$_4$ without any dopants were used as a core to shorten the distance between the NaGdF$_4$:49%Yb,1%Tm emissive layer and IR-806. The emission intensities of IR-806 grafted on Gd-CS$_1$S$_Y$S$_3$ and Gd-CS$_1$S$_Y$S$_3$S$_4$ increased 99 and 20 times, respectively, while the luminescence intensity in the UV region increased by 118 and 25 times and that in the visible region increased by 82 and 16 times, respectively. Moreover, the emission intensities of Y-CS$_1$S$_Y$S$_3$ and Y-CS$_1$S$_Y$S$_3$S$_4$ improved by 72 and 18 times after IR-806 loading. We also observed 81-fold and 22-fold enhancements in the UV spectral region and 63-fold and 14-fold enhancements in the visible region (Figure S10). These results are also consistent with our luminescence analysis, in that a significant enhancement in the UV luminescence of Gd-CS$_Y$S$_2$S$_3$ nanoparticles was observed compared to the visible range (Figure S11).

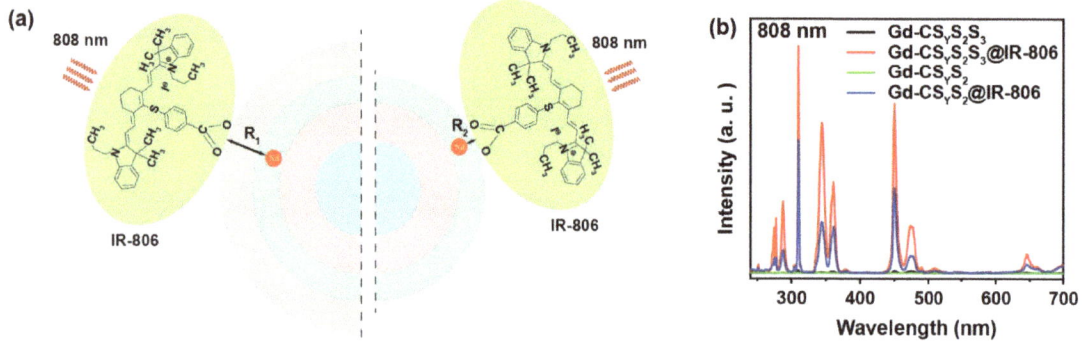

Figure 4. The effect of the distance between IR-806 and sensitizer Nd^{3+} on upconversion emission. (**a**) Schematic illustration of the nanostructural design to study the distance effect on upconversion emission. (**b**) The emission spectra of Gd-CS$_Y$S$_2$S$_3$, Gd-CS$_Y$S$_2$S$_3$@IR-806, Gd-CS$_Y$S$_2$, Gd-CS$_Y$S$_2$@IR-806 under 808 nm excitation.

3.6. Energy Transfer Mechanism

As shown in Scheme 2, IR-806 effectively absorbs the laser energy due to the absorption cross section under 808 nm excitation. To generate an efficient dye sensitization process, Nd^{3+} plays a critical role in bridging the energy transfer from the dye to the upconversion nanoparticles. Nd^{3+} ions trap the energy from the 808 nm laser and IR-806 mainly via the fluorescence–resonance energy transfer process and then gather photons at the $^4F_{5/2}$ energy state. Subsequently, relaxing to the $^4F_{3/2}$ energy state, Nd^{3+} transfers the energy to Yb^{3+} by an efficient energy transfer process. As an energy migrator, the excited Yb^{3+} populates the energy states of Tm^{3+} and gives rise to emission at 475 nm ($^1G_4\rightarrow{}^3H_6$), 450 nm ($^1D_2\rightarrow{}^3F_4$), 360 nm ($^1D_2\rightarrow{}^3H_6$), 345 nm ($^1I_6\rightarrow{}^3H_5$), and 290 nm ($^1I_6\rightarrow{}^3H_6$). Apart from emitting, Tm^{3+} serves as an energy donor donating energy to the Gd^{3+} ions via a five-photon process. Meanwhile, the six-photon upconversion process of 253 nm ($^6D_{9/2}\rightarrow{}^8S_{7/2}$) and the five-photon upconversion processes of 273 nm ($^6I_J\rightarrow{}^8S_{7/2}$), 276 nm ($^6I_J\rightarrow{}^8S_{7/2}$), 279 nm ($^6I_J\rightarrow{}^8S_{7/2}$), 306 nm ($^6P_{5/2}\rightarrow{}^8S_{7/2}$), and 310 nm ($^6P_{7/2}\rightarrow{}^8S_{7/2}$) are observed with the assistance of the appropriate energy matching of the following transition of $^2F_{5/2}\rightarrow{}^2F_{7/2}$ (9750 cm^{-1}, Yb^{3+}): $^6P_J\rightarrow{}^6D_J$ (~8750 cm^{-1}, Gd^{3+}). Notably, the utilization of an optically inert NaYF$_4$ host lattice with Yb^{3+} dopants as the interlayer plays a decisive role in protecting the energy by cooperative dye and Nd^{3+} sensitization from interior lattice defects, making it possible to effectively further increase UV via dye sensitizing.

3.7. Back Energy Transfer from Nanoparticles to IR-806

As well as increasing the luminescence intensity, a back energy transfer process from IR-806 to Gd-CS$_Y$S$_2$S$_3$ occurred. As depicted in Figure 5 and Figure S12, the lifetime of Gd^{3+} at 253, 276, and 310 nm, and Tm^{3+} at 290, 345, 475, and 650 nm slightly decreased after IR-806 loading, which can be ascribed to the nonradiative energy transfer from Gd^{3+} and Tm^{3+} to IR-806 [40–42].

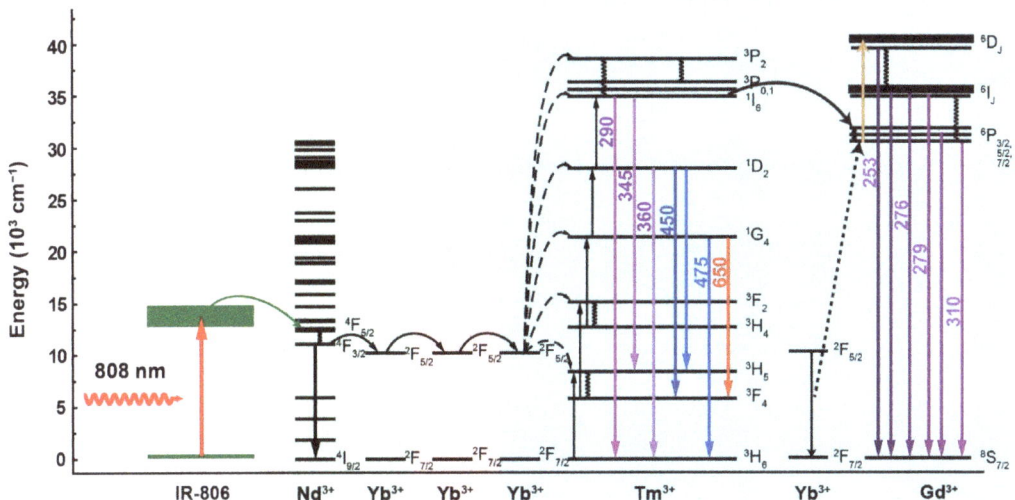

Scheme 2. Schematic illustration of the mechanism for cascade energy transfer in Gd-CS$_Y$S$_2$S$_3$@IR-806. Upon 808 nm laser excitation, IR-806 first absorbs excitation energy and transfers it to Nd^{3+}. Next, Yb^{3+} accepts the energy from Nd^{3+}, contributing to populating photons in the $^3P_2$ state of Tm^{3+} through a continuous five-photon energy transfer process and then relaxing to the $^1I_6$ state of Tm^{3+}. Trapping the energy from both five-photon upconversion from Tm^{3+} and one-photon upconversion from Yb^{3+}, six-photon and five-photon upconversion luminescence from ^6D_J, ^6I_J, and ^6P_J state of Gd^{3+} is observed.

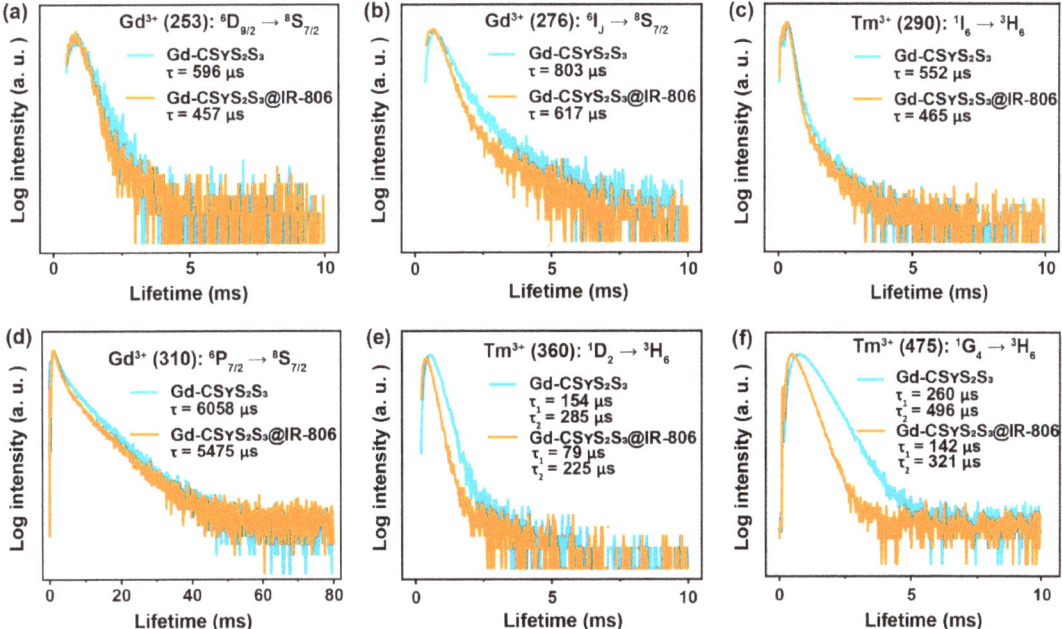

Figure 5. The decreased lifetime of Tm^{3+} and Gd^{3+} for Gd-CS$_Y$S$_2$S$_3$@IR-806. (**a**–**f**) The Tm^{3+} and Gd^{3+} lifetime decay curves of Gd-CS$_Y$S$_2$S$_3$ and Gd-CS$_Y$S$_2$S$_3$@IR-806 at 253, 276, 290, 310, 360, and 475 nm under 808 nm excitation, respectively.

4. Discussion

In this study, we developed a dye-sensitized heterogeneous lanthanide nanoparticle to regulate the energy transfer pathway for UV enhancement by 808 nm excitation. We systematically studied the influence of dye concentration, excitation wavelength, and distance between the dye and the sensitizer Nd^{3+} on upconversion emission, especially in the UV spectral region. Dye loading can improve the absorption of excitation light and thus improve the efficiency of energy-transfer-mediated upconversion. Moreover, our experimental results demonstrated a strengthened multiphoton upconversion process, which can be ascribed to the dominant effect of ligand loading on upconversion emission from high-lying energy states. The fundamentals gained from our investigations may provide insights into promoting the multiphoton upconversion process and the future design of organic–inorganic hybrid luminescent nanoparticles for applications in photocatalysis, biomedicine, environmental science, and more.

Supplementary Materials: The following are available online at https://www.mdpi.com/article/10.3390/nano11113114/s1. Figure S1: TEM and size distribution of Gd-CSYS2S3 nanoparticles; Figure S2: XRD of Gd-CSYS2S3 nanoparticles; Figure S3: EDX of Gd-CSYS2S3 nanoparticles; Figure S4: EDX lining analysis of Gd-CSYS2S3 nanoparticles; Figure S5: luminescence emission of NaGdF4:18%Yb, 2%Er@NaYF4:20%Yb@NaGdF4:10%Yb, 50%Nd@NaGdF4 with and without IR-806 loading; Figure S6: luminescence emission of Gd-CSYS2S3 nanoparticles with different Nd3+ doping before and after IR-806 loading; Figure S7: normalized intensity of luminescence spectra of Gd-CSYS2S3 with various contents of IR-806; Figure S8: TEM images of as-synthesized nanoparticles with different structures for distance effect studies; Figure S9: schematic illustration of five types of core–multishell structures including Gd-CSYS2S3, Y-CS1SYS3, Y-CS1SYS3S4, Gd-CS1SYS3, and Gd-CS1SYS3S4; Figure S10: luminescence spectra of as-synthesized nanoparticles with different structures for distance effect studies; Figure S11: normalized intensities of luminescence spectra of corresponding nanoparticles for distance effect studies; Figure S12: the lifetime decay of Tm3+ at 650 nm in Gd-CSYS2S3 and Gd-CSYS2S3@IR-806 nanoparticles under 808 nm excitation.

Author Contributions: Conceptualization, Q.S. and C.H; methodology, Q.S. and M.W.; validation, Q.S., C.H. and M.W.; investigation, M.W., Q.S., H.W. and S.W.; resources, Q.S.; data curation, M.W.; writing—original draft preparation, M.W.; writing—review and editing, Q.S. and C.H; visualization, M.W.; supervision, Q.S. and C.H.; project administration, Q.S.; funding acquisition, C.H. and Q.S. All authors have read and agreed to the published version of the manuscript.

Funding: This research was funded by the National Natural Science Foundation of China (Nos. 82002893 and 21701109).

Data Availability Statement: All of the relevant data are available from the correspondence authors upon reasonable request. Source data are provided with this paper.

Acknowledgments: The authors acknowledge Han for helpful discussions. The authors thank Jin and Guan for their help with the HAADF-STEM measurements.

Conflicts of Interest: The authors declare no conflict of interest.

References

1. Gu, Y.; Guo, Z.; Yuan, W.; Kong, M.; Liu, Y.; Liu, Y.; Gao, Y.; Feng, W.; Wang, F.; Zhou, J.; et al. High-sensitivity imaging of time-domain near-infrared light transducer. *Nat. Photon.* **2019**, *13*, 525–531. [CrossRef]
2. Zhu, X.; Su, Q.; Feng, W.; Li, F. Anti-stokes shift luminescent materials for bio-applications. *Chem. Soc. Rev.* **2017**, *46*, 1025–1039. [CrossRef]
3. Zhao, J.; Chu, H.; Zhao, Y.; Lu, Y.; Li, L. A NIR light gated DNA nanodevice for spatiotemporally controlled imaging of MicroRNA in cells and animals. *J. Am. Chem. Soc.* **2019**, *141*, 7056–7062. [CrossRef]
4. Wang, Y.-F.; Liu, G.-Y.; Sun, L.-D.; Xiao, J.-W.; Zhou, J.-C.; Yan, C.-H. Nd^{3+}-sensitized upconversion nanophosphors: Efficient In Vivo bioimaging probes with minimized heating effect. *ACS Nano* **2013**, *7*, 7200–7206. [CrossRef]
5. Xu, J.; Yang, P.; Sun, M.; Bi, H.; Liu, B.; Yang, D.; Gai, S.; He, F.; Lin, J. Highly emissive dye-sensitized upconversion nanostructure for dual-photosensitizer photodynamic therapy and bioimaging. *ACS Nano* **2017**, *11*, 4133–4144. [CrossRef]
6. Yang, D.; Ma, P.A.; Hou, Z.; Cheng, Z.; Li, C.; Lin, J. Current advances in lanthanide ion (Ln^{3+})-based upconversion nanomaterials for drug delivery. *Chem. Soc. Rev.* **2015**, *44*, 1416–1448. [CrossRef] [PubMed]

7. Idris, N.M.; Gnanasammandhan, M.K.; Zhang, J.; Ho, P.C.; Mahendran, R.; Zhang, Y. In Vivo photodynamic therapy using upconversion nanoparticles as remote-controlled nanotransducers. *Nat. Med.* **2012**, *18*, 1580–1585. [CrossRef] [PubMed]
8. Chen, G.; Qiu, H.; Prasad, P.N.; Chen, X. Upconversion nanoparticles: Design, nanochemistry, and applications in theranostics. *Chem. Rev.* **2014**, *114*, 5161–5214. [CrossRef] [PubMed]
9. Fan, W.; Bu, W.; Shi, J. On the latest three-stage development of nanomedicines based on upconversion nanoparticles. *Adv. Mater.* **2016**, *28*, 3987–4011. [CrossRef]
10. Dai, Y.; Xiao, H.; Liu, J.; Yuan, Q.; Ma, P.A.; Yang, D.; Li, C.; Cheng, Z.; Hou, Z.; Yang, P.; et al. In Vivo multimodality imaging and cancer therapy by near-infrared light-triggered trans-platinum pro-drug-conjugated upconverison nanoparticles. *J. Am. Chem. Soc.* **2013**, *135*, 18920–18929. [CrossRef]
11. Bansal, A.; Zhang, Y. Photocontrolled nanoparticle delivery systems for biomedical applications. *Acc. Chem. Res.* **2014**, *47*, 3052–3060. [CrossRef]
12. Zheng, W.; Zhou, S.; Xu, J.; Liu, Y.; Huang, P.; Liu, Y.; Chen, X. Ultrasensitive luminescent in vitro detection for tumor markers based on inorganic lanthanide nano-bioprobes. *Adv. Sci.* **2016**, *3*, 1600197. [CrossRef] [PubMed]
13. Chen, S.; Weitemier Adam, Z.; Zeng, X.; He, L.; Wang, X.; Tao, Y.; Huang Arthur, J.Y.; Hashimotodani, Y.; Kano, M.; Iwasaki, H.; et al. Near-infrared deep brain stimulation via upconversion nanoparticle–mediated optogenetics. *Science* **2018**, *359*, 679–684. [CrossRef]
14. Wu, X.; Zhang, Y.; Takle, K.; Bilsel, O.; Li, Z.; Lee, H.; Zhang, Z.; Li, D.; Fan, W.; Duan, C.; et al. Dye-sensitized core/active shell upconversion nanoparticles for optogenetics and bioimaging applications. *ACS Nano* **2016**, *10*, 1060–1066. [CrossRef] [PubMed]
15. Liu, J.; Zheng, X.; Yan, L.; Zhou, L.; Tian, G.; Yin, W.; Wang, L.; Liu, Y.; Hu, Z.; Gu, Z.; et al. Bismuth sulfide nanorods as a precision nanomedicine for in vivo multimodal imaging-guided photothermal therapy of tumor. *ACS Nano* **2015**, *9*, 696–707. [CrossRef] [PubMed]
16. Zhu, X.; Feng, W.; Chang, J.; Tan, Y.-W.; Li, J.; Chen, M.; Sun, Y.; Li, F. Temperature-feedback upconversion nanocomposite for accurate photothermal therapy at facile temperature. *Nat. Commun.* **2016**, *7*, 10437. [CrossRef]
17. Zuo, J.; Tu, L.; Li, Q.; Feng, Y.; Que, I.; Zhang, Y.; Liu, X.; Xue, B.; Cruz, L.J.; Chang, Y.; et al. Near infrared light sensitive ultraviolet–blue nanophotoswitch for imaging-guided "Off–On" therapy. *ACS Nano* **2018**, *12*, 3217–3225. [CrossRef]
18. Chan, M.-H.; Pan, Y.-T.; Chan, Y.-C.; Hsiao, M.; Chen, C.-H.; Sun, L.; Liu, R.-S. Nanobubble-embedded Inorganic 808 nm excited upconversion nanocomposites for tumor multiple imaging and treatment. *Chem. Sci.* **2018**, *9*, 3141–3151. [CrossRef]
19. Liu, C.; Liu, B.; Zhao, J.; Di, Z.; Chen, D.; Gu, Z.; Li, L.; Zhao, Y. Nd^{3+}-sensitized upconversion Metal–Organic frameworks for mitochondria-targeted amplified photodynamic therapy. *Angew. Chem. Int. Ed.* **2020**, *59*, 2634–2638. [CrossRef]
20. Zheng, K.; Qin, W.; Cao, C.; Zhao, D.; Wang, L. NIR to VUV: Seven-photon upconversion emissions from Gd^{3+} ions in fluoride nanocrystals. *J. Phys. Chem. Lett.* **2015**, *6*, 556–560. [CrossRef]
21. Chen, X.; Jin, L.; Kong, W.; Sun, T.; Zhang, W.; Liu, X.; Fan, J.; Yu, S.F.; Wang, F. Confining energy migration in upconversion nanoparticles towards deep ultraviolet lasing. *Nat. Commun.* **2016**, *7*, 10304. [CrossRef]
22. Zhao, C.; Kong, X.; Liu, X.; Tu, L.; Wu, F.; Zhang, Y.; Liu, K.; Zeng, Q.; Zhang, H. Li^+ Ion doping: An approach for improving the crystallinity and upconversion emissions of $NaYF_4$:Yb^{3+}, Tm^{3+} Nanoparticles. *Nanoscale* **2013**, *5*, 8084–8089. [CrossRef] [PubMed]
23. Shi, F.; Wang, J.; Zhang, D.; Qin, G.; Qin, W. Greatly enhanced size-tunable ultraviolet upconversion luminescence of monodisperse $β$-$NaYF_4$:Yb,Tm nanocrystals. *J. Mater. Chem.* **2011**, *21*, 13413–13421. [CrossRef]
24. Dawson, P.; Romanowski, M. Excitation modulation of upconversion nanoparticles for switch-like control of ultraviolet luminescence. *J. Am. Chem. Soc.* **2018**, *140*, 5714–5718. [CrossRef] [PubMed]
25. Wang, F.; Wang, J.; Liu, X. Direct evidence of a surface quenching effect on size-dependent luminescence of upconversion nanoparticles. *Angew. Chem. Int. Ed.* **2010**, *49*, 7456–7460. [CrossRef] [PubMed]
26. Wang, F.; Deng, R.; Wang, J.; Wang, Q.; Han, Y.; Zhu, H.; Chen, X.; Liu, X. Tuning upconversion through energy migration in core-shell nanoparticles. *Nat. Mater.* **2011**, *10*, 968–973. [CrossRef] [PubMed]
27. Sun, T.; Li, Y.; Ho, W.L.; Zhu, Q.; Chen, X.; Jin, L.; Zhu, H.; Huang, B.; Lin, J.; Little, B.E.; et al. Integrating temporal and spatial control of electronic transitions for bright multiphoton upconversion. *Nat. Commun.* **2019**, *10*, 1811. [CrossRef]
28. Su, Q.; Han, S.; Xie, X.; Zhu, H.; Chen, H.; Chen, C.-K.; Liu, R.-S.; Chen, X.; Wang, F.; Liu, X. The effect of surface coating on energy migration-mediated upconversion. *J. Am. Chem. Soc.* **2012**, *134*, 20849–20857. [CrossRef]
29. Xie, X.; Gao, N.; Deng, R.; Sun, Q.; Xu, Q.-H.; Liu, X. Mechanistic investigation of photon upconversion in Nd^{3+}-sensitized core–shell nanoparticles. *J. Am. Chem. Soc.* **2013**, *135*, 12608–12611. [CrossRef]
30. Su, Q.; Wei, H.-L.; Liu, Y.; Chen, C.; Guan, M.; Wang, S.; Su, Y.; Wang, H.; Chen, Z.; Jin, D. Six-photon Upconverted Excitation Energy Lock-in for Ultraviolet-C Enhancement. *Nat. Commun.* **2021**, *12*, 4367. [CrossRef]
31. Wen, S.; Zhou, J.; Schuck, P.J.; Suh, Y.D.; Schmidt, T.W.; Jin, D. Future and challenges for hybrid upconversion nanosystems. *Nat. Photon.* **2019**, *13*, 828–838. [CrossRef]
32. Zou, W.; Visser, C.; Maduro, J.A.; Pshenichnikov, M.S.; Hummelen, J.C. Broadband dye-sensitized upconversion of near-infrared light. *Nat. Photon.* **2012**, *6*, 560–564. [CrossRef]
33. Shao, Q.; Li, X.; Hua, P.; Zhang, G.; Dong, Y.; Jiang, J. Enhancing the upconversion luminescence and photothermal conversion properties of ~800 nm excitable core/shell nanoparticles by dye molecule sensitization. *J. Colloid Interface Sci.* **2017**, *486*, 121–127. [CrossRef] [PubMed]

34. Xue, B.; Wang, D.; Zhang, Y.; Zuo, J.; Chang, Y.; Tu, L.; Liu, X.; Yuan, Z.; Zhao, H.; Song, J.; et al. Regulating the color output and simultaneously enhancing the intensity of upconversion nanoparticles via a dye sensitization strategy. *J. Mater. Chem. C.* **2019**, *7*, 8607–8615. [CrossRef]
35. Wang, F.; Deng, R.; Liu, X. Preparation of core-shell NaGdF$_4$ nanoparticles doped with luminescent lanthanide ions to be used as upconversion-based probes. *Nat. Protoc.* **2014**, *9*, 1634–1644. [CrossRef] [PubMed]
36. Wang, S.; Shen, B.; Wei, H.-L.; Liu, Z.; Chen, Z.; Zhang, Y.; Su, Y.; Zhang, J.-Z.; Wang, H.; Su, Q. Comparative investigation of the optical spectroscopic and thermal effect in Nd^{3+}-doped nanoparticles. *Nanoscale* **2019**, *11*, 10220–10228. [CrossRef]
37. Xu, H.; Han, S.; Deng, R.; Su, Q.; Wei, Y.; Tang, Y.; Qin, X.; Liu, X. Anomalous upconversion amplification induced by surface reconstruction in lanthanide sublattices. *Nat. Photon.* **2021**, *15*, 732–737. [CrossRef]
38. Muhr, V.; Würth, C.; Kraft, M.; Buchner, M.; Baeumner, A.J.; Resch-Genger, U.; Hirsch, T. Particle-size-dependent förster resonance energy transfer from upconversion nanoparticles to organic dyes. *Anal. Chem.* **2017**, *89*, 4868–4874. [CrossRef]
39. Ray, P.C.; Fan, Z.; Crouch, R.A.; Sinha, S.S.; Pramanik, A. Nanoscopic optical rulers beyond the fret distance limit: Fundamentals and applications. *Chem. Soc. Rev.* **2014**, *43*, 6370–6404. [CrossRef]
40. Deng, R.; Wang, J.; Chen, R.; Huang, W.; Liu, X. Enabling förster resonance energy transfer from large nanocrystals through energy migration. *J. Am. Chem. Soc.* **2016**, *138*, 15972–15979. [CrossRef]
41. Tu, D.; Liu, L.; Ju, Q.; Liu, Y.; Zhu, H.; Li, R.; Chen, X. Time-Resolved FRET biosensor based on amine-functionalized lanthanide-doped NaYF$_4$ nanocrystals. *Angew. Chem. Int. Ed.* **2011**, *50*, 6306–6310. [CrossRef] [PubMed]
42. Kong, M.; Gu, Y.; Liu, Y.; Shi, Y.; Wu, N.; Feng, W.; Li, F. Luminescence lifetime–Based In Vivo detection with responsive rare earth–dye nanocomposite. *Small* **2019**, *15*, 1904487. [CrossRef] [PubMed]

Review

Encapsulation of Dyes in Luminescent Metal-Organic Frameworks for White Light Emitting Diodes

Zhihong Sun [1], Aaqib Khurshid [2], Muhammad Sohail [2], Weidong Qiu [1], Derong Cao [1,*] and Shi-Jian Su [1,*]

[1] State Key Laboratory of Luminescent Materials and Devices, School of Chemistry and Chemical Engineering, South China University of Technology, Guangzhou 510640, China; sunzhihong7@163.com (Z.S.); wdddqiu@foxmail.com (W.Q.)

[2] Department of Chemistry, University of Sargodha Sub Campus Bhakkar, Bhakkar 30000, Pakistan; chemistuos22@gmail.com (A.K.); msohail91147@gmail.com (M.S.)

* Correspondence: drcao@scut.edu.cn (D.C.); mssjsu@scut.edu.cn (S.-J.S.)

Abstract: The development of white light emitting diodes (WLEDs) holds great promise for replacing traditional lighting devices due to high efficiency, low energy consumption and long lifetime. Metal-organic frameworks (MOFs) with a wide range of luminescent behaviors are ideal candidates to produce white light emission in the phosphor-converted WLEDs. Encapsulation of emissive organic dyes is a simple way to obtain luminescent MOFs. In this review, we summarize the recent progress on the design and constructions of dye encapsulated luminescent MOFs phosphors. Different strategies are highlighted where white light emitting phosphors were obtained by combining fluorescent dyes with metal ions and linkers.

Keywords: metal–organic frameworks; organic dye; luminescence; white light emitting diodes

---

Citation: Sun, Z.; Khurshid, A.; Sohail, M.; Qiu, W.; Cao, D.; Su, S.-J. Encapsulation of Dyes in Luminescent Metal-Organic Frameworks for White Light Emitting Diodes. *Nanomaterials* **2021**, *11*, 2761. https://doi.org/10.3390/nano11102761

Academic Editor: Thomas Pons

Received: 10 September 2021
Accepted: 14 October 2021
Published: 18 October 2021

Publisher's Note: MDPI stays neutral with regard to jurisdictional claims in published maps and institutional affiliations.

Copyright: © 2021 by the authors. Licensee MDPI, Basel, Switzerland. This article is an open access article distributed under the terms and conditions of the Creative Commons Attribution (CC BY) license (https://creativecommons.org/licenses/by/4.0/).

1. Introduction

White light emitting diodes (WLEDs), as solid-state lighting sources, have attracted increasing attention in the past decades owing to their potential applications in displays and lighting [1]. WLEDs are energy saving and environmentally friendly, and have higher luminous efficiency than conventional incandescent and fluorescent lamps [2]. Moreover, WLEDs emit polychromatic light rather than monochromatic light that was emitted by traditional light emitting diodes (LEDs) [3]. It is well known that white light can be generated by mixing primary colors (red, green and blue) in appropriate proportions or using a pair of complementary colors [4]. Light sources with Commission International de l'Eclairage (CIE) coordinates (0.33, 0.33), color correlated temperature (CCT) between 2500 K and 6500 K, and color rendering indices (CRI) value above 80 are preferred for high-quality white light illumination [5]. Quantum yield (QY) is another important photophysical parameter, which refers to the ratio of photons emitted to the photons absorbed (unless otherwise specified, QY in this review is the absolute quantum yield). Currently, there are mainly two approaches to produce WLEDs: (1) multichip combination, in which three LEDs with primary colors are mixed appropriately to generate white light [6] and (2) phosphor-converted WLEDs (pc-WLEDs) approach, where phosphors are excited by a single-chip LED to produce white light. For pc-WLEDs, white light can often be obtained by a blue LED coated with a yellow-emitting phosphor or a ultraviolet (UV) LED coated with mixing phosphors [7]. Most commercially available WLEDs are pc-WLEDs due to the high cost and poor color stability of the color-mixed LEDs [8]. The first commercial WLED was developed by Nichia Chemical Co. in 1996 [9], which adopted a blue LED (InGaN) with yellow-emitting phosphor (YAG:Ce). Since then, tremendous progress has been made and the luminous efficacy has increased from 5 lm/W to over 300 lm/W [3]. Phosphors are of vital importance in determining the optical properties of WLEDs, including luminous efficiency, chromaticity coordinates, color temperature, lifetime and reliability. WLEDs phosphors should have the following properties: strong light absorption, broad excitation spectrum, useful emission spectrum, high quantum efficiency, optimal Stokes shift, high stability, etc. [4]. Current phosphors are almost all based on rare-earth metals and their self-quenching and absorption effects lower the phosphor performances [10]. Therefore, it is urgent to develop new phosphors, especially organic luminescent phosphors.

Metal-organic frameworks (MOFs) are a class of porous crystalline materials composed of inorganic and organic moieties via coordination bonds, which are known for tunable pore size, high surface areas, structure flexibility and multiple functionality. These extraordinary properties have made MOFs ideal candidates for catalysis, gas storage and separation, membranes, biomedical imaging and luminescence-based sensing and lighting [11,12]. Specially, MOFs offer a unique platform for the development of luminescent materials due to structural predictability, multifunctionality, nanoscale processability and well-defined environments for luminophores in crystalline states [13,14]. Luminescence in MOFs can arise from organic ligands, metal ions and charge transfers such as ligand-to-metal charge transfer (LMCT), metal-to-ligand charge transfer (MLCT), ligand-to-ligand charge transfer (LLCT) and metal-to-metal charge transfer (MMCT) [15]. In addition, some guests introduced into MOFs via supramolecular interactions can emit or induce luminescence, and white light can be easily obtained by rational structure design and luminescent guest selection. Overall, these various effects have naturally led to speculation that MOFs could find potential applications in WLEDs. The first attempt to obtain white light by using MOFs can be traced back to 2007 [16]. Since then, different color-emitting lanthanide metals, conjugated organic ligands and guest species such as dye molecules and quantum dots have been incorporated in MOFs to generate white light [17,18].

Encapsulation of emissive organic dyes is quite a simple way to obtain MOFs with multiple luminescence emissions [19]. Organic dyes are probably the most widespread fluorophores among the luminescent materials because of wide excitation band, large absorption coefficient, moderate-to-high quantum yields, short fluorescent lifetime and great availability [20]. However, there are two serious problems when directly applying organic dyes in WLEDs. One is the aggregation caused quenching (ACQ) effect induced by π-π stacking interactions of the organic dyes, which results in low fluorescence intensity in solid states in comparison with their bright solution states. Additionally, the other is the thermal and photo-stability of organic dyes [10]. MOFs are ideal supporting materials to prevent organic dyes aggregating in solid states [21,22], since MOFs are highly porous and able to encapsulate molecular dyes in confined pores, so they are capable of preventing aggregation-induced quenching and restricting internal molecular motions to inhibit non-radiative relaxation [23]. In addition, by carefully choosing fluorescent linkers and organic dyes, MOFs can serve as an antenna to transfer energy to the dyes. The emissions from encapsulated dyes can be easily adjusted by changing the component and content of dyes. Moreover, diverse luminescence properties can be achieved by engineering interactions between dyes and constituents of MOFs. Thus, encapsulation of dyes into MOFs is massively proposed as phosphor converters in white light emitting diodes [21].

There are three major methods to encapsulate organic dyes in MOFs [21]. The first is the two-step synthesis method, in which the pristine MOF is synthesized first and then immersed in a solution of fluorescent dyes. Despite the simplicity of this approach, the mismatch size between MOF aperture and organic dyes not only restricts the choice of dyes, but also causes guest leakage, which hiders the extensive application of this approach. The second is the in situ encapsulation method, where dyes are introduced during the crystal formation. Although this method is helpful in obtaining fluorescent MOFs with uniform distribution of fluorescent dyes, more factors including pore size, pore windows and structures of MOFs for desired organic dyes should be considered. The final method is to use fluorescent linkers incorporated in the frame of MOFs, in which permanent fluorescence can be easily obtained, although the steric hindrance caused by bulky ligands often reduces the yield of the fluorescent of MOFs. In practice, fluorescent ligands are often combined with dyes to induce dual emissions, and the ligand-to-dye energy transfer process can be controlled by changing the excitation energy.

MOFs materials with porosity, multifunctionality and crystallinity have aroused much interest since the debut of the "metal-organic frameworks" concept in 1995 [24], and the scope of this research has expanded from structure design and topology analysis to a wide range of applications in gas storage, catalysis and biomedicine [11,12,25–28]. A number of excellent reviews have summarized the properties and applications of luminescent metal–organic frameworks (LMOFs) [8,10,13–15,17,19,20,29–32], while the reviews that specifically and systemically discuss the encapsulation of organic dyes in MOFs (dye@MOFs) for WLEDs applications are still rare. This review mainly summarizes recent progress achieved in developing pc-WLEDs based on dye@MOFs, where white light can be generated by coating the dye encapsulated MOF hybrids on the corresponding blue-LED chip or UV-LED chip. The emphasis was put on the white light emitting phosphors fabrication. The origin of luminescence in dye@MOFs has been discussed to tune high-quality white light.

2. Phosphors Excited by Blue-LED Chip

The combination of a blue-LED chip with yellow phosphors belongs in a partial conversion. The blue light emitting from LED is partially absorbed by the phosphor and refurbished into yellow light, while the remaining part of blue light is transmitted through the phosphor [3]. The blue and yellow light, as a pair of complementary colors, mix together to generate white light. Generally, compared to the UV chip WLEDs, the blue LED chips have higher theoretical efficiency, better reproducibility and lower input energy, so they are quite attractive for low-cost bright white-light sources [33]. However, these WLEDs often show low CRI and high CCT caused by red emitting deficiency, which limits their indoor use. In the past decades, the design and synthesis of new blue-light-excitable single-phase phosphors have emerged as a hot research area, and much progress has been made in improving color-rendering properties, especially benefiting from the development of MOF materials. From a fundamental point of view, the abundant luminescent behaviors and ordered structures of MOFs allow for the fine-tuning of emission color across the CIE diagram and improve luminescent intensity simultaneously.

An effective way to improve color-rendering properties is to broaden the emission spectra. Qian et al. [34] simultaneously encapsulated green-emitting coumarin 6 (Cou-6), yellow-emitting rhodamine 6G (R6G) and red-emitting rhodamine 101 (R101) into a MOF crystal to synthesize a yellow broadband phosphor ZJU-28⊃Cou-6/R6G/R101 via ion exchange method. By coating the single-phase phosphor ZJU-28⊃Cou-6/R6G/R101 on commercial blue LED chips, the WLED lamp exhibits bright white light with luminous efficiency of 126 lm/W, CRI of 88 and CCT of 4446 K, and the total quantum yield (QY) can reach up to 82.9%. The good performance was ascribed to the high intrinsic quantum yields of dyes and fluorescence resonance energy transfer (FRET) process between them. In addition, the confinement effects of the MOFs can effectively inhibit the ACQ of dye molecules.

WLEDs can also be fabricated by combining blue chips with dye@MOFs and other commercialized phosphors [35]. Various concentration of rhodamine (Rh) dye was adopted to synthesize a series of Rh@bio-MOF-1 via cation exchange, and then the mixtures of the yellow-emitting Rh@bio-MOF-1, green $(Ba,Sr)_2SiO_4:Eu^{2+}$ and red $CaAlSiN_3:Eu^{2+}$ were coated on the blue LED chip to form phosphor film, which exhibits high luminous efficacy of 94–156 lm/W, CRI of 80–94 and excellent stability.

Unlike ion exchange, in situ encapsulation, in which fluorescent dyes are incorporated into the pores during the preparation of MOF crystals, have the advantage of uniform distribution of the fluorescent molecules, as long as the dyes can stand the synthesis conditions of MOFs. Li et al. [36] adopted the in situ encapsulation approach to avoid tedious ion-exchange synthesis and prevent dye leakage. Two yellow-emitting nanocomposites R6G@ZIF-8 and DBNT@UiO-66 with solubility compatibility and solution processability were synthesized, which can be excited by blue light to generate white light with absolute quantum yield of 63.1% and 22.7%, respectively. Similarly, Qian et al. [37] incorporated red, green and blue dyes into ZIF-8 by in situ self-assembly process to fabricate stable remote-type incandescent white-light device. They evaluated the thermostability and photostability of TPU-encapsulated ZIF-8⊃pm546/pm605/SRh101 phosphor in detail, and found the stability was greatly enhanced with TPU coating, which was ascribed to the protection against the oxygen and water invasion.

Guest species like carbon dots with strong resistance to irradiation and heat are preferable in order to improve the stability of phosphors. Li et al. [38] encapsulated both green-emitting carbon quantum dots (CQDs) and red-emitting rhodamine B (RhB) into ZIF-8, where RhB molecules can be sensitized by CQDs. Single-phase single-shell CQDs&RhB@ZIF-8^2 and single-phase multi-shell CQDs@ZIF-8^2@RhB@ZIF-8^2 were fabricated as yellow phosphors. Multi-shell CQDs@ZIF-8^2@RhB@ZIF-8^2 shows higher luminescence efficiency due to a large spatial distance that can suppress the FRET interactions between CQDs and dyes. Benefiting from the host-guest shielding effect, the stability of hybrid materials can be further improved. Tan et al. [39] conducted a long-term material stability test on dye-encapsulated MOF Gaq3@ZIF-8, and the results showed that after 8 months, not only the structure of ZIF-8 could remain stable with Gaq3 dye being encapsulated, but also the absolute QY of Gaq3@ZIF-8 (15%) was exactly the same as prepared, which demonstrates that when trapping Gaq3 in ZIF-8 pores, the host can act as a shield to protect Gaq3 from photodegradation. A WLED emitting uniform white light could be obtained by coating Gaq3@ZIF-8 on a blue LED.

3. Phosphors Excited by UV-LED Chip

For WLEDs based on UV-LED and phosphors, all radiation from UV LED is converted into red, green and blue (RGB) light, which refers to full conversion. The phosphors excited by UV-LED chip must emit white light, so RGB phosphors are often adopted. As mentioned before, pc-WLEDs fabricated by blue LED coated with yellow phosphors may suffer such weaknesses as poor CRI and low stability of color temperature, due to deterioration of the chip or the phosphors. By contrast, UV-LED combined with mixed phosphors is one of the best approaches to generate white light for both high luminous efficiency and high CRI, at the expense of poorer efficacy owing to higher wavelength-conversion losses. Recently, developing single-phase white light phosphors is of great significance and different strategies have been adopted to improve UV-LED luminous efficacy. In general, luminescence in MOFs can be obtained from linkers, framework metal ions, and absorbed guests [29].

3.1. Luminescence from Organic Dyes

In order to obtain a single-phase white light phosphor, Bu et al. [40] reported the encapsulation of RGB dyes into anionic MOF via dye exchange, as shown in Figure 1a. NKU-114 with abundant nitrogen sites can serve as an excellent host matrix to incorporate electron-deficient cationic dyes due to the donor–acceptor electrostatic interactions. DSM, AF and 9-AA, with strong red, green and blue light emissions, respectively, were used as selected cationic dyes (Figure 1b), because of suitable spectral overlap and proper molecule size. By carefully tuning the relative contents of three dyes, white-light-emitting NKU-114@DSM-AF/9-AA composite could be obtained, with CIE coordinates (0.34, 0.32), CRI and CCT values of 81 and 5101 K, respectively. The absolute quantum yield reaches a comparative high value of 42.07%, compared with some other reported dye-encapsulated system [41,42]. A WLED was assembled by coating NKU-114@DSM-AF/9-AA on the surface of a UV-LED chip. Under 50 mA, the WLED shows corresponding CIE coordinates, CRI, CCT and luminous efficiency values of (0.3402, 0.3365), 85.41, 5148 K and 2.4 lm/W, respectively.

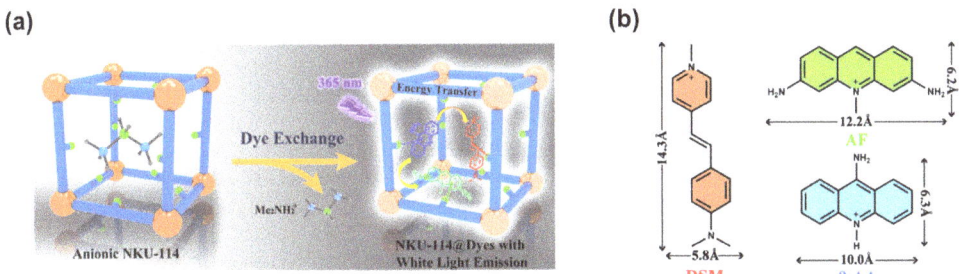

Figure 1. (**a**) Schematic representation of the incorporation of multicomponent dyes in an anionic MOF; (**b**) structures of cationic dyes. (Reproduced with permission from ref. [40]. Copyright © 2020, American Chemical Society).

In 2019, inspired by the extensive applications of core-shell structured MOFs, Gong [41] proposed a novel approach to construct core-shell structured cyclodextrin (CD) based MOFs by encapsulating different guests hierarchically into the framework. CD has the ability to improve fluorescence of organic dyes, because it can provide a confined hydrophobic cavity to change the stacking patterns of organic dyes and decrease the freedom of molecular motions [43]. CD-MOF⊃dyes, with γ-cyclodextrin (γ-CD) as organic ligands (Figure 2a), exhibit extremely high luminous intensity because of synergistic effect of CD and MOFs. Fluorescein (FL), RhB and 7-hydroxycoumarin (7-HCm) were chosen as encapsulates. The chemical structures of FL, RhB and fluorescence emission spectrum of CD-MOF⊃7-HCm are shown in Figure 2b–d. CD-MOF⊃RhB with longer emission wavelength was selected as core, and CD-MOF⊃7-HCm@FL@RhB was fabricated via epitaxial seeded growth (Figure 2e). The prepared core-shell crystals emit bright white light upon excitation of UV-LED, with CIE coordinate of (0.35, 0.32).

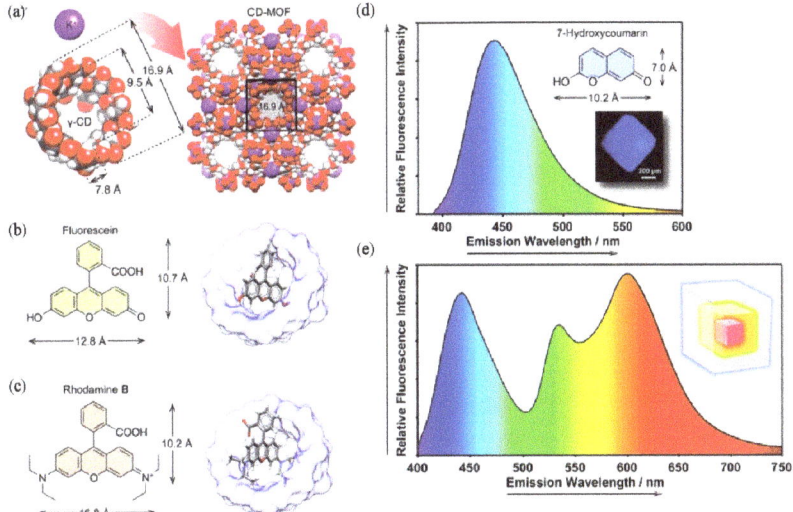

Figure 2. (**a**) γ-CD and formation of CD-MOF; (**b**) structure of FL; (**c**) structure of RhB; (**d**) fluorescence emission spectrum of CD-MOF⊃7-HCm; (**e**) fluorescence emission spectrum of CD-MOF⊃7-HCm@FL@RhB under an excitation wavelength of 365 nm. (Reproduced with permission from ref. [41]. Copyright © 2019, American Chemical Society).

Li et al. [44] reported high quality white-light-emitting dyes@ZIF-8 composites based on the three models (multiphase single-shell dye@ZIF-8, single-phase single-shell dyes@ZIF-

8, and single-phase multi-shell dyes@ZIF-8) (Figure 3a), in which dye locations are tunable. Red-emitting rhodamine B (RB), green-emitting fluorescein (F) and C-151 were chosen to match the pore structure of ZIF-8. Multiphase single-shell dye@ZIF-8^2 is solution-processable for device fabrication, and white light can be generated by optimizing the ratio of C-151@ZIF-8^2, F@ZIF8^2, and RB@ZIF-8^2, with CIE coordinates of (0.32, 0.34). Single-phase single-shell C-151&F&RB@ZIF-8^2 composite can be prepared by introducing C-151, F, and RB simultaneously into ZIF-8 via in situ encapsulation. By carefully tuning the content of red, green and blue emitting dyes, white light emitting C-151&F&RB@ZIF-8^2 composites with CIE coordinates (0.30, 0.34) and (0.34, 0.34) were obtained (Figure 3b). The efficiency decrease problem caused by FRET process can be solved by adopting model 3, a single-phase multi-shell dyes@ZIF-8. In model 3, RB, F and C-151 were encapsulated successively into ZIF-8 using shell-by-shell overgrowth, and the CIE chromaticity coordinates of multi-shell C-151@ZIF-8^2 @F@ZIF-8^2 @RB@ZIF-8^2 changed from (0.21, 0.26) to (0.32, 0.34) by tuning the concentration of RB (Figure 3c).

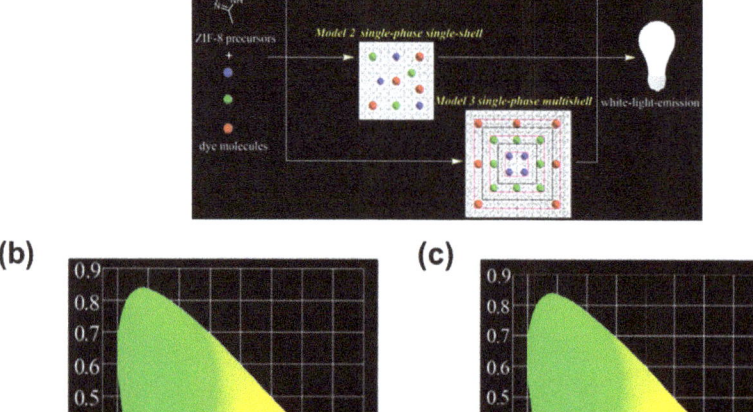

Figure 3. (a) Model 1 (multiphase single-shell dye@ZIF-8), Model 2 (single-phase single-shell dyes@ZIF-8), and Model 3 (single-phase multi-shell dyes@ZIF-8); (b) CIE coordinates of C-151&F&RB@ZIF-8^2 with different concentrations of dyes at λ_{ex} = 365 nm; (c) CIE chromaticity coordinates of C-151@ZIF-8^2@F@ZIF-8^2@RB@ZIF-8^2 with different concentrations of RB. (Reproduced with permission from ref. [44]. Copyright © 2019, American Chemical Society).

3.2. Luminescence from Dyes and Metals

During the development of luminescent MOFs, the lanthanide MOFs have aroused extensive interest from the very beginning owing to high luminescence quantum yield, large Stokes shifts and sharp line-emissions [30]. Since f–f transition is parity-forbidden, lanthanide ions are often sensitized by organic ligands due to antenna effect. Qian [45] fabricated a phosphor for WLED by encapsulating blue dye within lanthanide MOF. EuBPT, TbBPT and Eu_xTb_yBPT were synthesized by the solvothermal reaction. Owing to the energy transfer from BPT ligands to the lanthanide ions, the absolute quantum yields of red-emitting EuBPT and green-emitting TbBPT reached 37.11% and 73.68%, respectively. By optimizing the Eu^{3+}/Tb^{3+} ratio, $Eu_{0.05}Tb_{0.95}$BPT exhibits yellow light, and when combined with blue dye C460, white light emitting phosphor with absolute QY of 43.42% could be generated. The CRI and CCT values of the phosphors were estimated to be 90 and 6034 K, respectively. The WLED devices were fabricated by coating the prepared phosphor on a commercial UV-LED chip, and the luminous efficiency was measured to be 7.9 lm/W. Similarly, Saha [46] incorporated a single red emitting dye RhB into blue emitting gadolinium-based MOF to achieve perfect white light with high quantum yield.

Apart from the lanthanide, actinide can also be used to construct luminescent MOFs. Recently, inspired by the concept of 'molecular compartment' [47], Luo et al. synthesized a cage-based actinide MOF ECUT-300 [48]. Due to the trigonal building unit being constructed from the coordination of uranyl ions and carboxylate, ECUT-300 with mesopore A (2.8 nm), mesopore B (2.0 nm) and micropore C (0.9 nm) could be fabricated. Combining uranyl ions and 4,4′,4″,4‴-(ethene-1,1,2,2-tetrayl)tetrabenzoic acid as ligand, ECUT-300 with blue-green emission was observed upon excitation at 408 nm. Interestingly, RhB was encapsulated in the cage B of ECUT-300, and WLED device could be fabricated by coating RhB@ECUT-300 on an UV LED. While $[Fe(tpy)^2]^{3+}$ was encapsulated in cage C, which could be used to selectively adsorb C_2H_2 over CO_2. In addition, the incorporation of both RhB^+ and $[Fe(tpy)^2]^{3+}$ is helpful in stabilizing the framework structure.

3.3. Luminescence from Dyes and Organic Linkers

Combining the emissions from linkers and dyes to generate single-phase white light phosphors is a hot research topic in recently years. In 2015, Qian [49] first encapsulated two dyes simultaneously into blue-emitting anionic MOFs via ion exchange. ZJU-28 exhibits blue emission under excitation at 365 nm, which ascribes to the H_3BTB ligand. ZJU-28⊃DSM/AF, as white lighting phosphor, can be easily prepared by soaking ZJU-28 into the mixed solution of red-emitting DSM and green-emitting AF, exhibiting broadband white emission with CIE coordinates of (0.34, 0.32), CRI value up to 91% and CCT of 5327 K. Since the confinement of MOFs can effectively suppress ACQ, the absolute QY could be improved to 17.4%. Substituting the H_3BTB ligand with carbazole-based ligand 4,4′,4″-(9H-carbazole-3,6,9-triyl)-tribenzoic acid (H_3L) [50], a white-light-emitting phosphor with same CRI value could be obtained, while the quantum yield could reach up to 39.4%.

It is worth noting that efficient blue emission plays an important role in developing WLED, so strong blue fluorescent molecules are often introduced in company with red and green fluorescent molecules. Zhu [51] reported the incorporation of neutral and ionic RGB guest molecules into a neutral MOF HSB-W1 (Figure 4a). HSB-W1 exhibits blue emission upon excitation at 365 nm. HSB-W1⊃DCM, HSB-W1⊃C6 and HSB-W1⊃CBS-127 can be conveniently synthesized and exhibit red, green and blue emission, respectively, as shown in (Figure 4b). HSB-W1⊃DCM⊃C6⊃CBS-127 composite emits white-light with high quantum yield (up to 26%) and CRI (up to 92%). The results showed that incorporating RGB dyes into blue-emission MOFs is a useful strategy to design single-phase white-light phosphors.

The combination of blue and yellow emission can generate white light. Apart from H_3BTB and HSB ligands, 9,10-dibenzoate anthracene (DBA) is also an efficient blue emitter. A new phosphor for WLED was fabricated by encapsulating RhB into Al-DBA, and exhibits an emission lifetime of 1.8 ns and 5.4 ns for the blue and yellow light, respectively, enabling the WLED for visible light communication (VLC) [23].

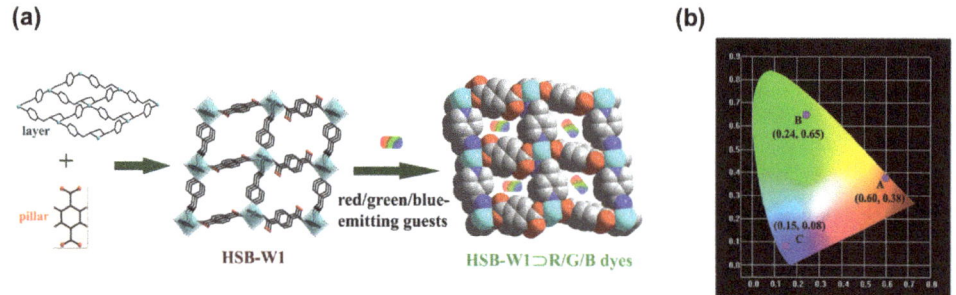

Figure 4. (**a**) Schematic synthesis of HSB-W1⊃R/G/B Dyes. (**b**) The CIE coordinates of HSB-W1⊃DCM (DCM, 0.31 wt%), HSB-W1⊃C6 (C6, 0.04 wt%), and HSB-W1⊃CBS-127 (CBS-127, 0.03 wt%) (λ_{ex} = 365 nm). (Reproduced with permission from ref. [51]. Copyright © 2019 Wiley—VCH Verlag GmbH & Co. KGaA, Weinheim).

There is a class of materials that are non-emissive in dilute solutions, but become highly luminescent after aggregation. This phenomenon is termed "aggregate induced emission" (AIE), which was proposed by Tang [52]. Tetraphenylethylene (TPE) as a typical AIE luminogen is of great interest in WLEDs [53]. Fu et al. [54] synthesized a TPE-based MOF with broadband green-yellow emission due to the energy transfer between dual linkers. White light can be generated by encapsulation of sulforhodamine 101 (SR101) into the MOF matrix, with corresponding QY, CRI and CCT values of 41.7%, 81.3 and 4527 K, respectively. Zhou et al. [55] investigated the dye encapsulation in TPE-based MOF for WLED. PCN-128W containing TPE-based ligand can be used to sensitize dye molecules through FRET, so DSM@PCN-128W shows dual emissions upon a single excitation. Compared with PCN-128W, the H_4ETTC ligand exhibits about 70 nm red-shift, because the confinement of MOFs increases the HOMO-LUMO energy gap of the linkers and generates high energy emission. In order to understand better, the typical molecular structures of the ligands are summarized in Table 1. By coating DSM@PCN-128W on UV LED chip, a WLED was obtained showing CIE chromaticity coordinates of (0.34, 0.33), CRI of 79.1, and CCT of 5525 K, and the absolute quantum yield of DSM@PCN128W was measured to be 21.2%. Similarly, by substituting H_4ETTC with H_8ETTB as a carboxylate ligand, Dong and Lei synthesized PCN-921 with a strong fluorescence emission at 447 nm [56]. The innovation of their work lies in the realization of room-temperature phosphorescence and white light emission by hierarchically encapsulating coronene and RhB dye. The results showed that by introducing guests into MOFs, coronene@PCN-91 exhibited a phosphorescence lifetime of 62.5 ns, and the hybrid material RhB@coronen@PCN-921 emitted bright white light by coating on a commercial UV LED. In addition, some TPE-based luminescent MOFs also exhibit piezofluorochromic behavior, and by combining organic dye encapsulation, white-light emission can be obtained. Adopting this strategy, Pan and coworkers constructed dual-emission luminescent MOF HNU-49 and generated relative pure white light by adjusting the pressure and the concentration of RhB [57]. The key parameters for white LEDs with dye-encapsulated MOFs as phosphors are summarized in Table 2 for easy comparison.

Table 1. Structural information about organic ligands of MOFs.

MOF	Organic Ligand	Ref.
ZJU-28	1,3,5-tris(4-carboxyphenyl)benzene	[34]
Bio-MOF-1	biphenyl-4,4'-dicarboxylic acid	[35]
UiO-66	terephthalic acid	[36]
ZIF-8	2-methylimidazole	[36–38]
NKU-114	bis(3,5-dicarboxyphenyl)methanol + 5-aminotetrazole	[40]
LIFM-WZ-6	tetrakis[4-(1-(4-carboxyphenyl)-1H-1,2,3-triazol-4-yl)phenyl]ethylene	[42]
EuBPT TbBPT	biphenyl-3,4',5-tricarboxylic acid	[45]

Table 1. Cont.

MOF	Organic Ligand	Ref.
$[Zn_4OL_2 \cdot xDMF]_n$		[50]
HSB-W1		[51]
$[Cd_7(SO_4)_6(tppe)_2]$ (2DMF·2H$_2$O) $[Zn_2(npd)_2(tppe)](2DMF·3H_2O)$		[54]
PCN-128W		[55]
PCN-921		[56]

Table 1. Cont.

MOF	Organic Ligand	Ref.
HNU-49	(tetrakis(4-carboxyphenyl)ethylene structure)	[57]

Table 2. Key parameters for white LEDs with dye-encapsulated MOFs as phosphors.

Dye-Encapsulated MOF Materials	CIE (x,y)	CCT(K)	CRI	QY (%)	Ref.
ZJU-28⊃Cou-6/R6G/R101	(0.34, 0.32)	4446	88	82.9	[34]
R6G@ZIF-8	-	-	–	63.1	[36]
ZIF-8⊃pm546/pm605/SRh101	(0.465, 0.413)	2642	85	-	[37]
NKU-114@DSM/AF/9-AA	(0.3402, 0.3365)	5148	85.41	-	[40]
CD-MOF⊃7-HCm@FL@RhB	(0.35, 0.32)	-	-	-	[41]
ZJU-28⊃DSM/ AF	(0.34, 0.32)	5327	91	17.4	[49]
[Zn4OL2·xDMF]n⊃DCM/C6	(0.32, 0.31)	6186	91	39.4	[50]
HSB-W1⊃DCM/C6a/CBS-127	(0.31, 0.32)	6638	90	26	[51]
DSM@PCN-128W	(0.34, 0.33)	5525	79.1	21.2	[55]

The luminescent MOFs mentioned above exhibit a large Stokes shift to prevent self-absorption, which are referred to as down-conversion materials. Another type of MOFs belongs to up-conversion materials, exhibiting an anti-Stokes shift luminescence character. Generally, two methods can be used to achieve up-conversion in MOFs, one exploiting energy transfer between lanthanide ions, the other triplet-triplet annihilation, which is based on ligand selection and design [20]. In 2019, Pan reported a series of dye-encapsulated MOFs exhibiting dual way (one-photon absorption (OPA) and two-photon absorption (TPA)) excited fluorescence (Figure 5a) [42]. LIFM-WZ-6 containing TPE ligand shows blue-green and strong green emission upon excitation at 365 nm and 730 nm, respectively. Electron-deficient cationic dyes RhB^+, $BR-2^+$, $BR-46^+$, DSM^+ and $APFG^+$ were chosen due to appropriate D-A interactions and molecule size. RhB^+@LIFM-WZ-6 was first synthesized and when excited at 365 nm, the corresponding CIE coordinates, CCT, CRI and absolute QY values were (0.33, 0.35), 4745 K, 77 and 9.8%, respectively. Similar results were obtained for another four dyes, confirming the universality of the OPA approach. WLED devices were fabricated by coating the prepared phosphors on the surface of the commercial UV LED chip. Compared with the typical OPA process, two-photon excited fluorescence emission (TPEF) is more complicated and often shows different colors. Through TPA process, white light emitting phosphors RhB^+@LIFM-WZ-6, $BR-2^+$WZ-6 and $APFG^+$@LIFM@LIFM-WZ-6 were obtained under the excitation of 800, 790 and 730 nm, respectively (Figure 5b–d). More importantly, the emissive color of dye@MOF can be adjusted by simply tuning the excitation wavelengths.

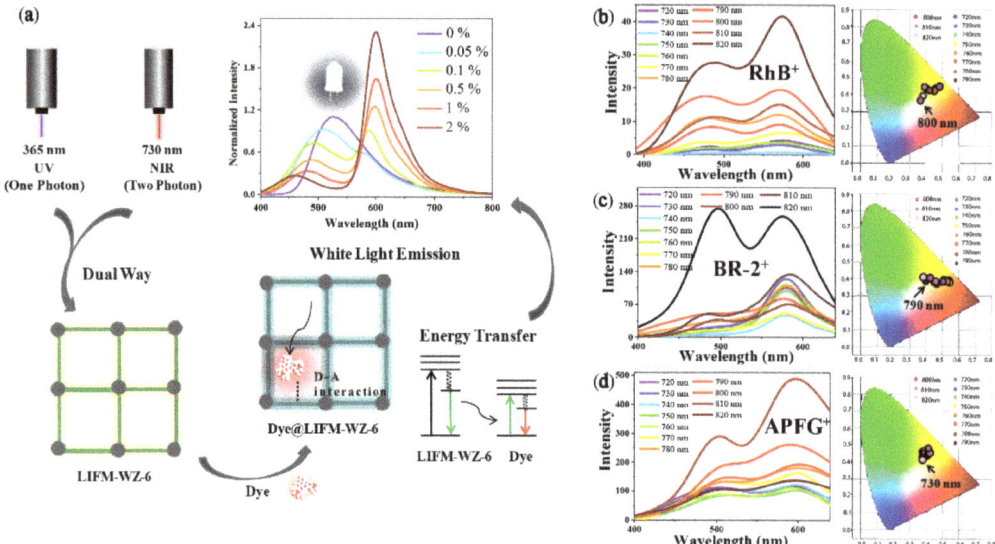

Figure 5. (**a**) Schematic illustration of OPA and TPA dual-way excited white-light emission in dye@LIFM-WZ-6; (**b**) TPEF spectra and CIE coordinate values of RhB+@LIFM-WZ-6 (0.05 wt%); (**c**) TPEF spectra and CIE coordinate values of BR-2+@LIFM-WZ-6 (1 wt%); (**d**) TPEF spectra and CIE coordinate values of APFG+ @LIFM-WZ-6 (0.05 wt%). (Reproduced with permission from ref. [42]. Copyright © 2019 Wiley—VCH Verlag GmbH & Co. KGaA, Weinheim).

3.4. Organic Dyes as Fluorescent Linkers

Inspired by substitutional solid solutions (SSS) concept applied in inorganic materials, Newsome [58] constructed luminescent MOFs by combining nonfluorescent linkers with dilute RGB fluorescent organic dyes, as shown in Figure 6a. Excited-state proton transfer (ESPT) dyes are of extensive interest due to unique photophysical properties caused by keto-enol tautomerism. They have enol tautomers in the ground state, but exists as a keto tautomers after excitation (Figure 6b). Multivariate MOFs are attractive for making multicolor emitting crystals, and the non-fluorescent link and ESPT dyes (Figure 6c) are chosen because of good stability, high quantum yield and color variability. Solid-state emission peaks centered at 430, 510, and 630 nm (Figure 6d) were seen after excitation at 365 nm for 10%-R, 10%-G and 10%-B, respectively. The keto emission in the MOFs is quite close to the ester forms of the RGB links solvated in toluene, as shown in dashed lines, suggesting that prepared MOFs exhibit solution-like properties. Finally, a series of $Zr_6O_4(OH)_4(R_xG_{1-2x}B_x)_yNF_{1-y}$ MOFs were synthesized. $(Zr_6O_4(OH)_4(R_{0.4}G_{0.2}B_{0.4})_{0.01}NF_{0.99})$ emitted combination of broadband emissions from RGB, with coordinates of (0.31, 0.33) on the CIE chromaticity diagram, an absolute QY of 4.3%, a CRI of 93 and a CCT of 6480 K. Other prepared MOFs also exhibited good fluorescence performance. These findings showed that substituting MOF linkers with fluorescent dyes are capable of obtaining both tunable emission chromaticity and accurate color rendering.

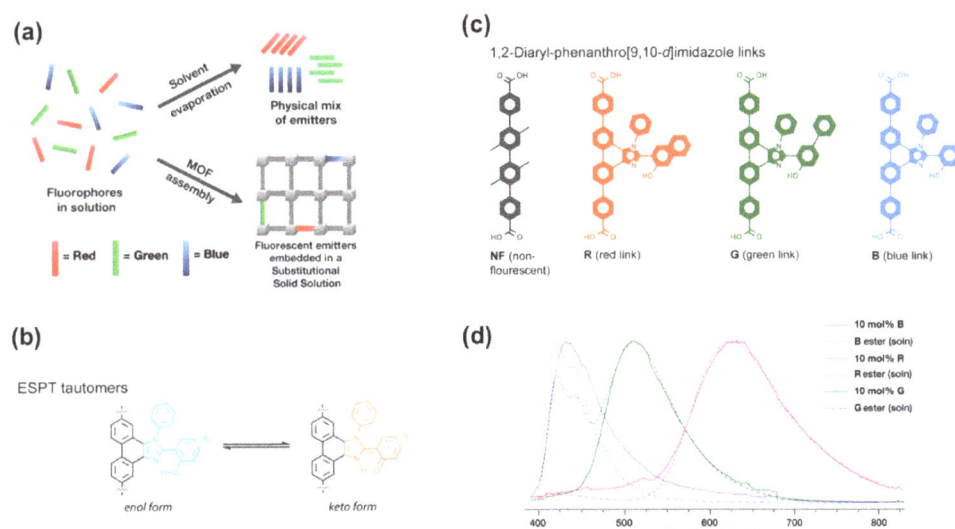

Figure 6. (**a**) Representation of luminescent MOFs based SSS. (**b**) Excited-state proton transfer enol and keto tautomer behavior of dyes. (**c**) Structure of organic linkers. (**d**) Solid-state emission of MOF with 10%-R, 10%-G and 10%-B peaks centered at 430, 510, and 630 nm. (Reproduced with permission from ref. [58]. Copyright © 2019, American Chemical Society).

Recently, Liu and Li applied a mixed-linker strategy to successfully synthesize a series of UiO-68 MOFs with full color emission by changing the ratios of chromophore and non-fluorescent linkers [59]. Obviously, introducing of non-fluorescent linkers is helpful in reducing the concentration of emissive linkers and increasing the spatial distances between fluorescent linkers, which effectively suppresses the π-π stacking interactions and thus enhances the emission efficiency. It is believed that this general approach is of great significance to overcome the challenge of ACQ, portending the potential application of luminescent MOFs in WLEDs [60].

4. Conclusions and Outlook

Luminescent MOFs materials offer a promising platform for light-emitting diodes, chemical sensing, bioimaging and anti-counterfeiting codes. In the past decades, much attention has been focused on design of linkers and encapsulation of guest molecules instead of lanthanide metal-based MOFs for environment consideration.

Encapsulation of organic dyes into MOFs is a feasible and ingenious approach to construct pc-WLEDs, which combines the benefits from dyes and MOF structures. The porosity and crystallinity of MOFs are helpful to suppressing ACQ of dye molecules and thus improving both fluorescent intensity and quantum yield. Meanwhile, the organic dyes enrich the luminescent behaviors without sacrificing the strength of MOFs. Although the warm white light can be generated by encapsulating fluorescent dyes in luminescent MOFs with high performance, the luminous efficiency is still low. In addition, organic dyes leakage, stability and unsuitable size hinder the extensive application of this method. According to the previous study, most reported dye@MOFs are synthesized based on currently available organic dyes or MOFs. From fundamental views, it is necessary to develop novel organic dyes and MOF structures considering factors such as topology, luminescence, charge transfer and stability. Moreover, in-depth research on mechanism behand should be devoted in order to provide instructions for material design and synthesis. For the purpose of industrialization and commercialization, the stability of phosphors, including photo-stability and thermal stability is of vital significance, while currently the reports on the stability of dye-encapsulated MOF phosphors are still rare. It is predictable that more effort will be devoted to investigate the stability of LMOF materials.

While the development of luminescent MOFs for WLEDs is still in infancy, it is evident that the future of WLEDs based on MOF is bright, not only because MOF structure provides various luminescence, but also because the low energy input can reduce the carbon footprint. There is still a long way ahead in achieving commercially available MOF-based WLEDs. In the coming decades, chemists and material scientists will work more closely to develop novel stable and high efficiency single phase white light emitting phosphors for WLED fabrication.

Author Contributions: Conceptualization, D.C. and S.-J.S.; literature investigation, Z.S., A.K. and M.S.; writing—original draft preparation, Z.S.; writing—review and editing, W.Q., D.C. and S.-J.S.; supervision, D.C. and S.-J.S.; project administration, D.C. and S.-J.S.; funding acquisition, D.C. All authors have read and agreed to the published version of the manuscript.

Funding: This work was supported by the National Natural Science Foundation of China (21772045, 22071066), the National Key Research and Development Program of China (2016YFA0602900), the Natural Science Foundation of Guangdong Province, China (2018B030311008), and the Foundation from the Guangzhou Science and Technology Project, China (201902010063).

Institutional Review Board Statement: Not applicable.

Informed Consent Statement: Not applicable.

Data Availability Statement: Not applicable.

Conflicts of Interest: The authors declare no conflict of interest.

References

1. Sun, C.Y.; Wang, X.L.; Zhang, X.; Qin, C.; Li, P.; Su, Z.M.; Zhu, D.X.; Shan, G.G.; Shao, K.Z.; Wu, H.; et al. Efficient and tunable white-light emission of metal-organic frameworks by iridium-complex encapsulation. *Nat. Commun.* **2013**, *4*, 2717–2724. [CrossRef]
2. Reineke, S.; Lindner, F.; Schwartz, G.; Seidler, N.; Walzer, K.; Lussem, B.; Leo, K. White organic light-emitting diodes with fluorescent tube efficiency. *Nature* **2009**, *459*, 234–238. [CrossRef]
3. Cho, J.; Park, J.H.; Kim, J.K.; Schubert, E.F. White light-emitting diodes: History, progress, and future. *Laser Photonics Rev.* **2017**, *11*, 1600147. [CrossRef]
4. Khanna, V.K. *Fundamentals of Solid-State Lighting LEDs, OLEDs, and their Applications in Illumination and Displays*, 1st ed.; CRC Press: Boca Raton, FL, USA, 2014.
5. D'Andrade, B.W.; Forrest, S.R. White organic light-emitting devices for solid-state lighting. *Adv. Mater.* **2004**, *16*, 1585–1595. [CrossRef]
6. Muthu, S.; Schuurmans, F.J.P.; Pashley, M.D. Red, green, and blue LEDs for white light illumination. *IEEE J. Sel. Top. Quantum Electron.* **2002**, *8*, 333–338. [CrossRef]
7. Gong, Q.; Hu, Z.; Deibert, B.J.; Emge, T.J.; Teat, S.J.; Banerjee, D.; Mussman, B.; Rudd, N.D.; Li, J. Solution processable MOF yellow phosphor with exceptionally high quantum efficiency. *J. Am. Chem. Soc.* **2014**, *136*, 16724–16727. [CrossRef] [PubMed]
8. Wang, M.S.; Guo, G.C. Inorganic-organic hybrid white light phosphors. *Chem. Commun.* **2016**, *52*, 13194–13204. [CrossRef] [PubMed]
9. Nakamura, S.; Fasol, G. *The Blue Laser Diode GaN Based Light Emitters and Lasers*; Springer: New York, NY, USA, 1997.
10. Mosca, M.; Macaluso, R.; Crupi, I. Hybrid inorganic-organic white light emitting diodes. In *Polymers for Light-Emitting Devices and Displays*; Wiley-Scrivener: Beverly, MA, USA, 2020; pp. 197–262.
11. Furukawa, H.; Cordova, K.E.; O'Keeffe, M.; Yaghi, O.M. The chemistry and applications of metal-organic frameworks. *Science* **2013**, *341*, 1230444. [CrossRef]
12. Cui, Y.; Li, B.; He, H.; Zhou, W.; Chen, B.; Qian, G. Metal-organic frameworks as platforms for functional materials. *Acc. Chem. Res.* **2016**, *49*, 483–493. [CrossRef]
13. Cui, Y.; Yue, Y.; Qian, G.; Chen, B. Luminescent functional metal-organic frameworks. *Chem. Rev.* **2012**, *112*, 1126–1162. [CrossRef]
14. Yan, B. Lanthanide-functionalized metal-organic framework hybrid systems to create multiple luminescent centers for chemical sensing. *Acc. Chem. Res.* **2017**, *50*, 2789–2798. [CrossRef]
15. Lustig, W.P.; Li, J. Luminescent metal–organic frameworks and coordination polymers as alternative phosphors for energy efficient lighting devices. *Coord. Chem. Rev.* **2018**, *373*, 116–147. [CrossRef]
16. Wang, M.-S.; Guo, G.-C.; Chen, W.-T.; Xu, G.; Zhou, W.-W.; Wu, K.-J.; Huang, J.-S. A White-light-emitting borate-based inorganic–organic hybrid open framework. *Angew. Chem. Int. Ed.* **2007**, *119*, 3983–3985. [CrossRef]

17. Tang, Y.; Wu, H.; Cao, W.; Cui, Y.; Qian, G. Luminescent metal–organic frameworks for white LEDs. *Adv. Opt. Mater.* **2020**, *2020*, 2001817. [CrossRef]
18. Lustig, W.P.; Shen, Z.; Teat, S.J.; Javed, N.; Velasco, E.; O'Carroll, D.M.; Li, J. Rational design of a high-efficiency, multivariate metal-organic framework phosphor for white LED bulbs. *Chem. Sci.* **2020**, *11*, 1814–1824. [CrossRef] [PubMed]
19. Yin, H.Q.; Yin, X.B. Metal-organic frameworks with multiple luminescence emissions: Designs and applications. *Acc. Chem. Res.* **2020**, *53*, 485–495. [CrossRef] [PubMed]
20. Nguyen, T.N.; Ebrahim, F.M.; Stylianou, K.C. Photoluminescent, upconversion luminescent and nonlinear optical metal-organic frameworks: From fundamental photophysics to potential applications. *Coord. Chem. Rev.* **2018**, *377*, 259–306. [CrossRef]
21. Ryu, U.; Lee, H.S.; Park, K.S.; Choi, K.M. The rules and roles of metal–organic framework in combination with molecular dyes. *Polyhedron* **2018**, *154*, 275–294. [CrossRef]
22. Yu, J.; Cui, Y.; Xu, H.; Yang, Y.; Wang, Z.; Chen, B.; Qian, G. Confinement of pyridinium hemicyanine dye within an anionic metal-organic framework for two-photon-pumped lasing. *Nat. Commun.* **2013**, *4*, 2719. [CrossRef]
23. Wang, Z.; Wang, Z.; Lin, B.; Hu, X.; Wei, Y.; Zhang, C.; An, B.; Wang, C.; Lin, W. Warm-white-light-emitting diode based on a dye-loaded metal-organic framework for fast white-light communication. *ACS Appl. Mater. Interfaces* **2017**, *9*, 35253–35259. [CrossRef]
24. Yaghi, O.M.; Li, G.; Li, H. Selective binding and removal of guests in a microporous metal–organic framework. *Nature* **1995**, *378*, 703–706. [CrossRef]
25. Zhou, H.C.; Long, J.R.; Yaghi, O.M. Introduction to metal-organic frameworks. *Chem. Rev.* **2012**, *112*, 673–674. [CrossRef]
26. Li, B.; Wen, H.M.; Cui, Y.; Zhou, W.; Qian, G.; Chen, B. Emerging multifunctional metal-organic framework materials. *Adv. Mater.* **2016**, *28*, 8819–8860. [CrossRef] [PubMed]
27. Ryu, U.; Jee, S.; Rao, P.C.; Shin, J.; Ko, C.; Yoon, M.; Park, K.S.; Choi, K.M. Recent advances in process engineering and upcoming applications of metal-organic frameworks. *Coord. Chem. Rev.* **2021**, *426*, 213544. [CrossRef]
28. Li, H.; Eddaoudi, M.; O'Keeffe, M.; Yaghi, O.M. Design and synthesis of an exceptionally stable and highly porous metal-organic framework. *Nature* **1999**, *402*, 276–279. [CrossRef]
29. Allendorf, M.D.; Bauer, C.A.; Bhakta, R.K.; Houk, R.J. Luminescent metal-organic frameworks. *Chem. Soc. Rev.* **2009**, *38*, 1330–1352. [CrossRef]
30. Cui, Y.; Chen, B.; Qian, G. Lanthanide metal-organic frameworks for luminescent sensing and light-emitting applications. *Coord. Chem. Rev.* **2014**, *273-274*, 76–86. [CrossRef]
31. Cui, Y.; Zhang, J.; He, H.; Qian, G. Photonic functional metal-organic frameworks. *Chem. Soc. Rev.* **2018**, *47*, 5740–5785. [CrossRef]
32. Guo, B.B.; Yin, J.C.; Li, N.; Fu, Z.X.; Han, X.; Xu, J.; Bu, X.H. Recent progress in luminous particle-encapsulated host–guest metal-organic frameworks for optical applications. *Adv. Opt. Mater.* **2021**, *2021*, 2100283. [CrossRef]
33. Pimputkar, S.; Speck, J.S.; DenBaars, S.P.; Nakamura, S. Prospects for LED lighting. *Nat. Photonics* **2009**, *3*, 180–182. [CrossRef]
34. Tang, Y.; Xia, T.; Song, T.; Cui, Y.; Yang, Y.; Qian, G. Efficient energy transfer within dyes encapsulated metal–organic frameworks to achieve high performance white light-emitting diodes. *Adv. Opt. Mater.* **2018**, *6*, 1800968. [CrossRef]
35. Chen, W.; Zhuang, Y.; Wang, L.; Lv, Y.; Liu, J.; Zhou, T.L.; Xie, R.J. Color-tunable and high-efficiency dye-encapsulated metal-organic framework composites used for smart white-light-emitting diodes. *ACS Appl. Mater. Interfaces* **2018**, *10*, 18910–18917. [CrossRef] [PubMed]
36. Liu, X.Y.; Li, Y.; Tsung, C.K.; Li, J. Encapsulation of yellow phosphors into nanocrystalline metal-organic frameworks for blue-excitable white light emission. *Chem. Commun.* **2019**, *55*, 10669–10672. [CrossRef]
37. Tang, Y.; Cao, W.; Yao, L.; Cui, Y.; Yu, Y.; Qian, G. Polyurethane-coated luminescent dye@MOF composites for highly-stable white LEDs. *J. Mater. Chem. C* **2020**, *8*, 12308–12313. [CrossRef]
38. Lu, Y.; Wang, S.; Yu, K.; Yu, J.; Zhao, D.; Li, C. Encapsulating carbon quantum dot and organic dye in multi-shell nanostructured MOFs for use in white light-emitting diode. *Microporous Mesoporous Mater.* **2021**, *319*, 111062. [CrossRef]
39. Gutiérrez, M.; Martín, C.; Van der Auweraer, M.; Hofkens, J.; Tan, J.C. Electroluminescent guest@MOF nanoparticles for thin film optoelectronics and solid-state lighting. *Adv. Opt. Mater.* **2020**, *8*, 2000670. [CrossRef]
40. Yin, J.C.; Chang, Z.; Li, N.; He, J.; Fu, Z.X.; Bu, X.H. Efficient regulation of energy transfer in a multicomponent dye-loaded MOF for white-light emission tuning. *ACS Appl. Mater. Interfaces* **2020**, *12*, 51589–51597. [CrossRef]
41. Chen, Y.; Yu, B.; Cui, Y.; Xu, S.; Gong, J. Core–shell structured cyclodextrin metal–organic frameworks with hierarchical dye encapsulation for tunable light emission. *Chem. Mater.* **2019**, *31*, 1289–1295. [CrossRef]
42. Wang, Z.; Zhu, C.Y.; Mo, J.T.; Fu, P.Y.; Zhao, Y.W.; Yin, S.Y.; Jiang, J.J.; Pan, M.; Su, C.Y. White-light emission from dual-way photon energy conversion in a dye-encapsulated metal-organic framework. *Angew. Chem. Int. Ed.* **2019**, *58*, 9752–9757. [CrossRef]
43. Yong, D.; Zhang, X.; She, W. Phosphorescence enhancement of organic dyes by forming β-cyclodextrin inclusion complexes: Color tunable emissive materials. *Dye. Pigms.* **2013**, *97*, 65–70. [CrossRef]
44. Liu, X.Y.; Xing, K.; Li, Y.; Tsung, C.K.; Li, J. Three models to encapsulate multicomponent dyes into nanocrystal pores: A new strategy for generating high-quality white light. *J. Am. Chem. Soc.* **2019**, *141*, 14807–14813. [CrossRef]
45. Song, T.; Zhang, G.; Cui, Y.; Yang, Y.; Qian, G. Encapsulation of coumarin dye within lanthanide MOFs as highly efficient white-light-emitting phosphors for white LEDs. *CrystEngComm* **2016**, *18*, 8366–8371. [CrossRef]
46. Mondal, T.; Bose, S.; Husain, A.; Ghorai, U.K.; Saha, S.K. White light emission from single dye incorporated metal organic framework. *Opt. Mater.* **2020**, *100*, 109706. [CrossRef]

47. Jiang, Z.; Xu, X.; Ma, Y.; Cho, H.S.; Ding, D.; Wang, C.; Wu, J.; Oleynikov, P.; Jia, M.; Cheng, J.; et al. Filling metal-organic framework mesopores with TiO$_2$ for CO$_2$ photoreduction. *Nature* **2020**, *586*, 549–554. [CrossRef]
48. Gu, S.F.; Xiong, X.H.; Gong, L.L.; Zhang, H.P.; Xu, Y.; Feng, X.F.; Luo, F. Classified encapsulation of an organic dye and metal-organic complex in different molecular compartments for white-light emission and selective adsorption of C$_2$H$_2$ over CO$_2$. *Inorg. Chem.* **2021**, *60*, 8211–8217. [CrossRef]
49. Cui, Y.; Song, T.; Yu, J.; Yang, Y.; Wang, Z.; Qian, G. Dye encapsulated metal-organic framework for warm-white LED with high color-rendering index. *Adv. Funct. Mater.* **2015**, *25*, 4796–4802. [CrossRef]
50. Xia, Y.-P.; Wang, C.-X.; An, L.-C.; Zhang, D.-S.; Hu, T.-L.; Xu, J.; Chang, Z.; Bu, X.-H. Utilizing an effective framework to dye energy transfer in a carbazole-based metal–organic framework for high performance white light emission tuning. *Inorg. Chem. Front.* **2018**, *5*, 2868–2874. [CrossRef]
51. Wen, Y.; Sheng, T.; Zhu, X.; Zhuo, C.; Su, S.; Li, H.; Hu, S.; Zhu, Q.L.; Wu, X. Introduction of red-green-blue fluorescent dyes into a metal-organic framework for tunable white light emission. *Adv. Mater.* **2017**, *29*, 1700778. [CrossRef] [PubMed]
52. Luo, J.; Xie, Z.; Lam, J.W.; Cheng, L.; Chen, H.; Qiu, C.; Kwok, H.S.; Zhan, X.; Liu, Y.; Zhu, D.; et al. Aggregation-induced emission of 1-methyl-1,2,3,4,5-pentaphenylsilole. *Chem. Commun.* **2001**, *18*, 1740–1741. [CrossRef] [PubMed]
53. Hong, Y.; Lam, J.W.; Tang, B.Z. Aggregation-induced emission. *Chem. Soc. Rev.* **2011**, *40*, 5361–5388. [CrossRef] [PubMed]
54. Zhao, Y.; Wang, Y.J.; Wang, N.; Zheng, P.; Fu, H.R.; Han, M.L.; Ma, L.F.; Wang, L.Y. Tetraphenylethylene-decorated metal-organic frameworks as energy-transfer platform for the detection of nitro-antibiotics and white-light emission. *Inorg. Chem.* **2019**, *58*, 12700–12706. [CrossRef]
55. Xing, W.; Zhou, H.; Han, J.; Zhou, Y.; Gan, N.; Cuan, J. Dye encapsulation engineering in a tetraphenylethylene-based MOF for tunable white-light emission. *J. Colloid Interface Sci.* **2021**, *604*, 568–574. [CrossRef] [PubMed]
56. Yuan, J.; Feng, G.; Dong, J.; Lei, S.; Hu, W. Dual-functional porous MOFs with hierarchical guest encapsulation for room-temperature phosphorescence and white-light-emission. *Nanoscale* **2021**, *13*, 12466–12474. [CrossRef]
57. Li, M.; Ren, G.; Yang, W.; Yang, Y.; Yang, W.; Gao, Y.; Qiu, P.; Pan, Q. Dual-emitting piezofluorochromic dye@MOF for white-light generation. *Chem. Commun.* **2021**, *57*, 1340–1343. [CrossRef]
58. Newsome, W.J.; Ayad, S.; Cordova, J.; Reinheimer, E.W.; Campiglia, A.D.; Harper, J.K.; Hanson, K.; Uribe-Romo, F.J. Solid state multicolor emission in substitutional solid solutions of metal-organic frameworks. *J. Am. Chem. Soc.* **2019**, *141*, 11298–11303. [CrossRef]
59. Wu, S.; Ren, D.; Zhou, K.; Xia, H.L.; Liu, X.Y.; Wang, X.; Li, J. Linker engineering toward full-color emission of UiO-68 type metal-organic frameworks. *J. Am. Chem. Soc.* **2021**, *143*, 10547–10552. [CrossRef]
60. Leith, G.A.; Martin, C.R.; Mayers, J.M.; Kittikhunnatham, P.; Larsen, R.W.; Shustova, N.B. Confinement-guided photophysics in MOFs, COFs, and cages. *Chem. Soc. Rev.* **2021**, *50*, 4382–4410. [CrossRef]

Article

Tuning Multicolor Emission of Manganese-Activated Gallogermanate Nanophosphors by Regulating Mn Ions Occupying Sites for Multiple Anti-Counterfeiting Application

Dangli Gao [1,*], Peng Wang [1], Feng Gao [1], William Nguyen [2] and Wei Chen [2,*]

1. College of Science, Xi'an University of Architecture and Technology, Xi'an 710055, China; wp2013141996@163.com (P.W.); gf@xauat.edu.cn (F.G.)
2. Department of Physics, The University of Texas at Arlington, Arlington, TX 76019-0059, USA; william.nguyen@uta.edu
* Correspondence: gaodangli@163.com (D.G.); weichen@uta.edu (W.C.)

Citation: Gao, D.; Wang, P.; Gao, F.; Nguyen, W.; Chen, W. Tuning Multicolor Emission of Manganese-Activated Gallogermanate Nanophosphors by Regulating Mn Ions Occupying Sites for Multiple Anti-Counterfeiting Application. *Nanomaterials* **2022**, *12*, 2029. https://doi.org/10.3390/nano12122029

Academic Editors: Yann Molard and Eleonore Fröhlich

Received: 10 May 2022
Accepted: 9 June 2022
Published: 13 June 2022

Publisher's Note: MDPI stays neutral with regard to jurisdictional claims in published maps and institutional affiliations.

Copyright: © 2022 by the authors. Licensee MDPI, Basel, Switzerland. This article is an open access article distributed under the terms and conditions of the Creative Commons Attribution (CC BY) license (https://creativecommons.org/licenses/by/4.0/).

Abstract: The ability to manipulate the luminescent color, intensity and long lifetime of nanophosphors is important for anti-counterfeiting applications. Unfortunately, persistent luminescence materials with multimode luminescent features have rarely been reported, even though they are expected to be highly desirable in sophisticated anti-counterfeiting. Here, the luminescence properties of $Zn_3Ga_2GeO_8$:Mn phosphors were tuned by using different preparation approaches, including a hydrothermal method and solid-state reaction approach combining with non-equivalent ion doping strategy. As a result, Mn-activated $Zn_3Ga_2GeO_8$ phosphors synthesized by a hydrothermal method demonstrate an enhanced red photoluminescence at 701 nm and a strong green luminescence with persistent luminescence and photostimulated luminescence at 540 nm. While Mn-activated $Zn_3Ga_2GeO_8$ phosphors synthesized by solid-state reactions combined with a hetero-valent doping approach only exhibit an enhanced single-band red emission. Keeping the synthetic method unchanged, the substitution of hetero-valent dopant ion Li^+ into different sites is valid for spectral fine-tuning. A spectral tuning mechanism is also proposed. Mn-activated $Zn_3Ga_2GeO_8$ phosphors synthesized by a hydrothermal approach with multimodal luminescence is especially suitable for multiple anti-counterfeiting, multicolor display and other potential applications.

Keywords: $Zn_3Ga_2GeO_8$:Mn phosphors; multicolor emission; hydrothermal approach; afterglow; anti-counterfeiting application

1. Introduction

The ability to tune the luminescent color of materials is essential for various applications, such as anti-counterfeiting, three-dimensional displays, information coding, bioimaging, optoelectronic devices and luminescent labeling [1–3]. Conventional approaches involve utilizing specially designed organic dyes or quantum dots upon ultraviolet or blue light excitation, where the emission color is modulated by tuning the wavelengths or the power density of excitation light [4,5], thus posing limitations in the resolution of imaging due to auto-fluorescence. Persistent luminescence (PersL) phosphors, which can emit luminescence lasting for hours after the stoppage of the excitation light [6], are particularly suitable for such imaging applications as they emit no background fluorescence [7,8].

The transition metal Mn with the multiple oxidation states, e.g., +2, +3 and +4, provides an opportunity for multi-color emission [9]. In recent years, Mn-doped $ZnGa_2O_4$ microcrystals with unique luminescence features have attracted much attention because a $ZnGa_2O_4$ host has two kinds of stable chemical coordination structure, including Ga^{3+} sites with octahedral coordination and Zn^{2+} sites with tetrahedral coordination [10]. Generally, a Mn^{2+} activation center occupying the tetrahedral-coordinated sites shows a green emission with long PersL, while a Mn^{4+} activation center with an octahedral-coordinated structure demonstrates a luminescence emission from red to deep red. Apparently, the Mn activator shows a green to deep red emission, and the emission color is determined by the coordination environment of Mn ions in the crystal structure.

In our work, Mn-activated $Zn_3Ga_2GeO_8$ phosphors are successfully prepared using a hydrothermal method and solid-state reaction approach. Interestingly, it was found that changing the preparation route was a more efficient method for the spectral tuning of Mn-activated $Zn_3Ga_2GeO_8$ phosphors relative to a non-equivalent ion doping strategy. Mn-activated $Zn_3Ga_2GeO_8$ phosphors show an enhanced red photoluminescence (PL) at 701 nm and a strong green emission at 540 nm with PersL and a green photostimulated luminescence (PSL) by Li^+ substituted for Zn^{2+} or Ga^{3+} sites under hydrothermal conditions. $Zn_3Ga_2GeO_8$ phosphors synthesized by a solid-state reaction only exhibit an enhanced pure-red broad-band luminescence. Particularly, $Zn_3Ga_2GeO_8$ phosphors synthesized by a hydrothermal approach exhibit multicolor and multimode luminescence properties, which are especially suitable for multiple anti-counterfeiting and have a great potential for multicolor display, anti-forgery, and other potential applications.

2. Experimental Section

2.1. Chemicals

Ga_2O_3 (99.999%), GeO_2 (99.999%), $LiNO_3$ (99.9%), $Mn(NO_3)_2 \cdot xH_2O$ (99.9% metals basis), $Zn(NO_3)_2 \cdot 6H_2O$ (99%), ZnO (99.99%), and MnO (99.99%) were obtained from Aladdin (Shanghai, China). Hydrochloric acid (AR, 36.0–38.0%), nitric acid (AR, 36.0–38.0%), sodium hydroxide, concentrated ammonium hydroxide, anhydrous ethanol and polyvinyl alcohol (PVA) with analytical grade are stocked from Sinopharm Chemical Reagent Co. Ltd. (Shanghai, China).

2.2. Sample Preparation

$Zn_3Li_{0.4}Ga_{1.6}GeO_8$:0.25% Mn phosphor was synthesized by using a hydrothermal method [8,11]. The synthesis procedure is described below. GeO_2 powders were dissolved in a sodium hydroxide solution to achieve 0.5 M Na_2GeO_3 solution. Then, 3.0 mL 0.5 M $Zn(NO_3)_2$, 0.0375 mL 0.1 M $Mn(NO_3)_2$ and 0.3 mL concentrated HNO_3 were together added slowly into 10 mL deionized water and then violently stirred. Subsequently, 0.4 mL 0.5 M $LiNO_3$, 1.6 mL 0.5 M $Ga(NO_3)_3$ and 1.1 mL 0.5 M Na_2GeO_3 were added to the above solution. After the solution was stirred vigorously for 1 h, concentrated ammonium hydroxide was added into the mixture of precursor solution to tune the pH of the mixed solution to 7.5, and then the mixed solution was stirred vigorously for 1 h at ambient temperature. The final colloid was transferred into a 50 mL polytetrafluoroethylene reactor, which was then placed in an oven and heated at 220 °C for 10 h. Finally, the resulting suspension was centrifuged and washed 3 times using deionized water. The collected products were dried at 70 °C, and were then annealed in a chamber-type electric resistance furnace in air at 1100 °C for 2 h.

The synthesis procedure of $Zn_{2.4}Li_{0.6}Ga_2GeO_8$:Mn phosphors by hydrothermal method was similar to that of $Zn_3Li_{0.4}Ga_{1.6}GeO_8$:0.25% Mn only via 0.6 mL 0.5 M $LiNO_3$ substituting for 20% $Zn(NO_3)_2$. According to our previous study [8,11], 20% Li substitution is an optimal doping concentration.

$Zn_3Ga_2GeO_8$:0.25% Mn phosphors were prepared via a solid-state reaction approach using ZnO, Ga_2O_3, GeO_2, and MnO as raw materials [12]. Raw materials were weighted on the basis of the formula of $Zn_3Ga_2GeO_8$:0.25%Mn and finely mixed in an agate. All the ground powders were then pre-calcined at 900 °C for 2 h. Subsequently, the pre-calcined products were sintered at 1100 °C for 2 h and then cooled down to ambient temperature.

The synthesis procedure of $Zn_{2.4}Li_{0.6}Ga_2GeO_8$:Mn phosphor synthesized by a solid state reaction approach was similar to that of $Zn_3Ga_2GeO_8$:0.25% Mn only via Li substituting for 20% ZnO.

2.3. Inks Preparation and Anti-Counterfeiting Patterns

Luminescence inks were fabricated based on a modified approach [13]. Typically, the prepared phosphors were dispersed into a mixed solution of 1.0 mL hydrochloric acid (0.2 mol/L) and 1.0 mL anhydrous C_2H_5OH, and then the mixed solution was centrifuged and washed with deionized water and anhydrous C_2H_5OH several times. Finally, the powders were mixed with PVA aqueous solution in a ratio of 1:1 to fabricate luminescent anti-counterfeiting inks. The luminescent patterns printed on the paper were achieved using screen printing method.

2.4. Characterization

A D/Max2400 X-ray diffractometer (XRD, Rigaku, Japan) with Cu Kα (40 kV, 100 mA) irradiation (λ = 1.5406 Å) was employed to characterize the crystal structures of phosphors. The shape and size of phosphors were characterized using a ZEISS Gemini 500 scanning electron microscopy (SEM) (Oberkochen, Germany). A spectrometer (PHI 5600ci ESCA, PerkinElmer, Waltham, MA, USA) with monochromatized Al K radiation was used to measure the X-ray photoelectron spectra (XPS) at room temperature. Thermoluminescence (TL) curves of products were measured using a TOSL-3DS TL spectrometer (Guangzhou, China) with a temperature range from 25 to 500 °C and a heating rate of 2 °C/s. A Horiba PTI QuantaMaster 8000 spectrofluorometer (Burlington, ON, Canada) equipped with a 75 W xenon lamp was used to study the optical properties of products. In addition, besides the xenon lamp as an irradiation source, two UV lamps (a 4-W 254 nm and a 5-W 365 nm) and two NIR laser diodes (0–2 W 808 nm and 0–5 W 980 nm) were also employed as excitation sources. When NIR lasers were used for irradiating the anti-counterfeiting patterns, the entire anti-counterfeiting pattern was programmatically scanned by the laser beam at a rate of 3 times per min. A Nikon EOS 60D camera (Tokyo, Japan) was used to take photographs of anti-counterfeiting patterns with suitable optical filters.

3. Results and Discussion

Figure 1 exhibits the XRD patterns of the four samples synthesized by different methods combined with a non-equivalent doping strategy. As shown in Figure 1b,e, the XRD patterns of the two samples (one is $Zn_3Li_{0.4}Ga_{1.6}GeO_8$:Mn synthesized by a hydrothermal approach, and the other is $Zn_3Ga_2GeO_8$:Mn prepared by a solid-state reaction method) are similar and both dominated by a spinel-structured solid solutions of cubic phase $ZnGa_2O_4$ (JCPDS No. 01-071-0843) and cubic phase Zn_2GeO_4 (JCPDS No. 01-007-9080) (Figure 1f,g), accompanied with a rhombohedral Zn_2GeO_4 (JCPDS No. 04-007-5691) secondary phase (Figure 1a). $Zn_{2.4}Li_{0.6}Ga_2GeO_8$:Mn samples synthesized by a hydrothermal approach, as shown in Figure 1c, are the compounds of cubic $ZnGa_2O_4$ (JCPDS No. 01-071-0843) with an $Fd\bar{3}m$ space group and rhombohedral Zn_2GeO_4 (JCPDS No. 04-007-5691) with an $R\bar{3}$ space group [6,14,15]. While the XRD patterns of $Zn_{2.4}Li_{0.6}Ga_2GeO_8$:Mn prepared by a solid-state reaction method in Figure 1d demonstrate spinel-structured solid solutions of $ZnGa_2O_4$ and Zn_2GeO_4 [16,17]. It is reported that $Zn_3Ga_2GeO_8$ solid solutions are formed by Ge^{4+} substituting for Ga^{3+} into a $ZnGa_2O_4$ lattice, where excessive Zn^{2+} is necessary to counterbalance the charge disequilibrium [6,17]. It was observed that a solid solution is easier to form when Li replaces Ga, as shown in Figure 1b, which corresponds to a reduced Ga to Zn ratio but an excess of Zn in the hydrothermal reaction process, relative to Li^+ replacing Zn^{2+} (Figure 1c). With the addition of Li^+, substituting for Zn^{2+} (Figure 1d) in the solid-state reaction, the phase purity becomes higher relative to the results shown in Figure 1e, indicating that the Ge^{4+} ions (53 pm for ionic radius) enter into octahedral Ga^{3+} (62 pm for ionic radius) sites [17] and Li^+ (76 pm for ionic radius) is substituted for Zn^{2+} (74 pm for ionic radius) sites [18] for charge compensation [19]. The results of XRD patterns indicate that different preparation approaches and non-equivalent doping strategies have a great influence on the crystal structure of the $Zn_3Ga_2GeO_8$ phosphor.

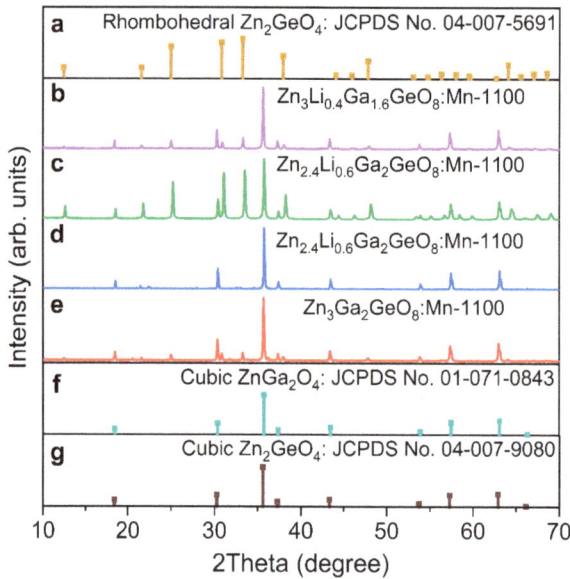

Figure 1. XRD patterns of $Zn_3Li_{0.4}Ga_{1.6}GeO_8$:Mn, $Zn_{2.4}Li_{0.6}Ga_2GeO_8$:Mn and $Zn_3Ga_2GeO_8$:Mn phosphors. (**a**) Standard data for rhombohedral Zn_2GeO_4 (JCPDS No. 04-007-5691); (**b,c**) $Zn_3Li_{0.4}Ga_{1.6}GeO_8$:Mn and $Zn_{2.4}Li_{0.6}Ga_2GeO_8$:Mn phosphors synthesized by hydrothermal approach; (**d,e**) $Zn_{2.4}Li_{0.6}Ga_2GeO_8$:Mn and $Zn_3Ga_2GeO_8$:Mn phosphors synthesized by solid-state reaction; (**f,g**) Standard data for cubic $ZnGa_2O_4$ (JCPDS No. 01-071-0843) and cubic Zn_2GeO_4 (JCPDS No. 04-007-9080).

SEM images (Figure 2) demonstrate that the morphology of these phosphors are irregular micro-particles with faceted surfaces. Their size remarkably changes from around 0.3 μm for $Zn_3Li_{0.4}Ga_{1.6}GeO_8$:Mn and $Zn_{2.4}Li_{0.6}Ga_2GeO_8$:Mn phosphors (Figure 2a,b) synthesized by a hydrothermal method to around 5 μm for $Zn_{2.4}Li_{0.6}Ga_2GeO_8$:Mn and $Zn_3Ga_2GeO_8$:Mn (Figure 2c,d) synthesized by a solid-state reaction. Elemental mappings of $Zn_3Li_{0.4}Ga_{1.6}GeO_8$:Mn phosphors synthesized by a hydrothermal method and $Zn_3Ga_2GeO_8$:Mn phosphors synthesized by a solid-state reaction (Figure 2e,f) show the even distributions of Zn, Ge, Ga, and O, indicating that there is no difference in element types and element distribution in these two kinds of samples. Clearly, the preparation method is one of the key factors that determines the crystal structure and morphology of these phosphors. The reasons for the differences in the crystal phase structure of the samples prepared by different methods could be complex. In a solid-state reaction system, a high temperature facilitates GeO_2 volatilization, leading to the formation of a solid solution. On the other hand, hydrothermal conditions may also be beneficial for the coexistence of rhombohedral Zn_2GeO_4 and cubic $ZnGa_2O_4$.

All phosphors used in the experiments were annealed at 1100 °C due to their more distinctive luminescence features than their as-synthesized counterparts. Manganese-activated gallogermanate-based phosphors synthesized by a hydrothermal approach via Li replacing 20% Ga^{3+} are referred to as $Zn_3Li_{0.4}Ga_{1.6}GeO_8$:Mn phosphor, and those synthesized via Li replacing 20% Zn^{2+} are referred to as $Zn_{2.4}Li_{0.6}Ga_2GeO_8$:Mn phosphor. The $Zn_3Li_{0.4}Ga_{1.6}GeO_8$:Mn phosphor was investigated first, and Figure 3 illustrates its excited wavelength-dependent luminescence features. Under a 254 nm UV lamp excitation, a blue–white broad-band spectrum in the range of 400–750 nm was achieved (Figure 3a) and could be ascribed to the transitions of matrix defect levels. When the excitation wavelength was switched to 357 nm, the strong red emission of Mn^{4+} ions from $^2E \rightarrow {}^4A_2$ (peaking at 701 nm) transitions could be obtained, accompanied with a weak green emission (540 nm) from $^4T_1(4G) \rightarrow {^6A_1}(6S)$ transitions of Mn^{2+} ions (Figure 3a) [20,21]. Finally, when the sample was irradiated with blue visible light at 467 nm, only a pure-red single band at 650–800 nm from $^2E \rightarrow {^4A_2}$ of Mn^{4+} was observed (Figure 3a) [21]. PL excitation (PLE) spectrum monitoring at 540 nm shows a typical Mn^{2+}–O^{2-} charge transfer band (CTB) centered at 283 nm (Figure 3b) [22]. PLE spectrum monitoring at 701 nm shows two broad bands (excitation band I at 200–400 nm and excitation band II at 400–620 nm) peaking at 357 nm and 467 nm in Figure 3b. The excitation band I peaking at 357 nm originated from $^4A_2 \rightarrow {^4T_1}$, while excitation band II peaking at 467 nm was assigned to the $^4A_2 \rightarrow {^4T_2}$ transitions of Mn^{4+} ions [23]. The profile and peak position on the PLE spectra monitored at 701 and 540 nm are different, which indicates that red and green emissions come from different Mn emission centers. These spectra indicate that Li^+ replacing Ga^{3+} provides an opportunity for Mn to occupy the octahedral Ga^{3+} sites. As a result, in $Zn_3Li_{0.4}Ga_{1.6}GeO_8$:Mn crystals, Mn occupies two kinds of positions: one is a tetrahedral Zn site, and the other is octahedral Ga site. These results are consistent with the above XRD analysis.

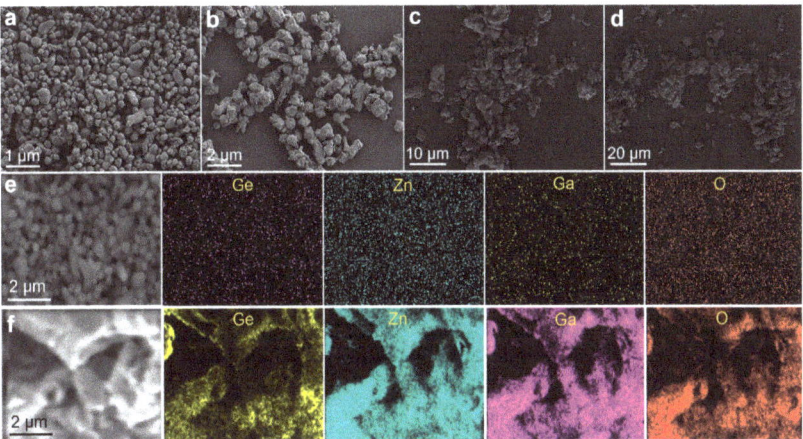

Figure 2. SEM images (**a–d**) and EDX mapping (**e,f**) of $Zn_3Ga_2GeO_8$ phosphors annealed at 1100 °C. (**a,b**) $Zn_3Li_{0.4}Ga_{1.6}GeO_8$:Mn phosphor in (**a**) and $Zn_{2.4}Li_{0.6}Ga_2GeO_8$:Mn in (**b**) synthesized by hydrothermal approach; (**c,d**) $Zn_{2.4}Li_{0.6}Ga_2GeO_8$:Mn in (**c**) and $Zn_3Ga_2GeO_8$:Mn in (**d**) synthesized by solid-state reaction; (**e**) EDX mapping of $Zn_3Li_{0.4}Ga_{1.6}GeO_8$:Mn synthesized by hydrothermal approach; and (**f**) EDX mapping of $Zn_3Ga_2GeO_8$:Mn synthesized by solid-state reaction.

In addition, the $Zn_3Li_{0.4}Ga_{1.6}GeO_8$:Mn phosphor can demonstrate superior green PersL properties. After the removal of UV radiation, the green luminescence (monitored at 540 nm) of the $Zn_3Li_{0.4}Ga_{1.6}GeO_8$:Mn phosphor shows a long PersL signal (Figure 3c). A PersL spectrum achieved at 7 s of the decay is shown in the inset of Figure 3b. The similar PersL profile indicated that the PersL could be attributed to the Mn^{2+} ions. From Figure 3c, it is evident that $Zn_3Li_{0.4}Ga_{1.6}GeO_8$:Mn phosphor exhibited a good green PersL. Moreover, the excitation wavelength-dependent PersL duration is shown in Figure 3c, which may originate from the pre-irradiated wavelength-dependent trapping and detrapping [24]. Aside from PL and PersL, the UV pre-irradiated $Zn_3Li_{0.4}Ga_{1.6}GeO_8$:Mn phosphor also exhibited superior PSL capabilities, peaking at 540 nm and 670 nm (Figure 3a) under the illumination of a 980 nm laser diode, which is consistent with previous reports [25,26]. We still found that the PL and PSL emission spectra were slightly different (Figure 3a). We know that green luminescence (with PersL feature) and red luminescence (without PersL characteristics) originate from the Mn^{2+} and Mn^{4+} luminescent centers, respectively. Theoretically, red PSL should not be observed due to no red PersL. Surprisingly, we could still observe red luminescence, but no green luminescence was achieved when we emptied the traps of $Zn_3Li_{0.4}Ga_{1.6}GeO_8$:Mn phosphor and then the sample was excited under 980 nm, suggesting that the red fluorescence may be derived from the upconversion fluorescence of Mn ions [27]. These results show that red and green PL with different luminescent features should derive from the luminescent center with different coordination environments.

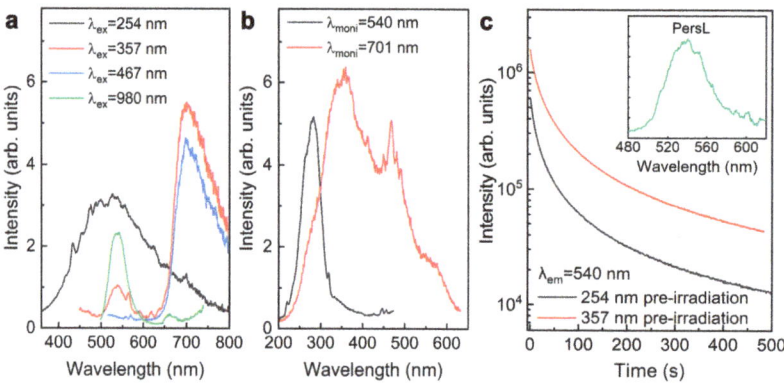

Figure 3. Excited wavelength-dependent fluorescence characteristics of $Zn_3Li_{0.4}Ga_{1.6}GeO_8$:Mn phosphors synthesized by hydrothermal approach. (**a**) PL spectra under various selective excitation and PSL upon 980 nm laser diode (0.5 W) irradiation; (**b**) PLE spectra (monitored at 540 or 701 nm); (**c**) PersL decay curves (monitoring at 540 nm) after irradiation by using 254 nm or 357 nm light for 5 min, the inset is the PersL emission spectrum achieved at 7 s of the decay.

Interestingly, the fluorescent colors of Mn-activated $Zn_3Ga_2GeO_8$ phosphor prepared by a hydrothermal approach could be further finely adjusted by substituting Zn^{2+} with Li^+ to tune the occupancy rate of Mn ions in Zn^{2+} and Ga^{3+} sites. Figure 4 shows the luminescence characteristics of $Zn_{2.4}Li_{0.6}Ga_2GeO_8$:Mn phosphor prepared by a hydrothermal approach. As expected, the fluorescence emission color changed from green to yellow to red in response to the excitation wavelength (Figure 4a). Under excitation at 254 nm, the sample showed a broad-band emission spanning a range of 400–800 nm wavelength (Figure 4a), which is similar to the emission of $Zn_3Li_{0.4}Ga_{1.6}GeO_8$:Mn phosphor. When the excitation wavelength was switched to 337 nm, the green emission (Figure 4a), which can be assigned to the $^4T_1(4G) \rightarrow {}^6A_1(6S)$ transition of Mn^{2+}, dominated the emission, accompanying the weak red emission peaking at 701 nm [26]. PLE spectrum monitored at 540 nm exhibited twin peaks in a 200–400 nm wavelength range (Figure 4b), which were assigned to the Mn^{2+}–O^{2-} CTB and the 5d→5d transition of Mn^{2+} with tetrahedral coordination [9,22]. Under the 413 nm excitation, the phosphors exhibited a narrowband red emission (Figure 4a) originating from the spin-forbidden $^4T_1(4G) \rightarrow {}^6A_1(6S)$ transition of Mn^{2+} in the strong octahedral crystal field, which is different from the deep red emission of Mn^{4+} (Figure 3a) of $Zn_3Li_{0.4}Ga_{1.6}GeO_8$:Mn phosphor. The PLE spectrum monitored at 701 nm exhibited triplet peaks at 276, 413 and 575 nm (Figure 4b), which were assigned to the Mn^{2+}–O^{2-} CTB (276 nm) and the 5d→5d Mn^{2+} transition (413 and 575 nm) [24]. Aside from wavelength-dependent PL properties, the UV pre-irradiated $Zn_{2.4}Li_{0.6}Ga_2GeO_8$:Mn phosphor also showed excellent green PersL (Figure 4c) and green PSL (Figure 4a) capability under the 980 nm laser diode illumination.

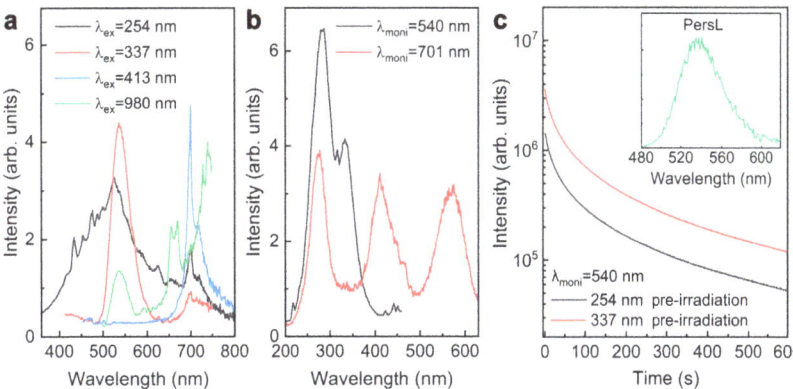

Figure 4. Excited wavelength-dependent fluorescence characteristics of $Zn_{2.4}Li_{0.6}Ga_2GeO_8$:Mn phosphors prepared by hydrothermal approach. (**a**) PL spectra under various selective excitation and PSL upon 980 nm laser diode (0.5 W) irradiation; (**b**) Excitation spectra monitoring at 540 and 701 nm; (**c**) PersL decay curves (monitoring at 540 nm) after irradiation by using 254 nm or 337 nm light for 5 min, the inset is the PersL emission spectrum achieved at 7 s of the decay.

$Zn_{2.4}Li_{0.6}Ga_2GeO_8$:Mn and $Zn_3Ga_2GeO_8$:Mn phosphors synthesized by solid-state reaction demonstrated completely different fluorescence characteristics. The PLE and PL spectra of these phosphors are illustrated in Figure 5. We found that $Zn_3Ga_2GeO_8$:Mn phosphor exhibited a broad-band red emission at 650–800 nm from $^2E \rightarrow {}^4A_2$ transition of Mn^{4+} ions, accompanied with the weak green color broad-band emission at 500–600 nm from matrix defect in Figure 5a. Meanwhile, $Zn_{2.4}Li_{0.6}Ga_2GeO_8$:Mn phosphor demonstrates a broad-band pure-red emission at 650–800 nm upon the 365 nm excitation, indicating that Li doping turns Mn^{4+} into the spin-allowed weak crystal field, leading to a broader red band emission. In addition, the perceptible color of luminescence from the two samples can be directly observed, as shown in the digital imaging signals (Figure 5b). Compared with the orange light of $Zn_3Ga_2GeO_8$:Mn phosphors without Li, the $Zn_{2.4}Li_{0.6}Ga_2GeO_8$:Mn phosphor exhibits an enhanced pure-red emission from Mn^{4+} ions at 701 nm. The profile and peak position of PLE spectra (Figure 5a) are similar to the PLE spectra monitored at 701 nm as shown in Figure 3b, which indicates that the broad-band red luminescence for three samples (including $Zn_{2.4}Li_{0.6}Ga_2GeO_8$:Mn and $Zn_3Ga_2GeO_8$:Mn phosphors synthesized by solid-state reaction and $Zn_3Li_{0.4}Ga_{1.6}GeO_8$:Mn phosphor synthesized by hydrothermal approach) come from the same Mn^{4+} emission centers.

Comparing the spectral characteristics of the four samples, we found that the emission peak at 701 nm and their PLE spectra show the same spectral profile and position in the three samples ($Zn_3Li_{0.4}Ga_{1.6}GeO_8$:Mn synthesized by a hydrothermal approach; $Zn_{2.4}Li_{0.6}Ga_2GeO_8$:Mn and $Zn_3Ga_2GeO_8$:Mn phosphors synthesized by solid-state reaction). Comparing the XRD patterns of the three samples, we can find that all three samples have the same spinel-structured solid solutions. Combined with the spectral features, the broad red emission bands peaking at 701 nm are easily ascribed to $^2E \rightarrow {}^4A_2$ transitions of Mn^{4+} [28,29], which occupied octahedral sites in a spinel-structured solid solutions of $ZnGa_2O_4$ and Zn_2GeO_4. While the emission spectrum peaking at 701 nm and its PLE spectrum of $Zn_{2.4}Li_{0.6}Ga_2GeO_8$:Mn synthesized by a hydrothermal approach (Figure 4a,b) are different from the other three samples. The spin-forbidden red narrow-band emission peaking at 701 nm can be ascribed to the $^4T_1(4G) \rightarrow {}^6A_1(6S)$ transitions of Mn^{2+}, which occupied the strong octahedral crystal field [30,31] in the cubic phase $ZnGa_2O_4$ lattice. Similarly, green luminescence with PersL and PSL features can be observed only in $Zn_3Li_{0.4}Ga_{1.6}GeO_8$:Mn and $Zn_{2.4}Li_{0.6}Ga_2GeO_8$:Mn phosphors with a rhombohedral Zn_2GeO_4 phase synthesized by a hydrothermal approach, indicating that the rhombohedral Zn_2GeO_4 is a good matrix for generating green afterglow of Mn^{2+}-occupied tetrahedral sites. The broad emission band (400–750 nm) from $Zn_3Li_{0.4}Ga_{1.6}GeO_8$:Mn and $Zn_{2.4}Li_{0.6}Ga_2GeO_8$:Mn phosphor synthesized by a hydrothermal approach could be assigned to matrix defects under a 254 nm UV lamp excitation, which can be further verified by the PL and PLE spectra of undoped $Zn_3Li_{0.4}Ga_{1.6}GeO_8$ phosphor, as shown in Figure 6a.

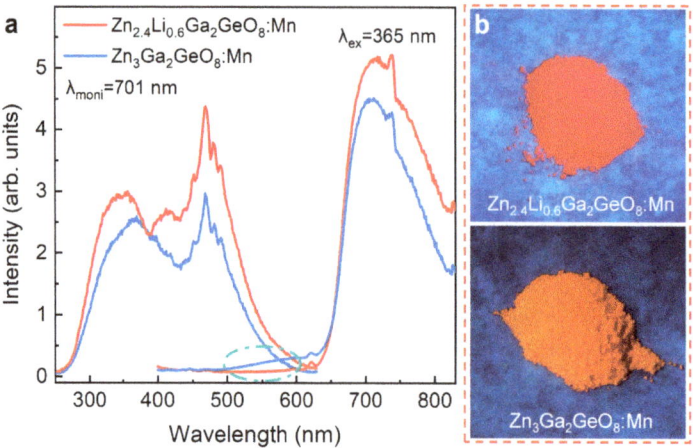

Figure 5. PLE and PL spectra and the corresponding digital PL images of $Zn_{2.4}Li_{0.6}Ga_2GeO_8$:Mn and $Zn_3Ga_2GeO_8$:Mn phosphors synthesized by solid-state reaction. (**a**) PLE spectra monitored at 701 nm and PL spectra under 365 nm UV lamp excitation; (**b**) Digital PL photographs under 365 nm irradiation. The camera parameters are manual/ISO 3200/1 s. The slight difference in PL spectra can be seen in the cyan dash dot ellipse.

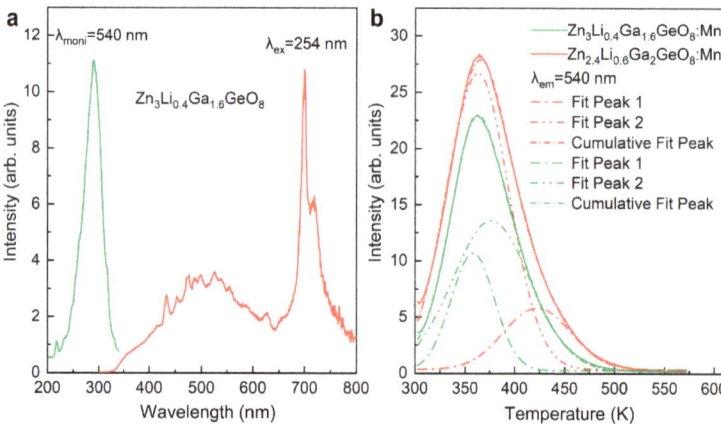

Figure 6. (**a**) PL (excitation at 254 nm) and PLE (monitoring at 540 nm) spectra of undoped $Zn_3Li_{0.4}Ga_{1.6}GeO_8$ phosphors synthesized by hydrothermal approach; (**b**) TL spectra obtained after 5 min illumination for $Zn_3Li_{0.4}Ga_{1.6}GeO_8$:Mn (green curve) and $Zn_{2.4}Li_{0.6}Ga_2GeO_8$:Mn phosphors (red curve) prepared by hydrothermal method.

As a result, the hydrothermal approach is an effective method for the preparation of afterglow phosphors. The stored energy of $Zn_3Li_{0.4}Ga_{1.6}GeO_8$:Mn and $Zn_{2.4}Li_{0.6}Ga_2GeO_8$:Mn PersL phosphors can also be triggered by heating, which helps to release the absorbed energy and provide insights into PersL mechanism. TL curves monitored at 540 nm are shown in Figure 6b. The TL curves for $Zn_3Li_{0.4}Ga_{1.6}GeO_8$:Mn (Figure 6b, green curve) and $Zn_{2.4}Li_{0.6}Ga_2GeO_8$:Mn (Figure 6b, red curve) are similar and can be divided into two broad bands at 358 K and 376 K, and at 363 K and 423 K, respectively. These results indicate the presence of a shallow trap and deep trap in the two matrix materials, and the existence of the deep traps provides the conditions for generating PersL. Compared to $Zn_3Li_{0.4}Ga_{1.6}GeO_8$:Mn phosphor, the traps of $Zn_{2.4}Li_{0.6}Ga_2GeO_8$:Mn phosphor are deeper, which may explain why $Zn_{2.4}Li_{0.6}Ga_2GeO_8$:Mn phosphor has a long green afterglow duration. The possible luminescence mechanisms, including green PL, PersL and PSL from Mn^{2+}, and the two red PL from Mn^{2+} and Mn^{4+} are proposed and shown in Figure 7 [8,22,32,33].

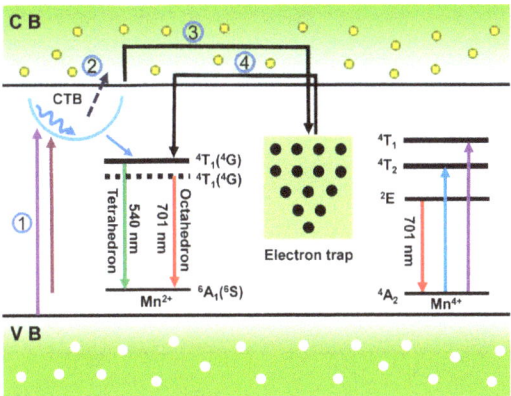

Figure 7. The proposed PL, PersL and PSL schematic diagram for green and red luminescence of Mn^{2+}/Mn^{4+} in the $Zn_3Ga_2GeO_8$ phosphors. Therein, ① UV light excitation, ② energy (or electron) transfer processes, ③ trapping, ④ release. The straight-line arrows and curved-line arrows stand for optical transitions and energy (or electron) transfer processes, respectively. Note that the solid and dashed lines represent the $^4T_1$ energy levels of Mn^{2+} in the weak tetrahedral crystal field and strong octahedral crystal field, respectively.

As shown in Figure 7, under UV excitation, electrons are first excited to the Mn^{2+}–O^{2-} CTB or the excited state of Mn ions, some electrons reach the conduction band through photoelectric separation and are trapped by traps, and some electrons are relaxed to $^4T_1$ and ^2E levels through lattice vibration or non-radiation, resulting in $^4T_1$ (4G)→$^6A_1$ (6S) (540 nm) transitions of the Mn^{2+}-occupied tetrahedral site, $^4T_1$ (4G)→$^6A_1$ (6S) (701 nm) transitions of Mn^{2+} at the octahedral site, and ^2E→$^4A_2$ (701 nm) transitions of Mn^{4+} at the octahedral site. When the excitation is stopped, the trapped electrons in the trap reach the conduction band through lattice thermal vibration and are released to the $^4T_1$ energy level of Mn^{2+} ions, resulting in green continuous fluorescence from $^4T_1$ (4G)→$^6A_1$ (6S) (540 nm).

The luminescent emission color response to excitation wavelength and excitation power is convenient for various applications [8,34,35]. To verify the feasibility of the phosphors in the anti-counterfeiting fields, we used these phosphors as inks printed the table lamp patterns in Figure 8. Figure 8a depicts the schematic diagram of these patterns. The lampshade and bulb are printed with $Zn_3Li_{0.4}Ga_{1.6}GeO_8$:Mn and $Zn_{2.4}Li_{0.6}Ga_2GeO_8$:Mn phosphor synthesized by a hydrothermal approach, respectively.

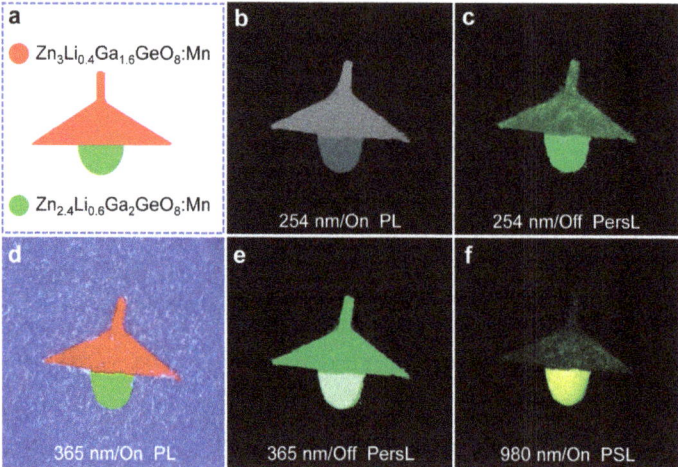

Figure 8. Digital photographs of a triple-mode (PL, PersL, and PSL modes) 'chandelier' pattern printed used $Zn_3Li_{0.4}Ga_{1.6}GeO_8$:Mn phosphor (lampshade) and $Zn_{2.4}Li_{0.6}Ga_2GeO_8$:Mn phosphor (light bulb) synthesized by hydrothermal approach. (**a**) The design of the 'chandelier' pattern; (**b**,**c**) PL and PersL images acquired upon and after 254 nm UV light excitation (for 5 min); (**d**,**e**) PL and PersL images acquired upon and after 365 nm UV light excitation (for 5 min); (**f**) PSL images upon a 980 nm laser diode irradiation (at 0.5 W). The PSL imaging was achieved until the disappearance of chandelier to the naked eye. The camera parameters are manual/ISO 3200/4 s.

Upon the irradiation of 254 nm UV light, the lampshade became blue–white due to the color mixture of the Mn PL and the defect fluorescence of matrix [36] (Figure 8b), while the bulb became blue–green (Figure 8b). After the stoppage of excitation light, both the lampshade (dark green) and the bulb (bright green) emitted green PersL (Figure 8c), while upon the irradiation of 365 nm UV light, the lampshade and the bulb emitted bright red PL and green PL, respectively (Figure 8d). After the stoppage of UV excitation, these two patterns both showed green PersL colors with different saturations (Figure 8e). After these two patterns disappeared, the entire 'desk lamp' pattern was illuminated by a 980 nm laser diode; the lampshade and the bulb were lit again (Figure 8f), with the lampshade emitting a dark green color and the bulb emitting a yellow–green color (for a 980 nm illumination at 0.5 W), which is consistent with spectral characterization. Therefore, Mn-activated $Zn_3Ga_2GeO_8$ phosphors hold a promise for multi-chrome (green, yellow and red) and multi-mode (PL, PersL, and PSL) anti-counterfeiting applications [37–39]. In addition, the PSL and PL of nanomaterials can be used for photodynamic activation [40], dosimetry [41], thermometry [42] and solid-state lighting [43].

Additionally, we need to point out that, for practical applications, the possibility of the power of the laser diodes (2 and 5 W for the 808 nm and 980 nm lasers, respectively) to generate even secondary heating effects on the samples should be considered. It is true that NIR lasers, such as 808 nm or 980 nm lasers, have been widely used for photothermal therapy (PTT) [44–46]. However, for PTT, the materials have strong absorptions for NIR, but they do not have luminescence. So, the energy absorbed is released as heat, while in the luminescence materials, the absorbed energy is released as luminescence [47,48]; therefore, heating is not a critical issue, even though it is unavoidable. Heating is always an issue for many applications with NIR lasers.

4. Conclusions

In summary, manganese-activated gallium germanate phosphors were successfully synthesized in two approaches, including a high-temperature solid-state reaction approach and a hydrothermal method. We found, for the first time, that these Mn-activated $Zn_3Ga_2GeO_8$ phosphors prepared by a hydrothermal method and solid-state reaction method have different crystal phase structures and spectral properties. The phosphors prepared by the hydrothermal method exhibited a double-peak emission, including green PL with a PersL feature and red PL without a PersL effect, due to simultaneously having two kinds of occupancy sites. While the Mn-activated $Zn_3Ga_2GeO_8$ phosphors prepared by a solid-state reaction method only exhibited red PL. In addition, Li^+ ions were selectively substituted for the tetrahedral Zn site or the octahedral Ga site, leading to green light emission at 540 nm ($^4T_1(G) \rightarrow {}^6A_1(S)$ transition of Mn^{2+}) with an afterglow and red light centered at 701 nm ($^2E \rightarrow {}^4A_2$ transition of Mn^{4+}) in the absence of afterglow. Particularly, Mn-activated $Zn_3Ga_2GeO_8$ phosphors synthesized by a hydrothermal approach via Li^+ replacing the Zn^{2+} or Ga^{3+} sites exhibit dynamic and multicolor emissions as luminescence labels for multiple anti-counterfeiting, thus revealing the great potential of these phosphors in multicolor display, anti-forgery, and other potential applications.

Author Contributions: Conceptualization, D.G. and W.C.; methodology, D.G. and P.W.; validation, P.W., F.G. and W.N.; formal analysis, D.G. and W.C.; investigation, P.W., F.G. and W.N.; resources, D.G. and W.C.; data curation, P.W.; writing—original draft preparation, D.G.; writing—review and editing, W.N., D.G. and W.C.; visualization, D.G. and P.W.; supervision, D.G.; project administration, D.G. and F.G.; funding acquisition, G.D and W.C. All authors have read and agreed to the published version of the manuscript.

Funding: This research was funded by National Natural Science Foundation of China, grant number 11604253 and Natural Science Foundation of Shaanxi Province of China, grant number 2018JM1036.

Data Availability Statement: Not applicable.

Acknowledgments: We would like to acknowledge the support from Guangxi Jialouyuan Medical Inc., Solgro, and the distinguished award from UT Arlington.

Conflicts of Interest: The authors declare no conflict of interest.

References

1. Thejas, K.K.; Abraham, M.; Kunti, A.K.; Tchernycheva, M.; Ahmad, S.; Das, S. Review on deep red-emitting rare-earth free germanates and their efficiency as well as adaptability for various applications. *Appl. Mater. Today* **2021**, *24*, 101094. [CrossRef]
2. Wang, T.; Wang, S.F.; Liu, Z.Y.; He, Z.Y.; Yu, P.; Zhao, M.Y.; Zhang, H.X.; Lu, L.F.; Wang, Z.; Wang, Z.; et al. A hybrid erbium(III)–bacteriochlorin near-infrared probe for multiplexed biomedical imaging. *Nat. Mater.* **2021**, *20*, 1571–1578. [CrossRef]
3. Zhang, Y.; Zhu, X.H.; Zhang, Y. Exploring heterostructured upconversion nanoparticles: From rational engineering to diverse applications. *ACS Nano* **2021**, *15*, 3709–3735. [CrossRef] [PubMed]
4. Cui, X.Y.; Shi, W.Y.; Lu, C. Control of multicolor and white emission by triplet energy transfer. *J. Phys. Chem. A* **2021**, *125*, 4209–4215. [CrossRef] [PubMed]
5. Shi, S.C.; Bai, W.H.; Xuan, T.T.; Zhou, T.L.; Dong, G.Y.; Xie, R.J. In situ inkjet printing patterned lead halide perovskite quantum dot color conversion films by using cheap and eco-friendly aqueous inks. *Small Methods* **2021**, *5*, 2000889. [CrossRef] [PubMed]
6. Pan, Z.W.; Lu, Y.Y.; Liu, F. Sunlight-activated long-persistent luminescence in the near-infrared from Cr^{3+}-doped zinc gallogermanates. *Nat. Mater.* **2012**, *11*, 58–63. [CrossRef] [PubMed]
7. Chuang, Y.J.; Zhen, Z.P.; Zhang, F.; Liu, F.; Mishra, J.P.; Tang, W.; Chen, H.M.; Huang, X.L.; Wang, L.C.; Chen, X.Y.; et al. Photostimulable near-infrared persistent luminescent nanoprobes for ultrasensitive and longitudinal deep-tissue bio-imaging. *Theranostics* **2014**, *4*, 1112–1122. [CrossRef] [PubMed]
8. Gao, D.L.; Gao, J.; Gao, F.; Kuang, Q.Q.; Pan, Y.; Chen, Y.F.; Pan, Z.W. Quintuple-mode dynamic anti-counterfeiting using multi-mode persistent phosphors. *J. Mater. Chem. C* **2021**, *9*, 16634–16644. [CrossRef]
9. Zhou, Q.; Dolgov, L.; Srivastava, A.M.; Zhou, L.; Wang, Z.L.; Shi, J.X.; Dramićanin, M.D.; Brik, M.G.; Wu, M.M. Mn^{2+} and Mn^{4+} red phosphors: Synthesis, luminescence and applications in WLEDs. A review. *J. Mater. Chem. C* **2018**, *6*, 2652–2671. [CrossRef]
10. Kanzaki, M. Crystal structures of Zn_2GeO_4 cubic/tetragonal spinel and Zn_2SiO_4 modified spinel phases. *J. Miner. Petrol. Sci.* **2018**, *113*, 41–46. [CrossRef]
11. Gao, D.L.; Ma, K.W.; Wang, P.; Zhang, X.Y.; Pang, Q.; Xin, H.; Zhang, Z.H.; Jiao, H. Tuning multicolour emission of Zn_2GeO_4: Mn phosphors by Li^+ doping for information encryption and anti-counterfeiting applications. *Dalton Trans.* **2022**, *51*, 553–561. [CrossRef] [PubMed]
12. Dong, L.P.; Zhang, L.; Jia, Y.C.; Xu, Y.H.; Yin, S.W.; You, H.P. $ZnGa_{2-y}Al_yO_4$: Mn^{2+}, Mn^{4+} thermochromic phosphors: Valence state control and optical temperature sensing. *Inorg. Chem.* **2020**, *59*, 15969–15976. [CrossRef]

13. Chen, X.; Yao, W.; Wang, Q.; Wu, W. Designing multicolor dual-mode lanthanide-doped NaLuF$_4$/Y$_2$O$_3$ composites for advanced anticounterfeiting. *Adv. Opt. Mater.* **2020**, *8*, 1901209. [CrossRef]
14. Wang, Y.J.; Włodarczyk, D.; Brik, M.G.; Barzowska, J.; Shekhovtsov, A.N.; Belikov, K.N.; Paszkowicz, W.; Li, L.; Zhou, X.J.; Suchocki, A. Effect of temperature and high pressure on luminescence properties of Mn^{3+} ions in Ca$_3$Ga$_2$Ge$_3$O$_{12}$ single crystals. *J. Phys. Chem. C* **2021**, *125*, 5146–5157. [CrossRef]
15. López, S.; Romero, A.H.; Rodríguez-Hernández, P.; Munoz, A. First-principles study of the high-pressure phase transition in ZnAl$_2$O$_4$ and ZnGa$_2$O$_4$: From cubic spinel to orthorhombic post-spinel structures. *Phys. Rev. B* **2009**, *79*, 214103. [CrossRef]
16. Liu, F.; Liang, Y.; Pan, Z. Detection of up-converted persistent luminescence in the near infrared emitted by the Zn$_3$Ga$_2$GeO$_8$: Cr^{3+}, Yb^{3+}, Er^{3+} phosphor. *Phys. Rev. Lett.* **2014**, *113*, 177401. [CrossRef]
17. Allix, M.; Chenu, S.; Véron, E.; Poumeyrol, T.; Kouadri-Boudjelthia, E.A.; Alahrache, S.; Porcher, F.; Massiot, D.; Fayon, F. Considerable improvement of long-persistent luminescence in germanium and tin substituted ZnGa$_2$O$_4$. *Chem. Mater.* **2013**, *25*, 1600–1606. [CrossRef]
18. Huang, K.; Dou, X.; Zhang, Y.F.; Gou, X.P.; Liu, J.; Qu, J.L.; Li, Y.; Huang, P.; Han, G. Enhancing light and X-ray charging in persistent luminescence nanocrystals for orthogonal afterglow anti-counterfeiting. *Adv. Funct. Mater.* **2021**, *31*, 2009920. [CrossRef]
19. Ma, Q.; Wang, J.; Zheng, W.; Wang, Q.; Li, Z.H.; Cong, H.J.; Liu, H.J.; Chen, X.Y.; Yuan, Q. Controlling disorder in host lattice by hetero-valence ion doping to manipulate luminescence in spinel solid solution phosphors. *Sci. China Chem.* **2018**, *61*, 1624–1629. [CrossRef]
20. Srivastava, B.B.; Gupta, S.K.; Li, Y.; Mao, Y. Bright persistent green emitting water-dispersible Zn$_2$GeO$_4$: Mn nanorods. *Dalton Trans.* **2020**, *49*, 7328–7340. [CrossRef]
21. Ma, Z.D.; Zhou, J.Y.; Zhang, J.C.; Zeng, S.H.; Zhou, H.; Smith, A.T.; Wang, W.X.; Sun, L.Y.; Wang, Z.F. Mechanics-induced triple-mode anticounterfeiting and moving tactile sensing by simultaneously utilizing instantaneous and persistent mechanoluminescence. *Mater. Horizons* **2019**, *6*, 2003–2008. [CrossRef]
22. Lin, S.S.; Lin, H.; Ma, C.G.; Cheng, Y.; Ye, S.Z.; Lin, F.L.; Li, R.F.; Xu, J.; Wang, Y.S. High-security-level multi-dimensional optical storage medium: Nanostructured glass embedded with LiGa$_5$O$_8$: Mn^{2+} with photostimulated luminescence. *Light Sci. Appl.* **2020**, *9*, 22. [CrossRef] [PubMed]
23. Dong, L.; Zhang, L.; Jia, Y.; Shao, B.; Lü, W.; Zhao, S.; You, H. Enhancing luminescence and controlling the Mn valence state of Gd$_3$Ga$_{5−x−\delta}$Al$_{x−y+\delta}$O$_{12}$:yMn phosphors by the design of the garnet structure. *ACS Appl. Mater. Interfaces* **2020**, *12*, 7334–7344. [CrossRef] [PubMed]
24. Du, J.R.; Li, K.; Deun, R.V.; Poelman, D.; Lin, H.W. Near-infrared persistent luminescence and trap reshuffling in Mn^{4+} doped alkali-earth metal tungstates. *Adv. Opt. Mater.* **2022**, *10*, 2101714. [CrossRef]
25. Liu, F.; Yan, W.Z.; Chuang, Y.-J.; Zhen, Z.P.; Xie, J.; Pan, Z.W. Photostimulated near-infrared persistent luminescence as a new optical read-out from Cr^{3+} doped LiGa$_5$O$_8$. *Sci. Rep.* **2013**, *3*, 1554. [CrossRef]
26. Zheng, S.H.; Shi, J.P.; Fu, X.Y.; Wang, C.C.; Sun, X.; Chen, C.J.; Zhuang, Y.X.; Zou, X.Y.; Li, Y.C.; Zhang, H.W. X-ray recharged long afterglow luminescent nanoparticles MgGeO$_3$: Mn^{2+}, Yb^{3+}, Li$^+$ in the first and second biological windows for long-term bioimaging. *Nanoscale* **2020**, *12*, 14037–14046. [CrossRef]
27. Chen, X.; Li, Y.; Huang, K.; Huang, L.; Tian, X.; Dong, H.; Kang, R.; Hu, Y.; Nie, J.; Qiu, J.; et al. Trap energy upconversion-like near-infrared to near-infrared light rejuvenateable persistent luminescence. *Adv. Mater.* **2021**, *33*, 2008722. [CrossRef]
28. Adachi, S. Review—Mn^{4+}-activated red and deep red-emitting phosphors. *ECS J. Solid State Sci. Technol.* **2020**, *9*, 016001. [CrossRef]
29. Brik, M.G.; Camardello, S.J.; Srivastava, A.M. Influence of covalency on the Mn^{4+} ^{2}E$_g$→^{4}A$_{2g}$ emission energy in crystals. *ECS J. Solid State Sci. Technol.* **2015**, *4*, R39. [CrossRef]
30. Brik, M.G.; Srivastava, A.M. Critical review—A review of the electronic structure and optical properties of ions with d^3 electron configuration (V^{2+}, Cr^{3+}, Mn^{4+}, Fe^{5+}) and main related misconceptions. *ECS J. Solid State Sci. Technol.* **2017**, *7*, R3079. [CrossRef]
31. Si, T.; Zhu, Q.; Xiahou, J.Q.; Sun, X.D.; Li, J.-G. Regulating Mn^{2+}/Mn^{4+} activators in ZnGa$_2$O$_4$ via Mg^{2+}/Ge^{4+} doping to generate multimode luminescence for advanced anti-counterfeiting. *ACS Appl. Electron. Mater.* **2021**, *3*, 2005–2016. [CrossRef]
32. Sun, X.Y.; Zhang, J.H.; Zhang, X.; Luo, Y.S.; Hao, Z.D.; Wang, X.J. Effect of retrapping on photostimulated luminescence in Sr$_3$SiO$_5$: Eu^{2+}, Dy^{3+} phosphor. *J. Appl. Phys.* **2009**, *105*, 013501. [CrossRef]
33. Ueda, J.; Maki, R.; Tanabe, S. Vacuum referred binding energy (VRBE)-guided design of orange persistent Ca$_3$Si$_2$O$_7$: Eu^{2+} phosphors. *Inorg. Chem.* **2017**, *56*, 10353–10360. [CrossRef] [PubMed]
34. Gao, D.L.; Gao, J.; Zhao, D.; Pang, Q.; Xiao, G.Q.; Wang, L.L.; Ma, K.W. Enhancing the red upconversion luminescence of hybrid porous microtubes via an in situ O-substituted reaction through heat treatment. *J. Mater. Chem. C* **2020**, *8*, 17318–17324. [CrossRef]
35. Gao, D.L.; Zhao, D.; Pan, Y.; Chai, R.P.; Pang, Q.; Zhang, X.Y.; Chen, W. Extending the color response range of Yb^{3+} concentration-dependent multimodal luminescence in Yb/Er doped fluoride microrods by annealing treatment. *Ceram. Int.* **2021**, *47*, 32000–32007. [CrossRef]
36. Wang, J.; Ma, Q.Q.; Zheng, W.; Liu, H.Y.; Yin, C.Q.; Wang, F.B.; Chen, X.Y.; Yuan, Q.; Tan, W.H. One-dimensional luminous nanorods featuring tunable persistent luminescence for autofluorescence-free biosensing. *ACS Nano* **2017**, *11*, 8185–8191. [CrossRef]

37. You, W.W.; Tu, D.T.; Li, R.F.; Zheng, W.; Chen, X.Y. "Chameleon-like" optical behavior of lanthanide-doped fluoride nanoplates for multilevel anti-counterfeiting applications. *Nano Res.* **2019**, *12*, 1417–1422. [CrossRef]
38. Ding, M.Y.; Dong, B.; Lu, Y.; Yang, X.F.; Yuan, Y.J.; Bai, W.F.; Wu, S.T.; Ji, Z.Z.; Lu, C.H.; Zhang, K.; et al. Energy manipulation in lanthanide-doped core-shell nanoparticles for tunable dual-mode luminescence toward advanced anti-counterfeiting. *Adv. Mater.* **2020**, *32*, 2002121. [CrossRef]
39. Zhang, Y.; Huang, R.; Li, H.L.; Lin, Z.X.; Hou, D.J.; Guo, Y.Q.; Song, J.; Song, C.; Lin, Z.W.; Zhang, W.X.; et al. Triple-mode emissions with invisible near-infrared after-glow from Cr^{3+}-doped zinc aluminum germanium nanoparticles for advanced anti-counterfeiting applications. *Small* **2020**, *16*, 2003121. [CrossRef]
40. Ma, L.; Zou, X.; Bui, B.; Chen, W.; Song, K.H.; Solberg, T. X-ray excited ZnS:Cu,Co afterglow nanoparticles for photodynamic activation. *Appl. Phys. Lett.* **2014**, *105*, 13702. [CrossRef]
41. Chen, W.; Wang, S.P.; Westcott, S.; Liu, Y. Dose Dependent X-Ray Luminescence in MgF_2: Eu^{2+}, Mn^{2+} Phosphors. *J. Appl. Phys.* **2008**, *103*, 113103. [CrossRef]
42. Wang, S.P.; Westcott, L.S.; Chen, W. Nanoparticle Luminescence Thermometry. *J. Phys. Chem. B* **2002**, *106*, 11203–11209. [CrossRef]
43. Chen, W.; Joly, A.; Malm, J.O.; Bovin, J.O.; Wang, S.P. Full-Color Emission and Temperature Dependence of The Luminescence In Poly-P-Phenylene Ethynylene-ZnS:Mn^{2+} Composite Particles. *J. Phys. Chem. B* **2003**, *107*, 6544–6551. [CrossRef]
44. Li, Y.; Lu, W.; Huang, Q.; Huang, M.; Li, C.; Chen, W. Copper sulfide nanoparticles for photothermal ablation of tumor cells. *Nanomedicine* **2010**, *5*, 1161–1171. [CrossRef] [PubMed]
45. Lakshmanan, S.B.; Zou, X.; Hossu, M.; Ma, L.; Yang, C.; Chen, W. Local field enhanced Au/CuS nanocomposites as efficient photothermal transducer agents for cancer treatment. *J. Biomed. Nanotechnol.* **2012**, *8*, 883–890. [CrossRef] [PubMed]
46. Chu, X.; Li, K.; Guo, H.; Zheng, H.; Shuda, S.; Wang, X.; Zhang, J.; Chen, W.; Zhang, Y. Exploration of Graphitic-C_3N_4 Quantum Dots for Microwave-Induced Photodynamic Therapy. *ACS Biomater. Sci. Eng.* **2017**, *3*, 1836–1844. [CrossRef]
47. Ma, L.; Chen, W. ZnS:Cu,Co water-soluble afterglow nanoparticles: Synthesis, luminescence and potential applications. *Nanotechnology* **2010**, *21*, 385604. [CrossRef]
48. Tang, X.; Yu, H.; Bui, B.; Wang, L.; Xing, C.; Wang, S.; Chen, M.; Hu, Z.; Chen, W. Nitrogen-doped fluorescence carbon dots as multi-mechanism detection for iodide and curcumin in biological and food samples. *Bioact. Mater.* **2020**, *6*, 1541–1554. [CrossRef]

Article

Self-Matrix N-Doped Room Temperature Phosphorescent Carbon Dots Triggered by Visible and Ultraviolet Light Dual Modes

Huiyong Wang [1], Hongmei Yu [1,*], Ayman AL-Zubi [2], Xiuhui Zhu [3,*], Guochao Nie [4], Shaoyan Wang [1] and Wei Chen [2,*]

1. School of Chemical Engineering, University of Science and Technology Liaoning, Anshan 114051, China; hywang@163.com (H.W.); aswsy64@163.com (S.W.)
2. Department of Physics, The University of Texas at Arlington, Arlington, TX 76019-0059, USA; ayman.alzubi@uta.edu
3. Department of Chemical Engineering, Yingkou Institute of Technology, Yingkou 115014, China
4. School of Physics and Telecommunication Engineering, Yulin Normal University, Yulin 537006, China; bccu518@ylu.edu.cn
* Correspondence: seesea0304@163.com (H.Y.); zhuxiuhui@126.com (X.Z.); weichen@uta.edu (W.C.)

Abstract: The synthesis of room temperature phosphorescent carbon dots (RTP-CDs) without any matrix is important in various applications. In particular, RTP-CDs with dual modes of excitation are more interesting. Here, we successfully synthesized matrix-free carbonized polymer dots (CPDs) that can generate green RTP under visible and ultraviolet light dual-mode excitation. Using acrylic acid (AA) and ammonium oxalate as precursors, a simple one-pot hydrothermal method was selected to prepare AA-CPDs. Here, acrylic acid is easy to polymerize under high temperature and high pressure, which makes AA-CPDs form a dense cross-linked internal structure. Ammonium oxalate as a nitrogen source can form amino groups during the reaction, which reacts with a large number of pendant carboxyl groups on the polymer chains to further form a cross-linked structure. The carboxyl and amino groups on the surface of AA-CPDs are connected by intermolecular hydrogen bonds. These hydrogen bonds can provide space protection (isolation of oxygen) around the AA-CPDs phosphor, which can stably excite the triplet state. This self-matrix structure effectively inhibits the non-radiative transition by blocking the intramolecular motion of CPDs. Under the excitation of WLED and 365 nm ultraviolet light, AA-CPDs exhibit the phosphorescence emission at 464 nm and 476 nm, respectively. The naked-eye observation exceeds 5 s and 10 s, respectively, and the average lifetime at 365 nm excitation wavelength is as long as 412.03 ms. In addition, it successfully proved the potential application of AA-CPDs in image anti-counterfeiting.

Keywords: carbon dots; room temperature phosphorescence; visible/ultraviolet light excitation; anti-counterfeiting

1. Introduction

Room temperature phosphorescence (RTP) has a longer life than fluorescence and has a wide range of applications in safety [1,2], optoelectronic devices [3,4], and biological imaging [5,6]. In fact, phosphorescence is difficult to achieve at room temperature due to spin prohibition, the probability of transitions between singlet (S_1) and triplet states (T_1) is extremely low, and triplet phosphorescence is easily quenched by oxygen [7,8]. However, recent studies have found that the possibility of adjusting the properties of afterglow emission through simple structural modifications [9–13] looks promising. CDs, as an emerging carbon-based luminescent nanomaterial, has attracted more and more attention for its RTP performance due to its low cost, convenient preparation, good stability, environmental friendliness, and low toxicity [14–16]. At present, the RTP phenomenon based on CDs is more through embedding in various substrates, including polyvinyl alcohol [13], polyacrylamide [17], polyurethane [18], urea/biuret [10], boric acid [19], layered double hydroxides [20], etc. Although the introduction of substrates can achieve the RTP of CDs, the inherent chemical and physical properties of substrates hinder the RTP properties and applications of CDs [21].

In recent years, matrix-free RTP CDs materials were obtained through hydrogen bonds formed by the internal functional groups of CDs, which can effectively inhibit intramolecular vibration and rotation [2,22,23]. Meanwhile, the complex preparation process, high toxicity, high cost, and potential environmental hazards caused by the method of embedding matrix are also solved [24]. Specifically, the introduction of -COOH and -OH groups facilitates the formation of hydrogen bonds, which can provide spatial protection around the phosphor and further stabilize triplet excitons of the RTP emission [25]. The doping of N, P, or halogen in CDs promotes the generation of n-π* transition, which promotes the transformation of excitons into triplet states through intersystem crossover (ISC) [1,2,22,23]. Moreover, the doped CDs can reduce energy gap (ΔE_{ST}) [1], which is conducive to the self-fixation of triplet excitons.

However, the vast majority of reported matrix-free RTP-CDs materials need to be excited under ultraviolet light, which greatly limits their application (such as biology related fields [26]). Compared with ultraviolet light, visible light is less phototoxic, more penetrating, and more likely to trigger RTP [21,27,28]. Hu [21] et al. reported that orange afterglow of CDs prepared from L-aspartic acid as raw material can be observed under commercial blue LED (420 nm), suggesting that L-aspartic acid could form a structure similar to crosslinked polymer at high temperature. In fact, the CDs obtained due to the incomplete carbonization of the polymer clusters should be classified as carbonized polymer dots (CPDs) [29]. CPDs exhibit a polymer/carbon hybrid structure, which not only have excellent optical properties, but also inherit the properties of polymers and special photoluminescence (PL) mechanism [30]. The CPDs formed by polymerization of certain structures may produce RTP emissions. This process is considered to be able to self-fix to excite the triplet state to form a more compact core structure, which is similar to the process of embedding in a solid matrix [31]. Of course, effective ISC is another key factor that must be considered to implement RTP.

In this work, we found that the CDs prepared with acrylic acid (AA) as carbon source and ammonium oxalate as nitrogen source were CPDs with self-matrix properties, namely AA-CPDs, which can be used as a solidified host or as a luminescent guest without any matrix doping. It has the property of generating green RTP emission under the dual-mode excitation of visible/ultraviolet light. This choice is based on the following considerations: (i) Acrylic acid can be polymerized and forms polymer at high temperature. The high-temperature and high-pressure conditions of the hydrothermal process increase the collision, entanglement, and crosslinking between polymer chains. Therefore, AA-CPDs have a tighter and better cross-linked internal structure. (ii) The -OH groups and the O atoms of C=O can be used as multiple reaction sites for cross-linking polymerization to form intramolecular hydrogen bonds with amino groups, respectively. Amino groups have been proven to be an effective sub-fluorophore for fluorescence emission [32]. (iii) The N atoms of amino groups are conducive to the n-π* transition, thereby promoting the effective filling of triplet excitons in ISC [33,34]. What is exciting is that solid AA-CPDs show a green afterglow after the visible/ultraviolet light is turned off. The afterglow is visible to the naked eye for more than 5 s and 10 s, respectively. The phosphorescence lifetime can reach 412.03 ms under 365 nm excitation, and it has stable phosphorescence performance. More importantly, AA-CPDs excited in the visible/ultraviolet light dual modes have great application potential in advanced anti-counterfeiting and hiding of complex patterns. This study provides a simple and rapid new method for preparing matrix-free RTP-CDs under the dual-mode excitation of visible/ultraviolet light.

2. Experimental

2.1. Chemicals

Acrylic acid (AA), ammonium oxalate, and oxalic acid were purchased from Aladdin (Shanghai, China). All reagents were of analytical grade and used as received without further purification. Ultrapure water was prepared by a Milli-Q ultrapure water system for all experiments.

2.2. Synthesis of AA-CPDs

Ammonium oxalate (0.2 g) and acrylic acid (3 mL) were dissolved in ultrapure water (12 mL). Then, the solution was subsequently transferred to a 50 mL poly(tetrafluoroethylene)-lined stainless steel autoclave, heated at 180 °C for 8 h, and cooled to room temperature. Thus, a dark-yellow suspension was obtained. The suspension was sonicated (at 40 kHz) for 10 min, filtered through a 0.22 μm filter membrane and centrifuged (at 10,000 r/min) for 15 min to remove large particles. The obtained solution was placed in a dialysis bag (Mw = 500 Da) to remove the unreacted reagents. The water in the dialysis bag should be changed frequently at the beginning of dialysis, and then changed every 4 h until it is clear and transparent for 72 h. The solution dialyzed was freeze-dried for 36 h to obtain AA-CPDs powders, then the powders were collected, sealed, and preserved for further usage.

2.3. Apparatus and Characterization

The morphology of AA-CPDs was measured by a JEM-ARM200F high-resolution transmission electron microscope (HRTEM) at an acceleration voltage of 200 kV (Hitachi, Tokyo, Japan). An X'Pert pro X-ray diffractometer (PANalytical, Amsterdam, The Netherlands) was used to determine the crystalline pattern of AA-CPDs. A Hyperion Fourier transform infrared (FTIR) spectrometer (Bruker, Leipzig, Germany) was used to record the FTIR spectrum of AA-CPDs. An ESCALAB 250Xi electronic spectrometer (Thermo, Waltham, MA, USA) was used to examine the X-ray photoelectron spectroscopy (XPS) spectra of AA-CPDs. A U-3900 UV–vis spectrophotometer (Hitachi, Japan) was used to measure the UV–vis absorption spectra of AA-CPDs. An FS5 spectrofluorometer (Edinburgh, England) was used to gain fluorescence and phosphorescence spectra of AA-CPDs. An Fluorolog 3-11 spectrofluorometer (Horiba, Japan) was used to determine the fluorescence lifetime of AA-CPDs. An FLS-1000 fluorescence spectrometer (Edinburgh, UK) was used to measure the phosphorescence lifetime of AA-CPDs.

3. Results and Discussion

3.1. Characterization of AA-CPDs

AA-CPDs with green phosphorescence under visible/ultraviolet light dual-mode excitation were prepared by one-step hydrothermal method using acrylic acid (AA) and ammonium oxalate as precursors without adding any acid, base, or metal ions (Figure 1a). The AA-CPDs powders show light yellow under sunlight and green phosphorescence under visible/ultraviolet dual-mode excitation (Figure 1b).

To understand the properties of the material, the morphology and particle size of the AA-CPDs were investigated and shown in Figure 2a,b. The high-resolution transmission electron microscopy (HRTEM) image (Figure 2a) of AA-CPDs reveals that AA-CPDs are nearly spherical and well-dispersed and the lattice spacing is 0.21 nm, which matches the in-plane lattice spacing of graphene (100 facet) [35]. Figure 2b shows the particle size distribution of AA-CPDs, with an average particle size of 4.7 ± 0.9 nm. It can be observed from the X-ray diffraction (XRD) pattern (Figure 2c) that the peak of AA-CPDs is located at $2\theta = 20.14°$, corresponding to the (100) crystal plane of graphite-like carbon [35]. The XRD pattern shows a wide diffraction band, which reveals the amorphous nature of AA-CPDs. FTIR analysis was performed to reveal information about the surface functional groups of AA-CPDs. The FTIR spectrum of AA-CPDs shown in Figure 2d exhibits that a strong and wide peak across the range of 3500–2500 cm^{-1} indicates the presence of polymerized carboxyl [36] functional groups on the surface of AA-CPDs. The peaks at 2850 cm^{-1} and 1760 cm^{-1} are the stretching vibrations of C-H [37] and C=O [38], respectively. In addition, the stretching vibration of C=N [39] (1440 cm^{-1}) and the bending vibration of N-H [40] (787 cm^{-1}) were detected in AA-CPDs, indicating that polyaromatic structures form in their skeletons. The results show that AA-CPDs have polymer/carbon hybridization structures consisting of hydroxyl, carbonyl, and imine groups. Notably, the existence of C=O/C=N functional groups has been reported to facilitate the generation of triple excitations through intersystem intersection (ISC) [2,41].

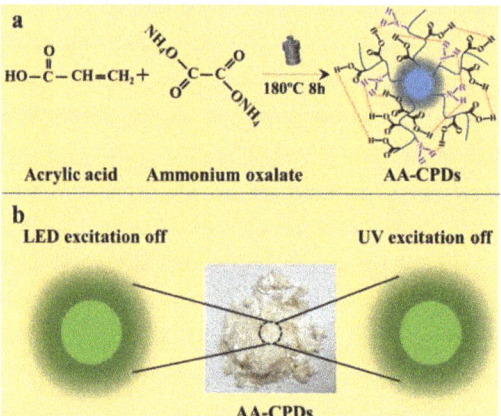

Figure 1. (a) Synthetic route of AA-CPDs. (b) Schematic illustration of AA-CPDs with WLED and UV excitation (365 nm).

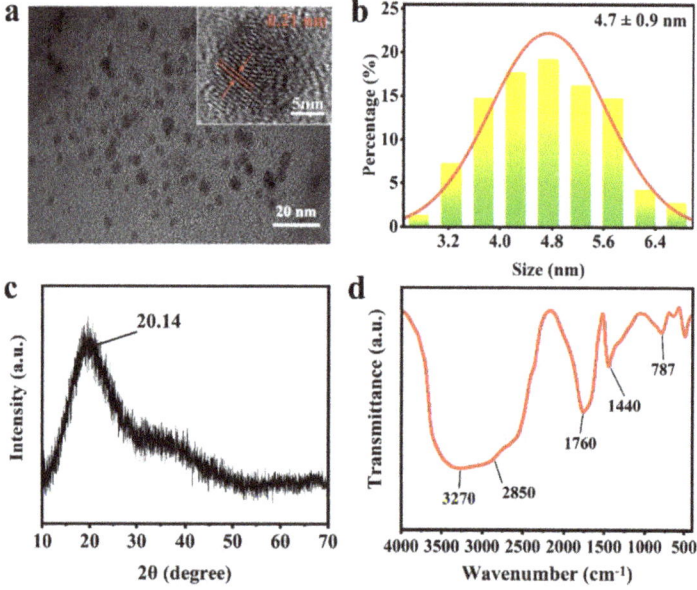

Figure 2. (a) HRTEM image, (b) Size distribution, (c) XRD pattern, and (d) FT − IR spectrum of AA-CPDs.

To further confirm the above FTIR analysis, the XPS spectra of the AA-CPDs were recorded in Figure 3a–d. The results show that AA-CPDs are composed of 73.03% C, 25.99%

O, and 0.98% N (Figure 3a). The C1s XPS spectrum (Figure 3b) has three peaks at 284.6 eV, 285.2 eV, and 288.8 eV, which are attributed to C–C or C=C [42], –COOH [43], and C–N [44], respectively. The N1s XPS spectrum (Figure 3c) consists of two components: pyrrolic N (400.2 eV) [40] and amino N (401.6 eV) [44]. The O1s XPS spectrum (Figure 3d) can be deconvoluted into two peaks at 531.9 eV and 533.3 eV, indicating the existence of C–OH and C=O [40], respectively. The wide O1s band facilitates RTP production [2,41].

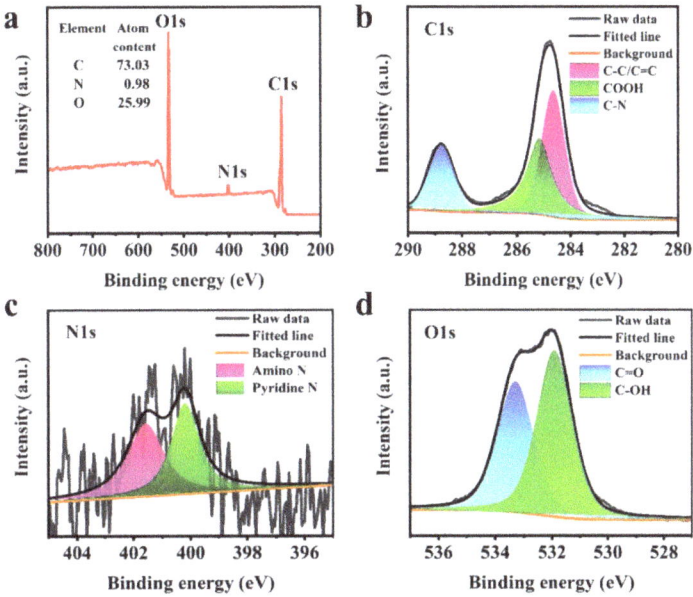

Figure 3. (a) XPS spectrum, (b–d) XPS fitting results for the C1s, N1s, and O1s spectra of AA-CPDs.

3.2. Fluorescence of AA-CPDs

The optical properties of AA-CPDs in aqueous solution and solid state were determined by UV–vis and fluorescence spectroscopy, respectively. The normalized UV–vis absorption, fluorescence excitation, and emission spectra of the AA-CPDs in aqueous solution are shown in Figure 4a. The UV–vis absorbance spectrum of the AA-CPDs (black line) showed one prominent peak centered around 210 nm. The peak was ascribed to the n–π^* transition of the C=O bonds of carboxyl groups and π–π^* transition of the C=C bonds comprising the aromatic network [45]. Notably, the presence of the C=O groups has previously been reported to facilitate the generation of intrinsic triplet excitons through ISC [10,41]. The emission wavelength of AA-CPDs was observed at 408 nm (blue line) under the optimal excitation wavelength of 330 nm (purple line). The excitation–emission map (Figure 4b) of the AA-CPDs in aqueous solution exhibits a single emission center (the maximum emission at 408 nm) ranging from 360 nm to 520 nm, which shows that fluorescence spectral characteristic is independent of excitation. The widespread absorption in UV regions (300–360 nm) is attributed to complex polymer/carbon hybrid structures [29,46,47]. The excitation-independent behavior of AA-CPDs in aqueous solution can be associated with the presence of one dominant fluorescence center. The best excitation peak of fluorescence excitation spectrum is 330 nm, which indicates that the C=O/C=N functional group of AA-CPDs should be responsible for fluorescence emission [21].

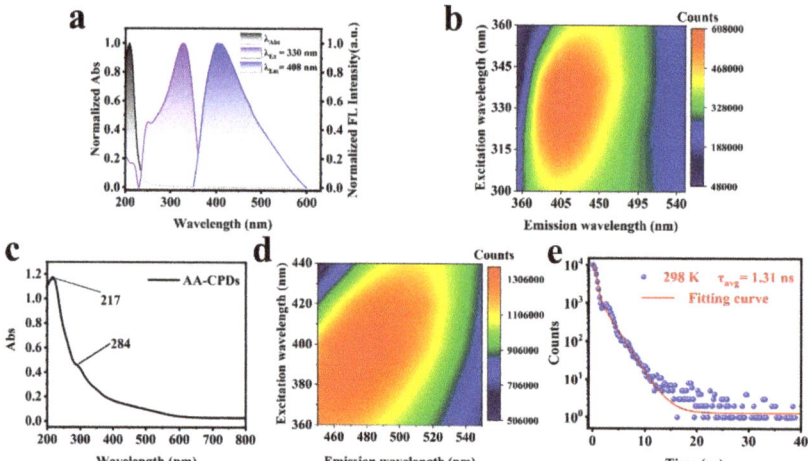

Figure 4. (**a**) The normalized UV–vis absorbance, FL excitation and emission spectra of AA-CPDs in water. (**b**) Fluorescence spectra of AA-CPDs in water. (**c**) UV–vis spectrum of the solid AA-CPDs. (**d**) Fluorescence spectra of the solid AA-CPDs. (**e**) FL decay curve of the AA-CPDs in water.

Further, the UV–vis absorption and fluorescence spectra of solid AA-CPDs were explored and shown in Figure 4c,d, respectively. As can be seen from Figure 4c, the UV–vis spectrum of solid AA-CPDs has two absorption peaks, including a strong absorption peak at 217 nm and a shoulder peak at 284 nm. The occurrence of acromion is due to the absorption peak formed by n-π* transition in amide functional groups. The UV–vis spectrum of solid AA-CPDs has two emission groups, which is consistent with their solid fluorescence spectrum. Fortunately, AA-CPDs powers were found to have solid fluorescence and no self-extinction [48,49]. Thanks to its inherent polymer structure, AA-CPDs successfully overcome the aggregation-induced annihilation problem faced by most solid CDs [50]. AA-CPDs, exhibit excellent solid-state fluorescence and unique RTP performance without the need for additional matrix or composite structures. The excitation and emission wavelengths of solid-state fluorescence are located at 400 nm and 478 nm, respectively. When the excitation wavelengths change from 360 nm to 440 nm, the maximum emission wavelengths of solid AA-CPDs change from 440 nm to 506 nm. It exhibits emission behavior associated with excitation wavelength, which is a common feature of CPDs [51]. The fluorescence emission wavelength of solid is red shifted compared with that of aqueous solution, which may be because the π-π* stacking interaction in solid enhances the energy transfer between AA-CPDs and generates new electronical states with emissions at long wavelength region [52–54]. The expansion and red shift of AA-CPDs photoluminescent peaks from aqueous solution to solid powder is attributed to the green RTP of solid powders [31].

The fluorescence decay profiles of AA-CPDs were studied at λex = 365 nm (Figure 4d). The decay curve of the AA-CPDs was fitted using a triadic exponential function with lifetimes (τ) of 0.6 ns (77.92%), 1.8 ns (21.83%) and 8.06 ns (0.25%). Average fluorescence life conforms to exponential function fit, based on Equation (1) [55]:

$$\tau_{avg} = \frac{\sum A_i \tau_i^2}{\sum A_i \tau_i} \tag{1}$$

The average fluorescence lifetime of AA-CPDs is calculated to be 1.31 ns. The absolute FL quantum yield (QY) of AA-CPDs measured under 330 nm excitation is 8.8%, showing the water-dispersed fluorescence properties.

3.3. Phosphorescence of AA-CPDs

The fluorescence and afterglow emission spectra of AA-CPDs at 365 nm and WLED as shown in Figure 5a. It can see that the AA-CPDs powders emit blue fluorescence at 443 nm (purple line) under UV light (365 nm). When the UV light was turned off, the AA-CPDs powders emit a green phosphorescence. The phosphorescence emission spectra have a broad peak from 400 to 600 nm, mainly centered at 476 nm (green line). Under the excitation of WLED, the phosphorescent emission wavelength of AA-CPDs is 464 nm (black line). The phosphorescent lifetime of AA-CPDs were measured at room temperature and shown in Figure 5b. The decay spectrum of AA-CPDs shows their long phosphorescent lifetime, which can be fitted with a three-exponential function. The lifetime components were determined to be 11.96 ms (16.04%), 98.40 ms (40.48%), and 476.08 ms (43.48%), respectively. These results suggest the AA-CPDs possess multiple decay channels. According to the following Equation (2) [56]:

$$\tau_{avg} = \frac{\sum B_i \tau_i^2}{\sum B_i \tau_i} \quad (2)$$

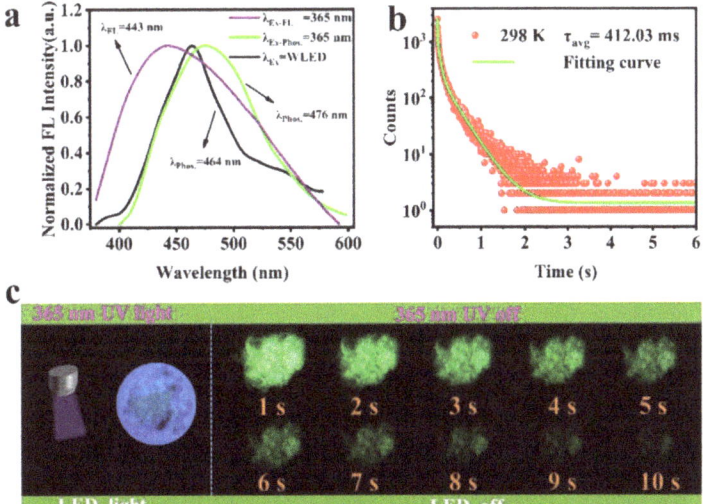

Figure 5. (a) FL and afterglow emission spectra of AA-CPDs at 365 nm and WLED. (b) Afterglow lifetime of AA-CPDs at 365 nm excitation. (c) Images of AA-CPDs powders under UV 365 nm and WLED on and off with time.

The average lifetime was calculated to be 412.03 ms (under excitation at 365 nm). RTP of AA-CPDs was determined and shown in Figure 5c. When the visible/ultraviolet light was removed, the phosphorescence signal of AA-CPDs can still be detected when t = 5 s or 10 s. By comparing the afterglow lifetimes of CDs reported in the literature (Table 1), it can be found that AA-CPDs have a great advantage in this work. Even if ground into powder, RTP performance is still shown, indicating that the crosslink structure is highly resistant to external damage.

Table 1. A comparison between the decay time to naked eyes and the lifetime of afterglow for the AA-CPDs with those of RTP carbon dots.

RTP Carbon Dots Materials	λ_{Em}/nm	Lifetime/ms	Decay Time to Naked Eyes/s	Reference
RTP C-dots	535	750	9	[1]
NCDs	519	459	6	[2]
AA-CDs	585	240.8	5	[21]
F-CDs	540	1390	10	[22]
P-CDs	525	1140	9	[25]
PCDs I-1	494	658.11	7	[37]
PCDs I-2	–	379.22	1.4	[37]
PCDs I-3	–	188.58	0.2	[37]
NCDs	515	747	2	[24]
AN-CPDs-150	485	373.5	–	[31]
AN-CPDs-180	490	436.1	–	[31]
AN-CPDs-200	494	466.5	–	[31]
AN-CPDs-230	500	257.8	–	[31]
AN-CPDs-250	518	174.8	–	[31]
AN-CPDs-280	532	117.2	–	[31]
AN-CPDs-300	558	61.4	–	[31]
AA-CPDs	476	412.03	10	This work

3.4. RTP Mechanism of AA-CPDs

In order to determine the source of RTP emission, the photophysical properties of the AA-CPDs were thoroughly investigated. Firstly, the AA-CPDs exhibit blue fluorescence (408 nm) in water without afterglow emission. The reason should be due to violent collisions caused by the free movement of molecules [57], causing hydrogen bonds inside AA-CPDs to be destroyed by the solvent effect of water molecules [24] (Figure 6 left). Secondly, the fluorescence spectrum of solid AA-CPDs shows a wide band with a centre wavelength of about 443 nm and a longer tail higher wavelength, which is caused by a phosphorescent phenomenon [1,19]. The AA-CPDs can be closely cross-linked by hydrogen bonds after water is removed, just as solid films are formed [34] (Figure 6 right). The AA-CPDs exhibit non-excitation-related fluorescence spectral characteristics in aqueous solutions; however, solids exhibit emission behaviour associated with excitation wavelengths, indicating that more emission centres are generated in solids that are responsible for the RTP characteristics of excitation dependence. Lastly, FT-IR and XPS analysis show that the AA-CPDs had a polymer/carbon hybrid structure with many groups including carbonyl and amino groups. According to previous reports, the C–N/C=N related groups in AA-CPDs have been confirmed to be the cause of the triplet state correlation emission [22,47,52]. Therefore, suppressing the quenching of triplet excitons generated in the C–N/C=N of CPDs is an effective way to achieve self-protected RTP. Hydrogen bonds effectively lock the emitted material and inhibit its intramolecular motion [13,41,58]. This is through nitrogen doping to achieve the self-protection RTP characteristics of AA-CPDs.

In order to prove that N doping is conducive to the generation of RTP. Therefore, we used oxalic acid instead of ammonium oxalate to prepare contrast CDs without N element, which were named AA-CPDs1. The excitation spectrum of AA-CPDs1 has the only peak located at 205 nm (Figure 7a). The best excitation and emission wavelengths of solid AA-CPDs1 are located at 360 nm and 426 nm, respectively (Figure 7b). Same as AA-CPDs, the solid AA-CPDs1 also has excitation wavelength dependency. The difference is that the fluorescence of AA-CPDs aqueous solution and solid has a red shift compared with that of AA-CPDs1, indicating that the increase of N element leads to the red shift of emission wavelength.

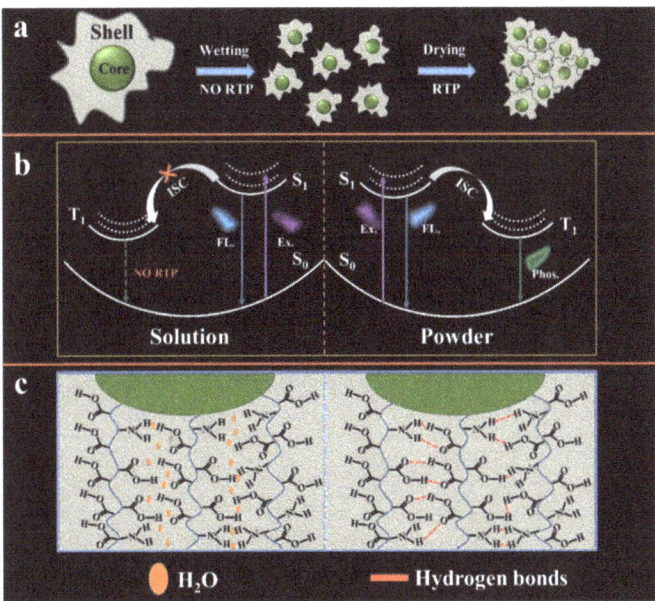

Figure 6. Mechanism of the photophysical processes of AA-CPDs. (**a**) Schematic of the process for achieving RTP; (**b**) The phosphorescence emission process of AA-CPDs; (**c**) Schematic diagram of hydrogen bond network interactions between AA-CPDs.

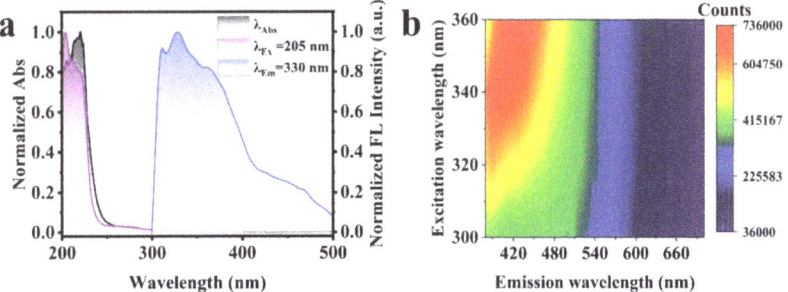

Figure 7. (**a**) The normalized UV absorbance, FL excitation and emission spectra of AA-CPDs1 in water. (**b**) Fluorescence spectra of the solid AA-CPDs1.

In addition, as can be seen from Figure 8, AA-CPDs1 has only 2 s of afterglow after turning off UV light, which is significantly less than the afterglow time of AA-CPDs. Under the excitation of WLED, AA-CPDs1 does not produce phosphorescence, only ultraviolet light single mode excitation produces phosphorescence. These images illustrate the large difference in RTP performance between the two materials. This result confirms that phosphorescent production is related to nitrogen element. This is consistent with previous reports [34,59] that nitrogen facilitates the n-π* transition, thereby facilitating the forbidden spin transfer from singlet to triplet state excitons through intersystem crossover.

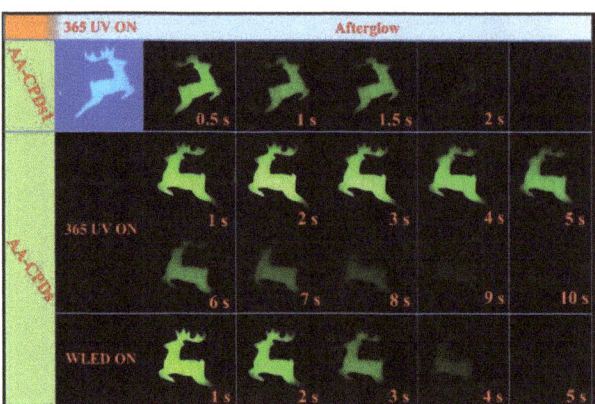

Figure 8. Image of AA-CPDs and AA-CPDs1 powders under UV 365 nm on and off with time.

In order to further illustrate that acrylic acid can form long polymer chains during the reaction process, this has been proved in several literatures [60,61]. It has been reported that acrylic acid can form a variety of oligomers, such as dimer, trimer, and tetramer, at high temperatures. Xia [29] et al. reported that carbonized polymer dots (CPDs) include polymer properties, such as rich functional groups, short polymer chains, and incomplete carbonization; meanwhile, a highly crosslinked network structure was formed through dehydration and carbonization [62]. Therefore, we believe that acrylic acid (AA) can form long chain polymers in 180 °C high pressure reactor, and the carbon dots prepared are CPDs. A large number of suspended carboxyl and amino groups on the surface of AA-CPDs can further form a cross-linked structure. The high degree of entanglement and cross-linking structure greatly limits the intramolecular space of AA-CPDs and at the same time shortens the distance between functional groups [50]. The existence of intermolecular interactions inside AA-CPDs can further reduce vibration and rotation, thereby inhibiting non-radiative relaxation [50]. Therefore, the self-matrix structure effectively inhibits the non-radiative transition by blocking the movement of the polymer chain, and can also strongly prevent oxygen quenching, presenting RTP at room temperature. In addition, the carbon dots can find applications in sensing [63–65], photodynamic activation on cancer treatment [66–68] and imaging [69–71], which are the topics for our research.

4. Conclusions

In summary, we successfully prepared RTP AA-CPDs using acrylic acid and ammonium oxalate as raw materials. AA-CPDs have the characteristics of producing green RTP under the dual-mode excitation of visible/ultraviolet light. Under environmental conditions, the phosphorescence lifetime of the AA-CPDs solid is 412.03 ms, and the afterglow of visible/ultraviolet light excitation is 5 s and 10 s for naked eye observation, respectively. The carboxyl and amino groups on the surface of AA-CPDs are connected by intermolecular hydrogen bonds. These hydrogen bonds may provide space protection (isolation of oxygen) around the AA-CPDs phosphor, which can stably excite the triplet state. In addition, through comparative experiments, we also verified that N doping promotes the generation of triplet excitons in AA-CPDs. More importantly, the prepared AA-CPDs can be used as a potential new type of smart hidden security material, which can be used in fields such as advanced anti-counterfeiting and the hiding of complex graphics. This research provides a method for simple and rapid preparation of metal-free RTP materials.

Author Contributions: H.W.: Experimental design, synthesis, and characterization; H.Y.: Data analysis, supervision, and funding; X.Z.: Characterization and funding; S.W.: Measurements and analysis; W.C. and H.Y.: Conceptualization, Supervision, and validation; H.W., H.Y., A.A.-Z., W.C., and G.N.: Writing and editing. All authors have read and agreed to the published version of the manuscript.

Funding: Financial supports from the XingLiao Talent Project Grants of Liaoning Province, China (No. XLYC1902076), the University of Science and Technology Liaoning Talent Project Grants (No. 601010302) and the Key Laboratory of Chemical Auxiliaries Synthesis and Separation of Liaoning Province, China (No. ZJNK2002, No. ZJNK2005) are highly appreciated. Nie GC would like to acknowledge the Guangxi Innovation Driven Development Major Project of Guangxi Province, China (No. Guike AA2030201) and Nanning Scientific Research and Technology Development Plan Project of Guangxi Province, China (No. RC20200001) for support. W.C. would like to thank the support from Solgro Inc. and UT Arlington for a distinguished award.

Institutional Review Board Statement: Not applicable.

Informed Consent Statement: Not applicable.

Data Availability Statement: Not applicable.

Conflicts of Interest: The authors declare no conflict of interest.

References

1. Qi, D.; Zhang, H.; Wu, X.; Tang, P.; Qian, M.; Wang, K.; Qi, P.H. Matrix-free and highly efficient room-temperature phosphorescence carbon dots towards information encryption and decryption. *Chem.-Asian J.* **2020**, *15*, 1281–1284. [CrossRef] [PubMed]
2. Li, H.; Ye, S.; Guo, J.Q.; Kong, J.T.; Song, J.; Kang, Z.H.; Qu, J.L. The design of room-temperature-phosphorescent carbon dots and their application as a security ink. *J. Mater. Chem. C* **2019**, *7*, 10605–10612. [CrossRef]
3. Zhao, F.F.; Zhang, T.Y.; Liu, Q.; Lü, C.L. Aphen-derived N-doped white-emitting carbon dots with room temperature phosphorescence for versatile applications. *Sens. Actuators B* **2019**, *304*, 127344. [CrossRef]
4. Miao, W.F.; Zou, W.S.; Zhao, Q.C.; Wang, Y.Q.; Chen, X.; Wu, S.B.; Liu, Z.M.; Xu, T.W. Coupling room-temperature phosphorescence carbon dots onto active layer for highly efficient photodynamic antibacterial chemotherapy and enhanced membrane properties. *J. Membr. Sci.* **2021**, *639*, 119754. [CrossRef]
5. Zhang, G.; Palmer, G.M.; Dewhirst, M.W.; Fraser, C.L. A dual-emissive-materials design concept enables tumour hypoxia imaging. *Nat. Mater.* **2009**, *8*, 747–751. [CrossRef]
6. Zhang, S.; Hosaka, M.; Yoshihara, T.; Negish, K.; Iida, Y.; Tobita, S.; Takeuchi, T. Phosphorescent light-emitting iridium complexes serve as a hypoxia-sensing probe for tumor imaging in living animals. *Cancer Res.* **2010**, *70*, 4490–4498. [CrossRef]
7. He, J.; Chen, Y.; He, Y.; Xu, X.; Lei, B.; Zhang, H.; Zhuang, J.; Hu, C.; Liu, Y. Anchoring carbon nanodots onto nanosilica for phosphorescence enhancement and delayed fluorescence nascence in solid and liquid states. *Small* **2020**, *16*, 2005228. [CrossRef]
8. Chen, K.C.; Liu, B. Enhancing the performance of pure organic room-temperature phosphorescent luminophores. *Nat. Commun.* **2019**, *10*, 2111.
9. Gou, H.; Liu, Y.; Zhang, G.; Liao, Q.; Huang, X.; Ning, F.; Ke, C.; Meng, Z.; Xi, K. Lifetime-tunable room-temperature phosphorescence of polyaniline carbon dots in adjustable polymer matrices. *Nanoscale* **2019**, *11*, 18311–18319.
10. Li, Q.; Zhou, M.; Yang, Q.; Wu, Q.; Shi, J.; Gong, A.; Yang, M. Efficient room-temperature phosphorescence from nitrogen-doped carbon dots in composite matrices. *Chem. Mater.* **2016**, *28*, 8221–8227. [CrossRef]
11. Li, Q.; Zhou, M.; Yang, M.; Yang, Q.; Zhang, Z.; Shi, J. Induction of long-lived room temperature phosphorescence of carbon dots by water in hydrogen-bonded matrices. *Nat. Commun.* **2018**, *9*, 734. [CrossRef] [PubMed]
12. Gao, Y.; Zhang, H.; Jiao, Y.; Lu, W.; Liu, Y.; Han, H.; Gong, X.; Shuang, S.; Dong, C. Strategy for activating room-temperature phosphorescence of carbon dots in aqueous environments. *Chem. Mater.* **2019**, *31*, 7979–7986. [CrossRef]
13. Deng, Y.; Zhao, D.; Chen, X.; Wang, F.; Song, H.; Shen, D. Long lifetime pure organic phosphorescence based on water soluble carbon dots. *Chem. Commun.* **2013**, *49*, 5751–5753. [CrossRef] [PubMed]
14. Baker, S.N.; Baker, G.A. Luminescent carbon nanodots: Emergent nanolights. *Angew. Chem.* **2010**, *49*, 6726–6744. [CrossRef] [PubMed]
15. Yuan, F.; Yuan, T.; Sui, L.; Wang, Z.; Xi, Z.; Li, Y.; Li, X.; Fan, L.; Tan, Z.; Chen, A.; et al. Engineering triangular carbon quantum dots with unprecedented narrow bandwidth emission for multicolored LEDs. *Nat. Commun.* **2018**, *9*, 2249. [CrossRef] [PubMed]
16. Wang, J.; Peng, F.; Lu, Y.; Zhong, Y.; Wang, S.; Xu, M.; Ji, X.; Su, Y.; Liao, L.; He, Y. Large-scale green synthesis of fluorescent carbon nanodots and their use in optics applications. *Adv. Opt. Mater.* **2015**, *3*, 103–111. [CrossRef]
17. Wang, H.Y.; Zhou, L.; Yu, H.M.; Tang, X.D.; Xing, C.; Nie, G.C.; Akafzade, H.; Wang, S.Y.; Chen, W. Exploration of room-temperature phosphorescence and new mechanism on carbon dots in a polyacrylamide platform and their applications for anti-counterfeiting and information encryption. *Adv. Opt. Mater.* **2022**, 202200678. [CrossRef]
18. Tan, J.; Zhang, J.; Li, W.; Zhang, L.; Yue, D. Synthesis of amphiphilic carbon quantum dots with phosphorescence properties and their multifunctional applications. *J. Mater. Chem. C* **2016**, *4*, 10146–10153. [CrossRef]
19. Li, W.; Zhou, W.; Zhou, Z.; Zhang, H.; Zhang, X.; Zhuang, J.; Liu, Y.; Lei, B.; Hu, C. A universal strategy for activating the multicolor room-temperature afterglow of carbon dots in a boric acid matrix. *Angew. Chem.* **2019**, *58*, 7278–7283. [CrossRef]

20. Bai, L.Q.; Xue, N.; Wang, X.R.; Shi, W.Y.; Lu, C. Activating efficient room temperature phosphorescence of carbon dots by synergism of orderly non-noble metals and dual structural confinements. *Nanoscale* **2017**, *9*, 6658–6664. [CrossRef]
21. Hu, S.; Jiang, K.; Wang, Y.; Wang, S.; Li, Z.; Lin, H. Visible-light-excited room temperature phosphorescent carbon dots. *Nanomaterials* **2020**, *10*, 464. [CrossRef] [PubMed]
22. Jiang, K.; Wang, Y.; Cai, C.; Lin, H. Conversion of carbon dots from fluorescence to ultralong room-temperature phosphorescence by heating for security applications. *Adv. Mater.* **2018**, *30*, 1800783. [CrossRef]
23. Long, P.; Feng, Y.; Cao, C.; Li, Y.; Han, J.; Li, S.; Peng, C.; Li, Z.; Feng, W. Self-protective room-temperature phosphorescence of fluorine and nitrogen Codoped carbon dots. *Adv. Funct. Mater.* **2018**, *28*, 1800791. [CrossRef]
24. Gao, Y.; Han, H.; Lu, W.; Jiao, Y.; Liu, Y.; Gong, X.; Xian, M.; Shuang, S.; Dong, C. Matrix-free and highly efficient room-temperature phosphorescence of nitrogen-doped carbon dots. *Langmuir* **2018**, *34*, 12845–12852. [CrossRef] [PubMed]
25. Lu, C.; Su, Q.; Yang, X. Ultra-long room-temperature phosphorescent carbon dots: pH sensing and dual-channel detection of tetracyclines. *Nanoscale* **2019**, *11*, 16036–16042. [CrossRef] [PubMed]
26. Cai, S.; Shi, H.; Li, J.; Gu, L.; Ni, Y.; Cheng, Z.; Wang, S.; Xiong, W.W.; Li, L.; An, Z.; et al. Visible-light-excited ultralong organic phosphorescence by manipulating intermolecular interactions. *Adv. Mater.* **2017**, *29*, 1701244. [CrossRef]
27. Wu, Z.; Liu, Z.; Yuan, Y. Carbon dots: Materials, synthesis, properties and approaches to long-wavelength and multicolor emission. *J. Mater. Chem. B* **2017**, *5*, 3794–3809. [CrossRef]
28. Pan, Z.; Lu, Y.Y.; Liu, F. Sunlight-activated long-persistent luminescence in the near-infrared from Cr^{3+}-doped zinc gallogermanates. *Nat. Mater.* **2012**, *11*, 58–63. [CrossRef]
29. Xia, C.; Zhu, S.; Feng, T.; Yang, M.; Yang, B. Evolution and synthesis of carbon dots: From carbon dots to carbonized polymer dots. *Adv. Sci.* **2019**, *6*, 1901316.
30. Wei, X.; Yang, J.; Hu, L.; Cao, Y.; Lai, J.; Cao, F.; Gu, J.; Cao, X. Recent advances in room temperature phosphorescent carbon dots: Preparation, mechanism, and applications. *J. Mater. Chem. C* **2021**, *9*, 4425–4443. [CrossRef]
31. Xia, C.; Zhu, S.; Zhang, S.; Zeng, Q.; Tao, S.; Tian, X.; Li, Y.; Yang, B. Carbonized polymer dots with tunable room-temperature phosphorescence lifetime and wavelength. *ACS Appl. Mater. Interfaces* **2020**, *12*, 38593–38601. [CrossRef] [PubMed]
32. Zhu, S.; Song, Y.; Zhao, X.; Shao, J.; Zhang, J.; Yang, B. The photoluminescence mechanism in carbon dots (graphene quantum dots, carbon nanodots, and polymer dots): Current state and future perspective. *Nano Res.* **2015**, *8*, 355–381. [CrossRef]
33. Zhao, J.; Wu, W.; Sun, J.; Guo, S. Triplet photosensitizers: From molecular design to applications. *Chem. Soc. Rev.* **2013**, *42*, 5323–5351. [CrossRef]
34. An, Z.; Zheng, C.; Tao, Y.; Chen, R.; Shi, H.; Chen, T.; Wang, Z.; Li, H.; Deng, R.; Liu, X.; et al. Stabilizing triplet excited states for ultralong organic phosphorescence. *Nat. Mater.* **2015**, *14*, 685–690. [CrossRef] [PubMed]
35. Li, H.; Zhang, M.; Song, Y.; Wang, H.; Liu, C.; Fu, Y.; Huang, H.; Liu, Y.; Kang, Z. Multifunctional carbon dot for lifetime thermal sensing, nucleolus imaging and antialgal activity. *J. Mater. Chem. B* **2018**, *6*, 5708–5717. [CrossRef]
36. Gao, D.; Zhang, Y.; Liu, A.; Zhu, Y.; Chen, S.; Wei, D.; Sun, J.; Guo, Z.; Fan, H. Photoluminescence-tunable carbon dots from synergy effect of sulfur doping and water engineering. *Chem. Eng. J.* **2020**, *388*, 124199. [CrossRef]
37. Das, P.; Ganguly, S.; Margel, S.; Gedanken, A. Immobilization of heteroatom-doped carbon dots onto nonpolar plastics for antifogging, antioxidant, and food monitoring applications. *Langmuir* **2021**, *37*, 3508–3520. [CrossRef]
38. Guo, D.; Wei, H.F.; Song, R.B.; Fu, J.; Lu, X.; Jelinek, R.; Min, Q.; Zhang, J.R.; Zhang, Q.; Zhu, J.J. N,S-doped carbon dots as dual-functional modifiers to boost bio-electricity generation of individually-modified bacterial cells. *Nano Energy* **2019**, *63*, 103875. [CrossRef]
39. Aswathy, A.O.; Anju, M.S.; Jayakrishna, J.; Vijila, N.S.; Anjali Devi, J.S.; Anjitha, B.; George, S. Investigation of heavy atom effect on fluorescence of carbon dots: NCDs and S, N-CDs. *J. Fluoresc.* **2020**, *30*, 1337–1344. [CrossRef]
40. Guo, Q.; Ma, Y.; Chen, T.; Xia, Q.; Yang, M.; Xia, H.; Yu, Y. Cobalt sulfide quantum dot embedded N/S-doped carbon nanosheets with superior reversibility and rate capability for sodium-ion batteries. *ACS Nano* **2017**, *11*, 12658–12667. [CrossRef]
41. Jiang, K.; Zhang, L.; Lu, J.; Xu, C.; Cai, C.; Lin, H. Triple-mode emission of carbon dots: Applications for advanced anti-counterfeiting. *Angew. Chem.* **2016**, *55*, 7231–7235. [CrossRef] [PubMed]
42. Saravanan, A.; Maruthapandi, M.; Das, P.; Luong, J.; Gedanken, A. Green synthesis of multifunctional carbon dots with antibacterial activities. *Nanomaterials* **2011**, *11*, 369. [CrossRef] [PubMed]
43. Joseph, J.; Anappara, A.A. Long life–time room–temperature phosphorescence of carbon dots in aluminum sulfate. *ChemistrySelect* **2017**, *2*, 4058–4062. [CrossRef]
44. Yang, Y.M.; Kong, W.Q.; Li, H.; Liu, J.; Yang, M.M.; Huang, H.; Liu, Y.; Wang, Z.Y.; Wang, Z.Q.; Sham, T.K.; et al. Fluorescent N-doped carbon dots as in vitro and in vivo nanothermometer. *ACS Appl. Mater. Interfaces* **2015**, *7*, 27324–27330. [CrossRef] [PubMed]
45. Sebastian, P.J.; Jun-Ray, M.; Alexia, M.; Alexandre, P.; Valeria, Q.; Isabelle, M.; Rafik, N. Effects of polydopamine-passivation on the optical properties of carbon dots and its potential usein vivo. *Phys. Chem. Chem. Phys.* **2020**, *22*, 16595–16605.
46. Fu, M.; Ehrat, F.; Wang, Y.; Milowska, K.; Reckmeier, C.; Rogach, A.; Stolarczyk, J.; Urban, A.; Feldmann, J. Carbon dots: A unique fluorescent cocktail of polycyclic aromatic hydrocarbons. *Nano Lett.* **2015**, *15*, 6030–6035. [CrossRef]
47. Song, Y.; Zhu, S.; Zhang, S.; Fu, Y.; Wang, L.; Zhao, X.; Yang, B. Investigation from chemical structure to photoluminescent mechanism: A type of carbon dots from the pyrolysis of citric acid and an amine. *J. Mater. Chem. C* **2015**, *3*, 5976–5984. [CrossRef]

48. Deng, Y.; Chen, X.; Wang, F.; Zhang, X.; Zhao, D.; Shen, D. Environment-dependent photon emission from solid state carbon dots and its mechanism. *Nanoscale* **2014**, *6*, 10388–10393. [CrossRef]
49. Liu, B.; Chu, B.; Wang, Y.L.; Hu, L.F.; Hu, S.; Zhang, X.H. Carbon dioxide derived carbonized polymer dots for multicolor light-emitting diodes. *Green Chem.* **2021**, *23*, 422–429. [CrossRef]
50. Tao, S.; Lu, S.; Geng, Y.; Zhu, S.; Redfern, A.T.; Song, Y.; Feng, T.; Xu, W.; Yang, B. Design of metal-free polymer carbon dots: A new class of room-temperature phosphorescent materials. *Angew. Chem.* **2018**, *57*, 2393–2398. [CrossRef]
51. Tang, G.; Zhang, K.; Feng, T.; Tao, S.; Han, M.; Li, R.; Wang, C.; Wang, Y.; Yang, B. One-step preparation of silica microspheres with super-stable ultralong room temperature phosphorescence. *J. Mater. Chem. C* **2019**, *7*, 8680–8687. [CrossRef]
52. Arshad, F.; Sk, M.P. Aggregation-induced red shift in N,S-doped chiral carbon dot emissions for moisture sensing. *New J. Chem.* **2019**, *43*, 13240–13248. [CrossRef]
53. Mai, X.D.; Phan, Y.T.H.; Nguyen, V.Q.; Manolakos, D.E. Excitation-independent emission of carbon quantum dot solids. *Adv. Mater. Sci. Eng.* **2020**, *2020*, 9643168. [CrossRef]
54. Yang, H.; Liu, Y.; Guo, Z.; Lei, B.; Zhuang, J.; Zhang, X.; Liu, Z.; Hu, C. Hydrophobic carbon dots with blue dispersed emission and red aggregation-induced emission. *Nat. Commun.* **2019**, *10*, 1789. [CrossRef]
55. Liu, Y.; Wu, P.; Wu, X.; Ma, C.; Luo, S.; Xu, M.; Li, W.; Liu, S. Nitrogen and copper (II) co-doped carbon dots for applications in ascorbic acid determination by non-oxidation reduction strategy and cellular imaging. *Talanta* **2020**, *210*, 120649. [CrossRef]
56. Jiang, K.; Wang, Y.; Gao, X.; Cai, C.; Lin, H. Facile, quick, and gram-scale synthesis of ultralong-lifetime room-temperature-phosphorescent carbon dots by microwave irradiation. *Angew. Chem.* **2018**, *57*, 6216–6220. [CrossRef]
57. Feng, T.; Zhu, S.; Zeng, Q.; Lu, S.; Tao, S.; Liu, J.; Yang, B. Supramolecular cross-link-regulated emission and related applications in polymer carbon dots. *ACS Appl. Mater. Interfaces* **2018**, *10*, 12262–12277. [CrossRef]
58. Chen, Y.; He, J.; Hu, C.; Zhang, H.; Lei, B.; Liu, Y. Room temperature phosphorescence from moisture-resistant and oxygen-barred carbon dot aggregates. *J. Mater. Chem. C* **2017**, *5*, 6243–6250. [CrossRef]
59. Tan, J.; Zou, R.; Zhang, J.; Li, W.; Zhang, L.; Yue, D. Large-scale synthesis of N-doped carbon quantum dots and their phosphorescence properties in a polyurethane matrix. *Nanoscale* **2016**, *8*, 4742–4747. [CrossRef]
60. Spychaj, T.; Hamielec, A.E. High-temperature continuous bulk copolymerization of styrene and acrylic acid: Thermal behavior of the reactants. *J. Appl. Polymer Sci.* **1991**, *42*, 2111–2119. [CrossRef]
61. Fujita, M.; Izato, Y.; Iizuka, Y.; Miyake, A. Thermal hazard evaluation of runaway polymerization of acrylic acid. *Process Saf. Environ. Prot.* **2019**, *129*, 339–347. [CrossRef]
62. Song, Y.; Zhu, S.; Shao, J.; Yang, B. Polymer carbon dots-a highlight reviewing their unique structure, bright emission and probable photoluminescence mechanism. *J. Polymer Sci. Part A Polymer Chem.* **2017**, *55*, 610–615. [CrossRef]
63. Tang, X.; Wang, H.; Yu, H.; Bui, B.; Zhang, W.; Wang, S.; Chen, M.; Hu, Z.; Chen, W. Exploration of Nitrogen-doped Grape Peels Carbon Dots for Baicalin Detection. *Mater. Today Phys.* **2022**, *22*, 100576.
64. Tang, X.; Yu, H.; Bui, B.; Wang, L.; Xing, C.; Wang, S.; Chen, M.; Hu, Z.; Chen, W. Nitrogen-doped fluorescence carbon dots as multi-mechanism detection for iodide and curcumin in biological and food samples. *Bioact. Mater.* **2021**, *6*, 1541–1554. [CrossRef] [PubMed]
65. Xu, Y.; Yu, H.; Chudal, L.; Pandey, N.; Amador, E.; Bui, B.; Wang, L.; Ma, X.; Deng, S.; Zhu, X.; et al. Striking Luminescence Phenomena of Carbon Dots and Their Applications as a Double Ratiometric Fluorescence Probes for H_2S Detection. *Mater. Today Phys.* **2021**, *17*, 100328. [CrossRef]
66. Ma, L.; Chen, W. ZnS:Cu,Co Water Soluble Afterglow Nanoparticles—Synthesis, Luminescence and Potential Applications. *Nanotechnology* **2010**, *21*, 385604. [CrossRef]
67. Chu, X.; Li, K.; Guo, H.; Zheng, H.; Shuda, S.; Wang, X.; Zhang, J.; Chen, W.; Zhang, Y. Exploration of Graphitic-C_3N_4 Quantum Dots for Microwave-Induced Photodynamic Therapy. *ACS Biomater. Sci. Eng.* **2017**, *3*, 1836–1844. [CrossRef]
68. Pandey, N.; Xiong, W.; Wang, L.; Chen, W.; Bui, B.; Yang, J.; Amador, E.; Chen, M.; Xing, C.; Athavale, A.; et al. Aggregation-induced emission luminogens for highly effective microwave dynamic therapy. *Bioact. Mater.* **2021**, *7*, 112–125. [CrossRef]
69. Chen, W.; Joly, A.; Roark, J. Photostimulated luminescence and dynamics of AgI and Ag nanoclusters in zeolites. *Phys. Rev. B* **2002**, *65*, 245404. [CrossRef]
70. Morgan, N.; English, S.; Chen, W.; Chernomordik, V.; Russo, A.; Smith, P.; Gandjbakhche, A. Real time in vivo non-invasive optical imaging using near-infrared fluorescent quantum dots. *Acad. Radiol.* **2005**, *12*, 313–323. [CrossRef]
71. Wang, S.; Westcott, S.; Chen, W. Nanoparticle luminescence thermometry. *J. Phys. Chem. B* **2002**, *106*, 11203–11209. [CrossRef]

Article

Novel Fluorescent Probe Based on Rare-Earth Doped Upconversion Nanomaterials and Its Applications in Early Cancer Detection

Zhou Ding [1], Yue He [1], Hongtao Rao [1], Le Zhang [1], William Nguyen [2], Jingjing Wang [1], Ying Wu [1], Caiqin Han [1], Christina Xing [2], Changchun Yan [1], Wei Chen [2,3,*] and Ying Liu [1,*]

[1] Jiangsu Key Laboratory of Advanced Laser Materials and Devices, School of Physics and Electronic Engineering, Jiangsu Normal University, Xuzhou 221116, China; 2020191044@jsnu.edu.cn (Z.D.); 2020201155@jsnu.edu.cn (Y.H.); 2020211255@jsnu.edu.cn (H.R.); zhangle@jsnu.edu.cn (L.Z.); jin16217@mail.ustc.edu.cn (J.W.); wuying@jsnu.edu.cn (Y.W.); hancq@jsnu.edu.cn (C.H.); yancc@jsnu.edu.cn (C.Y.)
[2] Department of Physics, The University of Texas at Arlington, Arlington, TX 76019-0059, USA; william.nguyen@uta.edu (W.N.); christina.xing@uta.edu (C.X.)
[3] Medical Technology Research Centre, Chelmsford Campus, Anglia Ruskin University, Chelmsford CM1 1SQ, UK
* Correspondence: weichen@uta.edu (W.C.); liuying@jsnu.edu.cn (Y.L.)

Abstract: In this paper, a novel rare-earth-doped upconverted nanomaterial NaYF$_4$:Yb,Tm fluorescent probe is reported, which can detect cancer-related specific miRNAs in low abundance. The detection is based on an upconversion of nanomaterials NaYF$_4$:Yb,Tm, with emissions at 345, 362, 450, 477, 646, and 802 nm, upon excitation at 980 nm. The optimal Yb^{3+}:Tm^{3+} doping ratio is 40:1, in which the NaYF$_4$:Yb,Tm nanomaterials have the strongest fluorescence. The NaYF$_4$:Yb, Tm nanoparticles were coated with carboxylation or carboxylated protein, in order to improve their water solubility and biocompatibility. The two commonly expressed proteins, miRNA-155 and miRNA-150, were detected by the designed fluorescent probe. The results showed that the probes can distinguish miRNA-155 well from partial and complete base mismatch miRNA-155, and can effectively distinguish miRNA-155 and miRNA-150. The preliminary results indicate that these upconverted nanomaterials have good potential for protein detection in disease diagnosis, including early cancer detection.

Keywords: upconversion nanomaterials; novel fluorescent probe; miRNA-155; miRNA-150

Citation: Ding, Z.; He, Y.; Rao, H.; Zhang, L.; Nguyen, W.; Wang, J.; Wu, Y.; Han, C.; Xing, C.; Yan, C.; et al. Novel Fluorescent Probe Based on Rare-Earth Doped Upconversion Nanomaterials and Its Applications in Early Cancer Detection. *Nanomaterials* **2022**, *12*, 1787. https://doi.org/10.3390/nano12111787

Academic Editor: Antonios Kelarakis

Received: 1 April 2022
Accepted: 20 May 2022
Published: 24 May 2022

Publisher's Note: MDPI stays neutral with regard to jurisdictional claims in published maps and institutional affiliations.

Copyright: © 2022 by the authors. Licensee MDPI, Basel, Switzerland. This article is an open access article distributed under the terms and conditions of the Creative Commons Attribution (CC BY) license (https://creativecommons.org/licenses/by/4.0/).

1. Introduction

MicroRNAs (miRNAs) are important regulators of cell proliferation, division, differentiation, and apoptosis. They can regulate the expression levels of various genes in DNA post-transcriptionally. The abnormal expression of miRNAs has been associated with many diseases (cancer, tumors, and diabetes) [1], and miRNAs can be obtained through blood, urine, etc. [2], which has the advantage of being non-invasive. Therefore, miRNAs are considered important biomarkers for tumors [3]. Patryk Krzeminski [4] et al. demonstrated that DNA methylation contributes to miRNA-155 expression, and the survival data of myeloma cells show a correlation between miR-155 expression and multiple myeloma outcomes. A number of studies have shown that miRNA-155 is closely related to MM, and the overexpression of miRNA 155 in blood is an important signal for the diagnosis of MM [5–7]. Therefore, the detection of miRNA-155 to diagnose early MM is an important and efficient method.

The concept of upconversion luminescence was first proposed by Auzel and Ovsyankin et al. [8]. The radiation process is a nonlinear anti-Stokes emission, excited by the effective absorption of two or more low-energy photons. Then, it transitions from the ground state to the excited state through a multi-step process, and finally, it returns to the ground state energy level in radiative transition, realizing the conversion of long-wavelength excitation light into short-wavelength emission light. The excitation wavelength (980 nm) of upconverted nanomaterials is in the "optical transmission window" of biological tissues [9–12] due to the least light absorbed by biological tissues [13–16]. This special luminescence process makes upconversion nanomaterials have incomparable advantages over traditional organic dyes, quantum dots, fluorescent proteins, and other biomolecular markers [17–19], such as near-infrared light as excitation light, resulting in less photo-damage, lower auto-fluorescence background of biological tissue, and a deep penetration depth [9–19]. In addition, rare-earth-doped upconversion nanoparticles have multi-wavelength emissions, high-fluorescence intensity and high stability, good water solubility, as well as good biocompatibility [20–31]. Moreover, rare-earth-doped nanoparticles also have strong scintillation luminescence that can be used for X-ray-induced photodynamic therapy, which is a well-captivated area, as this new therapy can be used for deep as well as skin cancer treatment [20–22], radiation dosimetry [23,24] and temperature sensing [25]. These irreplaceable advantages establish upconversion luminescent nanomaterials as a potential fluorescent marker in biological detection [28–31] and have developed rapidly. For example, Mao et al. [32] prepared $NaYF_4$:Yb,Er upconversion nanoparticles using the hydrothermal method, and discovered their specific detection of miRNA by using the principle of base stacking on the surface. Kowalik et al. [33] linked IgG antibodies to the PEG-NHS-modified $NaYF_4$:Yb,Tm@SiO_2 surface to achieve specific labeling to demonstrate the great potential of photodynamic targeted therapy.

The main body of rare-earth-doped upconversion nanoparticle materials is composed of host materials, sensitizers, and activators [34]. The commonly used host materials themselves have no fluorescence but can provide a suitable crystal field for the activated ions, so that the luminescent centers produce specific emissions. Y^{3+}, Gd^{3+}, Lu^{3+}, La^{3+} are usually selected as host elements [35]. The high-quality host material plays a decisive role in the entire upconversion luminescence efficiency. The spacing between rare-earth ions, the symmetry distribution, and the number of coordination ions would affect the crystal structure of the host material. The most common $NaYF_4$ is currently recognized as the most effective blue-green light upconversion luminescent host material [36,37]. Because the hexagonal $NaYF_4$ has a higher symmetry than the cubic $NaYF_4$, the luminous efficiency of the upconversion in hexagonal $NaYF_4$ is 10-times higher than in the cubic $NaYF_4$ [38]. Nd^{3+} and Yb^{3+} ions are the most common sensitizers in the upconversion systems [39,40]. Yb^{3+} has a large absorption cross-section at 950–1000 nm, which is related to the only excited state and can absorb the excited infrared photons as well as effectively obtain the energy. The energy is then transferred to the activators. Yb^{3+} is more often chosen as the sensitizer of nanomaterials because its upconversion luminescence efficiency is higher than that of Nd^{3+}. The activators are rare-earth ions, such as Er^{3+}, Tm^{3+}, Ho^{3+}, Nd^{3+}, Pr^{3+}, etc. [41–44], among which the Yb^{3+}/Er^{3+} (Tm^{3+}, Ho^{3+}) pairs are recognized as the most efficient [45].

The standard preparation methods of rare-earth-doped upconversion nanoparticles include the co-precipitation method, high-temperature thermal decomposition method, solvothermal method, microemulsion method, sol–gel method, etc. Nanoparticles with tunable morphology, high-fluorescence intensity, and good dispersion have advantages and disadvantages. For example, Gao et al. [46] compared the effects of factors, such as the concentration of reactants, the ratio of reactants and ligand solvents, and the types of ligand solvents on the synthesis of nanoparticles, respectively. Hydrothermal synthesis has been widely used for $NaYF_4$:Yb,Er nanoparticle synthesis. This method has a fast reaction and precipitation rate and crystallinity; however, the fluorescence efficiency of the nanoparticles synthesized is relatively low. Amphiphilic ligands can help the particles to be well dispersed in various polar solvents. However, the surface of the obtained upconversion nanoparticles is usually coated with hydrophobic organic ligands, such as oleic acid, oleylamine, and octadecene, which make them less water soluble and difficult for biological applications. To improve the water solubility and biocompatibility, the surface of upconvertion nanoparticles must be modified with bioactive ligands or biomolecules.

Surface ligand exchange, surface ligand oxidation, and surface ligand assembly are common surface modification methods [47–49]. Jiang et al. [48] used polymeric anhydride to interact with octadecene and oleic acid ligands on the surface of nanoparticles by coordination adsorption, and then cross linked with dicyclohexanetriamine to obtain stable water-soluble upconverted nanoparticles. The effective modification of upconverted materials has become a key factor in their applications in biological detections.

In this paper, high-temperature thermal decomposition was used to prepare NaYF$_4$:Yb^{3+}/Tm^{3+} upconverted nanoparticle materials and it was found that the optimal Yb^{3+}/Tm^{3+} ratio is 40:1, in terms of the luminescence efficiency. These new types of upconverted nanoparticles were modified by surface coating and tested for protein detection for the purpose of early cancer diagnosis.

2. Experimental Materials and Methods

2.1. Reagents and Instruments

Chemicals and Reagents: These chemicals were purchased from Shanghai Aladdin Biochemical Technology Co. Ltd. (Shanghai, China): yttrium chloride hexahydrate (YCl$_3$·6H$_2$O), ytterbium chloride hexahydrate (YbCl$_3$·6H$_2$O), thulium chloride hexahydrate (TmCl$_3$·6H$_2$O) gadolinium chloride hexahydrate (GdCl$_3$·6H$_2$O), oleic acid (OA), 1-octadecene (ODE), ammonium fluoride (NH$_4$F), cyclohexane (C$_6$H$_{12}$), methanol (MeOH). These Chemicals were purchased from Sinopharm Chemical Reagent Co., Ltd. (Shanghai, China): hydrochloric acid (HCL), succinic acid (C$_6$H$_{10}$O$_4$), sodium carbonate (Na$_2$CO$_3$), sodium bicarbonate (NaHCO$_3$), acetonitrile (C$_2$H$_3$N), and 1-ethyl-(3-Dimethylaminopropyl) carbonate diimide hydrochloride (EDC).

Anhydrous ethanol (C$_2$H$_6$O) and sodium hydroxide (NaOH) were purchased from Xilong Chemical Co., Ltd. (Guangzhou, Guangdong, China). Bovine serum albumin (BSA), N,N-dimethylformamide (DMF), and N-hydroxysuccinimide (NHS) were purchased from Shanghai Macklin Biochemical Co. Ltd. (Shanghai, China). Nitronium tetrafluoroborate (NOBF$_4$) was purchased from Hebei Bailingwei Superfine Materials Co., Ltd. (Langfang, Hebei, China), PBS buffer and DNA with amino and FAM were purchased from Sangon Biotech Co. Ltd. (Shanghai, China), corresponding base sequence is 5'-NH2-CCCCCCCCCCCCACCCCTATCACGATTAGCATTAA-6-FAM-3'. The ultrapure water was produced by a water purification system (H20BASIC-B, Sartorius, Germany).

Instrumentation: the intelligent digital magnetic stirring electric heating mantle (ZNCL-TS-250 mL, Shanghai Anchun Instrument Co., Ltd., Shanghai, China) was used to prepare upconverted nanomaterials and carboxylated proteins. Fluorescence was measured using a fluorescence spectrometer (F-4600, Hitachi, Hitachi, Ltd., Tokyo, Japan). Phase analysis of upconverted nanomaterials was carried out by X-ray powder diffractometer (X-ray Diffraction, XRD, D8 ADVANCE, Bruker, Germany). The morphology of the upconverted nanomaterials was characterized by field emission scanning electron microscope (Scanning Electron Microscope, SEM, SU8010, Hitachi Co., Ltd., Tokyo, Japan) and scanning transmission electron microscope (Transmission Electron Microscopy, TEM, FEI TECNAI G2 F20, Hitachi Co., Ltd., Tokyo, Japan). A Fourier transform infrared spectrometer measured infrared absorption (Tensor 27, Bruker, Germany). The luminescence was measured with a 980 nm fiber laser (BOT980-5W, Xi'an Leize Electronic Technology Co., Ltd., Xi'an, China).

2.2. Preparation of Upconverted Nanomaterials

Rare-earth-doped upconverted luminescent nanomaterials NaYF$_4$:Yb,Tm were synthesized by high-temperature thermal decomposition. The preparation of 2 mmol NaYF$_4$:20% Yb^{3+}, 0.5% Tm^{3+} nanomaterials was performed by first charging 1.39 mmol YCl$_3$·6H$_2$O, 0.6 mmol YbCl$_3$·6H$_2$O, and 0.01 mmol TmCl$_3$·6H$_2$O into a three-neck round-bottom flask. Then, oleic acid (12 mL) and octadecene (30 mL) were added, and the flask flowed with nitrogen gas for 10 min to ensure no oxygen in the flask. Next, under nitrogen protection with magnetic stirring, the mixture was heated to 160 °C and reacted for 1 h to obtain a pale-yellow solution. The mixture was cooled to 50 °C and then 10 mL of a methanol solution containing 8.0 mmol of ammonium fluoride and 5.0 mmol of sodium hydroxide was added dropwise to the mixture. The reaction was continuously stirred at 50 °C for 30 min to ensure a complete integration. Then, the temperature was raised to 80 °C to evaporate the methanol. During the evaporation of methanol, the solution continued to bubble, and the mixture was continuously heated to 120 °C for 30 min until no more bubbles were generated in the solution. Finally, the sample was heated to 300 °C for 90 min. After the reaction was over, the solution was naturally cooled to room temperature and cyclohexane was added to disperse the mixture. The mixture was centrifuged at 8000 rpm/min for 5 min to obtain a precipitate, which was then washed with cyclohexane. The above steps were centrifuged and washed three times, and the $NaYF_4$:Yb,Tm upconverted nanomaterial finally obtained was dispersed in cyclohexane and stored. In addition, the experimental steps for the preparation of $NaGdF_4$:Yb,Tm upconverted nanomaterials are the same as above. The $YCl_3 \cdot 6H_2O$ in the experimental material was replaced by $GdCl_3 \cdot 6H_2O$, and the experimental steps were repeated to obtain $NaGdF_4$:Yb,Tm upconverted nanomaterials.

2.3. Water-Soluble Upconverted Nanomaterials

Two mL of the cyclohexane solution of the upconverted nanomaterial prepared in the above experiment was prepared and ultrasonically dispersed for 5 min. Twenty mg of $NaBF_4$ was dissolved in two mL of acetonitrile solution and was added to the fully dispersed cyclohexane solution of upconverted nanomaterials by stirring at 1000 rpm/min for 30 min to obtain a mixed solution of water and oil separation. Then, the water-soluble $NaYF_4$:Yb, Tm upconverted nanomaterials were obtained by centrifuging at 8000 rpm/min for 15 min.

2.4. Preparation of Carboxylated Proteins

First, we dissolved 1 g of bovine serum albumin in 20 mL of ultrapure water, then excess oxalic acid was added, and we adjusted the pH to 7–8 with an aqueous sodium carbonate solution under magnetic stirring. Then, 5 mmol of EDC was added and the mixed solution was stirred overnight. A dialysis bag with a molecular-weight cut-off of 10k–30k Da was used for dialysis. The denatured proteins were put into the dialysis bag, clamped on both sides with dialysis clips to prevent leaking, and immersed into a sodium bicarbonate aqueous solution (1000 mL, 2 mmol) leaving it in a refrigerator at 4 °C. The sodium bicarbonate aqueous solution was replaced every 4–6 h to ensure that the protein was always in a slightly alkaline environment. The protein was dialyzed and purified under this condition for at least 72 h, and the carboxylated bovine serum albumin was obtained, which was divided into centrifuge tubes and stored in a −20 °C refrigerator.

2.5. Carboxylated Protein-Modified Upconverted Nanoparticles

Next, 0.5 mL of the prepared carboxylated protein, described above, was dissolved in 2 mL of DMF solution, then water-soluble upconverted nanomaterials were added to the DMF solution and the pH was adjusted to about 8 with aqueous sodium bicarbonate solution by stirring for 2 h. After the reaction was completed, the mixed solution was centrifuged at 10,000 rpm/min for 10 min. The precipitates were washed twice with DMF centrifugation, and finally dispersed with DMF for preservation. The above operations were all carried out at room temperature.

2.6. DNA Probes Linked to Carboxylated Protein-Modified Upconverting Nanoparticles

The pH of 1 mL of the carboxylated protein-modified upconverted nanomaterials obtained in the above steps was adjusted to about 6 with 20 μmol of hydrochloric acid aqueous solution, then 5 mg EDC and 5 mg NHS were added and stirred at room temperature for 25 min to let the carboxyl groups fully react with the nanoparticles. Next, the aminated DNA probe with FAM was added to a 40 μL PBS buffer to make a concentration of 100 μmol. An appropriate amount of the solution was immediately added to the carboxylated upconverted nanomaterial solution, and then put in a shaker to react for

30 min. The reaction was washed twice with PBS buffer at 10,000 rpm/min for 5 min each time, and the resulting precipitates were DNA probes with FAM linking to a carboxylated protein-modified upconverted nanomaterial fluorescent probe (DNA/dBSA /NaYF$_4$:Yb, Tm).

3. Experimental Results and Discussion

3.1. The Effect of Yb^{3+} Doping Concentration on the Luminescence of Upconverted Nanomaterials

Figure 1 shows the energy structure and the transitions of Yb^{3+} and Tm^{3+} ions in the upconverting luminescence process. It can be seen that upconversion luminescence is a complex multi-photon energy transfer and conversion process. The energy level transition of $^2F_{7/2} \to\ ^2F_{5/2}$ of the sensitizer Yb^{3+} ion matches the energy of the near-infrared photon at 980 nm, so it can continuously absorb the excitation energy and then transfer it to the adjacent luminescent center Tm^{3+}. The $^3H_5$, $^3F_2$ ($^3F_3$), and $^1G_4$ energy levels are from Tm^{3+} ions. Among them, there are three methods for upconversion luminescence: (1) the $^3H_6$ energy level absorbs three photons continuously and transitions to the $^1G_4$ energy level; (2) the $^3H_6$ energy first absorbs two photons continuously and then transitions to the $^3F_2$, then, through the cross-relaxation process $^3F_{2,3} + ^3H_4 \to\ ^3H_6 + ^1D_2$ to the $^1D_2$; (3) the non-radiation transitions from $^1G_4$, $^1D_2$ of Tm^{3+} to the lower energy levels $^3F_{2,3,4}$, $^3H_{4,5,6}$ to achieve upconversion luminescence. From the emission spectra of NaYF$_4$:Yb,Tm and NaGdF$_4$:Yb,Tm in Figure 3b,d, it can be seen that different host materials NaY(Gd)F$_4$ will not affect the position of the emission peak, but the doping concentration of Yb^{3+} ions affects the intensity of the emission peak. However, the change is not a simple linear increase or decrease with the increase or decrease in the doping ratio of Yb^{3+}.

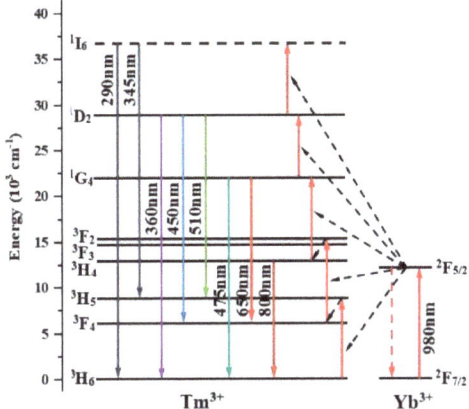

Figure 1. Schematic diagram of energy level transition of Yb^{3+} and Tm^{3+} ions.

3.2. The Effect of Rare-Earth Ion Doping Concentration on the Luminescence of Upconverted Nanomaterials

3.2.1. Effect of Yb^{3+} Doping Concentration on Upconverted Nanomaterials

NaYF$_4$:x%Yb^{3+}, 0.5%Tm^{3+} and NaGdF$_4$:x%Yb^{3+}, 0.5%Tm^{3+} (x = 5, 10, 20, 50, 80) nanomaterials were synthesized by high-temperature thermal decomposition under the same experimental conditions by varying the doping molar fraction of Yb^{3+} ions with a fixed Tm^{3+} ions molar fraction of 0.5%. Figures 2 and 3 show the multi-directional characterization results of the prepared NaYF$_4$:x%Yb^{3+}, 0.5%Tm^{3+} and NaGdF$_4$:x%Yb^{3+}, 0.5%Tm^{3+}. As shown in Figures 2a–j and 3a–j, all the nanomaterials exhibit the characteristics of high size dispersion and good crystallinity. It can be seen from Figures 2k and 3k that NaYF4: 20%Yb^{3+}, 0.5%Tm^{3+} and NaGdF$_4$:20%Yb^{3+}, 0.5%Tm^{3+} are uniform in morphology and size, forming a complete and regular hexagonal phase. Obviously, the doping concentration of Yb^{3+} ions does not have much effect on the morphology of the upconverted nanomaterials. The nanomaterials with different Yb^{3+} doping ratios can be synthesized with particle sizes between 25–38 nm, which lays a good foundation for the subsequent preparation of upconversion fluorescent probes. $NaYF_4$:x%Yb^{3+}, 0.5%Tm^{3+}(x = 5, 10, 20, 50, 80) and $NaGdF_4$:x%Yb^{3+}, 0.5%Tm^{3+} (x = 5, 10, 20, 50, 80) were subjected to phase analysis, as shown in Figures 2l and 3l, the XRD patterns were compared with standard card No.16-0994 (NaF_4), and the diffraction peaks obtained all corresponded to the standard card one by one, indicating that the samples obtained under this condition were all pure hexagonal $NaYF_4$:Yb, Tm and $NaGbF_4$:Yb, Tm.

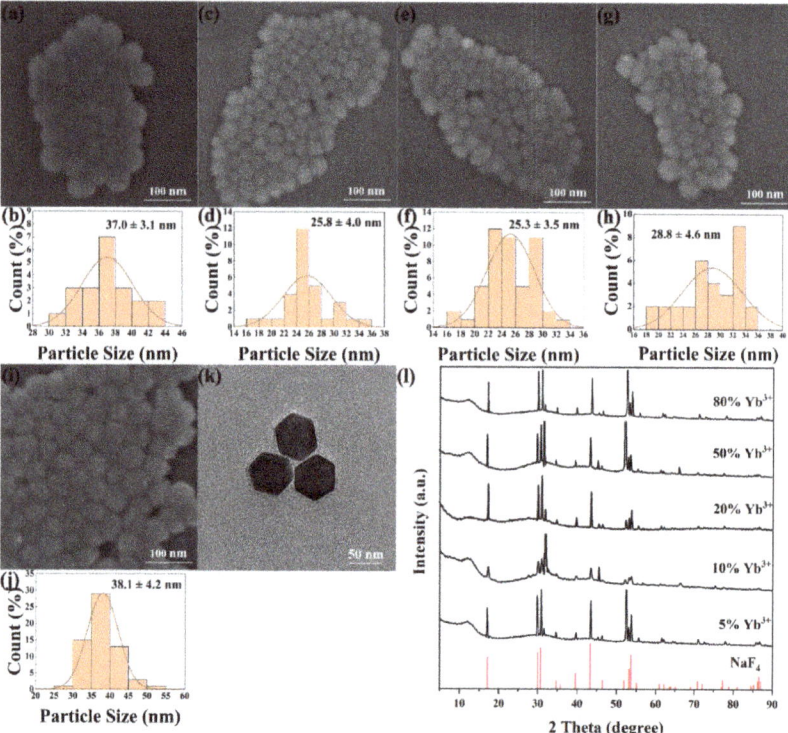

Figure 2. SEM, particle size analysis, TEM and XRD images of $NaYF_4$: x%Yb^{3+}, 0.5%Tm^{3+}. SEM: (**a**) x = 5, (**c**) x = 10, (**e**) x = 20, (**g**) x = 50, (**i**) x = 80; particle size analysis: (**b**) x = 5, (**d**) x = 10, (**f**) x = 20, (**h**) x = 50, (**j**) x = 80; (**k**) TEM image of $NaYF_4$: 20%Yb^{3+}, 0.5%Tm^{3+}; (**l**) XRD of $NaYF4$:x%Yb^{3+}, 0.5%Tm^{3+} (x = 5, 10, 20, 50, 80).

Figure 4 shows the fluorescence spectra of nanomaterials with different concentrations of Yb^{3+} doping. It can be seen from Figure 4a,c that the emission peak position is not affected by the host material $NaY(Gd)F_4$ or the doping concentration of Yb^{3+} ions. The intensity of the emission peak changes as the doping ratio of Yb^{3+} changes, but the change is not a simple linear increase or decrease with the increase or decrease in the doping ratio of Yb^{3+}.

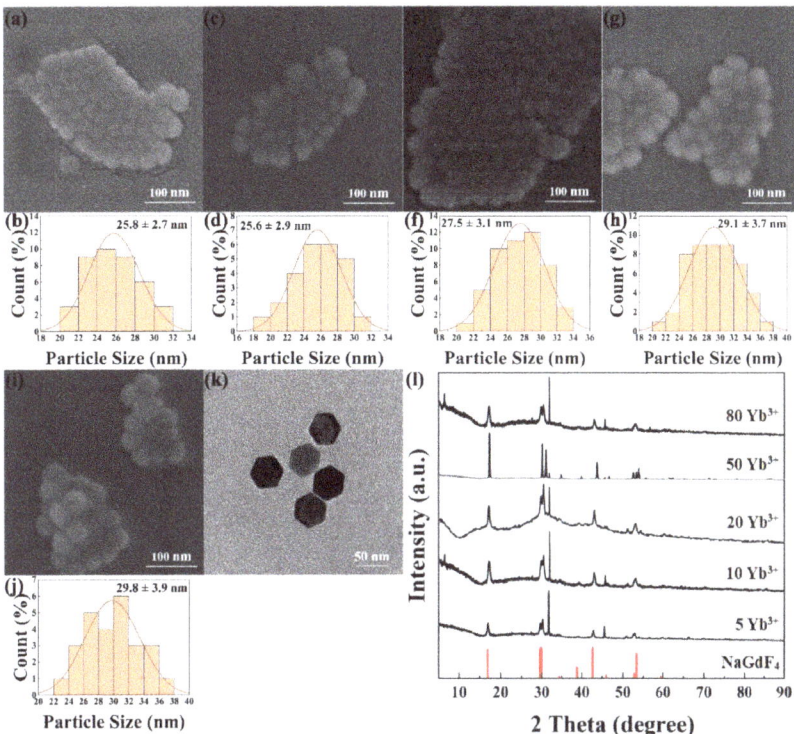

Figure 3. SEM, particle size analysis, TEM and XRD images of NaGdF$_4$: x%Yb^{3+}, 0.5%Tm^{3+}. SEM: (**a**) x = 5, (**c**) x = 10, (**e**) x = 20, (**g**) x = 50, (**i**) x = 80; particle size analysis: (**b**) x = 5, (**d**) x = 10, (**f**) x = 20, (**h**) x = 50, (**j**) x = 80; (**k**) TEM image of NaGdF$_4$: 20%Yb^{3+}, 0.5%Tm^{3+}; (**l**) XRD of NaGdF$_4$:x%Yb^{3+},0.5%Tm^{3+} (x = 5, 10, 20, 50, 80).

As shown in Figure 4, the luminescence intensity at 450, 477 and 646 nm increased gradually with the increase in Yb^{3+} concentration up to 20%, then the emission intensity is decreased with the increase in Yb^{3+} concentration (Figure 4b,d).

Upon Yb^{3+} doping, with the change in doping concentration, the number of photons absorbed at 980 nm increases, and the energy transferred to Tm^{3+} ions increases, so that its luminescence is enhanced. When the concentration of Yb^{3+} continues to increase, the photon energy absorbed by the Yb^{3+} ions will pass through the "bridge" between Yb^{3+} − Yb^{3+} and surface defects. Through energy resonance transfer, the energy will be transferred to the surface defects and organic vibration groups, through the free radiation process. According to the experimental results, when the optimal Yb^{3+} doping mole fraction is 20%, the luminescence reaches its peak.

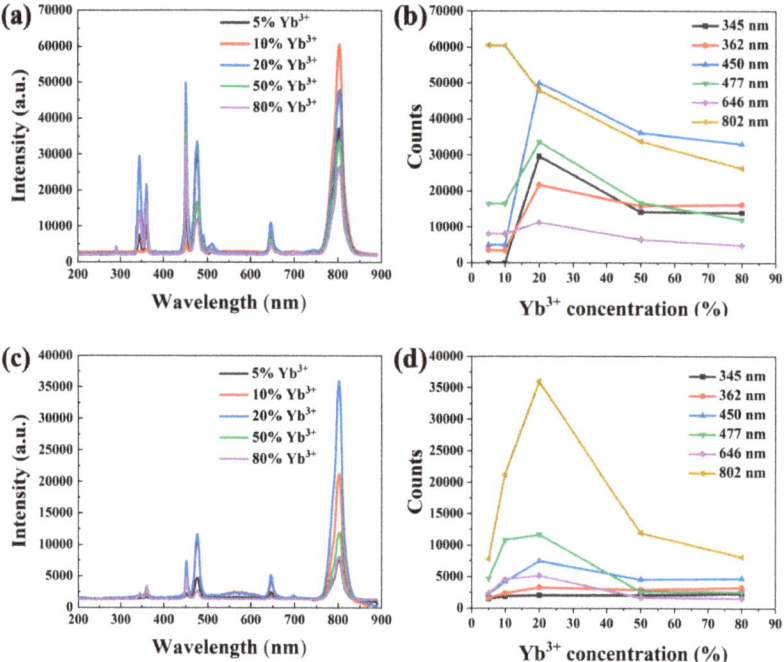

Figure 4. Fluorescence spectra of NaYF$_4$: Yb^{3+}, Tm^{3+} and NaGdF$_4$: Yb^{3+}, Tm^{3+} with different Yb^{3+} doping ratios. (**a**) Fluorescence spectra of NaYF$_4$:x%Yb^{3+}, 0.5%Tm^{3+} (x = 5, 10, 20, 50, 80) at 980 nm excitation. (**b**) Schematic diagram of the relationship between the maximum intensity of the luminescence peaks corresponding to NaYF$_4$: Yb^{3+} and Tm^{3+} with different doping ratios and the concentration change. (**c**) Fluorescence spectra of NaGdF$_4$:x%Yb^{3+}, 0.5%Tm^{3+} (x = 5, 10, 20, 50, 80) at 980 nm excitation. (**d**) Schematic diagram of the relationship between the maximum intensity of the luminescence peaks corresponding to NaGdF$_4$: Yb^{3+} and Tm^{3+} with different doping ratios and the concentration changes.

3.2.2. The Effect of Tm^{3+} Doping Concentration on Upconverted Nanomaterials

NaYF$_4$:20%Yb^{3+}, x%Tm^{3+} and NaGdF$_4$:20%Yb^{3+}, x%Tm^{3+} were synthesized by the same method and conditions by varying the concentration of Tm^{3+} (x = 0.2, 0.3, 0.5, 0.8, 1.0) while the concentration of Tb^{3+} was fixed at 20%. Figures 5 and 6 show the multi-directional characterization results of the prepared NaYF$_4$:20%Yb^{3+}, x%Tm^{3+} and NaGdF$_4$:20%Yb^{3+}, x%Tm^{3+}. As can be seen in Figures 5a–j and 6a–j, all the nanomaterials have high dispersibility in size distribution and good crystallinity. It can be seen from Figures 5k and 6k that the upconverted nanomaterials have uniform morphology and size, forming a regular hexagonal phase. Similarly, by changing the doping concentration of Tm^{3+}, the morphology of the upconverted nanomaterials does not change too much. The particle sizes of nanomaterials with different Tm^{3+} doping ratios can be between 21–43 nm. The XRD pattern of NaGdF$_4$: x%Yb^{3+}, 0.5%Tm^{3+} (x = 0.2, 0.3, 0.5, 0.8, 1) and NaGdF$_4$: 20%Yb^{3+}, x%Tm^{3+} (x = 0.2, 0.3, 0.5, 0.8, 1) nanomaterials was also compared with the standard card No.27-0699 (NaGdF$_4$), NaGdF$_4$: Yb, Tm is a pure hexagonal phase. As shown in Figures 5l and 6l, the diffraction peaks obtained all corresponded to the standard card, which means that the samples obtained under this condition were all pure hexagonal NaYF$_4$:Yb, Tm and NaGbF$_4$:Yb, Tm.

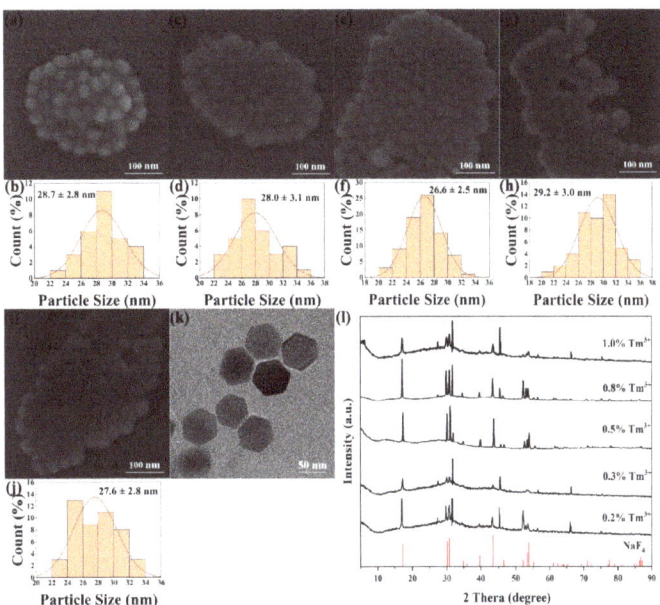

Figure 5. SEM, particle size analysis, TEM, and XRD images of NaYF$_4$: 20%Yb^{3+}, x%Tm^{3+}. (**a**) x = 0.2, (**c**) x = 0.3, (**e**) x = 0.5, (**g**) x = 0.8, (**i**) x = 1.0; particle size analysis: (**b**) x = 0.2, (**d**) x = 0.3, (**f**) x = 0.5, (**h**) x = 0.8, (**j**) x = 1.0; (**k**) TEM image of NaYF$_4$: 20%Yb^{3+}, 0.5%Tm^{3+}; (**l**) NaYF$_4$: 20%Yb^{3+}, x%Tm^{3+} (X = 0.2, 0.3, 0.5, 0.8, 1.0).

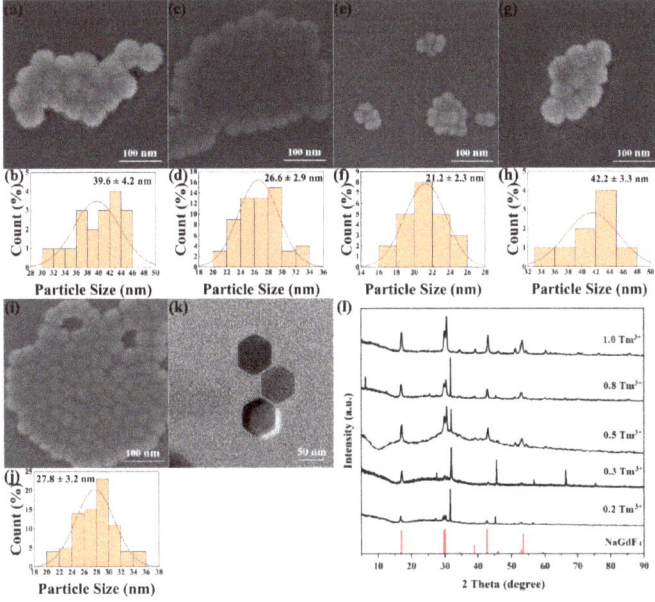

Figure 6. SEM, particle size analysis, TEM, and XRD images of NaGdF$_4$: 20%Yb^{3+}, x%Tm^{3+}. (**a**) x = 0.2, (**c**) x = 0.3, (**e**) x = 0.5, (**g**) x = 0.8, (**i**) x = 1.0; particle size analysis: (**b**) x = 0.2, (**d**) x = 0.3, (**f**) x = 0.5, (**h**) x = 0.8, (**j**) x = 1.0; (**k**) TEM image of NaGdF$_4$: 20%Yb^{3+}, 0.5%Tm^{3+}; (**l**) NaGdF$_4$: 20%Yb^{3+}, x%Tm^{3+} (X = 0.2, 0.3, 0.5, 0.8, 1.0).

The luminescence spectra of materials with different Tm^{3+} concentrations are shown in Figure 7. As the mole fraction of Tm^{3+} increases from 0.2% to 0.5%, the luminescence peaks at 450, 475, and 646 nm also change. As the mole fraction of Tm^{3+} increases from 0.5% to 1%, the emission intensities of these three luminescence peaks gradually decrease. When the mole fraction of Tm^{3+} is 0.2%, the luminescence intensity of the nanomaterial is not strong because there are not enough excitable Tm^{3+} ions in the nanomaterials. As the concentration of Tm^{3+} increases gradually, the number of excitable Tm^{3+} ions increases accordingly, and the luminescence of nanomaterials becomes stronger accordingly. When the mole fraction of Tm^{3+} reaches 0.5%, the upconversion luminescence intensity reaches the maximum, and then gradually becomes weaker. This is because the increase in Tm^{3+} ion concentration reduces the interionic distance and strengthens the interaction. Finally, the concentration quenching effect and cross-relaxation effect are observed, resulting in a decrease in luminescence intensity. At the same time, when the sample is excited with the same power of the 980 nm laser, since the total energy is fixed, the energy that each Tm^{3+} can receive will decrease with the increase in Tm^{3+}, leading to the weakening of the luminescence.

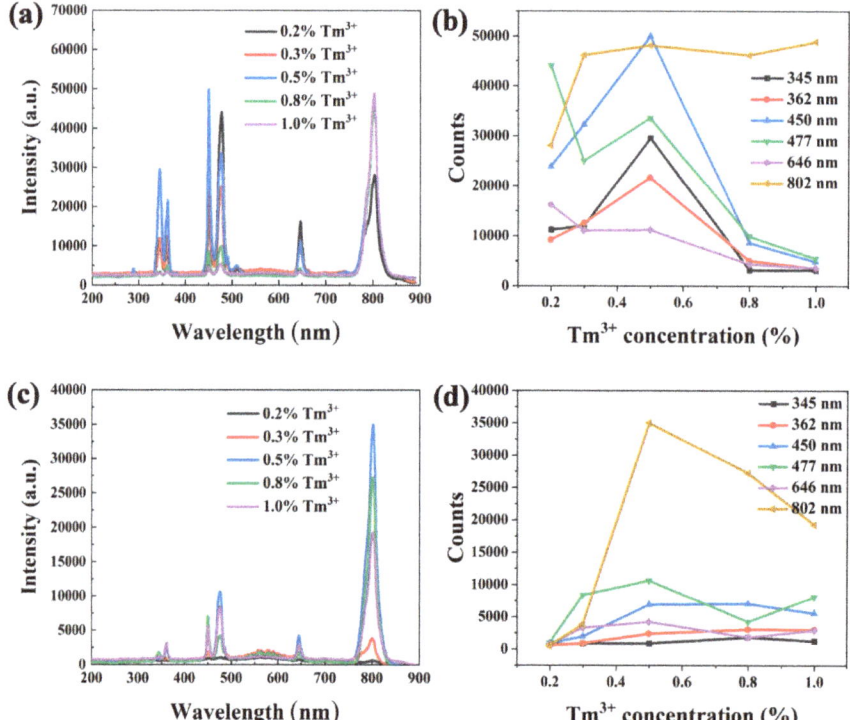

Figure 7. Fluorescence spectra of NaYF$_4$: Yb^{3+}, Tm^{3+} and NaGdF$_4$: Yb^{3+}, Tm^{3+} with different Tm^{3+} doping ratios. (**a**) Fluorescence spectra of NaYF$_4$:20%Yb^{3+}, x%Tm^{3+} (x = 0.2, 0.3, 0.5, 0.8, 1.0) at 980 nm excitation. (**b**) NaYF$_4$: Yb^{3+} with different Tm^{3+} doping ratios, the relationship between the maximum intensity of the luminescence peak corresponding to Tm^{3+} and the concentration change. (**c**) Fluorescence spectra of NaGdF$_4$:20%Yb^{3+}, x%Tm^{3+} (x = 0.2, 0.3, 0.5, 0.8, 1.0 at 980 nm excitation). (**d**) The corresponding doping ratios of NaGdF$_4$: Yb^{3+} and Tm^{3+} with different Tm^{3+} doping ratios.

3.2.3. Comparison of Luminescence Properties of Two Upconverted Nanomaterials with the Best Doping Ratio

From the above results, the luminescence intensity of nanomaterials is the strongest when the mole fraction of Yb^{3+} is 20% and the mole fraction of Tm^{3+} is 0.5%. Therefore, the optimal doping concentration ratio (40:1) is selected to prepare these two kinds of upconverting nanomaterials. Figure 8 shows the luminescence intensity comparison of $NaYF_4$:20%Yb^{3+},0.5%Tm^{3+} and $NaGdF_4$:20%Yb^{3+}, 0,5%Tm^{3+}. At 345, 362, 450, 477, 646, 802 nm, the luminescence intensity of $NaYF_4$:Yb,Tm is 4.4, 3.0, 4.2, 3.4, 2.3, and 2.7-times stronger than that of $NaGdF_4$:Yb,Tm, respectively. The luminescence intensity of $NaYF_4$:Yb, Tm is much higher than that of $NaGdF_4$:Yb, Tm under the same power of 980 nm laser excitation. Therefore, we chose $NaYF_4$:20%Yb^{3+},0.5%Tm^{3+} upconverted nanomaterials as the substrate to further investigate protein detections.

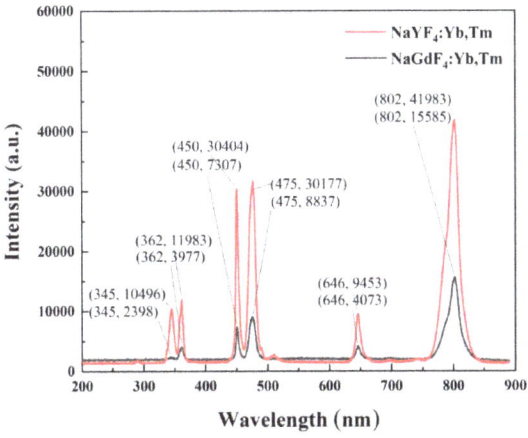

Figure 8. Comparison of fluorescence spectra of $NaYF_4$: 20% Yb^{3+}, 0.5% Tm^{3+} and $NaGdF_4$: 20% Yb^{3+}, 0.5% Tm^{3+} at 980 nm excitation.

3.3. Analysis of Fluorescence Characteristics Based on $NaYF_4$:Yb^{3+}, Tm^{3+} Biological Probes

After the carboxylated bovine serum albumin was added to the upconverted nanomaterial sample for reaction treatment, we carried out repeated centrifugal washing on the sample to remove the carboxylated bovine serum albumin, and then measured the infrared absorption spectrum of the remaining samples, as shown in Figure 9a (the red line). As shown in Figure 9a, the broad infrared (IR) absorption peak at 3295.86 cm^{-1} represents the stretching vibration peak of OH in the carboxyl functional group; the sharp IR absorption peak at 1650.82 cm^{-1} represents the C=O formed after the reaction between the carboxylated protein and NH_2-PEG group stretching vibration peak; 1024.05 cm^{-1} represents the stretching vibration absorption peak of -O- in PEG; 700.06 cm^{-1} broad absorption peak represents the out-of-plane rocking vibration absorption peak of NH in bovine serum albumin and NH_2-PEG. Repeated centrifugal washing can completely remove the free carboxylated bovine serum albumin in the solution, and the upconversion material is inorganic and will not absorb at these positions. Therefore, it is considered that the new characteristic peaks belonging to organic functional groups can only come from carboxylated bovine serum proteins that have been attached to the surface of upconverted nanomaterials, which cannot be washed away. The above results prove that the surface of $NaYF_4$:Yb^{3+}, Tm^{3+} nanomaterials contains many carboxyl functional groups, and the carboxylated bovine serum albumin has been successfully modified the surface of water-soluble nanomaterials. Figure 9b shows the excitation and emission spectra of DNA/dBSA/$NaYF_4$:Yb,Tm excited at 480 nm. The excitation spectrum is from 450 nm to 490 nm, and the emission spectrum is from 510 nm to 530 nm, which is mainly the contribution of FAM. As described in

Section 2.6, we carried out repeated centrifugal washing to fully wash the excess DNA, and then obtained the fluorescence spectrum in Figure 9c. As shown in Figure 9c, when the DNA/dBSA/NaYF$_4$:Yb,Tm fluorescent probes were excited at 480 nm, a strong emission peak is observed at 520 nm, which is consistent with the FAM fluorescence peak. It indicates that some DNA strands were not washed away due to their attachment to the carboxylated protein-modified upconversion material, namely DNA was attached successfully to the surfaces of UCNP via a protein and a FAM was added to the UCNP/DNA complex. Similarly, NaYF$_4$:Yb,Tm and DNA/dBSA/NaYF$_4$:Yb,Tm was excited at 980 nm, and the upconversion intensity of the luminescence is quite strong, as shown in Figure 9d. These nanocomposites have strong upconversion luminescence with good water solubility and biocompatibility, making them a new type of fluorescent probe.

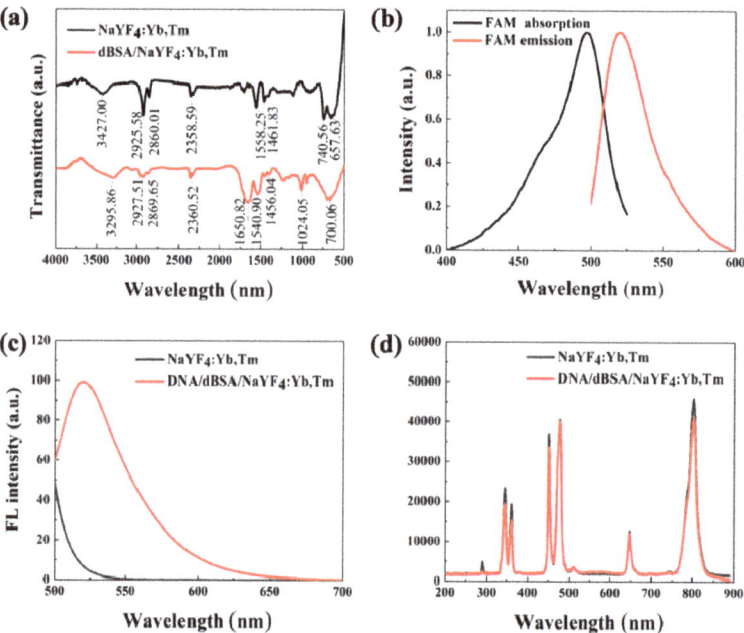

Figure 9. (a) Infrared absorption spectra of NaYF$_4$: Yb,Tm before and after modification with carboxylated bovine serum albumin. (b) Absorption and emission spectra of the FAM fluorophore. (c) Fluorescence spectra of NaYF$_4$:Yb,Tm and novel fluorescent probes under excitation at 480 nm. (d) Fluorescence spectra of NaYF$_4$:Yb,Tm and novel fluorescent probes under excitation at 980 nm.

3.4. Fluorescent Probes for the Detection of Different Proteins

To test the detection of these upconverted nanomaterials for protein detection, 100 pM solutions of miRNA-155, single-base mismatch of miRNA-155, double-base mismatch of miRNA-155, complete-base mismatch of miRNA-155 and miRNA-150 solution with the same concentration of 100 pM were prepared, respectively. The prepared nucleotide sequences of miRNAs and fluorescent probes are shown in Table 1. The prepared fluorescent probes of upconverted nanomaterials were tested with different miRNAs and mismatched miRNAs, and the samples were excited by a fiber laser with a wavelength of 980 nm and their fluorescence spectra were measured.

Table 1. Nucleotide sequences of miRNAs and fluorescent probes.

Name	Sequences (5'-3')
miRNA-155	UUAAUGCUAAUCGUGAUAGGGGU
miRNA-150	UCUCCCAACCCUUGUACCAGUG
miRNA-155 matched DNA strands	NH_2-CCCCCCCCCCCC-ACCCCTATCACGATTAGCATTAA-CGCTAT-FAM
miRNA-150 matched DNA strands	NH_2-CCCCCCCCCCCC-CACTGGTACAAGGGTTGGGAGA-CGCTAT-FAM
miRNA-155 single base mismatch	UUAAGGCUAAUCGUGAUAGGGGU
miRNA-155 double base mismatch	UUAAGGCUAAUAGUGAUAGGGGU
miRNA-155 complete base mismatch	AATTACGATTAGCACTATCCCCA

As shown in Figure 10, the fluorescence spectrum of the fluorescent probe after connecting different miRNA-155 changed significantly. In general, the fluorescence spectra of the four groups were very similar, and the fluorescence intensity decreased significantly (compared with Figure 9d). The sample added with miRNA-155 had the strongest fluorescence intensity. The more mismatched bases, the more obvious fluorescence quenching and the smaller spectral intensity. It is worth noting that the fluorescence quenching at 345 nm, 362 nm, 450 nm, 477 nm and 646 nm is more obvious than that at 802 nm.

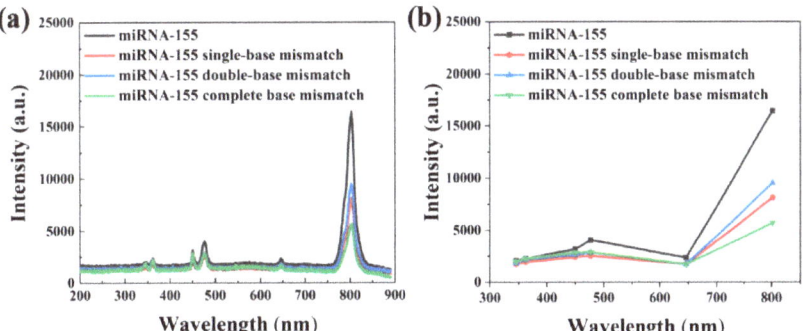

Figure 10. Fluorescence spectra of fluorescent probes connected to miRNA-155, miRNA-155 single-base mismatch, miRNA-155 double-base mismatch, and miRNA-155 complete-base mismatch. (**a**) Fluorescence spectra of mismatched sequences relative to miRNA-155 at 980 nm excitation. (**b**) MiRNA-155 with different sequences, the relationship between the maximum intensity of the luminescence peak corresponding to sequences change.

We divided the peak intensity at 802 nm (I_{802}) by the peak intensity at 345 nm (I_{345}), 362 nm (I_{362}), 450 nm (I_{450}), 477 nm (I_{477}) and 646 nm (I_{646}) to calculate a group of fluorescence peak ratios for further analysis of the differences in fluorescence spectra of samples with different miRNA-155s. The results are shown in Table 2. As can be seen in Table 2, the five peak ratios of the fluorescent probes are very close to those of the upconverted nanomaterials; then, the value of the completely mismatched miRNA155 is relatively close to that of the upconverted nanomaterials, and intact miRNA-155 had the greatest effect on all five peak ratios. It seems that the completely mismatched miRNA-155 has little effect on the peak ratios, and the intact miRNA-155 has the greatest effect on the peak ratios. This result may be that miRNAs with different sequences have different effects on different molecular bonds of upconverted nanomaterials. It is believed that these peak ratios can be used for specific recognition of miRNA-155. For fluorescent probes with multiple emission peaks, in addition to identifying the target substance by simply comparing the changes in peak intensity, the ratios between peaks can also be used for substance-specific identification. The fluorescence probe with multiple emission peaks provides more abundant optical information for the study of the characteristic changes of the detected object and has great application potential.

Table 2. The ratio of fluorescence peaks at 802 nm and 450 nm.

Fluorescent Substance	I_{802}/I_{345}	I_{802}/I_{362}	I_{802}/I_{450}	I_{802}/I_{477}	I_{802}/I_{646}
NaYF$_4$:20%Yb^{3+}, 0.5%Tm^{3+}	1.97	2.38	1.24	1.13	3.65
Fluorescent probes	2.09	2.71	1.23	1.01	3.37
FP + CmiRNA-155	2.87	2.52	2.01	1.96	3.32
FP + M2miRNA-155	5.19	4.38	3.61	3.29	5.5
FP + M1miRNA-155	4.66	4.21	3.38	3.22	4.69
FP + miRNA-155	7.92	7.28	5.19	4.06	6.99

Noting: FP + CmiRNA-155 represents fluorescent probe with completely mismatched miRNA155; FP+ M2miRNA-155 represents fluorescent probe with miRNA-155 double-base mismatch: FP+ M2miRNA-155 represents fluorescent probe with miRNA-155 single-base mismatch; FP + miRNA-155 represents fluorescent probe with miRNA-155.

As shown in Figure 11, when the fluorescent probes were connected to miRNA-155 and miRNA-150, respectively, the fluorescence intensity of miRNA-155 was higher than that of miRNA-150. The experimental results show that the fluorescent probe can effectively distinguish different types of miRNAs.

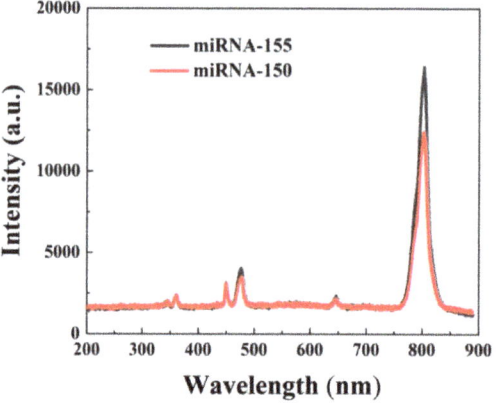

Figure 11. Fluorescence spectra of fluorescent probes connected to miRNA-155 and miRNA-150 under excitation at 980 nm.

4. Conclusions

In this paper, a novel NaYF$_4$:Yb,Tm surface-functionalized fluorescent probe was proposed based on upconverted nanomaterials. Quantitative analysis of the effects of Yb^{3+} and Tm^{3+} ion concentrations on the morphology, size, and luminescence properties of NaYF$_4$:Yb,Tm and NaGdF$_4$:Yb,Tm indicated that the optimal doping concentration ratio of Yb^{3+}:Tm^{3+} is 40:1. In the study, by comparing the fluorescence emission of NaYF$_4$: 20% Yb, 0.5% Tm and NaGdF$_4$: 20% Yb, 0.5% Tm, upconverted nanomaterials with better luminescence properties were obtained. A novel fluorescent probe was designed for the surface carboxylation of NaYF$_4$:Yb,Tm and the connection with amino group and DNA. The probe can be used for the detection of different specific biological miRNAs. When the fluorescent probes are used to detect different miRNAs, they can distinguish different miRNAs, especially miRNAs with base mismatches, from non-specific RNA molecular analytes. The preliminary studies indicate that the upconverted probes have a good potential for protein detection in early cancer diagnosis.

Author Contributions: Conceptualization, Instruction, Funding and Writing: Y.L., L.Z. and W.C.; Experimental design and processing: Z.D.; Material Characterization, data analysis and simulation: Y.H., H.R., J.W., W.N. and Y.W.; Data analysis and editing: C.H., C.X. and C.Y.; Z.D. and edited and formatted the figures: Z.D. and J.W.; Supervision and approval for submission: Y.L. and W.C. All authors have read and agreed to the published version of the manuscript.

Funding: This research was funded by National Natural Science Foundation of China grant number 61971207, Postgraduate Research & Practice Innovation Program of Jiangsu Province grant number (KYCX20_2213) and Natural Science Foundation of the Jiangsu Higher Education Institutes of China grant number 19KJA510003 and 20KJA430003.

Institutional Review Board Statement: Not applicable.

Informed Consent Statement: Not applicable.

Data Availability Statement: Not applicable.

Acknowledgments: W.C. would like to thank Solgro. Inc. and the University of Texas at Arlington for support.

Conflicts of Interest: The authors declare no conflict of interest.

References

1. Lv, Y.; Huang, Z.; Lin, Y.; Fang, Y.; Xu, Z. MiRNA expression patterns are associated with tumor mutational burden in lung adenocarcinoma. *OncoImmunology* **2019**, *8*, 1–8. [CrossRef] [PubMed]
2. Danese, E.; Minicozzi, A.M.; Benati, M.; Paviati, E.; Lima-Oliveira, G.; Gusella, M.; Pasini, F.; Salvagno, G.L.; Montagnana, M.; Lippi, G. Reference miRNAs for colorectal cancer: Analysis and verification of current data. *Sci. Rep.* **2017**, *7*, 8413. [CrossRef] [PubMed]
3. Wen, J.; Zhou, S.; Yu, Z.; Chen, J.; Yang, G.; Tang, J. Decomposable quantum-dots/DNA nanosphere for rapid and ultrasensitive detection of extracellular respiring bacteria. *Biosens. Bioelectron.* **2017**, *100*, 469–474. [CrossRef] [PubMed]
4. Krzeminski, P.; Sarasquete, M.E.; Misiewicz-Krzeminska, I.; Corral, R.; Corchete, L.A.; Martín, A.A.; García-Sanz, R.; San Miguel, J.F.; Gutiérrez, N.C. Insights into epigenetic regulation of microRNA-155 expression in multiple myeloma. *Biochim. Biophys. Acta—Gene Regul. Mech.* **2015**, *1849*, 353–366. [CrossRef] [PubMed]
5. Yébenes, V.G.D.; Bartolomé-Izquierdo, N.; Ramiro, A.R. Regulation of B-cell development and function by micrornas. *Immunol. Rev.* **2013**, *253*, 25–39. [CrossRef]
6. Spierings, D.C.; McGoldrick, D.; Hamilton-Easton, A.M.; Neale, G.; Murchison, E.P.; Hannon, G.J.; Green, D.R.; Withoff, S. Ordered progression of stage-specific miRNA profiles in the mouse B2 B-cell lineage. *Blood* **2011**, *117*, 5340–5349. [CrossRef]
7. Sacco, A.; Zhang, Y.; Maiso, P.; Manier, S.; Rossi, G.; Treon, S.P.; Ghobrial, I.M.; Roccaro, A.M. MicroRNA aberrations in waldenstrom macroglobulinemia. *Clin. Lymphoma Myeloma Leuk.* **2013**, *13*, 205–207. [CrossRef]
8. Auzel, F. Compteur quantique par transfert d'energie entre deux ions de terres rares dans un tungstate mixte et dans un verre. *C. R. Acad. Sci. Paris* **1966**, *262*, 1016–1019.
9. Zhan, Q.; Qian, J.; Liang, H.; Zhan, Q.; Somesfalean, G.; Andersson-Engels, S. Using 915 nm Laser Excited $Tm^{3+}/Er^{3+}/Ho^{3+}$-Doped $NaYbF_4$ Upconversion Nanoparticles for in Vitro and Deeper in Vivo Bioimaging without Overheating Irradiation. *Acs Nano* **2011**, *5*, 3744–3757. [CrossRef]
10. Chen, W. Manganese Doped Upconversion Luminescence Nanoparticles. U.S. Patent 7,008,559 B2, 29 June 2006.
11. Joly, A.G.; Chen, W.; McCready, D.E.; Malm, J.-O.; Bovin, J.-O. Upconversion luminescence of CdTe nanoparticles. *Phys. Rev. B* **2005**, *71*, 165304. [CrossRef]
12. Ouyang, J.; Ripmeester, J.A.; Wu, X.; Kingston, D.; Yu, K.; Joly, A.G.; Chen, W. Upconversion Luminescence of Colloidal CdS and ZnCdS Semiconductor Quantum Dots. *J. Phys. Chem. C* **2007**, *111*, 16261–16266. [CrossRef]
13. Morgan, N.Y.; English, S.W.; Chen, W.; Chernomordik, V.; Russo, A.; Smith, P.D.; Gandjbakhche, A. Real time in vivo non-invasive optical imaging using near-infrared fluorescent quantum dots. *Acad. Radiol.* **2005**, *12*, 313–323. [CrossRef] [PubMed]
14. Chen, W.; Zhang, X.; Huang, Y. Luminescence Enhancement of EuS Clusters in USY-Zeolite. *Appl. Phys. Lett.* **2000**, *76*, 2328–2330. [CrossRef]
15. Li, L.; Rashidi, L.H.; Yao, M.; Ma, L.; Chen, L.; Zhang, J.; Zhang, Y.; Chen, W. CuS Nanoagents for Photodynamic and Photothermal Therapies: Phenomena and Possible Mechanisms, Photodiagnosis and Photodynamic Therapy. *Photodign. Photody Ther.* **2017**, *19*, 5–14. [CrossRef] [PubMed]
16. Li, Y.; Lu, W.; Huang, Q.; Li, C.; Chen, W. In vitro Photothermal Ablation of Tumor Cells with CuS nanoparticles. *Nanomedicine* **2010**, *5*, 1161–1171. [CrossRef] [PubMed]
17. Zhu, G.; Zheng, J.; Song, E.; Donovan, M.; Zhang, K.; Liu, C.; Tan, W. Self-assembled, aptamer-tethered DNA nanotrains for targeted transport of molecular drugs in cancer theranostics. *Proc. Natl. Acad. Sci. USA* **2013**, *110*, 7998–8003. [CrossRef]
18. Dacosta, M.V.; Doughan, S.; Han, Y.; Krull, U.J. Lanthanide upconversion nanoparticles and applications in bioassays and bioimaging: A review. *Anal. Chim. Acta* **2014**, *832*, 1–33. [CrossRef]

19. Probst, C.E.; Zrazhevskiy, P.; Bagalkot, V.; Gao, X. Quantum dots as a platform for nanoparticle drug delivery vehicle design. *Adv. Drug Deliv. Rev.* **2013**, *65*, 703–718. [CrossRef]
20. Chen, W.; Zhang, J. Using Nanoparticles to Enable Simultaneous Radiation and Photodynamic Therapies for Cancer Treatment. *J. Nanosci. Nanotechnol.* **2006**, *6*, 1159–1166. [CrossRef]
21. Liu, F.; Chen, W.; Wang, S.P.; Joly, A.G. Investigation of Water-Soluble X-ray Luminescence Nanoparticles For Photodynamic Activation. *Appl. Phys. Lett.* **2008**, *92*, 43901. [CrossRef]
22. Liu, Y.F.; Chen, W.; Wang, S.P.; Joly, A.G.; Westcott, S.; Woo, B.K. X-ray Luminescence of $LaF_3:Tb$ and $LaF_3:Ce, Tb$ Water Soluble. *Nanoparticles J. Appl. Phys.* **2008**, *103*, 63105. [CrossRef]
23. Chen, W.; Westcott, S.L.; Zhang, J. Dose Dependence of X-ray Luminescence from $CaF_2:Eu^{2+}$, Mn^{2+} Phosphors. *Appl. Phys. Lett.* **2007**, *91*, 211103. [CrossRef]
24. Chen, W.; Westcott, S.L.; Wang, S.; Liu, Y. Dose Dependent X-Ray Luminescence in $MgF_2:Eu^{2+}$, Mn^{2+} Phosphors. *J. Appl. Phys.* **2008**, *103*, 113103. [CrossRef]
25. Wang, S.P.; Westcott, S.; Chen, W. Nanoparticle luminescence thermometry. *J. Phys. Chem. B* **2002**, *106*, 11203–11209. [CrossRef]
26. Chen, W.; Joly, A.G.; Zhang, J.Z. Up-Conversion Luminescence of Mn^{2+} in ZnS:Mn Nanoparticles. *Phys. Rev. B* **2001**, *64*, 412021–412024. [CrossRef]
27. Chen, X.; Liu, J.; Li, Y.; Pandey, N.K.; Chen, T.; Wang, L.; Amador, E.H.; Chen, W.; Liu, F.; Xiao, E.; et al. Study of copper-cysteamine based X-ray induced photodynamic therapy and its effects on cancer cell proliferation and migration in a clinical mimic setting. *Bioactive Materials* **2021**, *7*, 504–514. [CrossRef] [PubMed]
28. Gao, C.; Zheng, P.; Liu, Q. Recent Advances of Upconversion Nanomaterials in the Biological Field. *Nanomaterials* **2021**, *11*, 2474. [CrossRef]
29. Long, M.; Liu, Q.; Wang, D.; Wang, J.; Zhang, Y.; Tang, A.; Liu, N.; Buid, B.; Chen, W.; Yang, H. A New nanoclay-based bifunctional hybrid fiber membrane with hemorrhage control and wound healing for emergency self-rescue. *Mater. Today Adv.* **2021**, *12*, 100190. [CrossRef]
30. Pandey, N.K.; Xiong, W.; Wang, L.; Chen, W.; Lumata, L. Aggregation-induced emission luminogens for highly effective microwave dynamic therapy. *Bioact. Mater.* **2021**, *7*, 112–125. [CrossRef]
31. Wang, Y.; Alkhaldi, N.D.; Pandey, N.K.; Chudal, L.; Wang, L.Y.; Lin, L.W.; Zhang, M.B.; Yong, Y.X.; Amador, E.H.; Huda, M.N.; et al. A new type of cuprous-cysteamine sensitizers: Synthesis, optical properties and potential applications. *Mater. Today Phys.* **2021**, *19*, 100435. [CrossRef]
32. Mao, L.; Lu, Z.; He, N.; Zhang, L.; Deng, Y.; Duan, D. A new method for improving the accuracy of miRNA detection with $NaYF_4$:Yb,Er upconversion nanoparticles. *Sci. China Chem.* **2017**, *60*, 157–162. [CrossRef]
33. Kowalik, P.; Kaminska, I.; Fronc, K.; Borodziuk, A.; Sikora, B. The ROS-generating photosensitizer-free $NaYF_4$:Yb,Tm@SiO_2 upconverting nanoparticles for photodynamic therapy application. *Nanotechnology* **2021**, *32*, 475101. [CrossRef] [PubMed]
34. Etchart, I.M.; Bérard Laroche, M. Efficient white light emission by upconversion in Yb-, Er- and Tm-doped Y BaZnO. *Chem. Commun.* **2011**, *47*, 6263–6265. [CrossRef]
35. Zheng, K.; Liu, Z.; Zhao, D.; Zhang, D.; Qin, G.; Qin, W. Infrared to ultraviolet upconversion fluorescence of Gd^{3+} in β-$NaYF_4$ microcrystals induced by 1560 nm excitation. *Opt. Mater.* **2011**, *33*, 783–787. [CrossRef]
36. Shan, J.; Kong, W.; Wei, R.; Nan, Y.; Ju, Y. An investigation of the thermal sensitivity and stability of the β-$NaYF_4$:Yb,Er upconversion nanophosphors. *J. Appl. Phys.* **2010**, *107*, 937. [CrossRef]
37. SchäFer, H.; Ptacek, P.; Voss, B.; Eickmeier, H.; Haase, M. Synthesis and Characterization of Upconversion Fluorescent Yb^{3+}, Er^{3+} Doped RbY_2F_7 Nano and Microcrystals. *Cryst. Growth Des.* **2010**, *10*, 2202–2208. [CrossRef]
38. Li, C.X.; Quan, Z.W.; Yang, J.; Yang, P.P.; Lin, J. Highly uniform and monodisperse beta-$NaYF_4$:Ln($^{3+}$) (Ln = Eu, Tb, Yb/Er, and Yb/Tm) hexagonal microprism crystals: Hydrothermal synthesis and luminescent properties. *Inorg. Chem.* **2007**, *46*, 6329. [CrossRef]
39. Sch Fer, H.; Ptacek, P.; Zerzouf, O.; Haase, M. Synthesis and Optical Properties of KYF_4/Yb, Er Nanocrystals, and their Surface Modification with Undoped KYF_4. *Adv. Funct. Mater.* **2010**, *18*, 2913–2918. [CrossRef]
40. Zhang, F. *Photon Upconversion Nanomaterials*; Springer: Berlin/Heidelberg, Germany, 2015. [CrossRef]
41. Chien, H.W.; Tsai, M.T.; Yang, C.H.; Lee, R.H.; Wang, T.L. Interaction of $LiYF_4$:Yb^{3+}/Er^{3+}/Ho^{3+}/Tm^{3+}@$LiYF_4$:Yb^{3+} upconversion nanoparticles, molecularly imprinted polymers, and templates. *RSC Adv.* **2020**, *10*, 35600–35610. [CrossRef] [PubMed]
42. Gao, D.L.; Zheng, H.R.; Yu, Y.; Lei, Y.; Zhang, X.S. Spectroscopic properties of Tm^{3+} and Ln($^{3+}$ = Yb^{3+}, Er^{3+}, Pr^{3+}, Ho^{3+}, Eu^{3+}) co-doped fluoride nanocrystals. *Scientia Sinica (Phys. Mech. Astron.)* **2010**, *40*, 287–295.
43. Yun, R.; Luo, L.; He, J.; Wang, J.; Li, X.; Zhao, W.; Nie, Z. Tunable and white up-conversion emission from Tm^{3+}-Ho^{3+}-Yb^{3+}-/Nd^{3+} co-doped $GdVO_4$ phosphors under 808-nm excitation. *J. Mater. Sci. Mater. Electron.* **2021**, *32*, 8149–8156. [CrossRef]
44. Kasprowicz, D.; Brik, M.G.; Majchrowski, A.; Michalski, E.; Głuchowski, P. Up-conversion emission in $KGd(WO)$ single crystals triply-doped with Er /Yb /Tm, Tb /Yb /Tm and Pr /Yb /Tm ions. *Opt. Mater.* **2011**, *33*, 1595–1601. [CrossRef]
45. Li, Z.; Zhang, Y.; Hieu, L.; Zhu, R.; Ghida, E.B.; Wei, Y.; Han, G. Upconverting NIR Photons for Bioimaging. *Nanomaterials* **2015**, *5*, 2148–2168. [CrossRef] [PubMed]
46. Hou, Y.; Qiao, R.; Fang, F.; Wang, X.; Gao, M. $NaGdF_4$ Nanoparticle-Based Molecular Probes for Magnetic Resonance Imaging of Intraperitoneal Tumor Xenografts In Vivo. *Acs Nano* **2012**, *7*, 330–338. [CrossRef] [PubMed]

47. Gao, D.; Zhang, X.; Zheng, H.; Gao, W.; He, E. Yb^{3+}/Er^{3+} codoped β-$NaYF_4$ microrods: Synthesis and tuning of multicolor upconversion. *J. Alloy. Compd.* **2013**, *554*, 395–399. [CrossRef]
48. Jiang, G.; Pichaandi, J.; Johnson, N.; Burke, R.D.; Van Veggel, F.C.J.M. An Effective Polymer Cross-Linking Strategy To Obtain Stable Dispersions of Upconverting $NaYF_4$ Nanoparticles in Buffers and Biological Growth Media for Biolabeling Applications. *Langmuir* **2012**, *28*, 3239–3247. [CrossRef]
49. Huang, W.; Shen, J.; Lei, W.; Chang, Y.; Ye, M. Y_2O_3:Yb/Er nanotubes: Layer-by-layer assembly on carbon-nanotube templates and their upconversion luminescence properties. *Mater. Res. Bull.* **2012**, *47*, 3875–3880. [CrossRef]

Article

Microwave-Assisted Synthesis of Sulfur Quantum Dots for Detection of Alkaline Phosphatase Activity

Fanghui Ma [1], Qing Zhou [2], Minghui Yang [1,*], Jianglin Zhang [3,*] and Xiang Chen [4,*]

[1] Hunan Provincial Key Laboratory of Micro & Nano Materials Interface Science, College of Chemistry and Chemical Engineering, Central South University, Changsha 410083, China
[2] State Key Lab of Powder Metallurgy, Central South University, Changsha 410083, China
[3] Department of Dermatology, Shenzhen People's Hospital (The Second Clinical Medical College, Jinan University, The First Affiliated Hospital, Southern University of Science and Technology), Shenzhen 518020, China
[4] Department of Dermatology, Xiangya Hospital, Central South University, Changsha 410008, China
* Correspondence: yangminghui@csu.edu.cn (M.Y.); zhang.jianglin@szhospital.com (J.Z.); chenxiangck@126.com (X.C.)

Abstract: Sulfur quantum dots (SQDs) are a kind of pure elemental quantum dots, which are considered as potential green nanomaterials because they do not contain heavy metal elements and are friendly to biology and environment. In this paper, SQDs with size around 2 nm were synthesized by a microwave-assisted method using sulfur powder as precursor. The SQDs had the highest emission under the excitation of 380 nm and emit blue fluorescence at 470 nm. In addition, the SQDs had good water solubility and stability. Based on the synthesized SQDs, a fluorescence assay for detection of alkaline phosphatase (ALP) was reported. The fluorescence of the SQDs was initially quenched by Cr (VI). In the presence of ALP, ALP-catalyzed hydrolysis of 2-phospho-L-ascorbic acid to generate ascorbic acid. The generated ascorbic acid can reduce Cr (VI) to Cr (III), thus the fluorescence intensity of SQDs was restored. The assay has good sensitivity and selectivity and was applied to the detection of ALP in serum samples. The interesting properties of SQDs can find a wide range of applications in different sensing and imaging areas.

Keywords: sulfur quantum dots; nanomaterials; fluorescence sensing; alkaline phosphatase; ascorbic acid

1. Introduction

Fluorescent quantum dots (QDs) have been widely used in the fields of biosensing, cell imaging and biomedicine due to their unique size and superior optical properties [1–9]. However, heavy metal quantum dots such as CdS and CdTe are limited in practical applications due to their potential cytotoxicity and environmental hazard [10–13]. As a kind of pure elemental quantum dot, sulfur quantum dots (SQDs) have attracted much attention in recent years because of their low toxicity, good water solubility, stable optical properties and abundant raw materials [14–19]. So far, there have been reports on the application of SQDs in biosensing, cell imaging and antibacterial [20–24]. However, the complex and time-consuming synthesis process and the use of environmentally harmful substances make the synthesis of SQDs still a challenge, which limits its practical applications [25–29]. Therefore, it is necessary to develop green and time-saving methods to synthesize SQDs.

Alkaline phosphatase (ALP), as a membrane-bound enzyme widely exists in a variety of organisms, participating in the process of dephosphorylation in cells and hydrolyzing the phosphate groups of various substrates [30,31]. The normal level of ALP in adult blood is 40 to 190 U/L. Its abnormal expression is closely related to many diseases, such as bone diseases, liver dysfunction and various cancers. Serum ALP levels can be used as a reference for clinical diagnosis [32,33]. Therefore, it is of great significance to develop methods for effective detection of ALP. To date, surface-enhanced Raman scattering (SERS), electrochemical analysis, colorimetric detection, fluorescence analysis and other methods have been developed for the detection of ALP [34–38]. Among them, fluorescence detection technology has attracted significant attention because of its convenient operation, high throughput and high sensitivity.

Several fluorescent material-based chemical sensors have been developed for the detection and analysis of ALP [39]. For example, Li and co-workers designed a smartphone-based sensing strategy for ALP analysis using amino-functionalized copper (II)-based metal-organic frameworks (NH_2-Cu-MOFs), which have both oxidase mimetic properties and fluorescence properties. The catalytic activity of NH_2-Cu-MOFs was greatly inhibited due to the binding ability of Cu^{2+} with pyrophosphate (PPi). After the addition of ALP, the catalytic activity of NH_2-Cu-MOFs was recovered due to the hydrolysis of PPi into orthophosphate by ALP, and then o-phenylenediamine was further catalyzed to form 2,3-diaminophenazine, which constituted a ratiometric fluorescent probe for the detection of ALP. The method has been successfully applied to the determination of ALP in serum samples. Zhang et al. constructed a novel near infrared ratiometric fluorescent probe (APT), which can achieve a rapid response to ALP (Within 10 min) [40]. After adding ALP, the fluorescence spectrum showed a shift (from 580 to 650 nm), and the near-infrared fluorescence emission (650 nm) made it more suitable for biological detection. The method has been successfully applied to the determination of ALP in serum and the detection and imaging of endogenous ALP in cells. In addition, nanocomposites have also been used to construct fluorescent sensors to detect ALP. Li's group developed a simple hydrothermal method to construct "three-in-one" nanocomposites (Fef NCs) for the detection of ALP [41]. Fef NCs consist of three components, in which MnO_2 nanosheets (NSs) are assembled on Fe_3O_4 nanoparticles (NPs), and then CeO_2 NPs are modified. The nanometer material has various catalytic activity, and can realize label-free, ultrasensitive and selective detection of ALP by utilizing that characteristic. Duan et al. utilized WS_2 quantum dots and MnO_2 nanosheets to form a nanocomposite system to detect ALP [42]. MnO_2 nanosheets can quench the blue fluorescence of tungsten disulfide quantum dots (WS_2 QDs). However, in the presence of ALP and amifostine, their hydrolysis products triggered the decomposition of MnO_2 nanosheets. This results in the restoration of fluorescence. Based on this discovery, the researchers successfully used the switch principle to detect ALP and used it in the analysis of actual samples. However, the above methods have limitations such as use of inorganic substances which limit their application and are harmful to the environment, and cumbersome synthesis or construction processes. Therefore, it is necessary to develop green, simple methods for efficient and sensitive detection of ALP.

In this work, SQDs with stable optical properties were prepared by microwave-assisted heating using sublimed sulfur as a precursor and PEG-400 as a stabilizer. The synthesized SQDs was utilized for detecting ALP through a fluorescence "off-on" mechanism, as shown in Scheme 1. As the emission band of SQDs and the absorption band of Cr (VI) are well matched, there is a strong internal filtering effect (IFE) between them, so the addition of Cr (VI) can well quench the fluorescence of SQDs. However, ascorbic acid (AA) can reduce Cr (VI) to Cr (III), so the addition of AA can restore the fluorescence of the quenched SQDs. In addition, ALP can hydrolyze 2-phospho-L-ascorbic acid (AAP) to AA and phosphate ions; therefore, the activity of ALP can be detected via the recovery of the fluorescence of SQDs. Based on the above principle, the relationship between the fluorescence characteristics of sulfur quantum dots, Cr (VI), AA as well as ALP was explored, and the performance of SQDs for detection of ALP was studied in detail. The microwave synthesis of SQDs greatly shortens the synthesis time of SQDs and simplifies the operation steps, which have great reference significance for the exploration of the synthesis of SQDs. At the same time, the obtained SQDs are successfully used in the detection of ALP, which broadens the application of SQDs in biosensors.

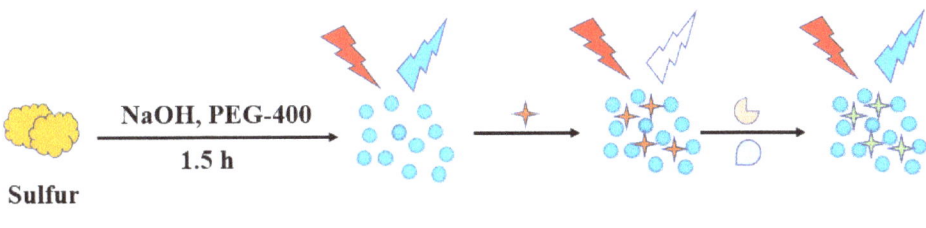

Scheme 1. Synthesis of SQDs and schematic diagram for determination of ALP.

2. Experimental Section

2.1. Materials and Apparatus

Potassium dichromate, sodium hydroxide, polyethylene glycol and sulfur powder were purchased from Aladdin (Shanghai, China). ALP and AA were obtained from Sigma-Aldrich (Shanghai, China). All reagents are of analytical grade and do not require further purification. The solutions used in the experiment were prepared by ultrapure water.

Transmission electron microscopy (TEM) images of the SQDs were obtained under a FEI Titan G2 60-300 microscope. Fluorescence measurements were performed on an F-7000 fluorescence spectrophotometer (Hitachi, Tokyo, Japan). UV-visible absorption spectra were recorded on a UV-2450 spectrophotometer (Hitachi, Tokyo, Japan).

2.2. Synthesis of SQDs

The synthesis of SQDs was based on previously reported methods with minor modifications [43–45]. Briefly, 1 g NaOH, 0.175 g sulfur powder, and 1.5 mL PEG-400 were mixed under stirring at 50 °C until the solution was clear. Then, the clear solution was placed in a microwave oven and reacted at 200 W for 1.5 h. Next, the obtained solution was centrifuged at 6000 rpm for 15 min, and the supernatant was centrifuged for two more times. Finally, the obtained supernatant was the SQDs solution, which was placed at 4 °C before use.

2.3. Determination of ALP

Initially, 20 µL of SQDs solution was mixed with various concentrations of aqueous $K_2Cr_2O_7$, then a PBS buffer (pH 7.4, 0.1 M) was added until the volume of the mixture reached 200 µL. The fluorescence spectrum of the solution was then measured. The excitation wavelength was maintained at 380 nm throughout the detection.

For the detection of ALP, 40 µL of solution containing different concentrations of ALP were mixed with 40 µL of 30 mM AAP. After mixing, the solution was incubated for 30 min in a water bath at 37 °C, then 10 µL of 10 mM $K_2Cr_2O_7$ was added. After incubating for another 10 min, 20 µL of SQDs was added into the mixture. Fluorescence spectra of the solution were then collected.

3. Results and Discussion

3.1. Preparation and Characterization of the SQDs

As shown in Scheme 1, using sulfur powder as a precursor, and PEG-400 as stabilizer, SQDs were synthesized by microwave-assisted heating. The whole synthesis process is easy to operate, the raw materials used being basically non-toxic to the environment, and the synthesis time was shortened to 90 min. In order to understand the size and morphology of SQDs, the morphology of SQDs was imaged by transmission electron microscope (TEM). As shown in Figure 1a, SQDs can be well dispersed into water. The particles display spherical shape, and the particle size is 2.27 ± 0.76 nm, which is similar to the fluorescent SQDs previously reported.

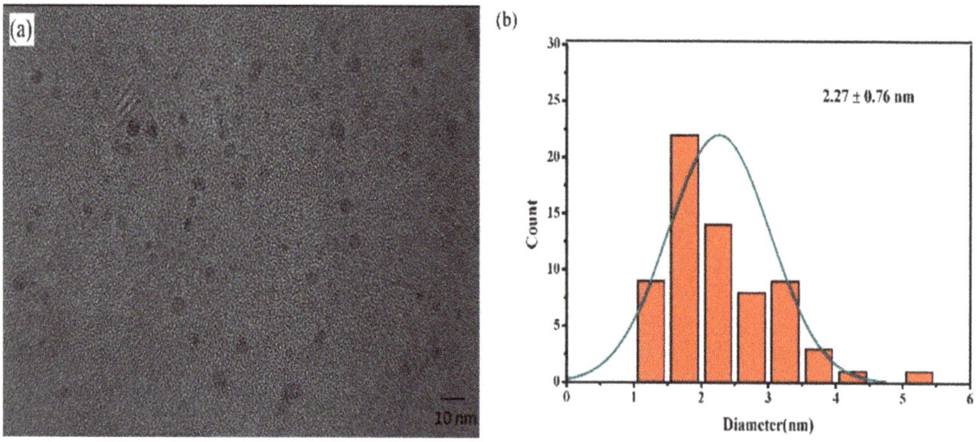

Figure 1. Morphological characterizations of the SQDs: (**a**) TEM images; (**b**) size distribution histogram.

The excitation and emission positions and fluorescence intensity of SQDs can be obtained by fluorescence tests. Fluorescence properties of the synthesized SQDs were studied (Figure S1). When excited under different wavelength, with the increase of wavelength from 320 nm to 380 nm, the emission intensity of SQDs increases with the increase of excitation wavelength. By further increasing the wavelength from 380 nm to 420 nm, the emission intensity decreases with the increase of excitation wavelength. In addition, the fluorescence emission peak is redshifted with the increase of excitation wavelength. Therefore, 380 nm was used as the optimal excitation wavelength in subsequent experiments.

3.2. Feasibility of the Assay for ALP Analysis

The feasibility of the assay for ALP analysis was tested. As described in Figure 2, SQDs has a strong fluorescence emission peak at about 470 nm. However, the fluorescence of the SQDs is quenched after adding a certain concentration of $K_2Cr_2O_7$, and can then recovered with further addition of AA. The recovery of fluorescence intensity is in accordance with the amount of AA added; this is because AA can reduce Cr (VI) to Cr (III). The fluorescence intensity of SQDs, mixed with a certain amount of chromium chloride solution, has no obvious difference with that of the single SQDs solution, which indicates that Cr (III) generated due to the reduction of $K_2Cr_2O_7$ by AA has no quenching effect on the fluorescence signal of the SQDs. Meanwhile, after the $K_2Cr_2O_7$ is added into the SQDs, the fluorescence signal is not changed by following addition of AAP, indicating AAP can not react with $K_2Cr_2O_7$ and AAP alone will not affect the fluorescence intensity of the system. However, in the presence of ALP, the fluorescence signal was restored, which verified that ALP can catalyze the hydrolysis of AAP to generate AA, thus achieving the same effect as adding AA. Based on the above experimental results, it can be seen that fluorescence sensing of ALP activity can be achieved based on SQDs.

UV-vis absorption spectra further verified the above experimental results (Figure 3). Cr (VI) has a strong absorption at 380 nm. In contrast, the absorption of SQDs is weak. The competion of Cr (VI) ion with SQDs for the absorption of 380 nm light resulted in the quenching of the fluorescence of SQDs. Adding SQDs to Cr (VI), the absorption at 380 nm is slightly enhanced compared to Cr (VI) alone which is due to the absorption of SQDs at 380 nm. After the addition of AA, the absorption at this point almost disappeared because AA reduced Cr (VI) to Cr (III). These results are consistent with the fluorescence data.

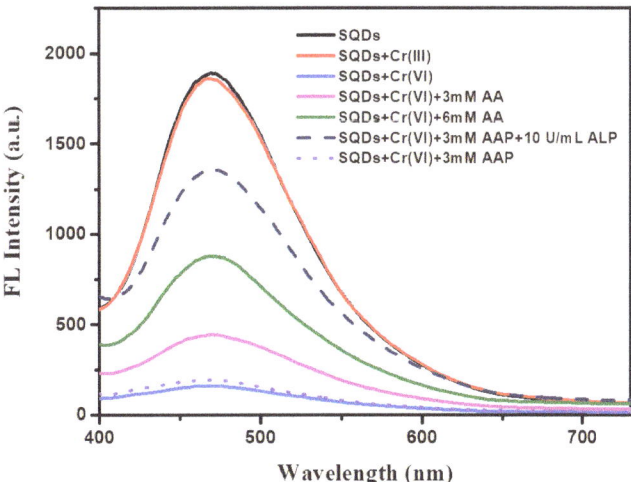

Figure 2. Feasibility analysis of the assay for ALP detection.

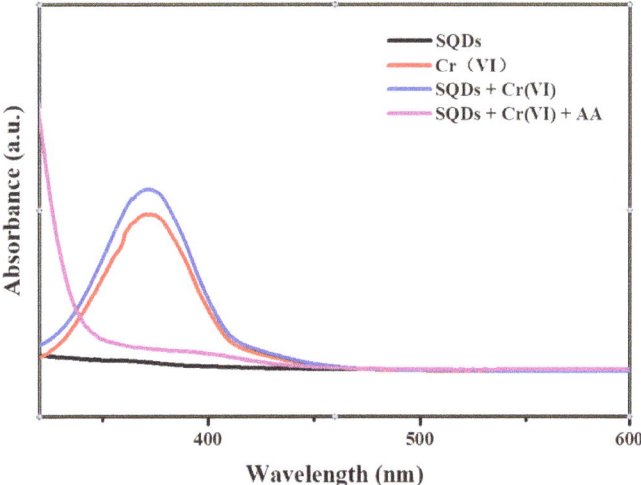

Figure 3. UV-vis absorption spectra of different solutions.

3.3. Detection of Cr (VI)

The quenching of the fluorescence of SQDs by Cr (VI) is due to fact that the excitation spectrum of SQDs overlaps well with the absorption band of Cr (VI). Therefore, a strong IFE occurred between SQDs and Cr (VI), because Cr (VI) can shield the excitation light of SQDs. As shown in Figure 4a, the fluorescence intensity of SQDs decreased gradually with the increase of the amount of Cr (VI) added. When the Cr (VI) concentration reaches 5 mM, the fluorescence of SQDs is almost completely quenched. The fluorescence intensity of SQDs was linearly correlated with Cr (VI) concentration in the range of 10–100 μM and has a good linear correlation coefficient ($R^2 = 0.997$) (Figure 4b).

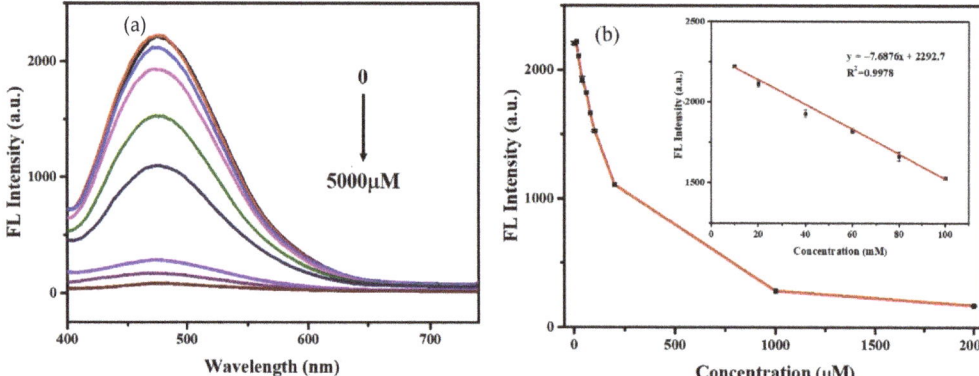

Figure 4. (**a**) Emission spectra of SQDs after adding different concentrations of Cr (VI) solution from 0 to 5 mM; (**b**) Fluorescence value of the SQDs in response to different concentrations of Cr (VI) in the range of 0 to 2 mM. Inset: the calibration curve to Cr (VI) in the linear range of 10–100 µM.

The selectivity of the synthesized SQDs for Cr (VI) was evaluated. Under the same experimental conditions, 2 mM Cr (VI) and 6 mM, other interfering ions, were added into SQDs and the fluorescence intensity was measured under 380 nm excitation. As described in Figure 5, when 2 mM Cr (VI) was added to the SQDs, the fluorescence intensity of the SQDs was reduced by more than 90%. However, even when 6 mM of other common metal ions were added, the fluorescence intensity of the SQDs was not affected significantly. These data show that the SQDs have good selectivity for Cr (VI).

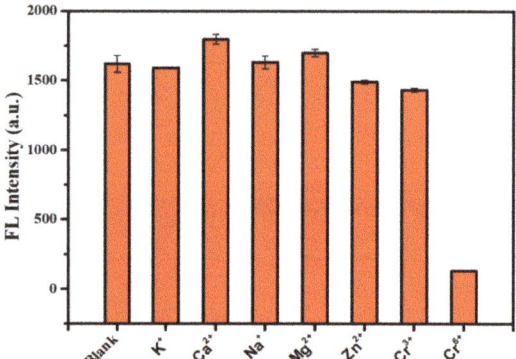

Figure 5. Response of SQDs to different ions.

3.4. Detection of ALP

The absorption peak of Cr (VI) at about 380 nm disappears after the addition of AA, which indicates that AA can reduce Cr (VI) to a low-valent Cr species, thereby eliminating the IFE between Cr (VI) and SQDs and restoring the fluorescence of SQDs. ALP can hydrolyze AAP to generate AA, which can also recover the fluorescence of SQDs-Cr (VI) system. Hence, the principle can be applied for detecting ALP. The selectivity of the system to AA was investigated. Other potential reductants including glutathione (GSH), cysteine (Cys), glucose and some common ions had little effect on the fluorescence of SQDs-Cr (VI) system, indicating that SQDs-Cr (VI) system had good selectivity for AA (Figure S2). As described in Figure 6a, the fluorescence intensity of SQDs increases with the increase of ALP activity. When the ALP activity is more than 10 U/mL, the fluorescence intensity of the system tends to be stable, and the fluorescence strength of SQDs was almost fully recovered. The fluorescence intensity of SQDs had a good linear relationship with ALP activity in the range of 1.5–5.0 U/mL, and the linear correlation coefficient R^2 was 0.992. Based on a signal–noise ratio of 3, the calculated LOD was 0.13 U/mL. Table 1 summarizes previously reported strategies for detecting ALP using different probes, indicating that the methods are comparable to those reported in other studies.

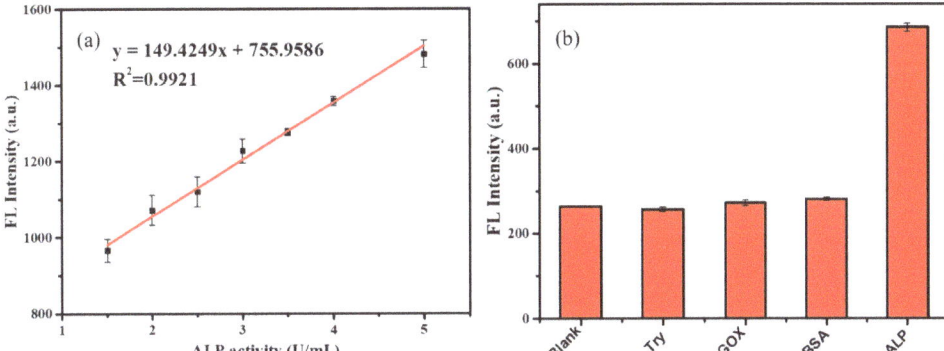

Figure 6. (**a**) Fluorescence response of the assay to ALP of different activities. Inset: linear response of SQDs to ALP in the activity range of 1.5–5.0 U/mL; (**b**) Selectivity of the assay to ALP.

Then, the selectivity of the assay for ALP was analyzed. Responses of the assay to 2 U/mL of alkaline phosphatase (ALP) and other enzymes with an activity of 5 U/mL were tested. As described in Figure 6b, when ALP is added, the fluorescence change is obvious, while the responses of the assay to other enzymes are negligible, similar to blanks. These data show that the assay has good selectivity to ALP sensing. In addition, the assay is used for the recovery testing of ALP in serum. With the serum signal as the background, ALP with the activity of 1, 2, 2.5 and 3.5 U/mL were added into the serum sample, and then the level of ALP was measured. The results are shown in Table 2. The recoveries for ALP in the serum were in the range of 102.01~119.61%, with the relative standard deviations (RSD) less than 5%. Therefore, the assay has potential application for clinical detection of ALP.

Table 1. Comparison employing fluorescent probes to detect ALP.

Probe	Linear Range/U·mL^{-1}	LOD/U·mL^{-1}	Ref.
Luminol–Tb–GMP CPNPs	0.00005–0.1	0.00002	[46]
Hydrogelator	0–2.8	0.06	[31]
NIR-Phos-1, NIR-Phos-2	0–1.0	10^{-5}–10^{-3}	[47]
5-bromo-4-chloro-3-indolyl phosphate	10–1000	0.87	[48]
CuNPs	0.1–40	0.05	[49]
SQDs	1.5–5.0	0.13	This work

Table 2. Recovery analysis of ALP in human serum.

Samples	Added (U/mL)	Found (U/mL)	Recovery (%)	RSD (%)
human serum	1	1.43 ± 0.02	-	0.301
	2	2.63 ± 0.09	119.61	0.495
	2.5	3.05 ± 0.20	108.15	2.497
	3.5	3.05 ± 0.20	102.01	1.181

4. Conclusions

In this work, by using sulfur powder as a precursor and polyethylene glycol as a stabilizer, SQDs were prepared successfully through a microwave-assisted method. The synthesis time of the SQDs was greatly shortened to only 90 min. The synthesized SQDs had good water solubility and good fluorescence stability. A sensitive and selective fluorescence assay based on IFE for detection of ALP was developed by using synthesized SQDs. In the presence of Cr (VI), the fluorescence of SQDs was quenched, and then recovered by the addition of AAP and ALP. The SQDs as fluorescence probes can find a wide range of applications in different sensing areas.

Supplementary Materials: The following supporting information can be downloaded at: https://www.mdpi.com/article/10.3390/nano12162787/s1, Figure S1: Fluorescence emission spectra of SQDs under different excitation wavelength, Figure S2: Selectivity of the SQDs-Cr (VI) system for AA.

Author Contributions: Conceptualization, M.Y., X.C. and J.Z.; investigation, F.M. and Q.Z.; methodology, F.M. and Q.Z.; formal analysis, F.M. and Q.Z.; resources, M.Y., X.C. and J.Z.; software, F.M. and Q.Z.; supervision, M.Y., X.C. and J.Z.; writing—original draft, F.M.; writing—review and editing, M.Y., X.C. and J.Z. All authors have read and agreed to the published version of the manuscript.

Funding: The authors are thankful for the support of this work by the National Natural Science Foundation of China (Grant No. 22174163), the Hunan Provincial Science and Technology Plan Project, China (No. 2019TP1001), and the Project of Intelligent Management Software for Multimodal Medical Big Data for New Generation Information Technology, Ministry of Industry and Information Technology of People's Republic of China (TC210804V).

Institutional Review Board Statement: Not applicable.

Informed Consent Statement: Not applicable.

Data Availability Statement: Data is contained within the article.

Conflicts of Interest: The authors declare no conflict of interest.

References

1. Lim, H.J.; Jin, H.; Chua, B.; Son, A. Clustered detection of eleven phthalic acid esters by fluorescence of graphene quantum dots displaced from gold nanoparticles. *ACS Appl. Mater. Interfaces* **2022**, *14*, 4186–4196. [CrossRef] [PubMed]
2. Ling, S.; Yang, X.; Li, C.; Zhang, Y.; Yang, H.; Chen, G.; Wang, Q. Tumor microenvironment-activated nir-ii nanotheranostic system for precise diagnosis and treatment of peritoneal metastasis. *Angew. Chem. Int. Ed.* **2020**, *59*, 7219–7223. [CrossRef] [PubMed]
3. Shi, R.; Feng, S.; Park, C.Y.; Park, K.Y.; Song, J.; Park, J.P.; Chun, H.S.; Park, T.J. Fluorescence detection of histamine based on specific binding bioreceptors and carbon quantum dots. *Biosens. Bioelectron.* **2020**, *167*, 112519. [CrossRef] [PubMed]
4. Wu, Y.; Chen, Z.; Yao, Z.; Zhao, K.; Shao, F.; Su, J.; Liu, S. Black phosphorus quantum dots encapsulated biodegradable hollow mesoporous mno$_2$: Dual-modality cancer imaging and synergistic chemo-phototherapy. *Adv. Funct. Mater.* **2021**, *31*, 2104643. [CrossRef]
5. Yang, H.; Huang, H.; Ma, X.; Zhang, Y.; Yang, X.; Yu, M.; Sun, Z.; Li, C.; Wu, F.; Wang, Q. Au-doped ag$_2$te quantum dots with bright nir-iib fluorescence for in situ monitoring of angiogenesis and arteriogenesis in a hindlimb ischemic model. *Adv. Mater.* **2021**, *33*, 2103953. [CrossRef]
6. Zhao, S.; Chen, X.; Zhang, C.; Zhao, P.; Ragauskas, A.J.; Song, X. Fluorescence enhancement of lignin-based carbon quantum dots by concentration-dependent and electron-donating substituent synergy and their cell imaging applications. *ACS Appl. Mater. Interfaces* **2021**, *13*, 61565–61577. [CrossRef]
7. Das, P.; Maruthapandi, M.; Saravanan, A.; Natan, M.; Jacobi, G.; Banin, E.; Gedanken, A. Carbon dots for heavy-metal sensing, ph-sensitive cargo delivery, and antibacterial applications. *ACS Appl. Nano Mater.* **2020**, *3*, 11777–11790. [CrossRef]
8. Wang, H.; Li, T.; Hashem, A.M.; Abdel-Ghany, A.E.; El-Tawil, R.S.; Abuzeid, H.M.; Coughlin, A.; Chang, K.; Zhang, S.; El-Mounayri, H.; et al. Nanostructured molybdenum-oxide anodes for lithium-ion batteries: An outstanding increase in capacity. *Nanomaterials* **2021**, *12*, 13. [CrossRef]
9. Das, P.; Ganguly, S.; Margel, S.; Gedanken, A. Tailor made magnetic nanolights: Fabrication to cancer theranostics applications. *Nanoscale Adv.* **2021**, *3*, 6762–6796. [CrossRef]
10. Yan, X.; Pei, Y.; Chen, H.; Zhao, J.; Zhou, Z.; Wang, H.; Zhang, L.; Wang, J.; Li, X.; Qin, C.; et al. Self-assembled networked pbs distribution quantum dots for resistive switching and artificial synapse performance boost of memristors. *Adv. Mater.* **2019**, *31*, 1805284. [CrossRef]

11. Hudson, M.H.; Chen, M.; Kamysbayev, V.; Janke, E.M.; Lan, X.; Allan, G.; Delerue, C.; Lee, B.; Guyot-Sionnest, P.; Talapin, D.V. Conduction band fine structure in colloidal hgte quantum dots. *ACS Nano* **2018**, *12*, 9397–9404. [CrossRef] [PubMed]
12. Zhang, X.; Li, L.; Sun, Z.; Luo, J. Rational chemical doping of metal halide perovskites. *Chem. Soc. Rev.* **2019**, *48*, 517–539. [CrossRef]
13. Pu, C.; Qin, H.; Gao, Y.; Zhou, J.; Wang, P.; Peng, X. Synthetic control of exciton behavior in colloidal quantum dots. *J. Am. Chem. Soc.* **2017**, *139*, 3302–3311. [CrossRef] [PubMed]
14. Pal, A.; Arshad, F.; Sk, M.P. Emergence of sulfur quantum dots: Unfolding their synthesis, properties, and applications. *Adv. Colloid Interface Sci.* **2020**, *285*, 102274. [CrossRef] [PubMed]
15. Shi, Y.E.; Zhang, P.; Yang, D.; Wang, Z. Synthesis, photoluminescence properties and sensing applications of luminescent sulfur nanodots. *Chem. Commun.* **2020**, *56*, 10982–10988. [CrossRef]
16. Song, Y.; Tan, J.; Wang, G.; Gao, P.; Lei, J.; Zhou, L. Oxygen accelerated scalable synthesis of highly fluorescent sulfur quantum dots. *Chem. Sci.* **2019**, *11*, 772–777. [CrossRef]
17. Arshad, F.; Sk, M.P. Luminescent sulfur quantum dots for colorimetric discrimination of multiple metal ions. *ACS Appl. Nano Mater.* **2020**, *3*, 3044–3049. [CrossRef]
18. Wang, C.; Wei, Z.; Pan, C.; Pan, Z.; Wang, X.; Liu, J.; Wang, H.; Huang, G.; Wang, M.; Mao, L. Dual functional hydrogen peroxide boosted one step solvothermal synthesis of highly uniform sulfur quantum dots at elevated temperature and their fluorescent sensing. *Sens. Actuators B Chem.* **2021**, *344*, 130326. [CrossRef]
19. Peng, X.; Wang, Y.; Luo, Z.; Zhang, B.; Mei, X.; Yang, X. Facile synthesis of fluorescent sulfur quantum dots for selective detection of p-nitrophenol in water samples. *Microchem. J.* **2021**, *170*, 106735. [CrossRef]
20. Zhang, C.; Zhang, P.; Ji, X.; Wang, H.; Kuang, H.; Cao, W.; Pan, M.; Shi, Y.E.; Wang, Z. Ultrasonication-promoted synthesis of luminescent sulfur nano-dots for cellular imaging applications. *Chem. Commun.* **2019**, *55*, 13004–13007. [CrossRef]
21. Zhang, X.; Chen, X.; Guo, Y.; Gu, L.; Wu, Y.; Bindra, A.K.; Teo, W.L.; Wu, F.G.; Zhao, Y. Thiolate-assisted route for constructing chalcogen quantum dots with photoinduced fluorescence enhancement. *ACS Appl. Mater. Interfaces* **2021**, *13*, 48449–48456. [CrossRef] [PubMed]
22. Qiao, G.; Liu, L.; Hao, X.; Zheng, J.; Liu, W.; Gao, J.; Zhang, C.C.; Wang, Q. Signal transduction from small particles: Sulfur nanodots featuring mercury sensing, cell entry mechanism and in vitro tracking performance. *Chem. Eng. J.* **2020**, *382*, 122907. [CrossRef]
23. Wang, Y.; Zhao, Y.; Wu, J.; Li, M.; Tan, J.; Fu, W.; Tang, H.; Zhang, P. Negatively charged sulfur quantum dots for treatment of drug-resistant pathogenic bacterial infections. *Nano Lett.* **2021**, *21*, 9433–9441. [CrossRef] [PubMed]
24. Lu, C.; Wang, Y.; Xu, B.; Zhang, W.; Xie, Y.; Chen, Y.; Wang, L.; Wang, X. A colorimetric and fluorescence dual-signal determination for iron (ii) and H_2O_2 in food based on sulfur quantum dots. *Food Chem.* **2022**, *366*, 130613. [CrossRef] [PubMed]
25. Sheng, Y.; Huang, Z.; Zhong, Q.; Deng, H.; Lai, M.; Yang, Y.; Chen, W.; Xia, X.; Peng, H. Size-focusing results in highly photoluminescent sulfur quantum dots with a stable emission wavelength. *Nanoscale* **2021**, *13*, 2519–2526. [CrossRef]
26. Rong, S.; Chen, Q.; Xu, G.; Wei, F.; Yang, J.; Ren, D.; Cheng, X.; Xia, Y.; Li, J.; Gao, M.; et al. Novel and facile synthesis of heparin sulfur quantum dots via oxygen acceleration for ratiometric sensing of uric acid in human serum. *Sens. Actuators B Chem.* **2022**, *353*, 131146. [CrossRef]
27. Arshad, F.; Sk, M.P.; Maurya, S.K.; Siddique, H.R. Mechanochemical synthesis of sulfur quantum dots for cellular imaging. *ACS Appl. Nano Mater.* **2021**, *4*, 3339–3344. [CrossRef]
28. Wang, H.; Wang, Z.; Xiong, Y.; Kershaw, S.V.; Li, T.; Wang, Y.; Zhai, Y.; Rogach, A.L. Hydrogen peroxide assisted synthesis of highly luminescent sulfur quantum dots. *Angew. Chem. Int. Ed. Engl.* **2019**, *58*, 7040–7044. [CrossRef]
29. Li, S.; Chen, D.; Zheng, F.; Zhou, H.; Jiang, S.; Wu, Y. Water-soluble and lowly toxic sulphur quantum dots. *Adv. Funct. Mater.* **2014**, *24*, 7133–7138. [CrossRef]
30. Niu, X.; Ye, K.; Wang, L.; Lin, Y.; Du, D. A review on emerging principles and strategies for colorimetric and fluorescent detection of alkaline phosphatase activity. *Anal. Chim. Acta* **2019**, *1086*, 29–45. [CrossRef]
31. Dong, L.; Miao, Q.; Hai, Z.; Yuan, Y.; Liang, G. Enzymatic hydrogelation-induced fluorescence turn-off for sensing alkaline phosphatase in vitro and in living cells. *Anal. Chem.* **2015**, *87*, 6475–6478. [CrossRef] [PubMed]
32. Haarhaus, M.; Brandenburg, V.; Kalantar-Zadeh, K.; Stenvinkel, P.; Magnusson, P. Alkaline phosphatase: A novel treatment target for cardiovascular disease in ckd. *Nat. Rev. Nephrol.* **2017**, *13*, 429–442. [CrossRef] [PubMed]
33. Chen, C.; Zhao, D.; Jiang, Y.; Ni, P.; Zhang, C.; Wang, B.; Yang, F.; Lu, Y.; Sun, J. Logically regulating peroxidase-like activity of gold nanoclusters for sensing phosphate-containing metabolites and alkaline phosphatase activity. *Anal. Chem.* **2019**, *91*, 15017–15024. [CrossRef] [PubMed]
34. Bekhit, M.; Blazek, T.; Gorski, W. Electroanalysis of enzyme activity in small biological samples: Alkaline phosphatase. *Anal. Chem.* **2021**, *93*, 14280–14286. [CrossRef]
35. Liu, X.; Mei, X.; Yang, J.; Li, Y. Hydrogel-involved colorimetric platforms based on layered double oxide nanozymes for point-of-care detection of liver-related biomarkers. *ACS Appl. Mater. Interfaces* **2022**, *14*, 6985–6993. [CrossRef]
36. Wang, X.; Zhou, S.; Chu, C.; Yang, M.; Huo, D.; Hou, C. Target-induced transcription amplification to trigger the trans-cleavage activity of crispr/cas13a (titac-cas) for detection of alkaline phosphatase. *Biosens. Bioelectron.* **2021**, *185*, 113281. [CrossRef]
37. Peng, J.; Han, X.X.; Zhang, Q.C.; Yao, H.Q.; Gao, Z.N. Copper sulfide nanoparticle-decorated graphene as a catalytic amplification platform for electrochemical detection of alkaline phosphatase activity. *Anal. Chim. Acta* **2015**, *878*, 87–94. [CrossRef]

38. Zhan, Y.; Yang, S.; Chen, L.; Zeng, Y.; Li, L.; Lin, Z.; Guo, L.; Xu, W. Ultrahigh efficient fret ratiometric fluorescence biosensor for visual detection of alkaline phosphatase activity and its inhibitor. *ACS Sustain. Chem. Eng.* **2021**, *9*, 12922–12929. [CrossRef]
39. Hou, L.; Qin, Y.; Li, J.; Qin, S.; Huang, Y.; Lin, T.; Guo, L.; Ye, F.; Zhao, S. A ratiometric multicolor fluorescence biosensor for visual detection of alkaline phosphatase activity via a smartphone. *Biosens. Bioelectron.* **2019**, *143*, 111605. [CrossRef]
40. Zhang, X.; Chen, X.; Liu, K.; Zhang, Y.; Gao, G.; Huang, X.; Hou, S. Near-infrared ratiometric probe with a self-immolative spacer for rapid and sensitive detection of alkaline phosphatase activity and imaging in vivo. *Anal. Chim. Acta* **2020**, *1094*, 113–121. [CrossRef]
41. Li, X.; Cai, M.; Shen, Z.; Zhang, M.; Tang, Z.; Luo, S.; Lu, N. "Three-in-one" nanocomposite as multifunctional nanozyme for ultrasensitive ratiometric fluorescence detection of alkaline phosphatase. *J. Mater. Chem. B* **2022**. [CrossRef] [PubMed]
42. Duan, X.; Liu, Q.; Su, X. Fluorometric determination of the activity of alkaline phosphatase based on a system composed of WS_2 quantum dots and MnO_2 nanosheets. *Microchim. Acta* **2019**, *186*, 839. [CrossRef] [PubMed]
43. Shen, L.; Wang, H.; Liu, S.; Bai, Z.; Zhang, S.; Zhang, X.; Zhang, C. Assembling of sulfur quantum dots in fission of sublimed sulfur. *J. Am. Chem. Soc.* **2018**, *140*, 7878–7884. [CrossRef] [PubMed]
44. Xiao, L.; Du, Q.; Huang, Y.; Wang, L.; Cheng, S.; Wang, Z.; Wong, T.N.; Yeow, E.K.L.; Sun, H. Rapid synthesis of sulfur nanodots by one-step hydrothermal reaction for luminescence-based applications. *ACS Appl. Nano Mater.* **2019**, *2*, 6622–6628. [CrossRef]
45. Fan, S.; Li, X.; Ma, F.; Yang, M.; Su, J.; Chen, X. Sulfur quantum dot based fluorescence assay for lactate dehydrogenase activity detection. *J. Photochem. Photobiol. A Chem.* **2022**, *430*, 113989. [CrossRef]
46. Tong, Y.J.; Yu, L.D.; Wu, L.L.; Cao, S.P.; Liang, R.P.; Zhang, L.; Xia, X.H.; Qiu, J.D. Aggregation-induced emission of luminol: A novel strategy for fluorescence ratiometric detection of alp and as(v) with high sensitivity and selectivity. *Chem. Commun.* **2018**, *54*, 7487–7490. [CrossRef]
47. Park, C.S.; Ha, T.H.; Kim, M.; Raja, N.; Yun, H.S.; Sung, M.J.; Kwon, O.S.; Yoon, H.; Lee, C.S. Fast and sensitive near-infrared fluorescent probes for alp detection and 3d printed calcium phosphate scaffold imaging in vivo. *Biosens. Bioelectron.* **2018**, *105*, 151–158. [CrossRef]
48. Mahato, K.; Chandra, P. Paper-based miniaturized immunosensor for naked eye ALP detection based on digital image colorimetry integrated with smartphone. *Biosens. Bioelectron.* **2019**, *128*, 9–16. [CrossRef]
49. Yang, D.; Guo, Z.; Tang, Y.; Miao, P. Poly(thymine)-templated selective formation of copper nanoparticles for alkaline phosphatase analysis aided by alkyne–azide cycloaddition "click" reaction. *ACS Appl. Nano Mater.* **2017**, *1*, 168–174. [CrossRef]

Article

Development of Ag-Doped ZnO Thin Films and Thermoluminescence (TLD) Characteristics for Radiation Technology

Hammam Abdurabu Thabit [1,*], Norlaili A. Kabir [2], Abd Khamim Ismail [1,*], Shoroog Alraddadi [3], Abdullah Bafaqeer [4] and Muneer Aziz Saleh [5]

1. Department of Physics, Faculty of Science, Universiti Teknologi Malaysia, UTM, Johor Bahru 81310, Malaysia
2. School of Physics, Universiti Sains Malaysia, Pulau Pinang 11800, Malaysia
3. Department of Physics, Umm AL-Qura University, Makkah 24382, Saudi Arabia
4. Chemical Reaction Engineering Group (GREG), School of Chemical and Energy Engineering, Universiti Teknologi Malaysia, UTM, Johor Bahru 81310, Malaysia
5. Office of Radiation Protection, Department of Health, Tumwater, WA 98501, USA
* Correspondence: hammam.tha@gmail.com or a.abdurabu@utm.my (H.A.T.); khamim@utm.my (A.K.I.)

Citation: Thabit, H.A.; Kabir, N.A.; Ismail, A.K.; Alraddadi, S.; Bafaqeer, A.; Saleh, M.A. Development of Ag-Doped ZnO Thin Films and Thermoluminescence (TLD) Characteristics for Radiation Technology. *Nanomaterials* **2022**, *12*, 3068. https://doi.org/10.3390/nano12173068

Academic Editors: Wei Chen and Derong Cao

Received: 1 August 2022
Accepted: 27 August 2022
Published: 3 September 2022

Publisher's Note: MDPI stays neutral with regard to jurisdictional claims in published maps and institutional affiliations.

Copyright: © 2022 by the authors. Licensee MDPI, Basel, Switzerland. This article is an open access article distributed under the terms and conditions of the Creative Commons Attribution (CC BY) license (https://creativecommons.org/licenses/by/4.0/).

Abstract: This work examined the thermoluminescence dosimetry characteristics of Ag-doped ZnO thin films. The hydrothermal method was employed to synthesize Ag-doped ZnO thin films with variant molarity of Ag (0, 0.5, 1.0, 3.0, and 5.0 mol%). The structure, morphology, and optical characteristics were investigated using X-ray diffraction (XRD), scanning electron microscope (SEM), energy-dispersive X-ray spectroscopy (EDX), photoluminescence (PL), and UV–vis spectrophotometers. The thermoluminescence characteristics were examined by exposing the samples to X-ray radiation. It was obtained that the highest TL intensity for Ag-doped ZnO thin films appeared to correspond to 0.5 mol% of Ag, when the films were exposed to X-ray radiation. The results further showed that the glow curve has a single peak at 240–325 °C, with its maximum at 270 °C, which corresponded to the heating rate of 5 °C/s. The results of the annealing procedures showed the best TL response was found at 400 °C and 30 min. The dose–response revealed a good linear up to 4 Gy. The proposed sensitivity was 1.8 times higher than the TLD 100 chips. The thermal fading was recorded at 8% for 1 Gy and 20% for 4 Gy in the first hour. After 45 days of irradiation, the signal loss was recorded at 32% and 40% for the cases of 1 Gy and 4 Gy, respectively. The obtained optical fading results confirmed that all samples' stored signals were affected by the exposure to sunlight, which decreased up to 70% after 6 h. This new dosimeter exhibits good properties for radiation measurement, given its overgrowth (in terms of the glow curve) within 30 s (similar to the TLD 100 case), simple annealing procedure, and high sensitivity (two times that of the TLD 100).

Keywords: dosimetry; Ag; ZnO; thermoluminescence; fading; linear response doses; sensitivity

1. Introduction

Thermoluminescence dosimeter, well-known as TLD, is a device that can keep radiation energy for a specific time and then read it out after being induced by heating. The process behind this device depends on the emission of light. TLD is a tool for measuring radiation doses in the clinical, radiotherapy, environmental, irradiated food, industrial, and quality assurance fields [1–8]. TLD remains the most powerful method to measure radiation doses due to its reliability, non-intricacy, and portability [1,3,9–11]. The thermoluminescence (TL) mechanism depends on impurities defects in the crystal structure, which increment the capacity of the materials to store radiation energy [6,11–13].

In recent years, many studies have been performed to establish novel high-performance TLDs that show a linear response at a wide range of doses, since most dosimeters exhibit nonlinear responses in a wide range of doses. TLD materials are microcrystalline powders or chips that can disperse light, where the near-surface light produced can approach the photon detector more than the light from within its depth. This result depends on the dosimeter's ability to keep light emitting through it. Most researchers focus on block form dosimeters such as pellets, chips, and discs [4,14–17]. In contrast, a few researchers have used limited efforts to investigate nanocomposites with a creative design of 2D thin films to serve as TL materials [18]. We see that it is of the utmost importance to consider modifying the dimensions of the dosimeter, to ensure that it is compatible with various applications and the critical regions for radiation measurements.

The development of thin films with dosimetric characteristics is of great importance in calculating low penetrating radiation doses, including in the study of dose distribution at interfaces. However, knowing the dose in the particular region and beyond clinics are significant. These are essential assets of knowledge to prevent unnecessary issues in the skin and treat the lymphatic system at a depth of 0.5 mm. According to the literature review, several attempts have been made to design two dimensional (2D) dosimeters [19–22]. Additionally, the 2D design of the dosimeter with a high resolution is essential in modern radiotherapy, for modalities where steep gradients of dose distributions occur. Moreover, overcoming the problems associated with measuring depth dosage distributions is desirable [14,23,24].

Hence, Zinc oxide, ZnO, is an excellent advantage in improving efficient TL phosphor tailored to be used in dosimetry applications, due to its wide bandgap of 3.37 eV, exciton binding energy of 60 MeV, and high transparency of 90% in the visible region [23]. Based on the characterization of ZnO, the surface area to volume ratio increases with the decrease in the nano-range of grain sizes, which changes the optical properties such as transmittance. Moreover, ZnO has inherent structures with different morphologies, i.e., nanoparticles, nanowires, and nanorods [24]. However, ZnO has weaknesses, such as a high electron-hole recombination rate. To address this limitation, doping with foreign atoms is crucial for modifying the characteristics and proposed uses of semiconductor nanocrystals. Adding impurities to ZnO will change the emission luminescence, as it creates defects in materials and increases the charge carriers. The results of ZnO doped by transition metals showed ZnO as a promising composite material in dosimetry, based on the TL glow curves, per the most recent literature review indicated in our previous work [25]. The introduction of Ag to ZnO caused the substitute of Ag^{2+} in the Zn^{2+} lattice, which increased the oxygen vacancies due to variation in the ionic radius between Ag (1.15 Å) and ZnO (.72 Å). These vacancies act as the sub-bandgap donor sites, producing traps; these sub-bandgap act as traps for the electrons during irradiation. Later by stimulating the nanocomposite, the trapped electrons tend to relax in the recombination center. A typical recombination center may be created by dislocating a negative ion that works as an electron trap; if this trap is shallow may be released by thermal vibrations of the lattice. On the other side, if the trap is deep (high activation energy), the electrons will recombine with the holes' trap at the recombination center, giving rise to light emission (TL).

Therefore, adding selective elements to ZnO offers a vital way to enhance and control optical and luminescence properties. According to the theory of valence control in oxide semiconductors, the Debye length (L_D) of ZnO is reduced when it is doped with acceptor elements such as Au, Cu, and Ag; Ag can promote the separation of spatially generated charge carriers. Furthermore, the unique interface interactions between Ag metal and ZnO may be related to the presence of the Schottky barrier, which promotes charge carrier separation. For instance, Huang observed that when Ag (NPs) is added to ZnO, electrons concentrate in ZnO along the Ag–ZnO interface until the electron-rich islands link. This difference in electron transport at Ag–ZnO is caused by the fact that the work function of ZnO (4.62 eV) is greater than that of Ag (4.24 eV), increasing electric conductivity up to 1000 times.

Furthermore, when the Ag content in the ZnO matrix increases, the electron concentration rises to 2.4×10^{20}/cm, causing the electron accumulation zones to overlap and forming a percolation channel for electron transport, without reducing electron mobility [26]. Similarly, Corro and their research group reported that adding Ag to pure ZnO increased the number of electrons in the conduction band (CB) of ZnO. This is caused by the interfacial electronic interactions between the metals and the ZnO. These impurities may increase the probability of localized electrons being trapped in de-traps (close to conduction band CB); this might occur because the electron transfer that emerged from Ag to ZnO caused the shifted absorption to a higher wavelength [27]. Saboor's findings demonstrated that the Ag-doping concentration significantly impacted the morphology, structure, and intrinsic defects of ZnO nanorods. As the modifier shifts the conduction band and Fermi level of the ZnO nanorods, it creates vacancies and forms ionic bonds with the oxygen atom rather than covalent bonds [28]. Likewise, incorporating silver into ZnO creates surface defects, which act as effective charge carrier traps to reduce the recombination rate of the photogenerated charge carriers. Thus, the addition of Ag to ZnO caused an increase in oxygen vacancies sites, due to the electron sensitization effect of Ag; as a result, an improvement in TL intensity can be obtained [29]. However, scholars are still making intensive efforts to improve the TL properties of these phosphors, either by preparing them in various ways, doping them with different impurities, or introducing new matrices, with ZnO being one of these new host materials. Although many researchers have been dedicated to developing a ZnO-based dosimeter, to our knowledge, no study has been reported that covers all of the features [30–34]. This study comprehensively investigated the dosimeter characteristics of Ag-doped ZnO thin films grown via the hydrothermal method.

2. Experimental Section

2.1. Materials

All of the compounds chosen for this research were conducted without additional purification. Zinc acetate dihydrate ($Zn(CH_3COO)_2 \cdot 2H_2O$) (BHD, Poole, UK), silver nitrate $AgNO_3$ (Sigma, Ronkonkoma, NY, USA), hexamethylenetetramine ($C_6H_{12}N$) (HMTA, Merck, Darmstadt, Germany). The ammonia (NH_3) and ionized water were available in the laboratory of School of Physics Universiti Sains Malaysia.

2.2. Preparation of Ag-Doped ZnO Thin Films and Measurements

The simple hydrothermal method was employed to synthesize Ag-doped ZnO thin film growth on the glass. The solution was prepared by mixing 0.1 mol of Zinc acetate dihydrate and 0.1 mol of hexamethylenetetramine in 100 mL of ionized water. After that, the mixture was vigorously stirred for half hour. $AgNO_3$ (0, 0.5, 1.0, 3.0, and 5.0 wt%) was introduced to the prepared solution and continuously stirred for 1 h to obtain a homogeneous solution. Then, NH_3 was added dropwise into the solution until the transparent solution was acquired. The solution was then poured into 150 mL Teflon autoclave, with the immersion of the glass substrate (slab glass that deposited the thin films on it), before it was closed tightly and heated at 180 °C in the furnace for 12 h, after which it was cooled to ambient temperature. The thin film was rinsed several times using deionized water. Finally, the sample was annealed at 400 °C for 1 h in the furnace for impurity removal. This procedure was performed several times for different concentrations of Ag, as previously mentioned.

Meanwhile, the structure analysis of pure ZnO and Ag-doped ZnO thin films was measured by X-ray diffraction (XRD) (Malvern PANalytical, Malvern, UK) and equipped with a Cu-K emission wavelength at (0.154 nm), for the range of 2θ at $20° \leq 2\theta \leq 80°$. Besides that, the morphology of the thin films was studied by scanning electron microscope (SEM) (JSM-6460LV SEM JEOL Ltd, Tokyo, Japan). The nanocomposite of the thin films was depicted via energy-dispersive X-ray (EDX) (Tokyo, Japan) analysis. In addition, photoluminescence (PL) (Jobin Yvon HR 800 UV Jobin Yvon, Kyoto, Japan) was investigated using a laser with a power of 0.18 mW, and the excitation source's wavelength was 325 nm. Ultraviolet visible (UV–vis) spectrophotometry (Model Cary 5000 UV-VIS-NIR Agilent Technologies, Santa Clara, CA, USA) was applied to measure the transmittance.

2.3. Thermoluminescence Measurements

In this study, the process of preparing the samples was carefully considered; the weighing by difference technique was applied to measure the mass of the thin films. After that, the thin film was cut into 5 × 5 mm². The samples were then labeled and placed in opaque containers. Before irradiating the samples, the annealing dosimeter procedure was first conducted to remove any traps. Annealing treatment is a method used to eliminate the residual signal, which may cause unwanted background readings in this work. The samples were kept under specific circumstances, to shield them against external physical and environmental effects such as dust and cleaning. Figure 1 shows the setting of the samples for irradiation using an X-ray machine (Toshiba KXO-50S) (Toshiba Medical Equipment, Tokyo, Japan) with its control panel setting, as mentioned in our previous study but with a slight modification, where the SSD and the field size were set at 80 cm and 10 × 10 cm², respectively [35]. In the present work, the temperature–time profile was preheated at 50 °C, and the maximum temperature was 400 °C. The heating rate was carried out within the intervals of (1, 3, 5, 7, and 10 °C/s) for all three compositions, where the samples were irradiated by X-ray radiation, and each data point is an average of three samples to estimate the average TL intensity and standard deviation. Win REMS application software (USA) for TLD readers involved two phases of preheating.

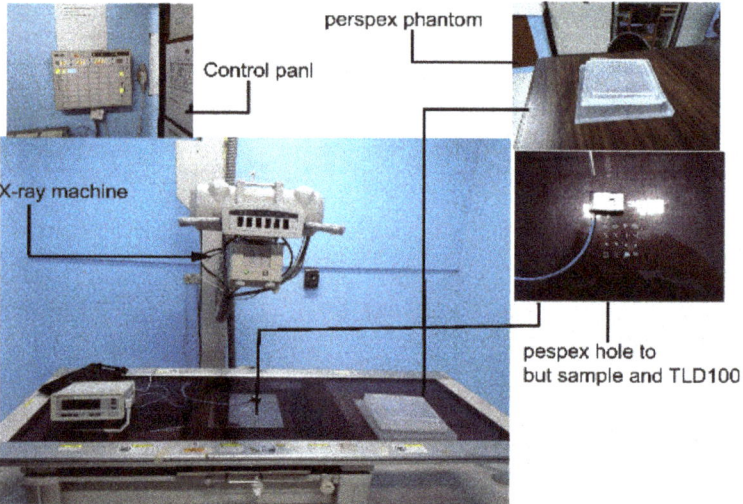

Figure 1. The schematic diagram of the experimental set-up for X-ray irradiation dose using Toshiba KXO-50S X-ray.

There are two steps in the readout (acquisition phase): collecting light emitted during the heating process and converting the light into an integrated value (display glow curve). In order to measure the percentage depth doses (PDD), 10 Perspex phantom slides were used, where the thickness of each slide was 1 cm. HARSHAW TLD Model 3500 (Thermo Fisher, Waltham, MA, USA) was utilized. Initially, the reader was warmed for 30 min, and nitrogen gas was turned on at a flow rate of 3.0 to 4.0 psi. The planchet's nitrogen flow improves the accuracy of low exposure reading and extends the planchet life by eliminating the oxygen in the planchet area. The time temperature profile (TTP) included the heating cycle parameters and was set in the Win REMS software. The TL charge was collected from each reading cycle in 200 data points.

3. Results and Discussion

3.1. Structural, Morphological, and Chemical Composition Investigations

Figure 2 displays the structure and crystallinity of undoped ZnO and Ag-doped ZnO thin films using XRD ($20° \leq 2\theta \leq 80°$). The results showed polycrystalline with a hexagonal wurtzite structure for the Ag-doped sample [36]. The diffraction peaks corresponding to

ZnO and Ag appeared to agree with the standard JCPDS data card 01-089-0511 and 03-065-2871, respectively. The peaks for all samples indicate that Ag-doped ZnO thin films have grown successfully. The peaks of 100, 002, and 101 planes correspond to 2 θ = 31.77°, 34.37°, and 36.35° of ZnO, respectively. The peak at (002) orientation is the highest, indicating the growth direction and c-axis orientation. In the meantime, the diffraction peak of 111 at 2 θ = 38.22° is consistent with Ag for Ag-doped ZnO thin films. No evidence of impurity peaks was observed from the XRD data [37].

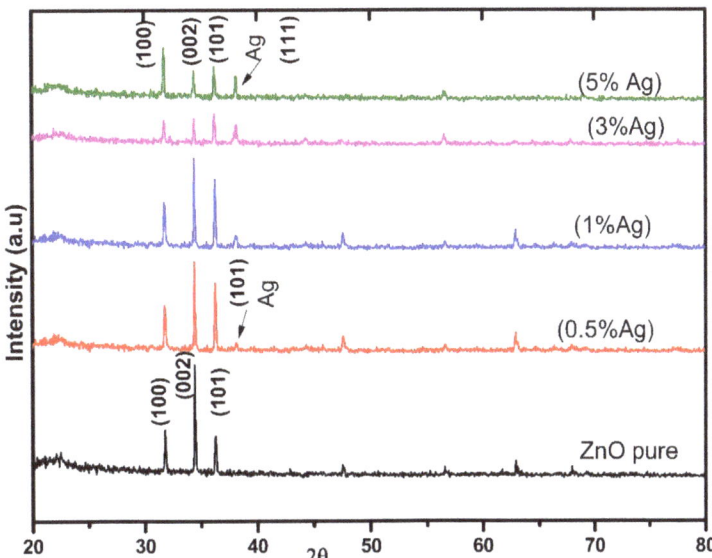

Figure 2. The X-ray diffraction pattern of undoped ZnO and Ag-doped ZnO thin films.

The results showed that the diffraction peaks connected to ZnO decreased along with the Ag dopant increment, due to the disorder formed by the Ag ions in the ZnO lattice structure; moreover, the dopant's chemical reactivity plays a role in crystal growth dynamics. As a result, the added Ag atoms cause a deformation in the crystalline structure of ZnO [38]. However, this deformity may be ascribed to Ag^+ having a considerably larger ionic radius (1.22 Å) than Zn^+ 0.72 Å, leading to the segregation of Ag atoms to the grain boundaries of ZnO crystal, indicating Ag cluster formation that appeared as separate Ag peaks. Consequently, a metallic phase of silver was produced on the surface of ZnO [28,39].

The crystallite size was estimated using Scherrer's formula and was tabulated in Table 1:

$$D = 0.89\lambda / \beta \cos\theta \quad (1)$$

As seen in Table 1, the TL intensity increases as the crystallite size decreases. The change of TL is consistent with that of the surface fluorescence, which increases as the surface area to volume ratio increases, as we mentioned in the introduction. When the nanocomposite is exposed to radiation, the electron and hole will be created and trapped in these metastable states. Hence, it is obvious that the TL of the nanoparticles is proportional to the surface defects. The surface-to-volume ratio increases when the size decreases, and the particles go to more easily obtainable carriers, which increase the holes and electrons for TL emission. There is still much to learn about the aspects of the upconversion of luminescence in these nanomaterials [40–43]. However, by increasing the dopant, the intensity decreased. Researchers ascribed the decrease in luminescence intensities to the quenching effect as an increasing dopant, which is consistent with this study [44,45].

Table 1. Calculation of the crystallite size of ZnO and Ag-doped ZnO thin films.

Samples	2 θ (°)	FWHM (rad)	Crystallite Size (nm)
Pure ZnO	31.80	0.147	36.21
1% Ag	31.79	0.1476	35.9
3% Ag	31.52	0.197	27.0
5% Ag	31.76	0.246	21.5

Figure 3's SEM images show the surface morphology of pure ZnO and Ag-doped ZnO thin films. One-dimension nanostructure hexagonal shapes were successfully synthesized through the hydrothermal method. The effect of the amount of Ag on the morphology of ZnO nanorods is observed. Based on the obtained results, the nanorods' length is 5 μm, as confirmed from the cross section (image inset of Figure 3b). The surface morphology of Ag-doped ZnO became deformed and flattened by increasing the Ag dopant. The morphologies of thin films are affected by adding Ag^+, which may be replaced with Zn^{2+}. Further increase in impurities caused the agglomeration of Ag atoms at the grain boundary of ZnO crystal, suggesting the creation of Ag clusters that appeared as separate Ag atoms, as can be seen in Figure 3c,d [46].

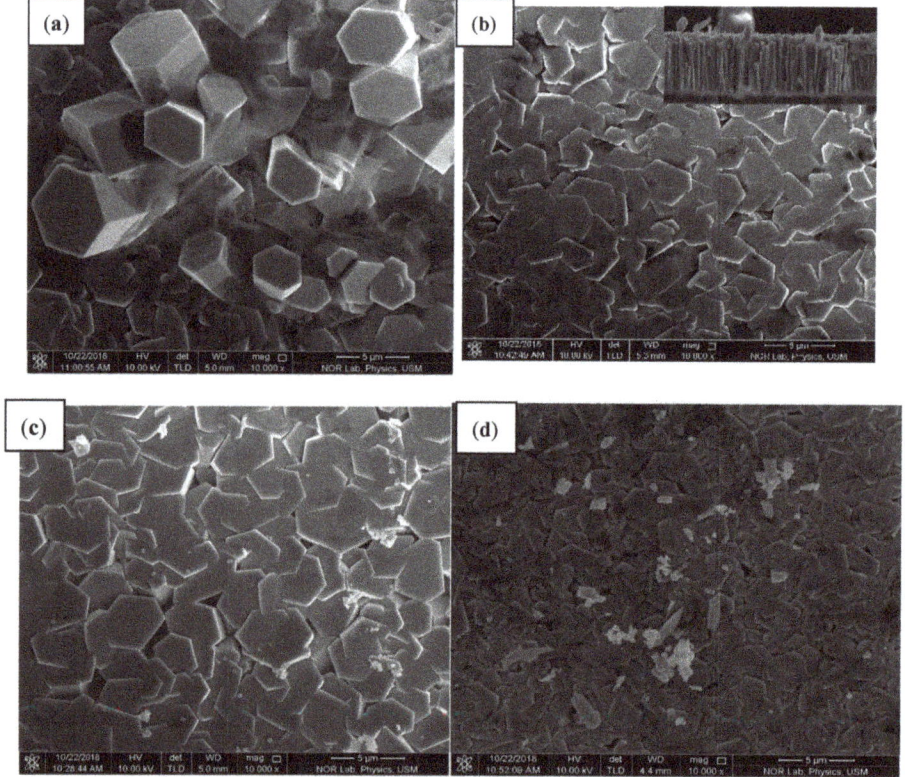

Figure 3. FE-SEM images of the (**a**) pure ZnO and silver dopant at (**b**) 0.5%; (**c**) 1%; and (**d**) 3%, respectively.

Energy-dispersive X-ray (EDX) spectroscopy was used to evaluate the chemical compositions of pristine ZnO and Ag-doped nanorods. The results showed that the sample comprises Zn and O elements; no additional elemental peaks were seen in the pristine ZnO analysis. Ag impurities have also been fully incorporated into the ZnO lattice. Figure 4 showed the elemental synthesis of ZnO and Ag-doped ZnO thin films via EDX and revealed a good agreement with the experimental composite [47]. The mapping images exhibited a uniform and homogeneous distribution of Ag$^+$ ions in ZnO.

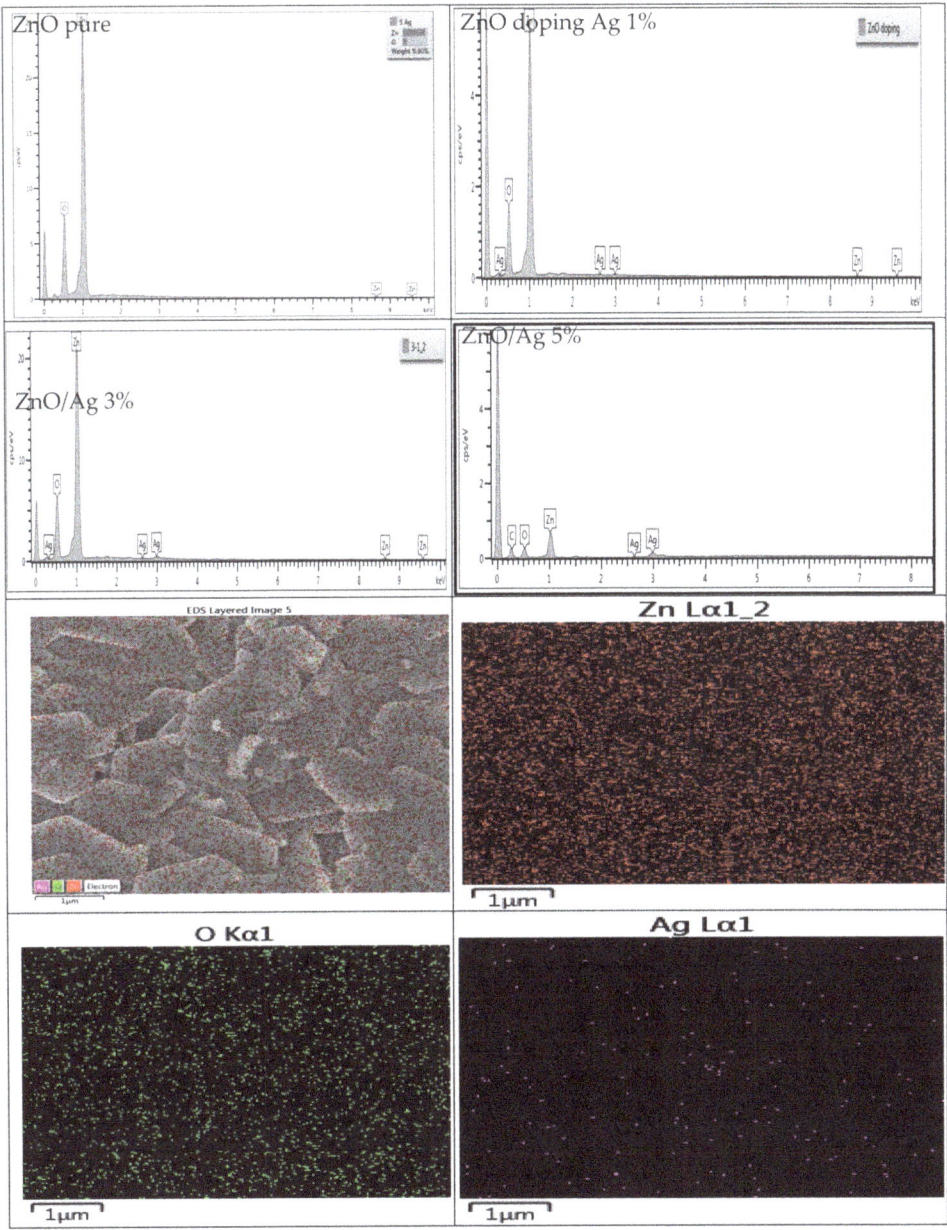

Figure 4. The EDX spectrum for undoped and Ag-doped ZnO thin films with typical mapping images.

3.2. Optical Properties Studies

The PL spectra of undoped ZnO and Ag-doped ZnO thin films at room temperature were recorded. Figure 5a shows the PL spectra of pure ZnO and Ag-doped ZnO as-synthesized by hydrothermal deposition at room temperature, after annealing at 400 °C for 2 h. The PL study was conducted at room temperature with an excitation wavelength of 325 nm xenon-lamp Laser. The obtained results revealed two peaks that included the near bandgap (NBE) (3.26 eV), due to the collision process of excitons. The other highest intensity peak in the visible region is defined as deep level emission (DLE) with a wide range (2.25 to 1.46 eV); this was mainly attributed to the electron-hole recombination, which gave rise to the surface and intrinsic defects in the crystalline structure lattice during the growth. In addition, the DLE is the result of various developmental defects, such as zinc interstitials (Zn_i), interstitial oxygen (O_i), zinc vacancies (V_{Zn}), and oxygen vacancies. These defects cause the DLE (V_O), and the red band at 1.6 E_v was observed due to the interface between O_i and V_{Zn} emissions [48–51].

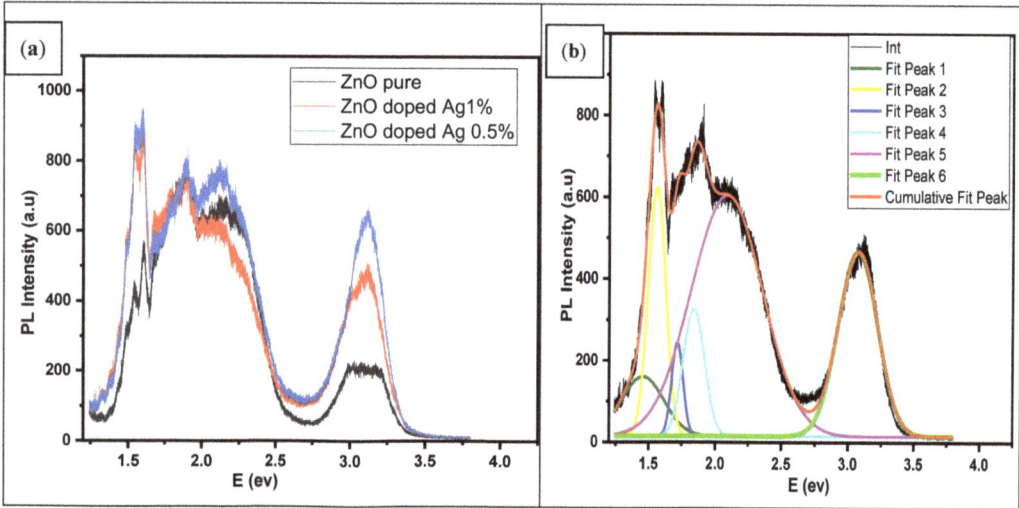

Figure 5. (a) Room temperature PL spectra of pure ZnO and Ag-doped ZnO 1%, 0.5% mol thin films, and (b) Gaussian deconvolution of Ag-doped ZnO thin film.

For the Ag-doped ZnO thin film, there is an increase in the intensity of NBE without shifting the peak's position. The enhancement in UV intensity after annealing may be attributed to excitonic recombination; we can explain it as due to the implanted Ag atoms; thermal annealing provides energy to occupy Zn atom sites in the lattice of ZnO. When no Ag atoms are stuck in non-equilibrium positions, the occupying probability could increase the temperature and gradually become steady at a specific value. The UV light in ZnO crystal can excite photocarriers [51,52]. However, the optimized intensity was 0.5 mol%, due to the electron-hole recombination as an electron transporting layer and the oxygen vacancies mechanism. With the increase in Ag molarity beyond 0.5% of the Ag amount, the PL intensity tended to reduce, which may be due to the surface plasmon resonance (SPR) of Ag, as reported in a prior study by [53].

The cause of the red region is still controversial. Due to the diverse array and complexity of defects existing in ZnO, some authors have attributed it to ZnO structure defects. In contrast, some other authors attributed it to an excess of oxygen impurities [54]. To understand the characteristics of the broad visible emission band, the deconvolution of the components band was applied via Gaussian fitting, see Figure 5b. The deconvolution peak at 2.10 (eV) may be ascribed to the V_{Zn}–V_O vacancies that other scholars assigned to positively charged oxygen vacancy V_{O++} [54–57]. Moreover, we observed other peaks at 1.84 (eV) and 1.7 (eV) assigned as the yellow and orange emissions in ZnO, respectively, attributed to neutral V_O and interstitial zinc atoms Zn_i [58–61]. The emission bands in the 1.7 (eV), 1.56 (eV), and 1.46 (eV) were caused by different types of defects, such as interstitial zinc, zinc vacancy, oxygen vacancy, and interstitial oxygen [56,62–64]. The emission bands have been improved by oxygen heat treatment and Ag-doping, suggesting that this emission band is due to Ag–O clusters [18].

Figure 6 shows the transmittance spectra of ZnO and Ag-doped ZnO over 200 to 800 nm. The results revealed that the transmittance decreased with the increment of Ag, which may be ascribed to the scattering of photons by crystal defects formed by the Ag dopant, agglomeration density, and adsorption of free carriers [65].

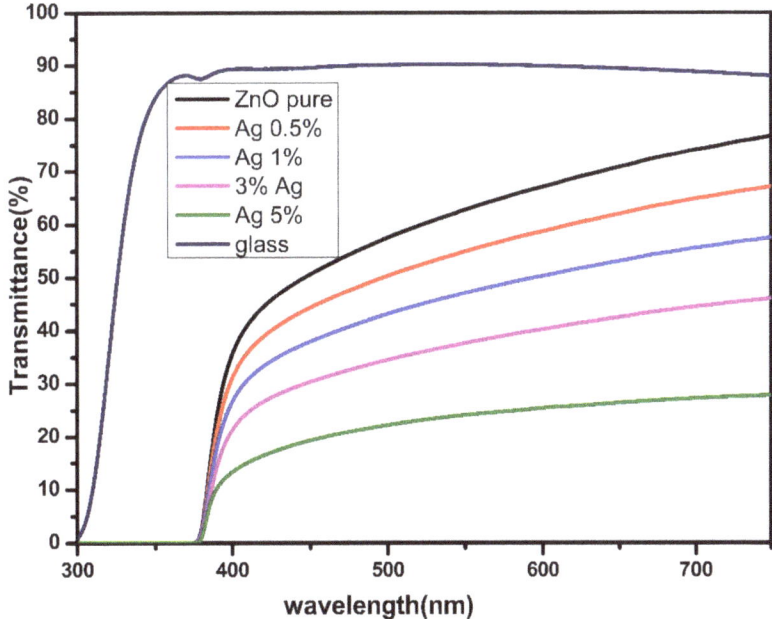

Figure 6. Optical transmittance for undoped ZnO and Ag-doped ZnO thin films.

Figure 7 shows the bandgap (E_g) of the samples. From the above transmittance, the absorption coefficient (α) was determined based on the following equation:

$$\alpha = -LnT/d \tag{2}$$

where T is the transmittance, and d is the thickness of the thin films. Meanwhile, the bandgap can be measured by extrapolating the linear portion curve of $(\alpha h\nu)^2$ versus the photon energy ($h\nu$), according to the following equation:

$$\alpha h\nu = (h\nu - Eg)^m \tag{3}$$

where E_g is the gap energy; $h\nu$ is the energy of the photon; α is the calculated absorption coefficient from the raw transmittance data; and $m = 1/2$ for the plotted direct transition bandgap $(\alpha h\nu)^2$ vis photon energy ($h\nu$).

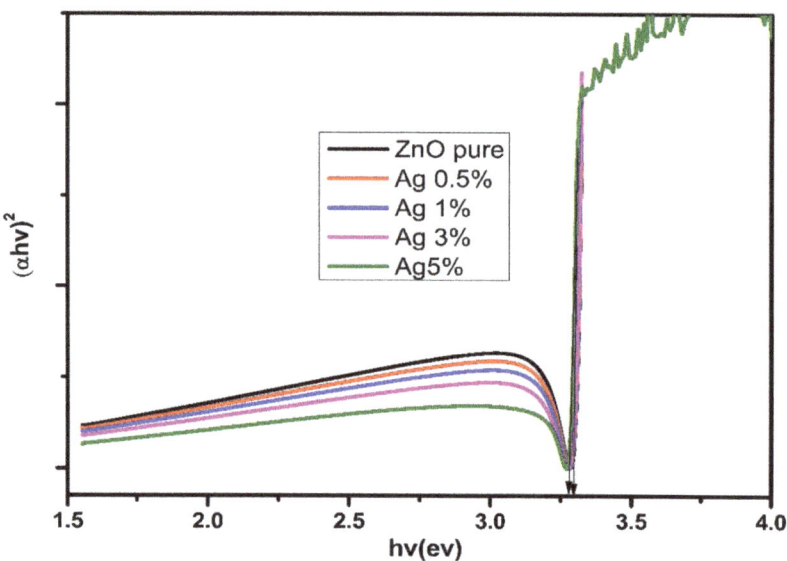

Figure 7. The bandgap for undoped ZnO and Ag-doped ZnO thin films.

The energy gap for ZnO recorded 3.2 eV, slightly decreasing with the addition of Ag. The decrease in E_g with the increase in Ag may signify Ag^+ was substituting for Zn^+ in the lattice [66]. The production of oxygen vacancies plays a vital role in reducing the bandgap, as they serve as trap centers that minimize the recombination of charge carriers by capturing the electrons. Furthermore, the addition of Ag may cause defects in the bandgap, which can broaden the spectrum and promote emission in the visible range, leading to improved luminescence characteristics of Ag-doped ZnO nanocomposites [67].

3.3. Dosimetry Characteristics

3.3.1. Sample Optimization

The samples were exposed to 3 Gy of X-ray radiation. Figure 8 displays the glow curve of the dosimetric peak for ZnO and Ag-doped ZnO thin films. The acquired glow curve exhibited a single peak sited at 240–270 °C with 5 °C/s of the heating rate. The results showed that the highest intensity of the thermoluminescence corresponded to a 0.5 mol% Ag-doped ZnO sample at 270 °C. This temperature peak is a desirable site for the dosimetry application because, if located at a low temperature, it will cause an increase in the fading and a loss of the signal, similar to if the dosimetric peak is situated at a high temperature, which leads to an interface with black body radiation as reported [25,68]. As the concentration of Ag decreased, the TL intensity appeared to increase, with a noticeable shift toward higher temperature.

The emission of TL intensity for Ag-doped ZnO increased when the dopant ratio reached 0.5%. Beyond this concentration, the Ag atoms became agglomerated into metallic Ag clusters, reducing the surface defects and, thus, quenching the TL intensity [69,70].

Another fact is that Ag-doped ZnO does not create a new TL peak but increases the trap center, increasing TL intensity. For more explanation of the TL mechanism of Ag-doped ZnO, when the samples were irradiated (the photon energy of X-ray was more significant than the bandgap of ZnO), the electrons in the valence band of this nanocomposite excited to the conduction band (e^-) simultaneously produce an equal number of holes (h^+) in the valence band. These electrons and holes will relax at the tarps created by the oxygen vacancies and Ag+ ions, preventing the immediate recombination of $(e^- - h^+)$ pairs. Consequently, the electrons and holes will stay trapped for a long time, depending on the lifetime of the levels trapped and the ambient temperature; later, when stimulated, the samples via the temperature, electrons, and holes will recombine at the recombination center, and emission photons will occur. The optimized sample of 0.5%Ag-doped ZnO was chosen for further investigations.

$$ZnO/Ag + + \text{radiation} \rightarrow ZnO/Ag\ (e_{cb}^- + h_{vb}^+)$$

$$(e_{cb}^- + h_{vb}^+) \xrightarrow{\text{recombined at LM center}} \text{emission light}$$

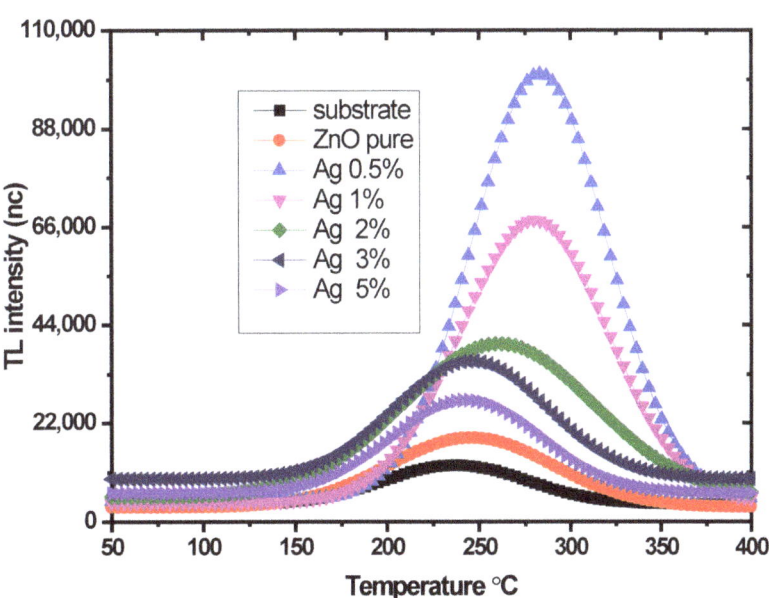

Figure 8. The glow curves of Ag-doped ZnO thin films with different concentrations of Ag percentage.

3.3.2. Heating Rate

The process of determining the heating rate is crucial in order to acquire the optimal recombination and sensitivity [71]. Figure 9 depicts the glow curve with a different heating rate of the optimum sample. For this study, each data value identified an average of three dosimeters samples to estimate the average TL intensity and a standard deviation. The prepared samples were automatically heated at 50 °C by the TLD reader before recording the glow curve, and the temperature was gradually incremented at the heating rate of 1 °C/s up to 10 °C/s. Ideally, the optimum heating rate should achieve the highest TL intensity with the lowest standard deviation [72]. As shown in Figure 10, the optimum heating rate corresponding with the highest TL intensity is set at 5 °C/s. It was observed that the intensity of the glow curve with the different heating rates of the Ag-doped ZnO film decreased when increasing the heating rate. The glow curve is connected to trap levels in the bandgap between the conduction bands and the valence bands of material at varying depths. These trap levels are distinguished by kinetic parameters such as activation energy, live time, and frequency factor. Besides, the reduction in intensity with an increase in heating rate can be attributed to the association between the time taken to de-trap the electrons trapped and the number of electrons de-trapped due to thermal stimulation [73].

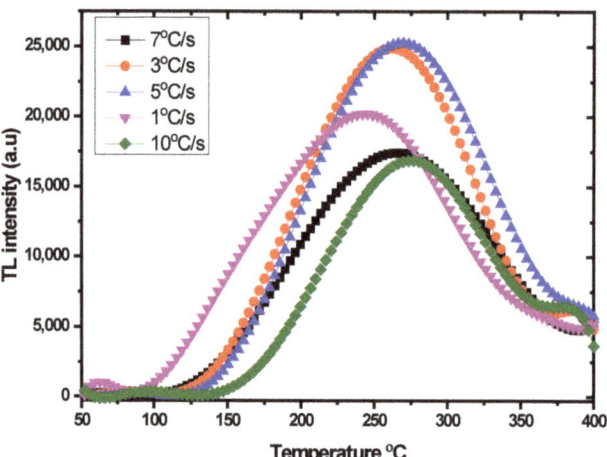

Figure 9. Glow curve of Ag-doped ZnO thin films as a function of heating rate.

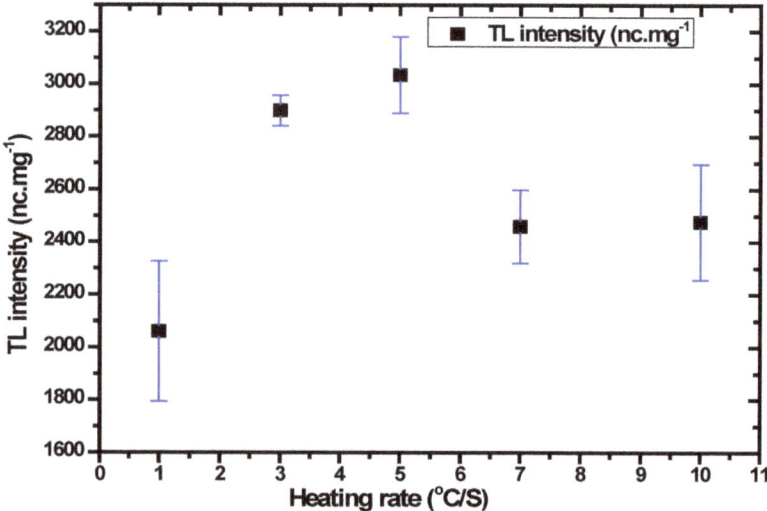

Figure 10. TL response of Ag-doped ZnO thin films as a function of heating rate.

3.3.3. Annealing Procedures

Annealing treatment is a method used to eliminate the residual signal of radiations, which may cause unwanted background readings; this process is essential to yielding a precise TLD outcome [74] The annealing procedure in the current study was carried out using different temperature values (100–400 °C) at a specific annealing time (1 h). In this investigation, each data value indicated an average of three dosimeter samples. Following that, the samples were exposed to 3 Gy of X-ray radiation.

Figure 11 displays the best TL response with a minimum standard deviation of the Ag-doped ZnO (0.5 mol%) at 400 °C. The samples were subjected to an annealing temperature at 400 °C with different annealing times (from 20 to 60 min), before being exposed to 3 Gy of X-ray radiation. As shown in Figure 12, the optimum TL response and low standard deviation were found at 30 min. These annealing processes demonstrated that all traps had been evacuated and restored the thermodynamic defect equilibrium before the irradiation [75]

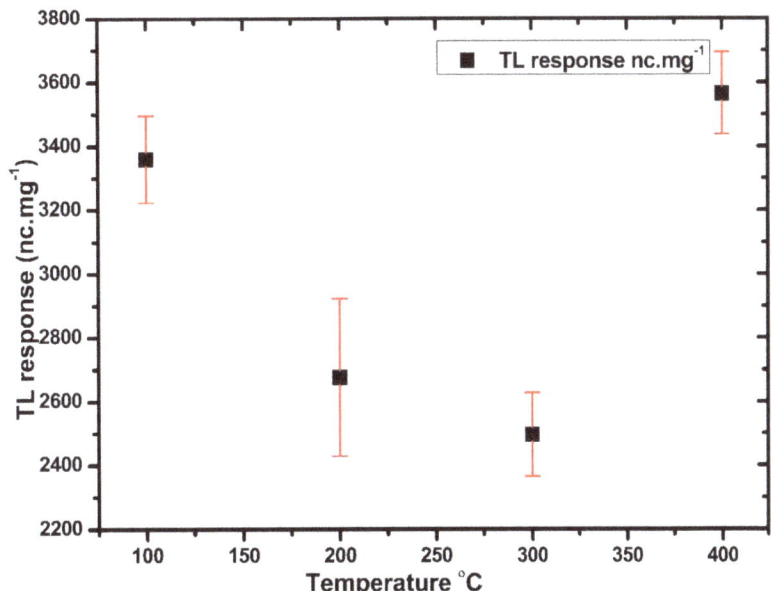

Figure 11. The TL response of Ag-doped ZnO thin films with different annealing temperatures.

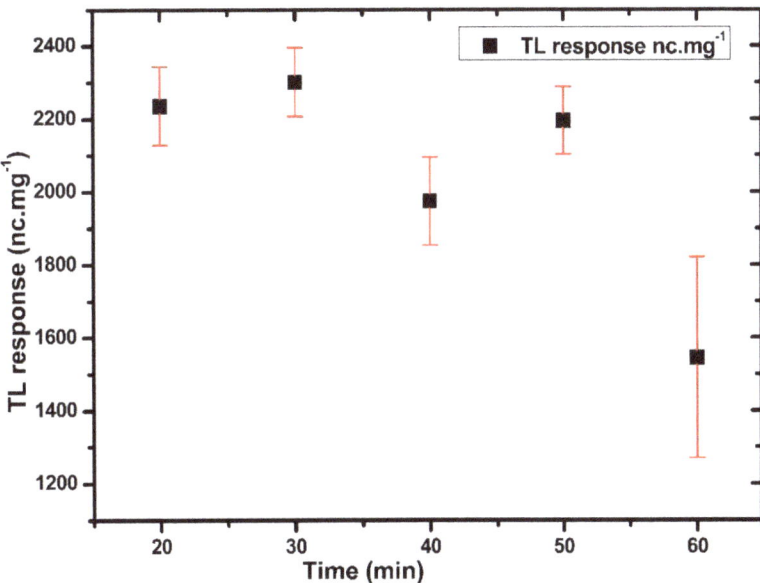

Figure 12. The TL response of Ag-doped ZnO thin films with different annealing times.

3.3.4. Dose-Response

The study investigated the TL response for the Ag-doped ZnO thin films. Each data point involved three samples; the average and standard deviations were taken. Figure 13 shows the linear response doses (0.1–4 Gy) for Ag-doped ZnO thin films. The study revealed an excellent linear response and correlation coefficients of 0.97 and 0.996 for Ag-doped ZnO thin films and TLD 100 chips, respectively. The results indicated the distribution of deep traps in Ag-doped (0.5 mol%) ZnO thin films located at different levels [76]. The Ag-doped ZnO thin films and TLD 100 chips were irradiated under the same condition for comparison (to the material host). The response of TLD 100 chips appeared linear within the range (0.1–4 Gy), as expected. This finding confirmed that the phosphorus established deep traps within the nanocomposites lattice, which is proportional to the various doses [77].

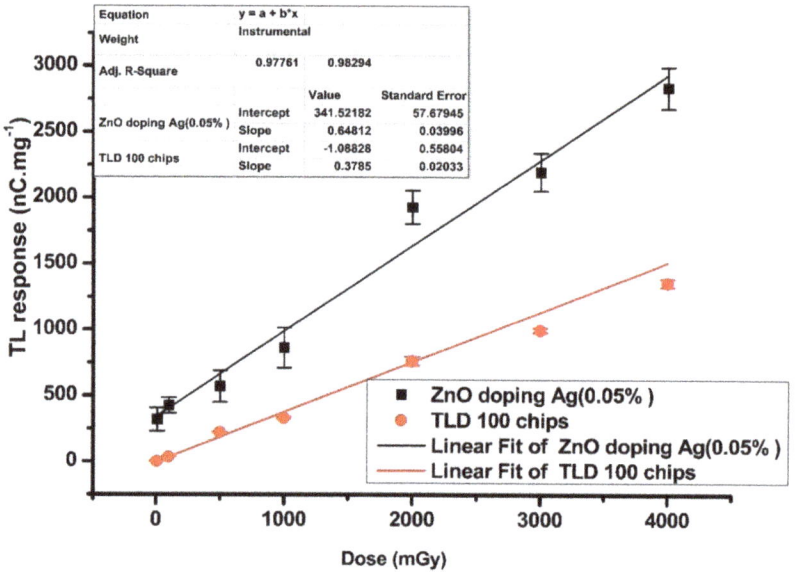

Figure 13. The linearity of response dose irradiated with X-ray (0.1–4) Gy.

3.3.5. Sensitivity

One of the essential characteristics of TLD properties is dosimeter sensitivity TL. Sensitivity is defined as the TL intensity per unit mass of the thin film and per unit dose of X-ray radiation (TL.mGy^{-1}.mg^{-1}), as shown in Equation (4).

$$S(D) = TL/m.D \qquad (4)$$

where TL is the intensity (expressed in nC), and $m.D$ is the dose (expressed in mGy).

Relative sensitivity, another term linked to the dosimeter material sensitivity, is often compared to the standard (TLD 100). The following equation expresses relative sensitivity [78].

$$R(D) = S(D)_{material}/S(D)_{TLD\ 100} \qquad (5)$$

The sensitivity could be obtained based on the slope of the graph of response doses (linearity) [79]. The sensitivity of 0.5 mol% Ag-doped ZnO appeared to be approximately two times that of the TLD 100 chips, which agreed with the past reported studies.

3.3.6. Reproducibility

Reproducibility is another significant aspect of TL material that should stay stable even after repeated usage. The samples were irradiated to 4 Gy and stored for 24 h at room temperature. The numerous cycles of TL reading were recorded. As a result, Figure 14 shows no considerable difference in the results after eight cycles and no changes in transparency or color. The results demonstrated remarkable stability of using the material as a dosimeter [80].

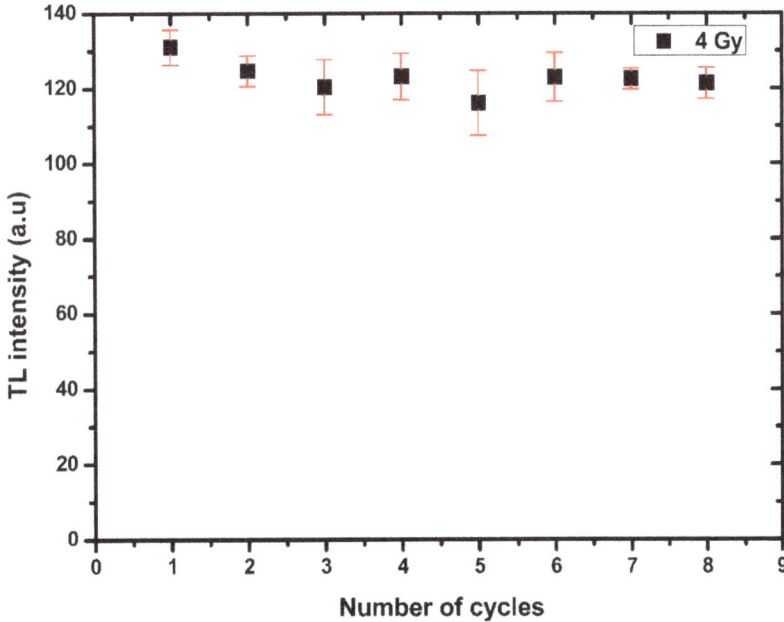

Figure 14. Reproducibility of Ag-doped ZnO thin films irradiated 4 Gy with X-ray.

3.3.7. Thermal Fading

Thermal fading is defined as the loss of stored thermoluminescence TL intensity during the storage after irradiation. Typically, the thermal stability of the dosimetry material is based on two categories: annealing, storage temperature, readout, radiation types, and time, which are discussed in the previous terms. Second, the thermal stability is strongly dependent on the nature of the material; in this context, the activation E energy of the depth trap should be $(E > kT)$, to liberate an electron from a trapping center.

The desirable dosimeters possess a glow curve that starts upward after 140 °C and reaches a maximum peak between 200–250 °C. However, commercial TLD materials possess fading. The fading may occur through defects or be due to their recombination in trapped holes [81–83]. The thermal fading property of the new host TL material in this work was examined. In this case, 0.5 mol% Ag-doped ZnO was exposed to 1 Gy and 4 Gy. The irradiated TL materials were divided into two groups, where each group had 35 samples and was stored in an opaque box (dark environment) at room temperature. The readout of TL started directly after 1 h and continued for 45 days.

Figure 15 displays the fading curve behavior of the samples. The signal loss was recorded at 8% for 1 Gy and 20% for 4 Gy in the first hour. After 45 days of irradiation, the signal loss was recorded at 32% and 40% for the cases of 1 Gy and 4 Gy, respectively. The residual signal noticeably decreased with the increase in the dose value of the X-ray radiation.

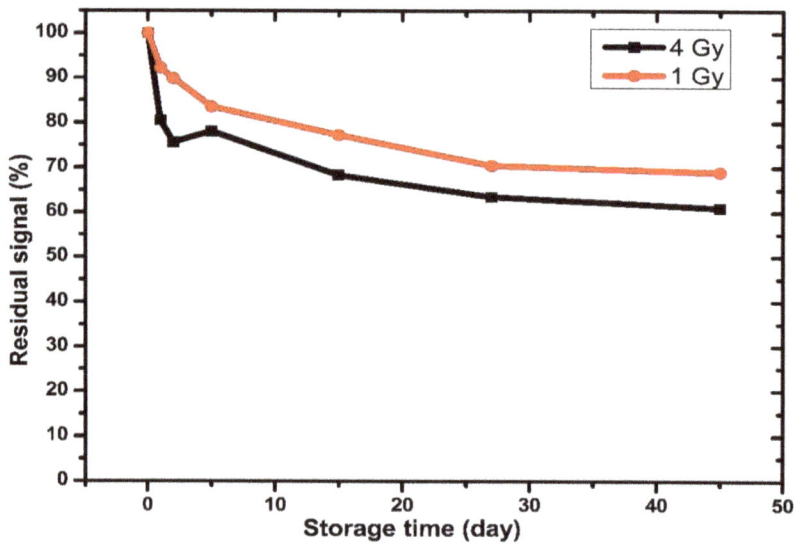

Figure 15. Thermal fading characteristics of Ag-doped ZnO thin films exposed to X-ray.

3.3.8. Optical Fading

Optical fading is another significant parameter for TLD materials; however, the optical fading dramatically depends on the light intensity, wavelength, and time exposure [83]. In the current study, optical fading was measured to estimate the sensitivity of the Ag-doped ZnO thin films to sunlight and room light. The samples were initially kept in a dark container during all the time measurements. Two groups of 30 samples were irradiated by 4 Gy of X-ray radiation and numbered, where each data point identified the outcomes of five samples. For six hours, the first group of samples was exposed to sunlight, while the second group was exposed to room light (by fluorescent lamp).

Figure 16 shows the behavior of the optical fading of Ag-doped ZnO thin films. The stored signals for the first group of samples yielded a loss of 53% after 1 h and 70% after 6 h of direct sunlight exposure in ambient conditions. As for the second group of samples, the stored signals yielded a loss of 30% after 1 h and 46% after 6 h. The trapped electrons or holes can be released optically at low temperatures, which suggests that further recombination could occur between the opposite charge carriers with an increasing absorbed temperature. Thus, a decrease in the TL intensity is expected when the proposed dosimeters are directly exposed to sunlight or room light [84]. This behavior has been observed in our previous works with multilayer thin films and nanopowder (pellets). The results revealed that the samples exposed to sunlight lost the signal more than the samples exposed to room light; this shrinking in the stored signals is attributed to the UV in sunlight. In light of this, the study concluded that the TL dosimeters should be stored in opaque containers when utilized [25].

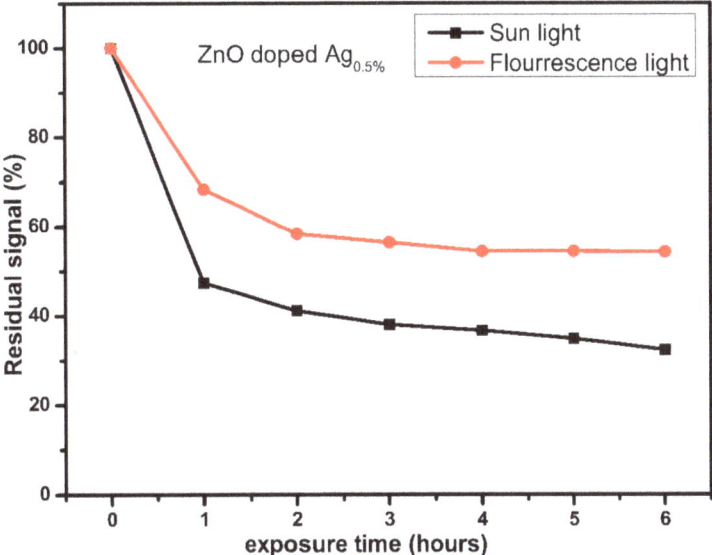

Figure 16. The optical fading of Ag-doped ZnO thin films exposed to sunlight and fluorescent light.

3.3.9. Minimum Detectable Dose (MDD)

For this study, the minimum detectable dose (MDD), also known as the lowest level detection, was calculated using Equation (6):

$$D_0 = (B^* + 2\sigma B)F \qquad (6)$$

where σB is the standard deviation of the background; B^* is the average background TL (zero dose reading); and F is the calibration factor expressed in Gy nC^{-1}.

In the current study, the five samples were read to measure the background before irradiation and exposed to 1 Gy of X-ray radiation. The average background and standard deviation for Ag-doped ZnO thin films were 0.414 and 0.099 nC, respectively. Additionally, the calibration factor recorded 0.0168 Gy nC^{-1}, substituting Equation (7). Consequently, the low detectable dose of the Ag-doped ZnO thin film was found to be 10.31 mGy. Table 2 shows the MMD for the Ag-doped ZnO thin film. The type of model of TLD reader is a significant factor in determining MMD [85].

$$F = Dose\,(Gy)/TL\,(nC) \qquad (7)$$

Table 2. The minimum detectable doses (MMD) for the Ag-doped ZnO thin film.

1 Gy X-ray Irradiated		Ag-Doped ZnO	
Samples	BG nC	TL Signal nC	TL–BG
1	0.27	63.27	63
2	0.36	59.62	59.26
3	0.47	48.94	48.47
4	0.52	65.28	64.76
5	0.451	61.85	61.399
Average	0.4142	59.792	59.3778
STDV	0.099238		
		F	0.016841311
		D_0	0.01031827

3.3.10. Percentage Depth Doses (PDD)

The percentage depth dose (PDD) is vital to determine the dose delivery, especially for cases that involve applying a narrow beam and a small field size [86]. PDD is given by the division of the absorbed dose at any depth, D, to the absorbed dose at a specific reference dose, which is along the central axis of the beam, as shown in Equation (8):

$$\%PDD = D_d/D_0 \qquad (8)$$

In this study, X-ray radiation (80 kVp, 100 Ams) was applied to determine the depth dose distribution and compare the use of a PTW Markus parallel plate chamber, TLD rods, and TLD 100 chips, a host material of Ag-doped ZnO thin films. As previously described in Figure 1, with a slight difference, the procedure was set up with a source-to-surface distance (SSD) of 80 cm and a field size of 20×20 cm^2. Furthermore, the phantom consisting of Perspex slabs included a slab with a thickness of 1 cm, where 10 slabs were used to get a maximum depth of 10 cm. The thin film set up in a slab with a thickness of 2 mm was placed in square holes, and the dose delivery was 3 Gy. The ionizing chamber, TLD rods, TLD 100 chips, and Ag-doped ZnO thin films were placed at different depths (from 0 to 10 cm).

Figure 17 shows the performance of the measurements at different depths, from the surface D_0 up to $D = 10$ cm. The depth doses (each data point representing five samples' outcomes) were normalized to $D_0 = D_{max}$. Ag-doped ZnO and all references recorded D_{max} at a depth of 0 cm. The PDD appeared to gradually decrease from the surface until the depth of 6 cm. The PDD values of Ag-doped ZnO at a depth of 5 cm were higher than the PPD values of the TLD 100 chips, ionization chamber (IC), and TLD rods, which were found to be 28.33%, 24.33%, 27.00%, and 25.00%, respectively. The variation in the values of depth dose is subjected to many factors, such as the location of effective point samples, field size (may be due to the scattering of radiation), measurement of detectors, presence of air gap between detectors and layers of Perspex phantom, and the Z_{eff} of the materials.

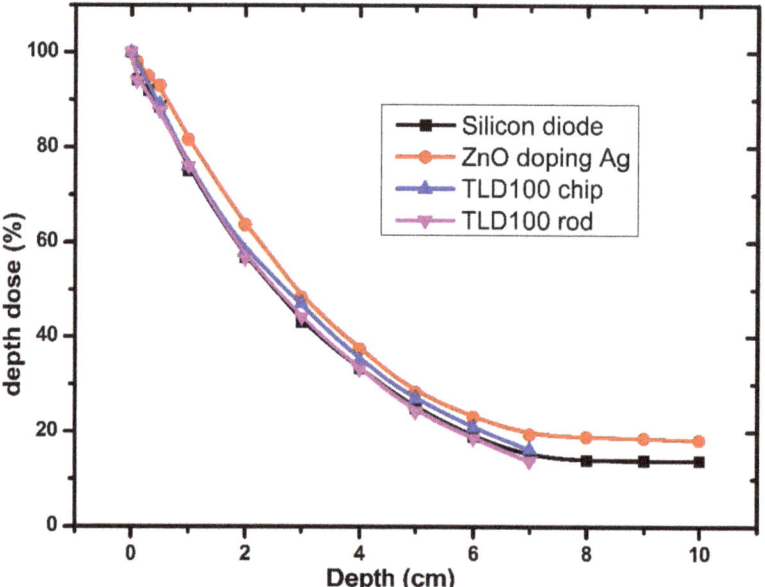

Figure 17. Percentage depth dose curve of the Perspex phantoms in 30 cm $\times$ 30 cm field size at 60 kVp of the X-ray energy with delivery dose 3 Gy.

4. Conclusions

The proposed dosimeter thin films were prepared successfully by the hydrothermal (autoclave) method. The structural and optical properties were studied. The PL spectra showed two peaks that included the near bandgap edge emission (3.26 eV) and high-intensity defect in the visible region with a wide range (2.25 to 1.55 eV). The dosimetric properties were investigated. TL optimum intensity was found at 0.5 mol% of Ag for a delivered dose of 4 Gy, with the glow curve showing a single peak at 270 °C. The TL sensitivity was normalized to the mass of the thin films and found to be two times that of the TLD 100 chips. Ag-doped ZnO thin films showed good reproducibility, and the results of PDD also confirmed an agreement with the case of TLD 100. However, it can be concluded that the advantage is that the glow curve's growth was acquired immediately with different heating rates and that the sensitivity is more than that of TLD 100. The limitations of this work are that the thin films' synthesizing conditions were a crucial effect on the shape and position of the glow curve and optical fading. These thin films were strongly affected by sunlight. The study demonstrated that the synthesized sample exhibits suitable TL properties for radiation monitoring.

Author Contributions: H.A.T., conceptualization and writing—original draft preparation; N.A.K., supervision, funding and review; A.K.I., supervision and funding acquisition; S.A., writing—review and editing; M.A.S., methodology; A.B., validation. All authors have read and agreed to the published version of the manuscript.

Funding: This research was funded by UTM Professional Development Research University (Postdoctoral Fellowship Ref. [No: PY/2022/03183] and an FRGS grant (vot num: [R.J130000.7854.5F236, Ref No: PY/2019/01269]. And also funded by Universiti Sains Malaysia (USM) [1001/PFIZIK/811345].

Data Availability Statement: The data presented in this study are available on request from the corresponding author.

Acknowledgments: The authors are thankful to UTM Professional Development Research University. The authors extend their appreciation to the Biophysics Laboratory, Medical Physics Laboratory and lab of Nano-Optoelectronics Research and Technology Laboratory (NOR Lab) staff members (https://fizik.usm.my/index.php/facilities/research-laboratory/nor-lab), USM, wherein this work was carried out.

Conflicts of Interest: The authors declare no conflict of interest.

References

1. Sinclair, S.; Pech-Canul, M. Development feasibility of TLD phosphors and thermoluminescent composite materials for potential applications in dosimetry: A review. *Chem. Eng. J.* **2022**, *443*, 136522. [CrossRef]
2. Bhatt, B.C.; Kulkarni, M. Thermoluminescent phosphors for radiation dosimetry. In *Defect and Diffusion Forum*; Trans Tech Publishing: Geneva, Switzerland, 2014; pp. 179–227.
3. Omanwar, S.; Koparkar, K.; Virk, H.S. Recent advances and opportunities in tld materials: A review. In *Defect and Diffusion Forum*; Trans Tech Publishing: Geneva, Switzerland, 2014; pp. 75–110.
4. Kry, S.F.; Alvarez, P.; Cygler, J.E.; DeWerd, L.A.; Howell, R.M.; Meeks, S.; O'Daniel, J.; Reft, C.; Sawakuchi, G.; Yukihara, E.G. AAPM TG 191: Clinical use of luminescent dosimeters: TLDs and OSLDs. *Med. Phys.* **2020**, *47*, e19–e51. [CrossRef] [PubMed]
5. Salah, N. Nanocrystalline materials for the dosimetry of heavy charged particles: A review. *Radiat. Phys. Chem.* **2011**, *80*, 1–10. [CrossRef]
6. Olko, P. Advantages and disadvantages of luminescence dosimetry. *Radiat. Meas.* **2010**, *45*, 506–511. [CrossRef]
7. Mohammed, B.; Jaafar, M.S.; Wagiran, H. Effect of Cu 2 O on the thermoluminescence properties of ZnO-B 2 O 3 –SiO 2 glass sample. *J. Lumin.* **2017**, *190*, 228–233. [CrossRef]
8. Alanazi, A. *A Study of Novel Forms of Thermoluminescent Media for Clinical Dosimetry*; University of Surrey: Surrey, UK, 2020.
9. Parauha, Y.R.; Dhoble, S.J. Synthesis and luminescence characterization of Eu(3+) doped Ca7 Mg2 (PO 4)6 phosphor for eco-friendly white LEDs and TL Dosimetric applications. *Luminescence* **2021**, *36*, 1837–1846.
10. Sadek, A.M.; Hassan, M.M.; Esmat, E.; Eissa, H.M. A new approach to the analysis of thermoluminescence glow-curve of TLD-600 dosimeters following Am-241 alpha particles irradiation. *Radiat. Prot. Dosim.* **2018**, *178*, 260–271. [CrossRef]
11. Biro, B.; Fenyvesi, A.; Timar-Gabor, A.; Simon, V. Thermoluminescence properties of 30Y2O3.30P2O5.40SiO2 vitroceramics in mixed neutron-gamma fields. *Appl. Radiat. Isot.* **2018**, *135*, 224–231. [CrossRef]

12. Al-Jawad, S.M.H.; Sabeeh, S.H.; Taha, A.A.; Jassim, H.A. Studying structural, morphological and optical properties of nanocrystalline ZnO:Ag films prepared by sol–gel method for antimicrobial activity. *J. Sol.-Gel Sci. Technol.* **2018**, *87*, 362–371. [CrossRef]
13. Yukihara, E.G.; McKeever, S.W.; Andersen, C.E.; Bos, A.J.; Bailiff, I.K.; Yoshimura, E.M.; Sawakuchi, G.O.; Bossin, L.; Christensen, J.B. Luminescence dosimetry. *Nat. Rev. Methods Primers* **2022**, *2*, 1–21. [CrossRef]
14. Grigorjeva, L.; Zolotarjovs, A.; Sokovnin, S.Y.; Millers, D.; Smits, K.; Il'ves, V.G. Radioluminescence, thermoluminescence and dosimetric properties of ZnO ceramics. *Ceram. Int.* **2017**, *43*, 6187–6191. [CrossRef]
15. Reddy, G.K.; Reddy, A.J.; Krishna, R.H.; Nagabhushana, B.M.; Gopal, G.R. Luminescence and spectroscopic investigations on Gd3+ doped ZnO nanophosphor. *J. Asian Ceram. Soc.* **2018**, *5*, 350–356. [CrossRef]
16. Singh, A.; Pandey, A.; Luthra, V. Modulating electrical, structural and thermoluminescence properties of γ- ray irradiated nanocrystalline Zn0.99M0.01O (M = Al/Gd). *Radiat. Phys. Chem.* **2018**, *152*, 69–74. [CrossRef]
17. Guckan, V.; Altunal, V.; Ozdemir, A.; Tsiumra, V.; Zhydachevskyy, Y.; Yegingil, Z. Calcination effects on europium doped zinc oxide as a luminescent material synthesized via sol-gel and precipitation methods. *J. Alloy. Compd.* **2020**, *823*, 153878. [CrossRef]
18. Thabit, H.A.; Kabir, N.A.; Ahmed, N.M. Synthesis & thermoluminescence characteristics & structural and optical studies of ZnO/Ag/ZnO system for dosimetric applications. *J. Lumin.* **2021**, *236*, 118097.
19. Jayaramaiah, J.R.; Nagabhushana, K.R.; Lakshminarasappa, B.N. Role of Li ion on luminescence performance of yttrium oxide thin films. *Dye. Pigment.* **2015**, *121*, 221–226. [CrossRef]
20. Montes-Gutiérrez, J.A.; Alcantar-Peña, J.J.; de Obaldia, E.; Zúñiga-Rivera, N.J.; Chernov, V.; Meléndrez-Amavizca, R.; Barboza-Flores, M.; Garcia-Gutierrez, R.; Auciello, O. Afterglow, thermoluminescence and optically stimulated luminescence characterization of micro-, nano-and ultrananocrystalline diamond films grown on silicon by HFCVD. *Diam. Relat. Mater.* **2018**, *85*, 117–124. [CrossRef]
21. Moradi, F.; Olatunji, M.A.; Sani, S.F.A.; Ung, N.M.; Forouzeshfar, F.; Khandaker, M.U.; Bradley, D.A. Composition and thickness dependence of TLD relative dose sensitivity: A Monte Carlo study. *Radiat. Meas.* **2019**, *129*, 106191. [CrossRef]
22. Moradi, F.; Ung, N.; Mahdiraji, G.; Khandaker, M.; Entezam, A.; See, M.; Taib, N.; Amin, Y.; Bradley, D. Angular dependence of optical fibre thermoluminescent dosimeters irradiated using kilo-and megavoltage X-rays. *Radiat. Phys. Chem.* **2017**, *135*, 4–10. [CrossRef]
23. Chen, H.; Qu, Y.; Sun, L.; Peng, J.; Ding, J. Band structures and optical properties of Ag and Al co-doped ZnO by experimental and theoretic calculation. *Phys. E Low-Dimens. Syst. Nanostructures* **2019**, *114*, 113602. [CrossRef]
24. Ahmad, K.S.; Jaffri, S.B. Phytosynthetic Ag doped ZnO nanoparticles: Semiconducting green remediators. *Open Chem.* **2018**, *16*, 556–570. [CrossRef]
25. Thabit, H.A.; Kabir, N.A.; Ahmed, N.M.; Alraddadi, S.; Al-Buriahi, M. Synthesis, structural, optical, and thermoluminescence properties of ZnO/Ag/Y nanopowders for electronic and dosimetry applications. *Ceram. Int.* **2021**, *47*, 4249–4256. [CrossRef]
26. Huang, P.-S.; Qin, F.; Lee, J.-K. Role of the interface between Ag and ZnO in the electric conductivity of Ag nanoparticle-embedded ZnO. *ACS Appl. Mater. Interfaces* **2019**, *12*, 4715–4721. [CrossRef] [PubMed]
27. Corro, G.; Flores, J.A.; Pacheco-Aguirre, F.U.P.; Olivares-Xometl, O. Effect of the Electronic State of Cu, Ag, and Au on Diesel Soot Abatement: Performance of Cu/ZnO, Ag/ZnO, and Au/ZnO Catalysts. *ACS Omega* **2019**, *4*, 5795–5804. [PubMed]
28. Saboor, A.; Shah, S.M.; Hussain, H. Band gap tuning and applications of ZnO nanorods in hybrid solar cell: Ag-doped verses Nd-doped ZnO nanorods. *Mater. Sci. Semicond. Processing* **2019**, *93*, 215–225. [CrossRef]
29. Wang, S.; Jia, F.; Wang, X.; Hu, L.; Sun, Y.; Yin, G.; Zhou, T.; Feng, Z.; Kumar, P.; Liu, B. Fabrication of ZnO nanoparticles modified by uniformly dispersed Ag nanoparticles: Enhancement of gas sensing performance. *ACS Omega* **2020**, *5*, 5209–5218. [CrossRef]
30. Razcón, J.L.I.; Vázquez, C.C.; Bernal, R.; Nuñez, H.A.B.; Castaño, V.M. Novel ZnO:Li phosphors for electronics and dosimetry applications. *Electron. Mater. Lett.* **2016**, *13*, 25–28. [CrossRef]
31. Borbón-Nuñez, H.A.; Iriqui-Razcón, J.L.; Cruz-Vázquez, C.; Bernal, R.; Furetta, C.; Chernov, V.; Castaño, V.M. Thermoluminescence kinetics parameters of ZnO exposed to beta particle irradiation. *J. Mater. Sci.* **2017**, *52*, 5208–5215. [CrossRef]
32. Isik, M.; Gasanly, N. Gd-doped ZnO nanoparticles: Synthesis, structural and thermoluminescence properties. *J. Lumin.* **2019**, *207*, 220–225. [CrossRef]
33. Prasad, A.R.; Anagha, M.; Shamsheera, K.; Joseph, A. Bio-fabricated ZnO nanoparticles: Direct sunlight-driven selective photodegradation, antibacterial activity, and thermoluminescence-emission characteristics. *New J. Chem.* **2020**, *44*, 8273–8279. [CrossRef]
34. Buryi, M.; Babin, V.; Artemenko, A.; Remeš, Z.; Děcká, K.; Mičová, J. Hydrothermally grown ZnO: Mo nanorods exposed to X-ray: Luminescence and charge trapping phenomena. *Appl. Surf. Sci.* **2022**, *585*, 152682. [CrossRef]
35. Thabit, H.A.; Kabir, N.A. The study of X-ray effect on structural, morphology and optical properties of ZnO nanopowder. *Nucl. Instrum. Methods Phys. Res. Sect. B: Beam Interact. Mater. At.* **2018**, *436*, 278–284. [CrossRef]
36. Xu, L.; Miao, J.; Chen, Y.; Su, J.; Yang, M.; Zhang, L.; Zhao, L.; Ding, S. Characterization of Ag-doped ZnO thin film for its potential applications in optoelectronic devices. *Optik* **2018**, *170*, 484–491.
37. Zakaria, M.A.; Menazea, A.; Mostafa, A.M.; Al-Ashkar, E.A. Ultra-thin silver nanoparticles film prepared via pulsed laser deposition: Synthesis, characterization, and its catalytic activity on reduction of 4-nitrophenol. *Surf. Interfaces* **2020**, *19*, 100438. [CrossRef]
38. Gurgur, E.; Oluyamo, S.; Adetuyi, A.; Omotunde, O.; Okoronkwo, A. Green synthesis of zinc oxide nanoparticles and zinc oxide–silver, zinc oxide–copper nanocomposites using Bridelia ferruginea as biotemplate. *SN Appl. Sci.* **2020**, *2*, 1–12. [CrossRef]

39. Zhou, F.; Jing, W.; Liu, P.; Han, D.; Jiang, Z.; Wei, Z. Doping Ag in ZnO Nanorods to Improve the Performance of Related Enzymatic Glucose Sensors. *Sensors* **2017**, *17*, 2214. [CrossRef]
40. Chen, W.; Wang, Z.; Lin, Z.; Lin, L. Thermoluminescence of ZnS nanoparticles. *Appl. Phys. Lett.* **1997**, *70*, 1465–1467. [CrossRef]
41. Chen, W.; Joly, A.G.; Roark, J. Photostimulated luminescence and dynamics of AgI and Ag nanoclusters in zeolites. *Phys. Rev. B* **2002**, *65*, 245404. [CrossRef]
42. Chen, W.; Westcott, S.L.; Zhang, J. Dose dependence of x-ray luminescence from Ca F 2: Eu 2+, Mn 2+ phosphors. *Appl. Phys. Lett.* **2007**, *91*, 211103. [CrossRef]
43. Chen, W.; Wang, Z.; Lin, L. Thermoluminescence of CdS clusters in zeolite-Y. *J. Lumin.* **1997**, *71*, 151–156. [CrossRef]
44. Johnson, N.J.; He, S.; Diao, S.; Chan, E.M.; Dai, H.; Almutairi, A. Direct evidence for coupled surface and concentration quenching dynamics in lanthanide-doped nanocrystals. *J. Am. Chem. Soc.* **2017**, *139*, 3275–3282. [CrossRef] [PubMed]
45. Wen, S.; Zhou, J.; Zheng, K.; Bednarkiewicz, A.; Liu, X.; Jin, D. Advances in highly doped upconversion nanoparticles. *Nat. Commun.* **2018**, *9*, 1–12.
46. Kumar, A.G.; Li, X.; Du, Y.; Geng, Y.; Hong, X. UV-photodetector based on heterostructured ZnO/(Ga, Ag)-co-doped ZnO nanorods by cost-effective two-step process. *Appl. Surf. Sci.* **2020**, *509*, 144770. [CrossRef]
47. Kandulna, R.; Choudhary, R.; Maji, P. Ag-doped ZnO reinforced polymeric Ag: ZnO/PMMA nanocomposites as electron transporting layer for OLED application. *J. Inorg. Organomet. Polym. Mater.* **2017**, *27*, 1760–1769. [CrossRef]
48. Torchynska, T.V.; Rodriguez, I.C.B.; el Filali, B.; Polupan, G.; Cano, A.I.D. Luminescence, structure and aging c-axis–Oriented silver doped ZnO nanocrystalline films. *Mater. Sci. Semicond. Processing* **2018**, *79*, 99–106. [CrossRef]
49. Kayani, Z.N.; Manzoor, F.; Zafar, A.; Mahmood, M.; Rasheed, M.; Anwar, M. Impact of Ag doping on structural, optical, morphological, optical and photoluminescent properties of ZnO nanoparticles. *Opt. Quantum Electron.* **2020**, *52*, 1–18. [CrossRef]
50. Wang, C.-C.; Shieu, F.-S.; Shih, H.C. Ag-nanoparticle enhanced photodegradation of ZnO nanostructures: Investigation using photoluminescence and ESR studies. *J. Environ. Chem. Eng.* **2021**, *9*, 104707. [CrossRef]
51. Zhao, Y.; Chen, X.; Fang, L.; Yang, L.; Li, H.; Gao, Y. Effects of Annealing on the Structural and Photoluminescent Properties of Ag-Doped ZnO Nanowires Prepared by Ion Implantation. *Plasma Sci. Technol.* **2013**, *15*, 817–820. [CrossRef]
52. Muthukumaran, S.; Gopalakrishnan, R. Structural, FTIR and photoluminescence studies of Cu doped ZnO nanopowders by co-precipitation method. *Opt. Mater.* **2012**, *34*, 1946–1953. [CrossRef]
53. Kuriakose, S.; Choudhary, V.; Satpati, B.; Mohapatra, S. Enhanced photocatalytic activity of Ag–ZnO hybrid plasmonic nanostructures prepared by a facile wet chemical method. *Beilstein J. Nanotechnol.* **2014**, *5*, 639–650. [CrossRef]
54. Uklein, A.; Multian, V.; Kuz'micheva, G.; Linnik, R.; Lisnyak, V.; Popov, A.; Gayvoronsky, V.Y. Nonlinear optical response of bulk ZnO crystals with different content of intrinsic defects. *Opt. Mater.* **2018**, *84*, 738–747. [CrossRef]
55. Dong, Y.; Tuomisto, F.; Svensson, B.G.; Kuznetsov, A.Y.; Brillson, L.J. Vacancy defect and defect cluster energetics in ion-implanted ZnO. *Phys. Rev. B* **2010**, *81*, 081201. [CrossRef]
56. Pal, S.; Gogurla, N.; Das, A.; Singha, S.; Kumar, P.; Kanjilal, D.; Singha, A.; Chattopadhyay, S.; Jana, D.; Sarkar, A. Clustered vacancies in ZnO: Chemical aspects and consequences on physical properties. *J. Phys. D Appl. Phys.* **2018**, *51*, 105107. [CrossRef]
57. Li, H.; Schirra, L.K.; Shim, J.; Cheun, H.; Kippelen, B.; Monti, O.L.; Bredas, J.-L. Zinc oxide as a model transparent conducting oxide: A theoretical and experimental study of the impact of hydroxylation, vacancies, interstitials, and extrinsic doping on the electronic properties of the polar ZnO (0002) surface. *Chem. Mater.* **2012**, *24*, 3044–3055. [CrossRef]
58. Rodnyi, P.; Khodyuk, I. Optical and luminescence properties of zinc oxide. *Opt. Spectrosc.* **2011**, *111*, 776–785. [CrossRef]
59. Peng, Y.; Wang, Y.; Chen, Q.-G.; Zhu, Q.; Xu, A.W. Stable yellow ZnO mesocrystals with efficient visible-light photocatalytic activity. *Cryst. Eng. Comm.* **2014**, *16*, 7906–7913. [CrossRef]
60. Kukreja, L.; Misra, P.; Fallert, J.; Phase, D.; Kalt, H. Correlation of spectral features of photoluminescence with residual native defects of ZnO thin films annealed at different temperatures. *J. Appl. Phys.* **2012**, *112*, 13525. [CrossRef]
61. Alvi, N.; Nur, O.; Willander, M. The origin of the red emission in n-ZnO nanotubes/p-GaN white light emitting diodes. *Nanoscale Res. Lett.* **2011**, *6*, 1–7. [CrossRef]
62. Pimpliskar, P.V.; Motekar, S.C.; Umarji, G.G.; Lee, W.; Arbuj, S.S. Synthesis of silver-loaded ZnO nanorods and their enhanced photocatalytic activity and photoconductivity study. *Photochem. Photobiol. Sci.* **2019**, *18*, 1503–1511. [CrossRef]
63. Duan, L.; Yu, X.; Ni, L.; Wang, Z. ZnO:Ag film growth on Si substrate with ZnO buffer layer by rf sputtering. *Appl. Surf. Sci.* **2011**, *257*, 3463–3467. [CrossRef]
64. Guidelli, E.J.; Baffa, O.; Clarke, D.R. Enhanced UV Emission From Silver/ZnO And Gold/ZnO Core-Shell Nanoparticles: Photoluminescence, Radioluminescence, And Optically Stimulated Luminescence. *Sci. Rep.* **2015**, *5*, 14004. [CrossRef] [PubMed]
65. Fayaz Rouhi, H.; Rozati, S. Synthesis and investigating effect of tellurium-doping on physical properties of zinc oxide thin films by spray pyrolysis technique. *Appl. Phys. A* **2022**, *128*, 1–8. [CrossRef]
66. Wang, C.; Wu, D.; Wang, P.; Ao, Y.; Hou, J.; Qian, J. Effect of oxygen vacancy on enhanced photocatalytic activity of reduced ZnO nanorod arrays. *Appl. Surf. Sci.* **2015**, *325*, 112–116. [CrossRef]
67. Zhang, Q.; Xu, M.; You, B.; Zhang, Q.; Yuan, H.; Ostrikov, K. Oxygen vacancy-mediated ZnO nanoparticle photocatalyst for degradation of methylene blue. *Appl. Sci.* **2018**, *8*, 353. [CrossRef]
68. Alajerami, Y.S. *Thermoluminescence and Optical Characteristics of Lithium Potassium Borate Glass for Radiation Therapy Dose Measurement*; Universiti Teknologi Malaysia: Johor Bahru, Malaysia, 2014.

69. Ahmad, M.; Ahmad, I.; Ahmed, E.; Akhtar, M.S.; Khalid, N.R. Facile and inexpensive synthesis of Ag doped ZnO/CNTs composite: Study on the efficient photocatalytic activity and photocatalytic mechanism. *J. Mol. Liq.* **2020**, *311*, 113326. [CrossRef]
70. Ignatovych, M.; Fasoli, M.; Kelemen, A. Thermoluminescence study of Cu, Ag and Mn doped lithium tetraborate single crystals and glasses. *Radiat. Phys. Chem.* **2012**, *81*, 1528–1532. [CrossRef]
71. Saidu, A.; Wagiran, H.; Saeed, M.A.; Obayes, H.K.; Bala, A.; Usman, F. Thermoluminescence response of rare earth activated zinc lithium borate glass. *Radiat. Phys. Chem.* **2018**, *144*, 413–418. [CrossRef]
72. Townsend, P.D.; Finch, A.A.; Maghrabi, M.; Ramachandran, V.; Vázquez, G.V.; Wang, Y.; White, D.R. Spectral changes and wavelength dependent thermoluminescence of rare earth ions after X-ray irradiation. *J. Lumin.* **2017**, *192*, 574–581. [CrossRef]
73. İflazoğlu, S.; Yılmaz, A.; Kafadar, V.E.; Topaksu, M.; Yazıcı, A. Neutron+Gamma response of undoped and Dy doped MgB4O7 thermoluminescence dosimeter. *Appl. Radiat. Isot.* **2019**, *147*, 91–98. [CrossRef]
74. Singh, R.; Kainth, H.S. Effect of heating rate on thermoluminescence output of LiF: Mg, Ti (TLD-100) in dosimetric applications. *Nucl. Instrum. Methods Phys. Res. Sect. B Beam Interact. Mater. At.* **2018**, *426*, 22–29. [CrossRef]
75. Mohammed, B.; Jaafar, M.S.; Wagiran, H. Thermoluminescence dosimetry properties and kinetic parameters of zinc borate silica glass doped with Cu2O and co-doped with SnO2. *J. Lumin.* **2018**, *204*, 375–381. [CrossRef]
76. Trindade, N.M.; Kahn, H.; Yoshimura, E.M. Thermoluminescence of natural BeAl2O4: Cr3+ Brazilian mineral: Preliminary studies. *J. Lumin.* **2018**, *195*, 356–361. [CrossRef]
77. Ozdemir, A.; Guckan, V.; Altunal, V.; Kurt, K.; Yegingil, Z. Thermoluminescence in MgB4O7: Pr, Dy dosimetry powder synthesized by solution combustion synthesis method. *J. Lumin.* **2021**, *230*, 117761. [CrossRef]
78. Prabhu, N.S.; Sharmila, K.; Somashekarappa, H.; GLakshminarayana, S. Thermoluminescence features of Er3+ doped BaO-ZnO-LiF-B2O3 glass system for high-dose gamma dosimetry. *Ceram. Int.* **2020**, *46*, 19343–19353. [CrossRef]
79. Hashim, S.; Omar, S.S.C.; Ibrahim, S.A.; Hassan, W.M.S.W.; Ung, N.M.; Mahdiraji, G.A.; Bradley, D.A.; Alzimami, K. Thermoluminescence response of flat optical fiber subjected to 9MeV electron irradiations. *Radiat. Phys. Chem.* **2015**, *106*, 46–49. [CrossRef]
80. Rammadhan, I.; Taha, S.; Wagiran, H. Thermoluminescence characteristics of Cu 2 O doped Calcium Lithium borate glass irradiated with the cobalt-60 gamma rays. *J. Lumin.* **2017**, *186*, 117–122. [CrossRef]
81. Glennie, G.D. A comparison of TLD dosimeters: LiF: Mg, Ti and LiF: Mg, Cu, P for measurement of radiation therapy doses. *Med. Phys.* **2003**, *30*, 3262. [CrossRef]
82. Mohammed, B. Development of undoped, doped and codoped boron silicate composite as thermoluminescent dosimeters for medium and high dose levels. In *School of Physics*; University of Southern Mississippi: Hattiesburg, MS, USA, 2017.
83. Salama, E.; Soliman, H.A.; Youssef, G.M.; Hamad, S. Thermoluminescence properties of borosilicate glass doped with ZnO. *J. Lumin.* **2017**, *186*, 164–169. [CrossRef]
84. Laopaiboon, R.; Thumsa-ard, T.; Bootjomchai, C. The thermoluminescence properties and determination of trapping parameters of soda lime glass doped with erbium oxide. *J. Lumin.* **2018**, *197*, 304–309. [CrossRef]
85. Bakhsh, M.; Abdullah, W.S.W.; Mustafa, I.S.; al Musawi, M.S.A.; Razali, N.A.N. Synthesis, characterisation and dosimetric evaluation of MgB4O7 glass as thermoluminescent dosimeter. *Radiat. Eff. Defects Solids* **2018**, *173*, 446–460. [CrossRef]
86. Tousi, E.T.; Aboarrah, A.; Bauk, S.; Hashim, R.; Jaafar, M.S. Measurement of percentage depth dose and half value layer of the Rhizophora spp. particleboard bonded by Eremurus spp. to 60, 80 and 100 kVp diagnostic X-rays. *MAPAN* **2018**, *33*, 321–332. [CrossRef]

Article

Construction of Novel Nanocomposites (Cu-MOF/GOD@HA) for Chemodynamic Therapy

Ya-Nan Hao [1,†], Cong-Cong Qu [1,†], Yang Shu [1,*], Jian-Hua Wang [1,*] and Wei Chen [2,3,*]

1. Department of Chemistry, College of Sciences, Northeastern University, Shenyang 110819, China; neuhaoyanan@163.com (Y.-N.H.); qcc1531486377@163.com (C.-C.Q.)
2. Departments of Physics, University of Texas at Arlington, Arlington, TX 76019, USA
3. Medical Technology Research Centre, Chelmsford Campus, Anglia Ruskin University, Chelmsford CM1 1SQ, UK
* Correspondence: shuyang@mail.neu.edu.cn (Y.S.); jianhuajrz@mail.neu.edu.cn (J.-H.W.); weichen@uta.edu (W.C.)
† These authors contributed equally to this work.

Abstract: The emerging chemodynamic therapy (CDT) has received an extensive attention in recent years. However, the efficiency of CDT is influenced due to the limitation of H_2O_2 in tumor. In this study, we designed and synthesized a novel core-shell nanostructure, Cu-metal organic framework (Cu-MOF)/glucose oxidase (GOD)@hyaluronic acid (HA) (Cu-MOF/GOD@HA) for the purpose of improving CDT efficacy by increasing H_2O_2 concentration and cancer cell targeting. In this design, Cu-MOF act as a CDT agent and GOD carrier. Cu(II) in Cu-MOF are reduced to Cu(I) by GSH to obtain Cu(I)-MOF while GSH is depleted. The depletion of GSH reinforces the concentration of H_2O_2 in tumor to improve the efficiency of CDT. The resultant Cu(I)-MOF catalyze H_2O_2 to generate hydroxyl radicals (·OH) for CDT. GOD can catalyze glucose (Glu) to supply H_2O_2 for CDT enhancement. HA act as a targeting molecule to improve the targeting ability of Cu-MOF/GOD@HA to the tumor cells. In addition, after loading with GOD and coating with HA, the proportion of Cu(I) in Cu-MOF/GOD@HA is increased compared with the proportion of Cu(I) in Cu-MOF. This phenomenon may shorten the reactive time from Cu-MOF to Cu(I)-MOF. The CDT enhancement as a result of GOD and HA effects in Cu-MOF/GOD@HA was evidenced by in vitro cell and in vivo animal studies.

Keywords: Fenton reaction; hydroxyl radicals; GSH depletion; hydrogen peroxide; glucose oxidase

1. Introduction

Chemodynamic therapy (CDT) is an emerging cancer treatment, which depends on the Fenton or Fenton-like reactions to obtain highly toxic hydroxyl radicals (·OH) for killing cancer cells [1]. The Fe/hydrogen peroxide (H_2O_2) system is defined as the Fenton reagent, and the others (e.g., Co, Cd, Cu, Ag, Mn, Ni) are called Fenton-like reagents [2–6]. Compared with normal cells, the tumor microenvironment is characterized by overexpression of H_2O_2, glutathione (GSH) and weak acidity [7,8]. Up to now, most of the metal ions used to construct CDT nanotherapeutics include Mn, Fe and Cu. Fe ions in particular are the most commonly used for CDT [9]. However, the Fe-based Fenton reaction is only effective under strongly acidic conditions (pH 2–4) [10]. Therefore, the Fenton-reaction of Fe will be limited under the neutral and weakly acidic microenvironment conditions encountered in tumors. Mn(II) used in Fenton-like reactions only remains stable when pH < 4. In contrast, the efficiency of Cu(I) is not influenced by the pH. Even under the best reaction conditions, the reaction rate of Cu(I) is 160 times than that of Fe(II) [11–13]. In addition, a large amount of GSH in tumor cells can react with reactive oxygen species and affect the concentration of H_2O_2 and ·OH. GSH can be depleted during the reduction of the Cu(II) to Cu(I), which is beneficial for CDT [14]. Therefore, Cu-based nanotherapeutic agents for CDT have attracted a lot of attention.

The relatively higher concentration of H_2O_2 in tumor cells than in normal cells is the basis for CDT due to Fenton/Fenton-like reactions with H_2O_2. However, the H_2O_2 concentration in tumor cells is still limited and this limitation actually influences the efficiency of CDT [15]. Different kinds of strategies have been proposed to overcome this drawback. The introduction of extraneous enzyme (glucose oxidase (GOD), superoxide dismutase and so on) into the cell interior may facilitate the production of H_2O_2, which ensures the continuous and effective treatment of tumor cells [16]. GOD can catalyze the reaction of glucose (Glu) to produce gluconic acid and H_2O_2 [16,17]. Therefore, combining GOD and a CDT agent is a promising strategy. GOD consumes a large amount of Glu needed for physiological activities, which achieves a "cell starving" effect therapy during this process. This phenomenon also supplies H_2O_2 for CDT [18].

Metal-organic frameworks (MOFs) are organic-inorganic hybrid materials formed by the combination of inorganic metal ions or metal clusters and organic ligands. MOFs have been widely used in biomedical imaging and therapy [19], biosensing [20], catalysis and as drug carriers [21] due to their high specific surface area, porosity and diversified structures. MOFs can be easily modified with appropriate treatments during or after their synthesis [22]. Therefore, Cu-MOFs are suitable as modifiable CDT agents.

Hyaluronic acid (HA) is a natural acid mucopolysaccharide present in the synovial fluid and extracellular matrix [23]. HA has been widely used in tissue engineering, drug delivery and molecular imaging which all benefit from its biocompatibility and biodegradability [23–25]. More importantly, HA can specifically target CD44 overexpressed in various cancer cells and be decomposed by the intracellular hyaluronidase [26]. Therefore, HA is often used to bounding various drug-loaded nanoparticles as a targeting moiety for enhanced cancer therapy [27–30].

In this work, Cu-MOF are chosen as a cascade nanoreactor for CDT. As shown in Figure 1, GOD are first loaded onto Cu-MOF and Cu-MOF/GOD composites are thus obtained. HA acts as a shell on the Cu-MOF/GOD to avoid GOD leakage and as a targeting molecule to tumor site. The resulting Cu-MOF/GOD@HA nanocomposites are activated by GSH in tumors and catalyze H_2O_2 to produce ·OH for CDT for cancer treatment as illustrated in Figure 1.

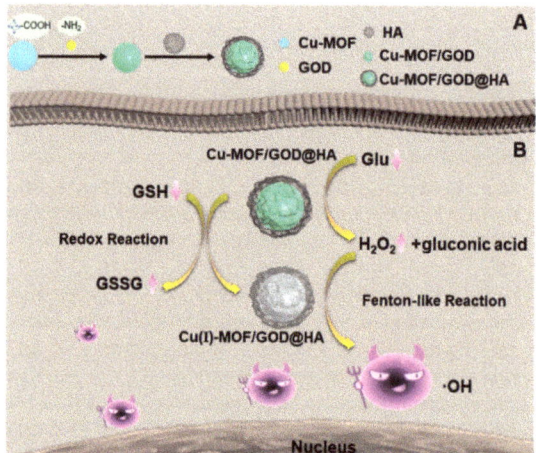

Figure 1. Schematic of the Cu-MOF/GOD@HA preparation process (**A**) and the Cu-containing nanoformulation mediated CDT (**B**).

2. Materials and Methods

2.1. The Preparation of Cu-MOF/GOD@HA Nanocomposites

2.1.1. Cu-MOF/GOD

Cu-MOFs (HKUST-1) were prepared according to a previously reported approach [31]. Thus, 10 mg of Cu-MOF were dissolved in 1 mL of absolute ethanol. Next 10 mg of coupling agent (1-(3-dimethylaminopropyl)-3-ethylcarbodiimide hydrochloride/N-hydroxysuccinimide) (EDC/NHS) were dissolved in 6 mL of DI water. The above Cu-MOF and EDC/NHS solutions were mixed with vigorous stirring for 30 min. Then 2 mg of GOD was dissolved in 3 mL of DI water and added dropwise into the above mixture under stirring for 4 h. Finally, the resultant Cu-MOF/GOD was washed with DI water three times.

2.1.2. Cu-MOF/GOD@HA

The above Cu-MOF/GOD nanoparticles (final concentration: 2 mg mL^{-1}) mixed with HA (final concentration: 0.5 mg mL^{-1}) under sonication for 30 min. Finally, the resultant Cu-MOF/GOD@HA was washed with DI water 3 times.

2.2. Extracellular ·OH Generation under Catalysis by Cu-MOF

2.2.1. Reaction between Cu-MOF and GSH

Two hundred μL of Cu-MOF solution (1 mg mL^{-1}) was mixed with 800 μL GSH solution (10 mmol L^{-1}) under vortex mixing until the appearance of a white precipitate (Cu(I)-MOF). The precipitated intermediate products were characterized via UV-vis absorption spectroscopy (Hitachi, Hitachi, Japan).

2.2.2. Extracellular ·OH Generation under Catalysis by Cu-MOF

A 5 μg mL^{-1} methylene blue (MB) solution was mixed with 5 mmol L^{-1} H_2O_2 and 200 μg mL^{-1} Cu(I)-MOF or Cu-MOF under vortex mixing. After different time intervals of reaction, ·OH-induced MB degradation was evaluated based on the change in the absorption at the maximum absorption wavelength 630 nm.

2.2.3. Gluconic Acid Generation under Catalysis by GOD

(1) A 100 μg mL^{-1} Cu-MOF or Cu-MOF/GOD solution was mixed with 500 μg mL^{-1} Glu under vortex mixing. After different times of reaction, the generation of gluconic acid was monitored by the acidity change of the solution. (2) A 100 μg mL^{-1} Cu-MOF or Cu-MOF/GOD solution was mixed with different concentrations of Glu under vortex mixing for 24 h. The generation of gluconic acid was monitored by the acidity change of the solution.

2.2.4. The H_2O_2 Generation under Catalysis by GOD

The fluorescence signals of Ampliflu Red are influenced by the change of H_2O_2 concentration. Thus, Ampliflu Red was used to investigate the H_2O_2 generation. Specific steps were as follows: (1) A 600 μg mL^{-1} Ampliflu Red solution was mixed with 500 μg mL^{-1} Glu and 100 μg mL^{-1} Cu-MOF/GOD under vortex mixing. After different time intervals (0, 5, 10, 25, 50 min), the fluorescence intensity at ~585 nm under 530 nm wavelength excitation was measured. (2) A 600 μg mL^{-1} Ampliflu Red solution was mixed with 100 μg mL^{-1} Cu-MOF/GOD and different concentrations of Glu (0, 20, 50, 70, 100 μg mL^{-1}) under vortex mixing. After 30 min of reaction, the fluorescence intensity at ~585 nm under 530 nm wavelength excitation was measured.

2.3. Cell Experiments

2.3.1. Cytotoxicity Assay

Briefly, 10^4 cells/well were seeded in 96-well plates and cultured overnight at 37 °C in 5% CO_2 atmosphere. The pending test cells were then further incubated with different concentrations of Cu-MOF, Cu-MOF/GOD or Cu-MOF/GOD@HA (10, 20, 30, 50, 70, 100 μg mL^{-1}) under different conditions (GSH: 2.5, 5.0 mmol L^{-1}, H_2O_2: 100 μmol L^{-1}

or trypan blue: 10 μg mL^{-1}) for 24 h. The cytotoxicity of Cu-MOF, Cu-MOF/GOD and Cu-MOF/GOD@HA was evaluated through an MTT assay.

2.3.2. Intracellular ·OH Generation Capability of Cu-MOF/GOD@HA

For the evaluation of ·OH generation capability, MCF-7 cells were cultured with 50 μg mL^{-1} of Cu-MOF, Cu-MOF/GOD and Cu-MOF/GOD@HA for 4 h, the cellular-ROS stress levels were then evaluated by using a ROS assay kit. After washing with PBS for twice, MCF-7 cells were further incubated with 2′,7′-dichlorodihydrofluorescein diacetate (DCFH-DA) (10 μmol L^{-1}) for 20 min followed by washing for 3 times with PBS. Finally, ROS associated signals in the cells were observed by fluorescence microscopy.

3. Results

3.1. Preparation and Characterization of Cu-MOF/GOD@HA

The preparation process of Cu-MOF/GOD@HA was shown in Figure 1A as described in the experimental section. GOD was firstly loaded into Cu-MOF by amide reaction between –COOH on the Cu-MOF and –NH$_2$ on GOD. Then, Cu-MOF/GOD was coated by HA to avoid the leakage of GOD and to improve its biocompatibility as well as the targeting ability. The CDT mechanism of Cu-MOF/GOD@HA was shown in Figure 1B. After endocytosis into MCF-7 cells, i.e., the tumor cells, CDT process based on Cu-MOF/GOD@HA was triggered sequentially by GSH "AND" H$_2$O$_2$ in the cancer cell interior. In addition, GOD in Cu-MOF catalysis Glu to supply H$_2$O$_2$ to improve CDT efficacy.

Transmission electron microscope (TEM) images in Figure 2A,B illustrate the Cu-MOF with diameter of 63.32 ± 9.12 nm. SEM image and Elemental mapping of C, N, O and Cu of Cu-MOF confirmed the successful preparation (Figure S1A,B). From the EDS spectra from SEM, Cu-MOF were consisted of 1% Cu, 2% N, 5% O and 92% C. As shown in Figure 2C, the shells of loaded GOD and coated HA were 5 nm. Compared with Cu-MOF, the FT-IR spectra of Cu-MOF/GOD, Cu-MOF/GOD@HA and GOD showed bands at 1550 cm^{-1}, attributed to vibrational stretches characteristic of GOD, indicating the successful loading of GOD (Figure S2A). Furthermore, the TGA analysis of Cu-MOF and Cu-MOF/GOD@HA further confirmed the GOD loading and HA coating (Figure S2B). From the TGA analysis, Cu-MOF/GOD@HA were consisted with 15.32% HA, 21.56% GOD and 63.12% Cu-MOF.

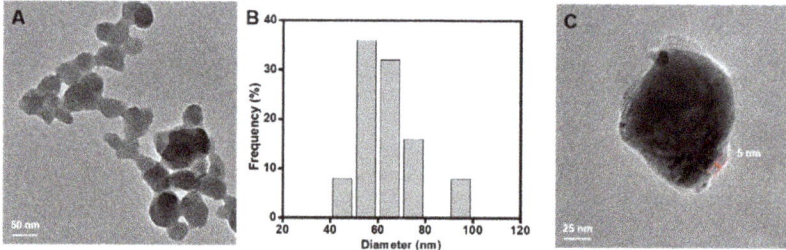

Figure 2. (**A**) TEM images of Cu-MOF. Scale bar: 50 nm. (**B**) The size distribution of Cu-MOF as measured by TEM image. (**C**) TEM images of Cu-MOF/GOD@HA (Scale bar: 25 nm).

In order to confirm the successful preparation of Cu-MOF/GOD@HA, the UV-vis absorption spectra of the reaction mixture were recorded. The absorption-change of Cu-MOF, Cu-MOF/GOD and Cu-MOF/GOD@HA in the 600–900 nm region, indicating the successful assembly of Cu-MOF/GOD@HA is clearly seen in Figure 3A. The color of Cu-MOF and Cu-MOF/GOD@HA changed from sky blue to green, also indicating successful loading of GOD and coating of HA. Zeta potential values of the corresponding nanocomposites obtained in each step are shown in Figure 3B, where the Cu-MOF aqueous solution exhibits a potential of −10.03 ± 0.16 mV. The loading of GOD reduces the zeta potential to −4.84 ± 0.19 mV, implying the deprotonation of the –NH$_2$ of the GOD. The increase of zeta potential to −7.84 ± 0.28 mV due to the presence of the –OH of HA [32,33].

Furthermore, the X-ray diffraction (XRD) spectra of Cu-MOF, Cu-MOF/GOD and Cu-MOF/GOD@HA indicates that the loading of GOD and coating of HA did not influence the crystal structure of Cu-MOF (Figure S3).

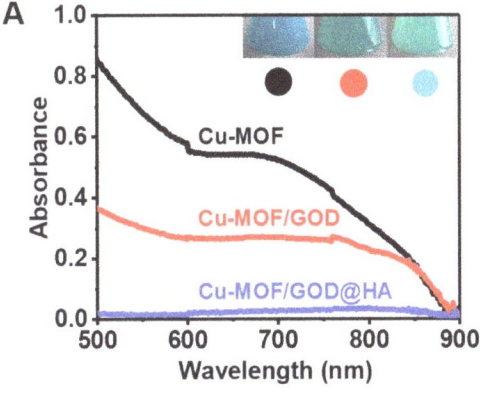

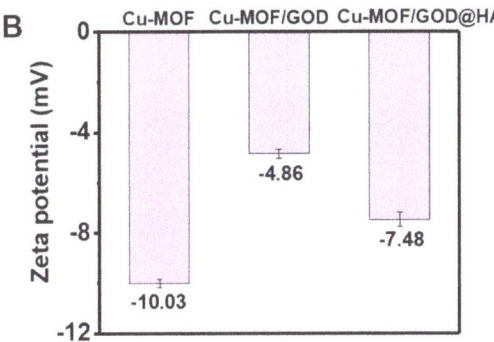

Figure 3. (**A**) UV-vis absorption spectra of Cu-MOF, Cu-MOF/GOD and Cu-MOF/GOD@HA (The concentrations of pending test samples were 100 μg mL^{-1}). Inset in (**A**) shows the photographs of the solutions containing Cu-MOF, Cu-MOF/GOD and Cu-MOF/GOD@HA. (**B**) Zeta potential values of Cu-MOF, Cu-MOF/GOD and Cu-MOF/GOD@HA. (The concentrations of pending test samples are 200 μg mL^{-1}).

3.2. Depletion of GSH and Generation of H_2O_2

In order to confirm the oxidation-reduction reactions between Cu-MOF and GSH, molar equivalents of Cu-MOF and GSH were mixed in an aqueous solution. It is clearly illustrated in Figure S4 that the absorption of Cu-MOF completely disappeared after reaction of Cu-MOF and GSH for ca. 2 min along with the formation of Cu(I)-MOF, and the color of the mixture of Cu-MOF and GSH turned from sky blue to white. This observation obviously demonstrated that the incubation of Cu-MOF with GSH facilitates the reduction of divalent Cu(II) in the Cu-MOF by GSH. In addition, compared with TEM image of Cu-MOF, Cu-MOF was collapsed after treatment with 10 mmol L^{-1} GSH (Figure S5), proving the reaction between GSH and Cu-MOF.

X-ray photoelectron spectroscopy (XPS) was also used to investigate the oxidation-reduction reactions between Cu-MOF and GSH. XPS is a surface sensitive technique. Compared with Cu-MOF (Figure S6A), an obvious N1s peak was observed in the XPS spectra of Cu-MOF/GOD and Cu-MOF/GOD@HA (Figure S6B,C), further demonstrating the occurrence of GOD loading. After treatment by GSH, an obvious N1s peak appears in Cu-MOF due to the residues of GSH (Figure S6D). Furthermore, after treatment of GSH, the XPS analysis of Cu(I)-MOF only shows Cu(I) (932.7, 952.4 eV) peaks (Figure S6E), also confirming the reduction of Cu-MOF by GSH. Moreover, Figure 3A also shows that after loading with GOD and coating with HA, the UV-vis absorption of between 600–900 nm is significantly reduced. Interestingly, the proportion of Cu(I) on the surface of Cu-MOF have increased after loading with GOD and coating with HA. As shown in Figure S6F–H, the valence state of Cu is "+1" or "+2", since the Cu 2p XPS spectra displayed peaks on Cu(II) (934.4 eV, 954.2 eV)/Cu(I) (932.7 eV, 952.4 eV). The existence of Cu(II)/Cu(I) redox pair provide a great potential for Fenton-like reactions. In the paramagnetic chemical state (Figure S6F–H), the Cu $2p_{3/2}$ XPS spectrum is quantitatively analyzed, and the Cu(II)/Cu(I) ratio of Cu-MOF is 1.2872. The Cu(II)/Cu(I) ratios on the surface of Cu-MOF/GOD and Cu-MOF/GOD@HA in Cu $2p_{3/2}$ XPS spectra are 1.0376 and 0.8744, respectively. Obviously, after loading with GOD and coating with HA, the proportion of Cu(I) on the surface of Cu-MOF increased, indicating that Cu(II) was reduced to Cu(I) during the assembly process on the surface of Cu-MOF. This phenomenon may shorten the reaction time from Cu-MOF to Cu(I)-MOF.

After reduction of Cu-MOF by GSH, Cu(I) moiety catalyzes H_2O_2 to generate ·OH via a Fenton-like reaction. The Cu(I)-MOF-involved CDT processes are investigated and demonstrated by observing the degradation methylene blue (MB) [34]. ·OH induced degradation of MB which may result in a significant decrement on the absorption of MB at the maximum wavelength of 663 nm. Figure 4A shows that the degradation results of MB under different conditions by incubation for 40 min. In the presence of only Cu(I)-MOF or H_2O_2, virtually no degradation of MB was observed. On the contrary, a remarkable degradation of MB was recorded by introducing H_2O_2 into the mixture containing MB and Cu(I)-MOF. In addition, in the presence of free Cu(I) and H_2O_2, no remarkable degradation of MB was observed within 40 min. Thus, compared with free Cu(I), Cu(I) activated from Cu-MOF has a higher catalytic performance. Moreover, Figure S7 shows that Cu-MOF could not react with H_2O_2 to generate ·OH to reduce the degradation of MB. This provides a clear evidence for the occurrence of the Cu-MOF involving in the cascade reactions in CDT process. First, divalent Cu(II) in Cu-MOF was activated by GSH to convert to monovalent Cu(I) in Cu(I)-MOF. Subsequently, Cu(I)-MOF catalyzes the generation of ·OH from H_2O_2. The absorption spectra of MB illustrating its time-dependent degradation is shown in Figure 4B. It is obvious that MB was completely degraded after incubation for 40 min.

The depletion of GSH by Cu-MOF and the generation of H_2O_2 by GOD were investigated. First of all, Cu-MOF were treated by excess GSH. The residual GSH was indicated by 5,5′-dithiobis (2-nitrobenzoic acid) (DTNB), a kind of sulphydryl (–SH) indicator at different time points [35]. With the extended time, the characteristic absorbance at ~412 nm decreased, indicating the depletion of GSH. As shown in Figure S8A, the depletion capacity of GSH exhibits a concentration-dependent model. The fluorescence signals of Ampliflu Red were influenced by the change of H_2O_2 concentration [36]. Therefore, Ampliflu Red can be used to investigate the generation of H_2O_2. As shown in Figure S8B,C, the generation of H_2O_2 exhibits a time-dependent model and a Glu-concentration-dependent model. The catalytic reaction of GOD with glucose to produce gluconic acid and H_2O_2 will cause the pH of the aqueous solution to change. Therefore, after mixing Cu-MOF or Cu-MOF/GOD and Glu for different time, the pH change can reflect the catalytic performance of GOD in Cu-MOF/GOD. As shown in Figure S9, the pH values show a time-dependent mode and a concentration-dependent mode (Glu). These results indicate that GOD loaded in Cu-MOF/GOD still can catalyze Glu to produce H_2O_2.

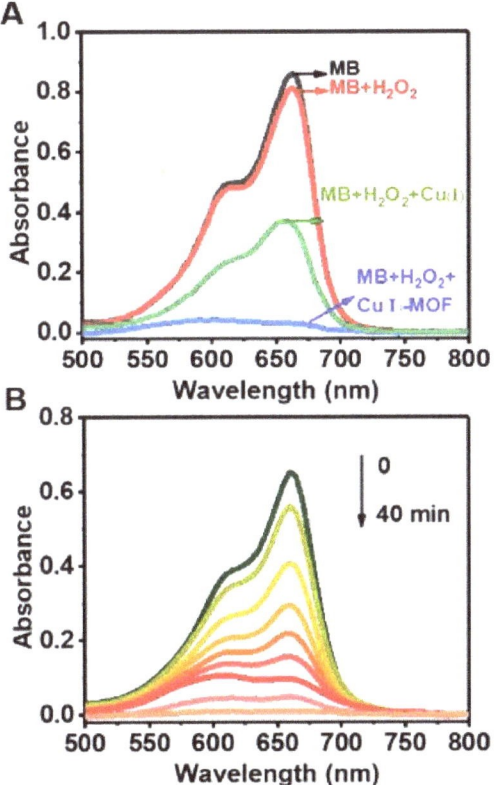

Figure 4. Experimental demonstration of the mechanism for the CDT process with Cu-MOF as the nano-therapeutic agent. OH generated from the Fenton-like reaction between Cu(I)-MOF and H_2O_2 leads to the degradation of MB. (**A**) MB degradation after incubation for 40 min in the presence of Cu(I)-MOF, H_2O_2, H_2O_2+Cu(I)-MOF and H_2O_2+free Cu(I), respectively. (**B**) The time-dependent degradation process of MB in the presence of Cu(I)-MOF. MB: 5 µg mL^{-1}, H_2O_2: 10 mmol L^{-1}, Cu(I)-MOF: 200 µg mL^{-1} (Cu(I): 0.494 µg mL^{-1}), free Cu(I): 0.494 µg mL^{-1}, time interval: 5 min.

3.3. Cell Experiment

After Cu-MOF/GOD@HA endocytosis into tumor cells, Cu(II) in Cu-MOF were reduced to Cu(I) by GSH in tumor. Cu(I) catalyzes H_2O_2 to generate a large number of ·OH. The excessive ·OH oxidize signal molecules, cytokines, proteins, nucleic acids, carbohydrates, lipids and etc to promote apoptosis of tumor cells [1,2,4,5,7].

MTT experiments were used to evaluate the lethality of Cu-MOF, Cu-MOF/GOD and Cu-MOF/GOD@HA [37–40]. The viabilities of MCF-7 cells treated with Cu-MOF, Cu-MOF/GOD and Cu-MOF/GOD@HA at several concentrations (0, 10, 20, 30, 50, 70, 100 µg mL^{-1}) levels were calculated, as illustrated in Figure 4A. The quantitative results showed that 10 µg mL^{-1} the Cu-MOF/GOD@HA nanocomposites gave rise to an MCF-7 cell viability of 49.8%, while the same concentration of Cu-MOF and Cu-MOF/GOD produced an MCF-7 cell viability of 87.9% and 83.1%, respectively. This could be attributed to the targeting ability of HA. 20 µg mL^{-1} the Cu-MOF nanocomposites gave rise to an MCF-7 cell viability of 84.5%, while the same concentration of Cu-MOF/GOD and Cu-MOF/GOD@HA produced an MCF-7 cell viability of 36.5% and 19.9%, respectively. This could be due to the loading of GOD. GOD catalyzes Glu to supply more H_2O_2 for CDT. The 50% inhibitory concentration (IC$_{50}$) values of Cu-MOF, Cu-MOF/GOD and Cu-MOF/GOD@HA for MCF-7 cells were found to be 70.8, 17.5, 9.7 µg mL^{-1}. In general,

Cu-MOF/GOD@HA improves its own tumor lethality in two aspects, one is the catalytic ability of GOD, and the other is the targeting ability of HA. When MCF-7 cells were incubated with lower concentration Cu-MOF/GOD@HA, the targeting performance of HA plays a dominant role. When MCF-7 cells were incubated with higher concentration Cu-MOF/GOD@HA, the catalysis performance of GOD plays a dominant role. Moreover, trypan blue was used to observe the lethality of Cu-MOF/GOD@HA more intuitively as illustrated in Figure 5B. It is clearly shown that the incubation of MCF-7 cells with Cu-MOF, Cu-MOF/GOD and Cu-MOF/GOD@HA for 4 h leads to the significant decrease on the number of cells and the morphology of MCF-7 cells became irregular. Catalytic ability of GOD and targeting ability of HA improve the lethality of Cu-MOF/GOD@HA.

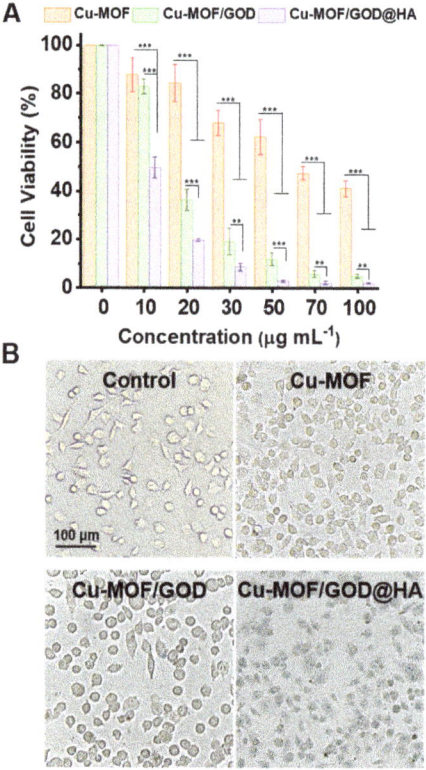

Figure 5. Cytotoxicity of different nanocomposites under different conditions. (**A**) MCF-7 cells were incubated with Cu-MOF, Cu-MOF/GOD and Cu-MOF/GOD@HA at concentrations of 0, 10, 20, 30, 50, 70, 100 µg mL^{-1}. (**B**) Microscopy images of MCF-7 cells with trypan blue staining. MCF-7 cells were incubated with Cu-MOF, Cu-MOF/GOD and Cu-MOF/GOD@HA for 4 h at the levels of 20 µg mL^{-1}. Scale bar: 100 µm. Values of $p < 0.05$ were considered statistically significant, with *, **, *** represent $p < 0.05$, $p < 0.01$ and $p < 0.001$, respectively.

3.4. In Vivo Antitumor Efficacy

The above discussions about the Cu-MOF/GOD@HA mediated CDT mechanisms indicate that the CDT process is significantly sensitive to GSH, which to a large extent determines the efficiency of CDT. In the present case, the variation of cell viability is evaluated by regulating the level of GSH in the cancer cell interior. For this purpose, MCF-7 cells were pre-incubated with GSH at the concentration of 2.5 and 5.0 mmol L^{-1}, respectively. Then, the cytotoxicity of Cu-MOF was assessed. As shown in Figure 6A, the increase of GSH level in tumor cell environment leads to a significant increase on the lethality of Cu-MOF to MCF-7 cells. The corresponding IC$_{50}$ values of Cu-MOF were 70.8, 24.0 and 19.1 µg mL^{-1}, at GSH levels of 0, 2.5 and 5.0 mmol L^{-1}, respectively. Moreover, the variation of cell viability was evaluated by regulating the level of H$_2$O$_2$ in the cancer cell interior. For this purpose, MCF-7 cells were pre-incubated with H$_2$O$_2$ at the concentration of 100 µmol L^{-1}. Then, the cytotoxicity of Cu-MOF was assessed. As shown in Figure 6B, the increase of H$_2$O$_2$ level in the tumor cell environment leads to a significant increase in the lethality of Cu-MOF to MCF-7 cells. The corresponding IC$_{50}$ values were 70.8 and 20.3 µg mL^{-1}, at H$_2$O$_2$ levels of 0 and 100 µmol L^{-1}. Furthermore, in order to investigate the influence of Cu-MOF, Cu-MOF/GOD and Cu-MOF/GOD@HA on the production of intracellular ROS, ROS detection kit was used for cell staining after MCF-7 cells were incubated with 50 µg mL^{-1} of Cu-MOF, Cu-MOF/GOD and Cu-MOF/GOD@HA for 4 h. The results in Figure S10 indicate that the loading of GOD and targeting ability of HA are beneficial for the generation of ·OH.

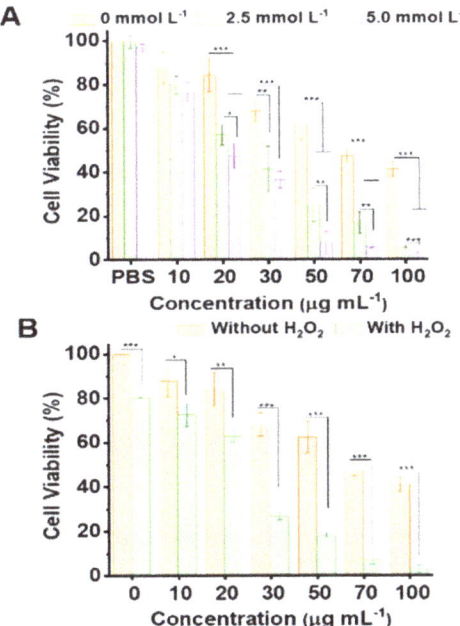

Figure 6. (**A**) The dependence of MCF-7 cell viability on the concentration of GSH (0, 2.5 and 5.0 mmol L^{-1}) by incubation with Cu-MOF at the levels of 10, 20, 30, 50, 70, 100 µg mL^{-1}. (**B**) The MCF-7 cell viability on H$_2$O$_2$ (100 µmol L^{-1}) by incubation with Cu-MOF at the levels of 10, 20, 30, 50, 70, 100 µg mL^{-1}. Values of $p < 0.05$ were considered statistically significant, with *, **, *** represent $p < 0.05, p < 0.01$, and $p < 0.001$, respectively.

The in vivo antitumor efficacy of the Cu-MOF, Cu-MOF/GOD and Cu-MOF/GOD@HA was investigated in nude mice bearing MCF-7. Once the tumors had grown to approximately 90 mm^3, the mice were intertumorally injected with either PBS (control, 10 mmol L^{-1}), Cu-MOF, Cu-MOF/GOD and Cu-MOF/GOD@HA (The concentrations were all 2.5 mg kg^{-1}) in 10 mmol L^{-1} PBS every 2 days (n = 3). A total of seven injections were performed over 2 weeks. During the treatment process, the mice in control (PBS) and experimental groups exhibit virtually no difference in their body weight, indicating negligible systemic toxicity of Cu-MOF, Cu-MOF/GOD and Cu-MOF/GOD@HA itself (Figure S11A). As shown in Figure S11B, tumor sizes in the PBS group were evidently increased from ~100 mm^3 to ~774 mm^3. the tumor sizes in the groups by injecting 2.5 mg kg^{-1} Cu-MOF, Cu-MOF/GOD were increased from ~90 mm^3 to ~238, 95 mm^3, respectively. On the contrary, the tumor sizes in the groups by injecting 2.5 mg kg^{-1} Cu-MOF/GOD@HA were decreased from ~90 mm^3 to 45 mm^3. At the end of the treatment (15 days), the tumor tissues were excised from the mice and weighed. As shown in Figure S11C, the tumor sizes in the experimental and groups (PBS, Cu-MOF, Cu-MOF/GOD and Cu-MOF/GOD@HA) decreased by 7.72, 2.98, 1.08 and 0.57 with respect to the initial-tumor volumes. The tumor sizes were obviously increased in the PBS group. In contract, the tumor sizes for the experimental group were remarkably decreased (Figure S11D).

In the concentration range of 10–100 μg mL^{-1}, Cu-MOF/GOD@HA exhibited no obvious hemolytic effect, and the hemolytic rate at each concentration level was less than 4%. This result well indicated the excellent blood compatibility of the Cu-MOF/GOD@HA. Figure S11E illustrated that after H&E staining of the major organs of mice including heart, liver, spleen, lung and kidney, no obvious pathological changes in these tissues were observed in the presence of Cu-MOF/GOD@HA. In order to avoid the possible hemolysis or blood cell aggregation after intravenous injection, the hemolysis experiment of Cu-MOF/GOD@HA was carried out (Figure S12).

4. Conclusions

In summary, Cu-metal organic framework (Cu-MOF)/glucose oxidase (GOD)@hyaluronic acid (HA) (Cu-MOF/GOD@HA) were prepared for chemodynamic therapy (CDT) with the aims to increase H_2O_2, targeting and efficacy. Cu-MOF/GOD@HA were activated by intracellular GSH and catalyzed intracellular H_2O_2 reactions to generate ·OH. During these processes, a large amount of ·OH were generated which well facilitates CDT for cancer destruction. The loading of GOD and coating of HA can improve the efficacy of CDT. The present study provides a useful route for the development of nanotherapeutic agents for the treatment of tumor by taking advantage of the intracellular ingredients.

Supplementary Materials: The following are available online at https://www.mdpi.com/article/10.3390/nano11071843/s1, Figure S1: (A) SEM image of Cu-MOF. Scar bar: 50 nm. (B) Elemental mapping of C, N, O and Cu of Cu-MOF, Figure S2. (A) FT-IR spectra of Cu-MOF, Cu-MOF/GOD, Cu-MOF/GOD@HA and GOD. (B) TGA curves of Cu-MOF and Cu-MOF/GOD@HA, Figure S3. XRD spectra of Cu-MOF, Cu-MOF/GOD and Cu-MOF/GOD@HA and simulated XRD pattern of Cu-MOF, Figure S4. UV-vis absorption spectra of Cu-MOF and Cu(I)-MOF (The concentrations of pending test samples were 100 μg mL^{-1}). Inset image shows the photographs of the solutions containing Cu-MOF and Cu(I)-MOF, Figure S5. TEM image of Cu-MOF after treatment by GSH, Figure S6. XPS pattern of Cu-MOF (A), Cu-MOF/GOD (B), Cu-MOF/GOD@HA (C), Cu(I)-MOF (D). Cu 2p high-resolution XPS pattern of Cu(I)-MOF (E), Cu-MOF (F), Cu-MOF/GOD (G), Cu-MOF/GOD@HA (H). Figure S7. UV-vis spectra of Cu-MOF+H_2O_2+MB and MB. Figure S8. (A) GSH depletion after incubation for 1 h in the presence of DTNB (720 μg mL^{-1}) and Cu-MOF (0, 5, 15, 25, 50 μg mL^{-1}). (B) The time-dependent reaction of Ampliflu Red solution (600 μg mL^{-1}) with Glu (500 μg mL^{-1})+Cu-MOF/GOD (100 μg mL^{-1}). (C) The concentration-dependent reaction of Ampliflu Red solution (600 μg mL^{-1}) with Glu+Cu-MOF/GOD (100 μg mL^{-1}) for 30 min ($\lambda_{ex}/\lambda_{em}$ = 530/585 nm). Figure S9. (A) The pH values of Cu-MOF+Glu and Cu-MOF/GOD+Glu aqueous solution in different time intervals. Cu-MOF and Cu-MOF/GOD: 100 μg mL^{-1}, Glu: 500 μg mL^{-1}. (B) The pH values of Cu-MOF and Cu-MOF/GOD with different concentration of Glu aqueous solution after 24 h. Cu-MOF/GOD: 100 μg mL^{-1}, Glu: 0, 20, 50, 100, 500, 750, 1000 μg mL^{-1}. Figure S10. ROS staining in MCF-7 cells after incubation with Cu-MOF, Cu-MOF/GOD and Cu-MOF/GOD@HA for 4 h at the concentration of 50 μg mL^{-1}, scale bar: 10 μm. Figure S11. In vivo CDT treatment for MCF-7 cancer cell-bearing mice with different nanocomposites (PBS, Cu-MOF, Cu-MOF/GOD and Cu-MOF/GOD@HA 2.5 mg kg^{-1}). (A) The changes of body weight for the KunMing mice during the process of therapy. (B) The variation of tumor size for the KunMing mice during the process of therapy. (C) The average relative mass excised from MCF-7 tumor-bearing mice after the treatment. (D) Photographs showing the tumor size after the treatment. Scale bar, 1 cm. (E) H&E staining of the major organs/tissues of mice after CDT process. Scale bar, 50 μm. Values of $p < 0.05$ were considered statistically significant, with *, **, *** represent $p < 0.05$, $p < 0.01$, and $p < 0.001$, respectively. 1, 2, 3 and 4 represent the group of PBS, Cu-MOF, Cu-MOF/GOD and Cu-MOF/GOD@HA, respectively. Figure S12. Hemolysis percentage of RBCs by Cu-MOF/GOD@HA nanodots at various concentration levels (10–100 µg mL^{-1}). Inset: the photographs for direct observation of hemolysis.

Author Contributions: Y.-N.H.: Conceptualization, Writing-Original Draft, Writing-Review & Editing. C.-C.Q.: Formal analysis, Investigation, Writing-Review & Editing. Y.S.: Conceptualization, Resources, Project administration, Writing-Review & Editing, Funding acquisition. J.-H.W.: Conceptualization, Resources, Writing-Review & Editing, Supervision, Funding acquisition. W.C.: Conceptualization, Resources, Writing-Review & Editing, Supervision. All authors have read and agreed to the published version of the manuscript.

Funding: This research received no external funding.

Data Availability Statement: The details regarding where data supporting reported results can be obtained from the authors.

Acknowledgments: This work is financially supported by the Natural Science Foundation of China (21974018, 21727811 and 22074011), Fundamental Research Funds for the Central Universities (N2005015, and N2005027), and Liaoning Revitalization Talents Program (XLYC1907191, XLYC1802016).

Conflicts of Interest: The authors declare no conflict of interest.

References

1. Hao, Y.N.; Zhang, W.X.; Gao, Y.R.; Wei, Y.N.; Shu, Y.; Wang, J.H. State-of-the-art advances of copper-based nanostructures in the enhancement of chemodynamic therapy. *J. Mater. Chem. B* **2021**, *9*, 250–266. [CrossRef] [PubMed]
2. Tian, Q.; Xue, F.; Wang, Y.; Cheng, Y.; An, L.; Yang, S.; Chen, X.; Huang, G. Recent advances in enhanced chemodynamic therapy strategies. *Nano Today* **2021**, *39*. [CrossRef]
3. Li, S.; Jiang, P.; Jiang, F.; Liu, Y. Recent advances in nanomaterial-based nanoplatforms for chemodynamic cancer therapy. *Adv. Funct. Mater.* **2021**, *31*. [CrossRef]
4. Lin, L.-S.; Huang, T.; Song, J.; Ou, X.-Y.; Wang, Z.; Deng, H.; Tian, R.; Liu, Y.; Wang, J.-F.; Liu, Y.; et al. Synthesis of copper peroxide nanodots for H$_2$O$_2$ self-supplying chemodynamic therapy. *J. Am. Chem. Soc.* **2019**, *141*, 9937–9945. [CrossRef]
5. Chang, M.; Wang, M.; Wang, M.; Shu, M.; Ding, B.; Li, C.; Pang, M.; Cui, S.; Hou, Z.; Lin, J. A multifunctional cascade bioreactor based on hollow-structured Cu$_2$MoS$_4$ for synergetic cancer chemo-dynamic therapy/starvation therapy/phototherapy/immunotherapy with remarkably enhanced efficacy. *Adv. Mater.* **2019**, *31*. [CrossRef]
6. Fu, L.-H.; Hu, Y.-R.; Qi, C.; He, T.; Jiang, S.; Jiang, C.; He, J.; Qu, J.; Lin, J.; Huang, P. Biodegradable manganese-doped calcium phosphate nanotheranostics for traceable cascade reaction-enhanced anti-tumor therapy. *ACS Nano* **2019**, *13*, 13985–13994. [CrossRef]
7. Tang, Z.; Liu, Y.; He, M.; Bu, W. Chemodynamic therapy: Tumour microenvironment-mediated Fenton and Fenton-like reactions. *Angew. Chem. Int. Ed.* **2018**, *58*, 946–956. [CrossRef]
8. Chudal, L.; Pandey, N.K.; Phan, J.; Johnson, O.; Lin, L.; Yu, H.; Shu, Y.; Huang, Z.; Xing, M.; Liu, J.P.; et al. Copper-cysteamine nanoparticles as a heterogeneous Fenton-like catalyst for highly selective cancer treatment. *ACS Appl. Bio Mater.* **2020**, *3*, 1804–1814. [CrossRef]
9. Zeng, L.; Cao, Y.; He, L.; Ding, S.; Bian, X.W.; Tian, G. Metal-ligand coordination nanomaterials for radiotherapy: Emerging synergistic cancer therapy. *J. Mater. Chem. B* **2021**, *9*, 208–227. [CrossRef]
10. Zhang, C.; Bu, W.; Ni, D.; Zhang, S.; Li, Q.; Yao, Z.; Zhang, J.; Yao, H.; Wang, Z.; Shi, J. Synthesis of iron nanometallic glasses and their application in cancer therapy by a localized Fenton reaction. *Angew. Chem.* **2016**, *128*, 2141–2146. [CrossRef]
11. Bokare, A.D.; Choi, W. Review of iron-free Fenton-like systems for activating H$_2$O$_2$ in advanced oxidation processes. *J. Hazard. Mater.* **2014**, *275*, 121–135. [CrossRef] [PubMed]
12. Chudal, L.; Pandey, N.K.; Phan, J.; Johnson, O.; Li, X.; Chen, W. Investigation of PPIX-Lipo-MnO$_2$ to enhance photodynamic therapy by improving tumor hypoxia. *Mater. Sci. Eng.* **2019**, *104*. [CrossRef] [PubMed]
13. Yao, M.; Ma, L.; Li, L.; Zhang, J.; Lim, R.X.; Chen, W.; Zhang, Y. A new modality for cancer treatment—nanoparticle mediated microwave induced photodynamic therapy. *J. Biomed. Nanotechnol.* **2016**, *12*, 1835–1851. [CrossRef]
14. Fu, L.; Wan, Y.; Qi, C.; He, J.; Li, C.; Yang, C.; Xu, H.; Lin, J.; Huang, P. Nanocatalytic theranostics with glutathione depletion and enhanced reactive oxygen species generation for efficient cancer therapy. *Adv. Mater.* **2021**, *33*. [CrossRef] [PubMed]
15. Chen, Q.; Liang, C.; Sun, X.; Chen, J.; Yang, Z.; Zhao, H.; Feng, L.; Liu, Z. H$_2$O$_2$-responsive liposomal nanoprobe for photoacoustic inflammation imaging and tumor theranostics via in vivo chromogenic assay. *Proc. Natl. Acad. Sci. USA* **2017**, *114*, 5343. [CrossRef]
16. Liu, Y.; Wu, J.; Jin, Y.; Zhen, W.; Wang, Y.; Liu, J.; Jin, L.; Zhang, S.; Zhao, Y.; Song, S.; et al. Copper(I) phosphide nanocrystals for in situ self-generation magnetic resonance imaging-guided photothermal-enhanced chemodynamic synergetic therapy resisting deep-seated tumor. *Adv. Funct. Mater.* **2019**, *29*. [CrossRef]

17. Wang, Y.; Song, M. pH-responsive cascaded nanocatalyst for synergistic like-starvation and chemodynamic therapy. *Colloids Surf. B* **2020**, *192*. [CrossRef]
18. Fan, W.; Lu, N.; Huang, P.; Liu, Y.; Yang, Z.; Wang, S.; Yu, G.; Liu, Y.; Hu, J.; He, Q.; et al. Glucose-responsive sequential generation of hydrogen peroxide and nitric oxide for synergistic cancer starving-like/gas therapy. *Angew. Chem. Int. Ed.* **2016**, *56*, 1229–1233. [CrossRef]
19. Wang, H.; Chen, Y.; Wang, H.; Liu, X.; Zhou, X.; Wang, F. DNAzyme-loaded, metal-organic frameworks (MOFs) for self-sufficient gene therapy. *Angew. Chem. Int. Ed.* **2019**, *58*, 7380–7384. [CrossRef]
20. Zhang, K.; Meng, X.; Yang, Z.; Dong, H.; Zhang, X. Enhanced cancer therapy by hypoxia-responsive copper metal-organic frameworks nanosystem. *Biomaterials* **2020**, *258*. [CrossRef]
21. Wang, Y.; Wu, W.; Mao, D.; Teh, C.; Wang, B.; Liu, B. Metal-organic framework assisted and tumor microenvironment modulated synergistic image-guided photo-chemo therapy. *Adv. Funct. Mater.* **2020**, *30*. [CrossRef]
22. Ding, S.-S.; He, L.; Bian, X.-W.; Tian, G. Metal-organic frameworks-based nanozymes for combined cancer therapy. *Nano Today* **2020**, *35*. [CrossRef]
23. Zhou, J.; Li, M.; Hou, Y.; Luo, Z.; Chen, Q.; Cao, H.; Huo, R.; Xu, C.; Sutrisno, C.; Hao, L.; et al. Engineering of a nanosized biocatalyst for combined tumor starvation and low-temperature photothermal therapy. *ACS Nano* **2018**, *12*, 2858–2872. [CrossRef] [PubMed]
24. Phua, S.Z.F.; Yang, G.; Lim, W.Q.; Verma, A.; Chen, H.; Thanabalu, T.; Zhao, Y. Catalase-integrated hyaluronic acid as nanocarriers for enhanced photodynamic therapy in solid tumor. *ACS Nano* **2019**, *13*, 4742–4751. [CrossRef] [PubMed]
25. Jia, H.R.; Zhu, Y.X.; Liu, X.; Pan, G.Y.; Gao, G.; Sun, W.; Zhang, X.; Jiang, Y.W.; Wu, F.G. Construction of dually responsive nanotransformers with nanosphere-nanofiber-nanosphere transition for overcoming the size paradox of anticancer nanodrugs. *ACS Nano* **2019**, *13*, 11781–11792. [CrossRef] [PubMed]
26. Zhang, X.; He, F.; Xiang, K.; Zhang, J.; Xu, M.; Long, P.; Su, H.; Gan, Z.; Yu, Q. CD44-targeted facile enzymatic activatable chitosan nanoparticles for efficient antitumor therapy and reversal of multidrug resistance. *Biomacromolecules* **2018**, *19*, 883–895. [CrossRef] [PubMed]
27. Choi, K.Y.; Chung, H.; Min, K.H.; Yoon, H.Y.; Kim, K.; Park, J.H.; Kwon, I.C.; Jeong, S.Y. Self-assembled hyaluronic acid nanoparticles for active tumor targeting. *Biomaterials* **2010**, *31*, 106–114. [CrossRef] [PubMed]
28. Choi, K.Y.; Yoon, H.Y.; Kim, J.H.; Bae, S.M.; Park, R.W.; Kang, Y.M.; Kim, I.S.; Kwon, I.C.; Choi, K.; Jeong, S.Y.; et al. Smart nanocarrier based on PEGylated hyaluronic acid for cancer therapy. *ACS Nano* **2011**, *5*, 8591–8599. [CrossRef]
29. Lv, Y.; Xu, C.; Zhao, X.; Lin, C.; Yang, X.; Xin, X.; Zhang, L.; Qin, C.; Han, X.; Yang, L.; et al. Nanoplatform assembled from a CD44-targeted prodrug and smart liposomes for dual targeting of tumor microenvironment and cancer cells. *ACS Nano* **2018**, *12*, 1519–1536. [CrossRef]
30. Mu, J.; Lin, J.; Huang, P.; Chen, X. Development of endogenous enzyme-responsive nanomaterials for theranostics. *Chem. Soc. Rev.* **2018**, *47*, 5554–5573. [CrossRef]
31. Li, Y.; Li, X.; Guan, Q.; Zhang, C.; Xu, T.; Dong, Y.; Bai, X.; Zhang, W. Strategy for chemotherapeutic delivery using a nanosized porous metal-organic framework with a central composite design. *Int. J. Nanomed.* **2017**, *12*, 1465–1474. [CrossRef]
32. Ming, J.; Zhu, T.; Yang, W.; Shi, Y.; Huang, D.; Li, J.; Xiang, S.; Wang, J.; Chen, X.; Zheng, N. Pd@Pt-GOx/HA as a novel enzymatic cascade nanoreactor for high-efficiency starving-enhanced chemodynamic cancer therapy. *ACS Appl. Mater. Interfaces* **2020**, *12*, 51249–51262. [CrossRef] [PubMed]
33. Špadina, M.; Gourdin-Bertin, S.; Dražić, G.; Selmani, A.; Dufrêche, J.F.; Bohinc, K. Charge properties of TiO_2 nanotubes in $NaNO_3$ aqueous solution. *ACS Appl. Mater. Interfaces* **2018**, *10*, 13130–13142. [CrossRef] [PubMed]
34. Lin, L.; Wang, S.; Deng, H.; Yang, W.; Rao, L.; Tian, R.; Liu, Y.; Yu, G.; Zhou, Z.; Song, J.; et al. Endogenous labileiron pool-mediated free radical generation for cancer chemodynamic therapy. *J. Am. Chem. Soc.* **2020**, *142*, 15320–15330. [CrossRef] [PubMed]
35. Zhong, X.; Wang, X.; Cheng, L.; Tang, Y.; Zhan, G.; Gong, F.; Zhang, R.; Hu, J.; Liu, Z.; Yang, X. GSH-depleted $PtCu_3$ nanocages for chemodynamic-enhanced sonodynamic cancer therapy. *Adv. Funct. Mater.* **2019**, *30*. [CrossRef]
36. Morlock, L.K.; Böttcher, D.; Bornscheuer, U.T. Simultaneous detection of NADPH consumption and H_2O_2 production using the Amplifu™ Red assay for screening of P450 activities and uncoupling. *Appl. Microbiol. Biotechnol.* **2017**, *102*, 985–994. [CrossRef]
37. Pandey, N.K.; Xiong, W.; Wang, L.; Chen, W.; Bui, B.; Yang, J.; Amador, E.; Chen, M.; Xing, C.; Athavale, A.A.; et al. Aggregation-induced emission luminogens for highly effective microwave dynamic therapy. *Bioact. Mater.* **2021**. [CrossRef]
38. Chen, X.; Liu, J.; Li, Y.; Pandey, N.K.; Chen, T.; Wang, L.; Amador, E.H.; Chen, W.; Liu, F.; Xiao, E.; et al. Study of copper-cysteamine based X-ray induced photodynamic therapy and its effects on cancer cell proliferation and migration in a clinical mimic setting. *Bioact. Mater.* **2021**. [CrossRef]
39. Wang, Y.; Alkhaldi, N.; Pandey, N.; Chudal, L.; Wang, L.; Lin, L.; Zhang, M.; Yong, Y.; Amador, E.; Huda, M.; et al. A new type of cuprous-cysteamine sensitizers: Synthesis, optical properties and potential applications. *Mater. Today Phys.* **2021**. [CrossRef]
40. Zhang, Q.; Guo, X.; Cheng, Y.; Chudal, L.; Pandey, N.K.; Zhang, J.; Ma, L.; Xi, Q.; Yang, G.; Chen, Y.; et al. Use of copper-cysteamine nanoparticles to simultaneously enable radiotherapy, oxidative therapy and immunotherapy for melanoma treatment. *Signal Transduct. Target. Ther. (Nat.)* **2020**, *5*. [CrossRef]

Article

Synthesis of a Two-Dimensional Molybdenum Disulfide Nanosheet and Ultrasensitive Trapping of *Staphylococcus Aureus* for Enhanced Photothermal and Antibacterial Wound-Healing Therapy

Weiwei Zhang [1,†], Zhao Kuang [1,†], Ping Song [1], Wanzhen Li [1], Lin Gui [2], Chuchu Tang [1], Yugui Tao [1,*], Fei Ge [1,*] and Longbao Zhu [1,*]

[1] School of Biological and Food Engineering, Anhui Polytechnic University, Wuhu 241000, China; zwwjcf0908@163.com (W.Z.); kz827478890@163.com (Z.K.); songping1987@foxmail.com (P.S.); liwanzhen129@126.com (W.L.); tangcc2333@126.com (C.T.)
[2] Department of Microbiology and Immunology, Wannan Medical College, Wuhu 241002, China; guilin729@126.com
* Correspondence: swgctaoyg@126.com (Y.T.); gerrylin@126.com (F.G.); lbzhu2008@126.com (L.Z.)
† These authors contributed equally to this work.

Abstract: Photothermal therapy has been widely used in the treatment of bacterial infections. However, the short photothermal effective radius of conventional nano-photothermal agents makes it difficult to achieve effective photothermal antibacterial activity. Therefore, improving composite targeting can significantly inhibit bacterial growth. We inhibited the growth of *Staphylococcus aureus* (*S. aureus*) by using an extremely low concentration of vancomycin (Van) and applied photothermal therapy with molybdenum disulfide (MoS_2). This simple method used chitosan (CS) to synthesize fluorescein 5(6)-isothiocyanate (FITC)-labeled and Van-loaded MoS_2-nanosheet hydrogels (MoS_2-Van-FITC@CS). After modifying the surface, an extremely low concentration of Van could inhibit bacterial growth by trapping bacteria synergistically with the photothermal effects of MoS_2, while FITC labeled bacteria and chitosan hydrogels promoted wound healing. The results showed that MoS_2-Van-FITC@CS nanosheets had a thickness of approximately 30 nm, indicating the successful synthesis of the nanosheets. The vitro antibacterial results showed that MoS_2-Van-FITC with near-infrared irradiation significantly inhibited *S. aureus* growth, reaching an inhibition rate of 94.5% at nanoparticle concentrations of up to 100 μg/mL. Furthermore, MoS_2-Van-FITC@CS could exert a healing effect on wounds in mice. Our results demonstrate that MoS_2-Van-FITC@CS is biocompatible and can be used as a wound-healing agent.

Keywords: MoS_2 nanosheet; antibacterial activity; photothermal therapy; wound-healing

1. Introduction

Bacterial infections are consistently ranked as one of the leading causes of human mortality, with infection rates, mortality rates, and hospitalization costs increasing annually [1–3]. The high prevalence of bacterial infections has led to the misuse of antibiotics, resulting in the emergence of superbugs as bacteria become resistant to treatment. Unfortunately, superbugs are arising at a rate much faster than that of new antibiotic discovery, thereby leading to a growing threat of untreatable bacterial infections [4–8]. Therefore, new treatments are urgently needed.

With the rapid development of technologies in the fields of modern nanotechnology and biomedicine, numerous antibacterial inorganic nanoparticles (NPs), such as silver, gold, copper nanoparticles, alumina, zinc oxide, magnesium oxide, silica titanium dioxide, and graphene oxide NPs, as well as their composites, have been used in antibacterial therapy [3,9–11]. For example, two-dimensional (2D) graphene-based nanocomposites and analogues have been demonstrated to have wide application prospects because of their multi-functional antibacterial mechanisms, which include physical and chemical damage to bacterial cells [12,13]. Molybdenum disulfide (MoS_2), a transition metal dichalcogenide, is similar to graphene. Currently, MoS_2 has several potential applications in the field of biomedicine due to its unique electronic, optical, mechanical, and chemical properties, and MoS_2 has been demonstrated to have unique optical properties in that it can convert itself into heat by absorbing light energy when irradiated with near-infrared (NIR) light. Through this property, MoS_2 can kill bacteria [14–16].

Free MoS_2 without irradiation does not have significant antibacterial effects. To improve the antibacterial ability of MoS_2, we selected vancomycin (Van). Van is a peptide antibiotic that can kill bacteria, such as *Staphylococcus aureus* (*S. aureus*), residing on a wound's surface [17,18]. Van has an attractive property in that it can target bacteria through hydrogen bonding to the terminal D-Ala-D-Ala sequence of the cytosolic peptide unit of Gram-positive bacteria. In other words, Van recognizes the D-Ala-D-Ala sequence on the cell surface of bacteria, increasing the half-life and effective working radius of the nano-antibacterial composites and enhancing the antibacterial effect [19–22]. Together, MoS_2 and Van can target Gram-positive bacteria, which can further enhance the thermal response of MoS_2 and promote bacterial growth inhibition while applying antibiotic therapy. Van has a primary amine group that binds covalently to fluorescein 5(6)-isothiocyanate (FITC), and we selected FITC-labeled Van because it can maintain both the fluorescent properties of FITC and the ability of Van to bind to the bacterial cell wall, thereby facilitating subsequent experimental validation.

To promote wound healing, it is important not only to clean the wound of germs but also to maintain an environment that is suitable for healing. A moist and clean environment accelerates the migration of epidermal cells, which facilitates skin cell granulation and division [23–25]. Polymer hydrogels are one of the most practiced soft-wet materials used for biomedical applications [26]. They can provide this type of environment for wounds, and thermosensitive hydrogels are easy to prepare. Chitosan (CS) is a biocompatible and weakly immunogenic material that can be degraded by enzymes in vivo, and the degradation products, namely oligosaccharides, are non-toxic. In addition, CS can enhance drug penetration by affecting the tightness between epithelial cells, which makes CS a valuable entity in the biomedical field [27]. As such, we prepared a temperature-sensitive hydrogel excipient by wrapping our synthesized MoS_2-Van-FITC nanomaterials, which possess photothermic and chemotherapeutic properties, in a hydrogel. By physically mixing CS with a sodium β-glycerophosphate (β-GP) solution, a sol-gel phase transition was achieved at temperatures higher than 37 °C due to the enhanced hydrogen bonding, electrostatic attraction, and hydrophobic interactions between CS and β-GP [28].

In this study (Figure 1), Van not only efficiently inhibited the growth of bacteria on the wound surface but also enhanced the antibacterial effect by targeting bacteria through hydrogen bonding with the terminal D-Ala-D-Ala sequence of the cytosolic peptide unit of Gram-positive bacteria, increasing the half-life of the composite and the effective working radius of the photothermal treatment. Therefore, when MoS_2 is combined with Van, the effective working radius of MoS_2 PTT can be further increased to better inhibit bacterial growth while applying anti-infective therapy. The covalent binding of primary amine groups using FITC with MoS_2-Van confers the nanocomposite to maintain the fluorescent properties of FITC, which can be observed for real-time detection. Additionally, photothermal treatment combined with chemotherapy cleaned the wound surface of pathogenic bacteria, thereby accelerating wound healing.

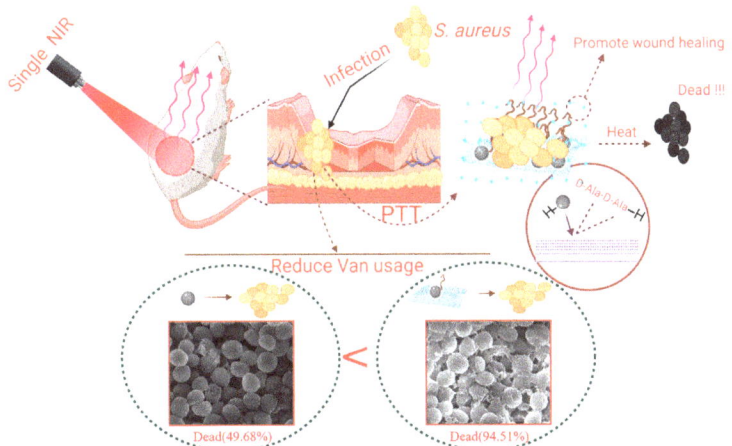

Figure 1. Schematic illustration of the synergism between the two components of MoS$_2$ and Van in MoS$_2$-Van-FITC@CS, which was constructed and applied to wound healing in vivo with their highly efficiently antibacterial properties.

2. Materials and Methods

2.1. Materials

Raw MoS$_2$ (99.5%), FITC (90%, mixture of 5- and 6-isomers), CS (≥99.8%), and Van HCL were purchased from Aladdin Chemical Reagent Co., Ltd. (Shanghai, China). The beef extract peptone agar medium, *Staphylococcus aureus* ATCC 25,923 cells (*S. aureus*, Gram-positive), and *Escherichia coli* ATCC 11,303 cells (*E. coli*, Gram-negative) were obtained from Anhui Polytechnic University.

2.2. Synthesis of MoS$_2$ Nanosheets

MoS$_2$ nanosheets were synthesized by liquid ultrasonic stripping [29–32]. In brief, 0.5 g of MoS$_2$ was combined with 50 mL of 45% ethanol to form the dispersion system of MoS$_2$ (concentration, 10 μg/mL), which was sonicated in a water bath for 12 h. The solution was centrifuged at 12,000 rpm for 15 min, and the supernatant was collected and processed by a rotary evaporator to yield a thin film. The film was weighed, and MoS$_2$ was resuspended in deionized water to form the MoS$_2$ nanosheet aqueous solution.

2.3. Synthesis of MoS$_2$-Van-FITC

Van is a heptapeptide-containing glycopeptide antibiotic with a primary amine moiety that binds covalently to FITC [33]. In brief, 4 mL of the MoS$_2$ nanosheet aqueous solution (concentration, 0.002 g/mL) was aliquoted, 0.2 mg of Van was added to the solution, and the volume was adjusted to 20 mL. The solution was mixed on a magnetic stirrer at 250 rpm for 6 h and centrifuged at 12,000 rpm for 15 min. The precipitate was resuspended in 20 mL of pure water, and 1 mL of FITC was added. The solution was further mixed for 12 h. MoS$_2$-Van-FITC was obtained by centrifugation.

2.4. Synthesis of the MoS$_2$-Van-FITC@CS Hydrogel

To prepare the hydrogel, 300 mg of CS was added to 18 mL of 0.1 mol HCl solution and mixed on a magnetic stirrer until the CS solution was clarified. Thereafter, 2 mg of sodium β-glycerophosphate was added to 2 mL of 0.001 g/mL MoS$_2$-Van-FITC solution, dissolved completely, and mixed for 4 h [27,34,35]. The MoS$_2$-Van-FITC@CS hydrogel was obtained by heating in a water bath at 37 °C for 1 h.

2.5. Characterization

Ultraviolet-visible (UV–vis) absorption spectra were recorded by UV–vis spectroscopy (model S-3100, Scinco Co., Daejeon, Korea). Fluorescence spectra of MoS_2-Van-FITC NPs were recorded by fluorescence spectrophotometry (model RF-5301PC, Shimadzu Corp., Kyoto, Japan). Fourier transform infrared (FT-IR) spectroscopy was performed with an FT-IR spectrometer (model Nicolette is50, Thermo Fisher Scientific, Waltham, MA, USA). The ultrastructural characteristics of synthesized MoS_2 were observed via scanning electron microscopy (SEM; model S-4800, Hitachi, Tokyo, Japan). The Brookhaven Zeta Pals instrument was used to obtain zeta potential measurements and to characterize the optical properties of MoS_2-Van-FITC (Brookhaven Instruments Corp., Holtsville, NY, USA). NPs of different concentrations were illuminated by using NIR irradiation at 808 nm at different power densities for 15 min, and the temperature was detected with an infrared camera with an accuracy of 0.1 °C [36,37].

2.6. In Vitro Antibacterial Assays

The minimum inhibitory concentration (MIC) of MoS_2-Van-FITC NPs was determined using the 96-well microtitration plate dilution method. In brief, 100 μL of LB medium was added to each well of a sterile 96-well plate, and 100 μL of the drug solution was added to the first well, mixed, and diluted in multiples until the last well was mixed. Thereafter, 100 μL of the mixture was discarded, followed by the addition of 100 μL of the bacterial diluent to each well (final density of bacteria, 1×10^6 CFU/mL). The positive controls were kanamycin and ampicillin. The cells of the NIR group were NIR irradiated (1.5 W/cm^2) for 6 min and cultured at 37 °C for 12 h. The OD value was measured with a microplate reader [38]. The three independent measurements were averaged, and each treatment group had three wells.

Log-phase *S. aureus* cultures were inoculated (1:40) into medium containing MoS_2-Van-FITC (100 μg/mL) or MoS_2 (100 μg/mL) and incubated in a shaking incubator at 37 °C for 12 h. The irradiation power densities were 0.5 W/cm^2, 1 W/cm^2, and 1.5 W/cm^2, and the irradiation times were 0 s, 150 s, and 300 s. Thereafter, the cells were coated on solid medium. The number of live bacteria was calculated using the CFU counting method.

2.7. Cellular Uptake Assays

To confirm the effect of Van in capturing *S. aureus* cells, we performed cellular uptake assays. Log-phase *S. aureus* cultures were incubated with different concentrations (15, 30, 45 μg/mL) of MoS_2-Van-FITC NPs for 12 h. The cells treated with PBS served as the control. *S. aureus* cells were harvested and stained with 4′-6-diamidino-2-phenylindole (DAPI; concentration, 5 μg/mL) for 15 min. Thereafter, the cells were washed twice with PBS, and the red and blue channels were examined under a fluorescence microscope.

2.8. LIVE–DEAD Assays

To further investigate the antibacterial ability of the drug, we conducted LIVE–DEAD assays. Log-phase *S. aureus* cultures were inoculated (1:40) in medium containing MoS_2-Van-FITC (100 μg/mL), MoS_2 (100 μg/mL), or Van (1 μg/mL) and incubated in a shaking incubator at 37°C for 12 h. The irradiation power density was 1.5 W/cm^2, and the irradiation time was 300 s. Thereafter, the cells were stained with SYTO9 and propidium iodide for 30 min in the dark, washed twice with PBS, and red and green channels were examined under a fluorescence microscope. The number of non-viable bacterial cells was determined by the CytoFLEX system (Beckman Coulter, Brea, CA, USA) [2,39].

2.9. Cell Integrity Assays

Log-phase *S. aureus* cultures were treated with different concentrations (25, 50, 100 μg/mL) of MoS_2-Van-FITC NPs. Van (1 μg/mL) and MoS_2 (100 μg/mL) served as the controls. The PBS-treated group served as the blank. The irradiation power density was 1.5 W/cm^2, and the irradiation time was 6 min. The cells were collected by centrifugation at 4000 rpm for

10 min, washed twice with PBS, and fixed with 2.5% glutaraldehyde at 4 °C for 12 h. The cells were subjected to gradient dehydration with different concentrations of ethanol (35%, 50%, 70%, 80%, 95%, 100%) for 20 min each time and then placed into acetone [40,41]. The specimens were observed under a scanning electron microscope.

2.10. Establishment of the Wound Mice Model

Kunming female mice (~25 g body weight; 4–5 weeks old) were obtained from the Model Animal Research Center of Nanjing University (Nanjing, China) and housed under standard environmental conditions (temperature, 22 ± 3 °C; humidity, $55 \pm 5\%$; 12-h dark/12-h light cycles). All animals were housed according to the guidelines in the "Guide for the Care and Use of Laboratory Animals". All animal studies were approved by Suzhou University (Suzhou, China) (approval number: 202010A415). After 7 days of acclimatization, an oval wound of approximately 1.5 cm in length was made by shaving the back of each mouse under anesthesia and adding 100 µL of activated S. aureus cells (1×10^6 CFU/mL) dropwise to each wound for two consecutive days. The wound was treated after inflammation [14,42].

2.11. Wound Healing Assays

The mice whose wounds were infected with S. aureus were randomly divided into five groups (n = 5–7 mice) as follows: blank group, MoS_2 group, MoS_2-Van-FITC@CS group, NIR MoS_2 group, and NIR MoS_2-Van-FITC@CS group. The wound site of the blank group was treated with 100 µL of PBS, that of the MoS_2 group was treated with 100 µL of MoS_2 each day, that of the MoS2-Van-FITC@CS group was treated with 100 µL of MoS_2-Van-FITC@CS hydrogel each day, that of the NIR MoS_2 group was treated with MoS_2, followed by irradiation at 1.5 W/cm^2 for 5 min, and that of the NIR MoS_2-Van-FITC@CS group was treated with MoS_2-Van-FITC@CS, followed by irradiation at 1.5 W/cm^2 for 5 min. The mice were photographed daily for eight consecutive days. The mice were euthanized, and their epidermises were harvested for hematoxylin–eosin staining [43].

2.12. Safety Evaluation of MoS_2-Van-FITC@CS

To examine the toxic effects of NIR and MoS_2-Van-FITC@CS on the heart, liver, spleen, lungs, and kidneys of mice, the wounded mice were divided into four groups as follows: control group, NIR irradiation group, MoS_2-Van-FITC@CS group, and MoS_2-Van-FITC@CS + NIR group. The irradiation power density was 1.5 W/cm^2, the wavelength was 808 nm, and the irradiation time was 6 min. The mice were treated until their wounds healed, and they were euthanized 30 days after the end of treatment. Their organs were collected, and the cross-sections were stained with hematoxylin–eosin [44].

3. Results and Discussion
3.1. Characterization of MoS_2-Van-FITC@CS

The nanocarriers were exfoliated by liquid phase ultrasound. The composite nanomaterials conjugated to FITC-labeled Van showed strong antibacterial effects. After irradiation, MoS_2 converted light energy into heat energy, thereby killing the bacteria by actively trapping the cells through Van. MoS_2-Van-FITC had good antibacterial and wound-promoting abilities through the temperature-sensitive hydrogel formed with chitosan (Figure 2A). To determine the synthesis of two-dimensional MoS_2 nanosheets, the morphology and thickness of MoS_2 NPs were observed via SEM (Figure 2B). The results of SEM showed that the flake NPs, which had a thickness of approximately 40 nm, were uniformly distributed. Zeta potentiometry can be used to determine the solid–liquid interfacial electrical properties of dispersed systems of particulate matter, so we can determine the successful synthesis of materials by using the potential changes of nanomaterials. The zeta potential of MoS_2 was -21.4 ± 1.2 mV, whereas the loading of Van resulted in a potential of -8.3 ± 1.9 mV and a zeta potential of -33 ± 0.8 mV after labeling with FITC (Figure 2C). The synthesis of MoS_2-Van-FITC@CS causes changes in the structure of a single component as a result of electron leaps between electronic energy levels in the valence and molecular orbitals, which can be observed by UV-vis spectrum (Figure 2D). In the UV-vis spectrum, MoS$_2$ NPs were observed to have an absorption peak near 808 nm, showing a longitudinal surface plasmon resonance band, which indicated photothermal effects, whereas Van did not show an absorption peak near 808 nm. Van was adsorbed on MoS$_2$, after which this characteristic peak significantly shifted. After FITC was decorated on the Van surface, the characteristic peak of MoS$_2$-Van-FITC was significantly shifted, and a new characteristic peak was observed near 450 nm. MoS$_2$-Van-FITC@CS also showed a shift in the characteristic peak compared to MoS$_2$. FTIR spectroscopy allows the observation of the functional groups and chemical bonds contained in the material, in order to be able to determine the successful synthesis of MoS$_2$-Van-FITC nanocomposites. FT-IR spectral analysis showed that the presence of Van resulted in an amino peak near 3000 cm^{-1}, and a distinct peak at 1000 cm^{-1} in the fingerprint region after the loading of FITC (Figure 2E). A significant change was also found in the fingerprint region after the wrapping of chitosan. The fluorescence properties of MoS$_2$-Van-FITC@CS are shown in Figure 2F,G. MoS$_2$-Van-FITC@CS emitted fluorescence under UV light irradiation at 465 nm, and the fluorescence properties of the nanomaterials were further confirmed by the fluorescence spectra. Taken collectively, these findings indicate that MoS$_2$, which has photothermal properties, was successfully synthesized, Van was successfully loaded, and the surface was modified by FITC. The morphological features of the MoS$_2$-Van-FITC@CS hydrogel were indicative of its successful synthesis.

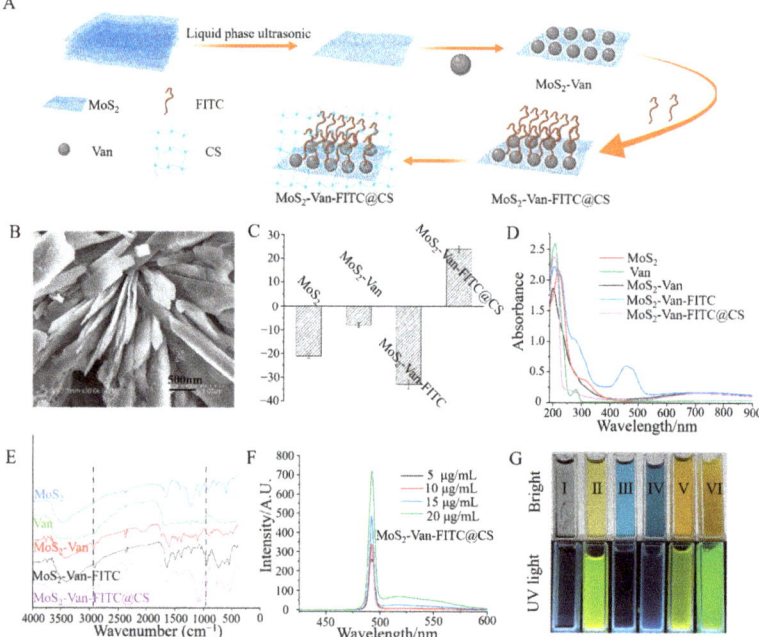

Figure 2. Synthesis and characterization of MoS$_2$-Van-FITC@CS. (**A**) Synthesis illustration of MoS$_2$-Van-FITC@CS. (**B**) SEM images of MoS$_2$ NPs. (**C**) Zeta potential of MoS$_2$, MoS$_2$-Van, MoS$_2$-Van-FITC, and MoS$_2$-Van-FITC@CS. (**D**) UV-vis absorption spectra of MoS$_2$, Van, MoS$_2$-Van, MoS$_2$-Van-FITC, and MoS$_2$-Van-FITC@CS. (**E**) FT-IR spectrometry of MoS$_2$, Van, MoS$_2$-Van, MoS$_2$-Van-FITC, and MoS$_2$-Van-FITC@CS. (**F**) Digital images of I (Milli-water), II (FITC), III (MoS$_2$), IV (MoS$_2$-Van), V (MoS$_2$-Van-FITC), and VI (MoS$_2$-Van-FITC@CS) under bright and UV light. (**G**) Fluorescence emission spectra of MoS$_2$-Van-FITC@CS.

3.2. In Vitro Photothermal Efficiency

MoS$_2$ is a photothermal agent that produces a large amount of heat to kill bacteria in the NIR region of 808 nm [45,46]. Therefore, we measured its photothermal conversion efficiency to understand the photothermal properties of MoS$_2$. As expected, MoS$_2$-Van-FITC acted as a good photothermal nanomaterial under NIR irradiation in that it converted light energy into heat energy, resulting in rapid warming (Figure 3A,B). Furthermore, the MoS$_2$-Van-FITC concentration was 400 μg/mL, m_{PBS} (m_D) was 1.0 g, C_{H_2O} (C_D) was 4.2 J/g/°C, ΔT_{max} was 47.3 °C (Figure 3C), I was 2 W, and τ_s was 296 s (Figure 3D). Thus, the photothermal conversion efficiency (η) of MoS$_2$-Van-FITC was 52%. Figure S1 shows the thermal images of MoS$_2$-Van-FITC at different concentrations under an NIR irradiation at 808 nm. In summary, MoS$_2$-Van-FITC has good photothermal conversion efficiency and can be used as a photothermal nanomaterial for killing bacteria.

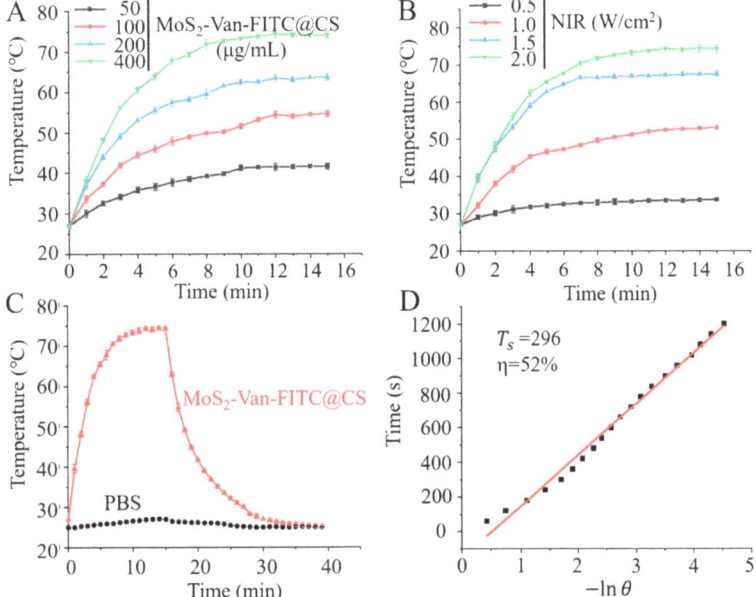

Figure 3. In vitro photothermal efficiency. (**A**) The photothermal responses of MoS$_2$-Van-FITC@CS. Different concentrations (50, 100, 200 and 400 μg/mL) were exposed to NIR irradiation (808 nm; 2 W/cm^2). PBS served as a control. (**B**) The photothermal responses of MoS$_2$-Van-FITC@CS (400 μg/mL), which was exposed to different power density of NIR irradiation (808 nm; 0.5, 1, 1.5 and 2 W/cm^2). (**C**) The heating and cooling curves of MoS$_2$-Van-FITC@CS (400 μg/mL) and PBS (laser irradiation at 808 nm). (**D**) The linear regression between the cooling period and −ln(θ) of the driving force temperature. Results shown are mean ± SD, n = 3.

3.3. In Vitro Antibacterial Activity

We used *S. aureus* and *E. coli* as Gram-positive and Gram-negative bacteria, respectively, in subsequent experiments (Table 1). In the MIC test, we found that MoS$_2$-Van-FITC + NIR had the highest killing effect against *S. aureus*, which may have been related to the fact that Van is a narrow-spectrum antibiotic that is only effective against Gram-positive bacteria. MoS$_2$-Van-FITC + NIR showed the effects of common antibiotics at low doses, and the inhibition of growth made it difficult for bacteria to develop resistance (Table 1). Thermal images of the test in MIC were showed in Figure S2. Figure 4 shows the thermogram of the solution temperature increase after NIR irradiation. We found that our nanomaterials had a stronger growth inhibition ability against *S. aureus*. In the CFU test, we screened the power density and light time using power densities of 0.5, 1, and 1.5 W/cm^2, and light times of 0, 150, and 300 s. We observed that a power density of 1.5 W/cm^2 and a light time of 300 s inhibited bacterial growth. As shown in Figure 4, the inhibition rate of bacteria in the absence of MoS$_2$ was low, and the survival rate of bacteria was 89%. However, the survival rate of bacteria gradually decreased after NIR irradiation, and the survival rate of bacteria was only 4.2% after treatment with MoS$_2$ (100 µg/mL, 1.5 W/cm^2, 300 s). After treatment with MoS$_2$-Van-FITC + NIR (100 µg/mL), the survival rate of bacteria was 0.9% after an irradiation time of 300 s at a power density of 1.5 W/cm^2. In the in vitro antibacterial test, the nanomaterials inhibited the growth of *S. aureus* cells, which played a role in the elimination of bacteria from the wound, thereby speeding up wound healing.

Table 1. Antibacterial activities with MICs values test (µg/mL).

Materials	Bacteria	
	S. aureus	*E. coli*
MoS$_2$	>128	>128
Van	2	>128
MoS$_2$-Van-FITC	64	64
MoS$_2$-Van-FITC@CS	64	64
MoS$_2$ + NIR	64	128
MoS$_2$-Van-FITC + NIR	36	128
MoS$_2$-Van-FITC@CS + NIR	36	128
Kanamycin	6	12
Ampicillin	18	24

Data are average values of at least three replicates.

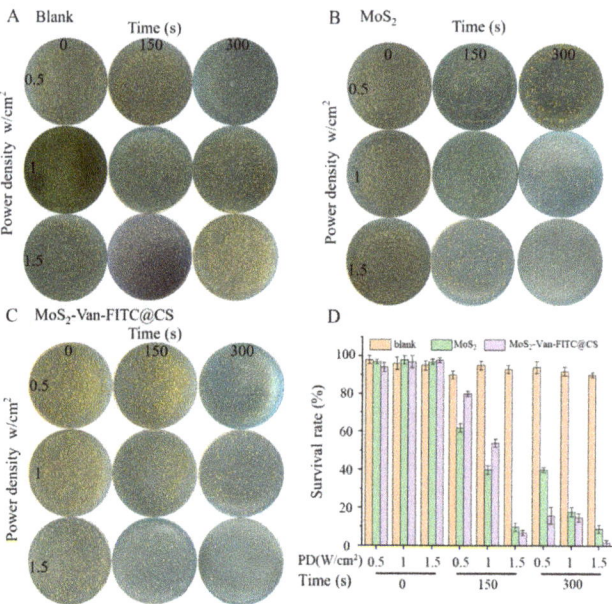

Figure 4. In vitro antibacterial activity. (**A**) The results of the CFU assay for the blank group. PBS as a blank group. (**B**) The results of the CFU assay for MoS$_2$ (100 µg/mL) for different times (0, 150 and 300 s) at different power densities (0.5, 1 and 1.5 W/cm^2). (**C**) The results of the CFU assay for MoS$_2$-Van-FITC (100 µg/mL) for different times (0, 150 and 300 s) at different power densities (0.5, 1 and 1.5 W/cm^2). (**D**) Quantitative statistical results of (**A–C**) (PD = Power Density). graphed using the Origin software. Results shown are mean ± SD, n = 3.

3.4. Cellular Uptake Assays

Van targets Gram-positive bacteria by binding to the hydrogen bond of the terminal D-Ala-D-Ala sequence of the cytosolic peptide of bacteria. It is also a heptapeptide-containing glycopeptide antibiotic with a primary amine moiety that binds covalently to FITC [46–48]. Van can label FITC on the bacterial surface; therefore, we verified the targeting of Van by DAPI staining. When MoS$_2$-Van-FITC was used at a concentration of 15 μg/mL, most of the bacteria were labeled, similar to higher concentrations of 30 μg/mL and 45 μg/mL. However, as the concentration increased, the number of bacteria decreased, showing the excellent antibacterial ability of MoS$_2$-Van-FITC (Figure 5). The results of cellular uptake assays indicated that MoS$_2$-Van-FITC successfully targeted bacteria and showed excellent antibacterial ability, thereby achieving our goal of using combined chemotherapy and photothermal therapy to inhibit bacterial growth.

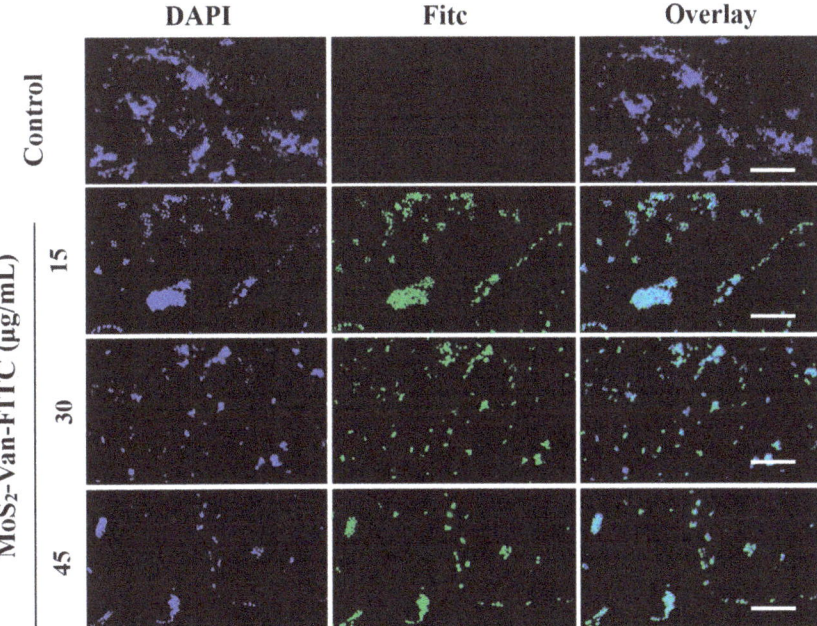

Figure 5. Fluorescence microscopy images of *S. aureus* cultures treated with MoS$_2$-Van-FITC at various concentrations (15, 30, 45 μg/mL) for 12 h. PBS served as a control for the blank group. The cells stained with DAPI for 30 min were fluorescent blue, and those stained with MoS$_2$-Van-FITC were fluorescent green. Scale bar = 15 μm. (DAPI, Ex = 358 nm and Em = 461 nm; FITC, Ex = 490 nm and Em = 525 nm).

3.5. Fluorescent Staining Analysis of Antibacterial Activity

The CFU assay can detect only viable bacteria to determine the antibacterial activity of nanomaterials. To further examine the antibacterial activity of nanomaterials, we determined the number of viable and non-viable cells using the LIVE–DEAD assay to examine the antibacterial activity of MoS$_2$-Van NPs. Viable bacterial cells were stained green, whereas non-viable bacterial cells were stained red (Figure 6A). No cell death was observed in the blank group. However, cell death was observed in the Van group, indicating that Van can inhibit bacterial growth. Similar results were obtained for the cells treated with MoS$_2$ + NIR and MoS$_2$-Van, with the MoS$_2$-Van + NIR group exhibiting stronger inhibition of bacterial growth after NIR irradiation.

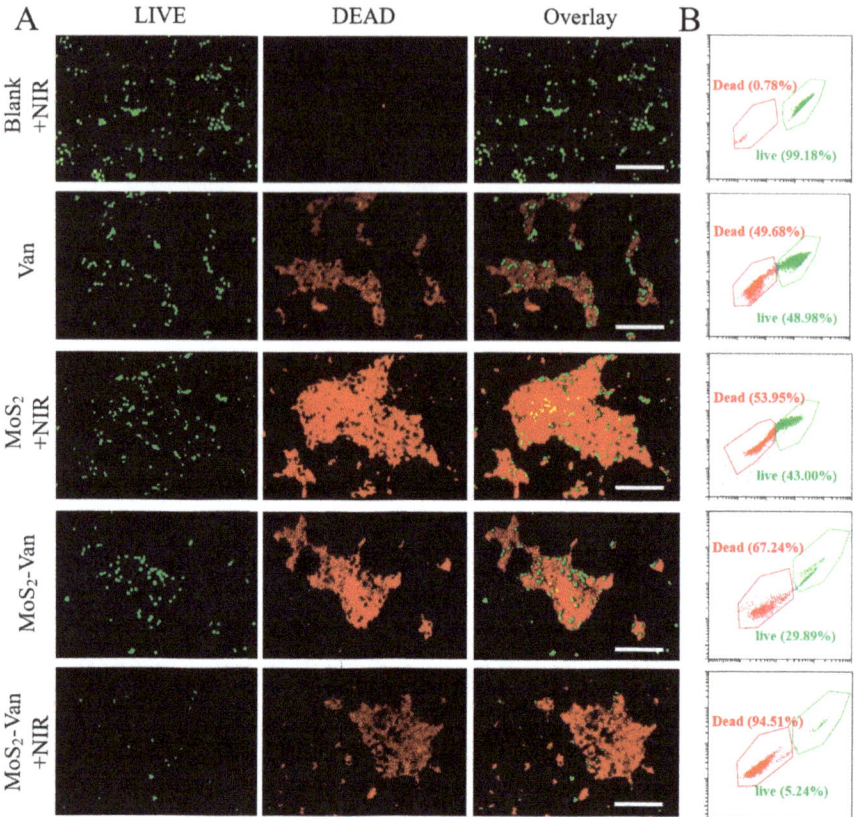

Figure 6. Confocal fluorescence microscopy assay (**A**). *S. aureus* cultures after treatment with MoS$_2$-Van NPs + NIR (100 µg/mL). Van solution (1 µg/mL), MoS$_2$ NPs + NIR (100 µg/mL) and MoS$_2$-Van (100 µg/mL) served as control groups. PBS served as a blank group. The cells were stained with SYTO 9 (green fluorescence) and PI (red fluorescence) for 30 min. The cells underwent NIR irradiation at 808 nm (1.5 W/cm^2, 6 min). The results of the apoptotic assay by flow cytometry analysis were statistically analyzed by CytExpert software (version 2.4.0.28) (**B**). Scale bar = 15 µm. The data are expressed as mean ± SD (*n* = 3).

To further confirm that MoS$_2$-Van+NIR reduced the survival of bacteria, the number of viable and non-viable bacterial cells was quantified via flow cytometry. As shown in Figure 6B, the apoptotic rate of the blank + NIR group was 0.78%. The apoptotic rate of MoS$_2$ + NIR (100 µg/mL) after NIR irradiation was 53.95%. When the concentration of Van was 1 µg/mL, the apoptotic rate was 49.68%. The apoptotic rate of MoS$_2$-Van (100 µg/mL) was 67.24% without irradiation, which was mainly due to the effects of Van, but MoS$_2$ also played its own role after NIR irradiation, and the apoptotic rate was 94.51%. The increased antibacterial activity of MoS$_2$-Van NPs was further confirmed by the quantitative analysis of viable and non-viable cells via flow cytometry.

3.6. Cell Integrity Study

Based on our findings, MoS$_2$-Van NPs + NIR showed efficient antibacterial activity against *S. aureus*. Because photothermal action mainly targets the bacterial cell surface, and the cell surface is also the site of action of Van, we speculate that changes in cell integrity may be the main mechanism behind the induction of apoptosis in bacteria. As such, we investigated the effects of MoS$_2$-Van NPs + NIR on the cellular integrity of *S. aureus* by SEM.

The integrity of bacterial cells was examined via SEM, as shown in Figure 7. *S. aureus* cells in the blank group had normal cell morphology, including intact cell membranes. The results showed that NIR irradiation alone did not affect the structure of cells. The rupture and shrinkage of cells could be clearly seen after treatment with Van and MoS$_2$ in the control group, and the enlarged area showed that the cells did not have intact cell membranes. In addition, there was cell leakage. In the MoS$_2$-Van-FITC group, we observed more severe cell damage, even at a concentration of 25 µg/mL, compared to the blank group. As the concentration increased, the cell damage increased, indicating that MoS$_2$-Van-FITC had a stronger antibacterial effect. The results from the elemental analysis chart showed that the bacterial surface did contain elemental sulfur and molybdenum, indicating that the bacterial surface contained MoS$_2$. Taken collectively, these findings indicate that MoS$_2$-Van-FITC NPs have very effective antibacterial activity compared to MoS$_2$ NPs and Van alone.

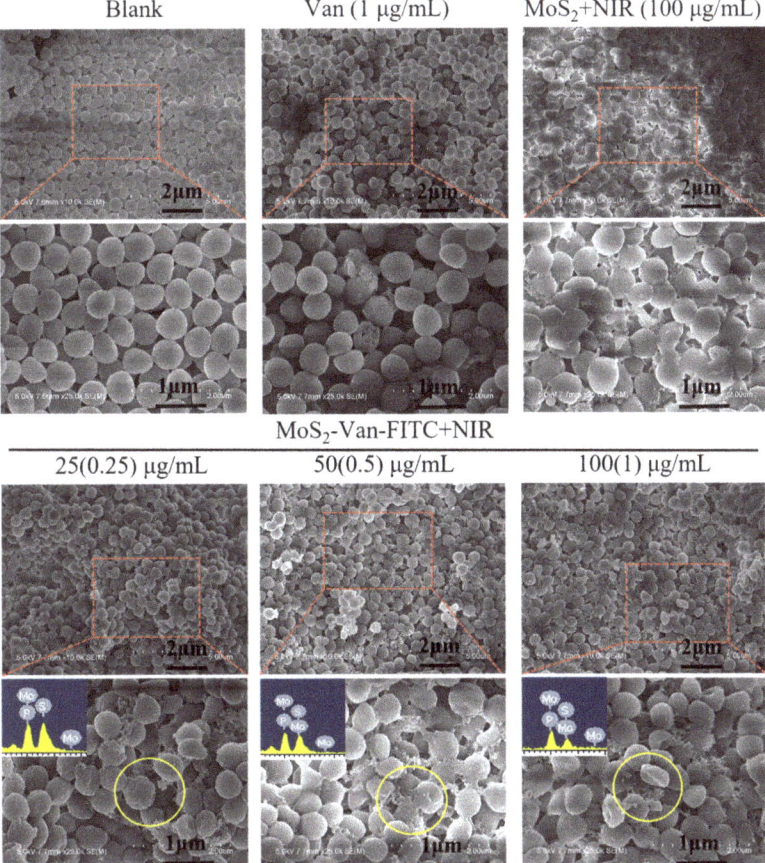

Figure 7. SEM images of *S. aureus* cells. The bacterial cultures were treated with MoS$_2$-Van-FITC NPs at various concentrations (25, 50 and 100 µg/mL). The bacterial cultures treated with Van (1 µg/mL) and MoS$_2$ NPs + NIR (100 µg/mL) served as control groups. PBS + NIR served as a blank group. The red squares indicate the enlarged regions. The blue area is the elemental analysis chart. The cells underwent NIR irradiation at 808 nm (1.5 W/cm^2, 6 min).

3.7. In Vivo Wound Healing Evaluation

We established a wound-healing mouse model and examined the pro-wound healing effects by directly applying NPs combined with irradiation. As shown in Figure 8, the MoS$_2$-Van-FITC@CS hydrogel alone was slightly therapeutic, and the rate of wound healing increased with irradiation (Figure 8A). Heating, which increased the temperature of the nanomaterial to 50 °C by irradiation at 1.5 W/cm^2 for 6 min, achieved a good therapeutic effect. The MoS$_2$ group had the largest relative wound area, which did not heal. However, MoS$_2$ decreased the relative wound area after NIR irradiation. MoS$_2$-Van-FITC@CS hydrogel + NIR had the best therapeutic effect after NIR irradiation, where the relative wound area was reduced to 20.8%. The weight of mice in all groups decreased and then increased. The reason for the decrease in weight was likely due to the appearance of wounds, but as treatment progressed, the mice recovered and regained the weight. MoS$_2$ was least effective, and the healing of mice treated with MoS$_2$ was similar to that of the controls (Figure 8B). Immunohistochemical analysis was performed to examine the epidermis from the different groups of mice, and the abnormal histological features could be clearly seen in the stained sections (Figure 8D). No changes in morphology were observed in the epidermises of mice in the control group and the other four groups. The epidermises showed no significant damage compared to those of normal mice, indicating that the elevated temperature of the nanosheets during the photothermal treatment did not cause significant damage to the wounds. Therefore, the therapeutic effect of the MoS$_2$-Van-FITC@CS hydrogel combined with NIR irradiation showed that there was no harm to the mice and their wound healing was accelerated. The preliminary photothermal imaging of mice also demonstrated the thermal response of NPs. In summary, the MoS$_2$-Van-FITC@CS hydrogel combined with photothermolysis significantly promoted wound healing in mice.

3.8. Biosafety Evaluation

The evaluation of toxicity is important for each new drug. As such, we evaluated the safety of our hydrogels in vivo. We applied the hydrogels to the wounds of healthy mice, followed by NIR irradiation, and performed a routine histological analysis of the major organs (Figure 9). Compared with the controls, no significant organ damage, tissue edema, cell death, or inflammatory cell infiltration was observed in the examined organs after treatment with NIR irradiation, MoS$_2$-Van-FITC@CS, and MoS$_2$-Van-FITC@CS + NIR, indicating that MoS$_2$-Van-FITC@CS + NIR was non-toxic in mice.

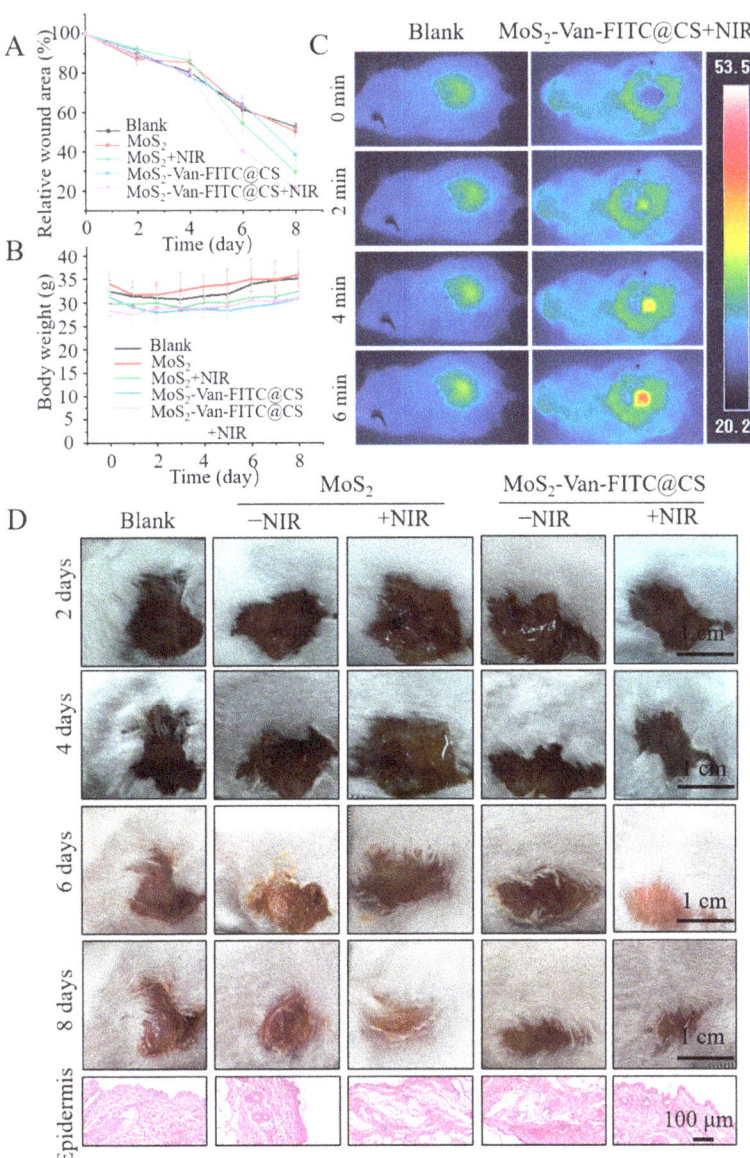

Figure 8. In vivo wound healing analysis. (**A**) The relative wound area curve shows the MoS$_2$-Van-FITC@CS hydrogel had a significant positive effect on wound healing. (**B**) The body weights of different groups after treatment. (**C**) The thermal imaging of mice after treatment. The cells underwent NIR irradiation at 808 nm (1.5 W/cm^2, 6 min). (**D**) The photographs of wounds were taken every 2 days after treatment. MoS$_2$-Van-FITC@CS hydrogel + NIR (100 μg/mL), MoS$_2$ NPs (100 μg/mL), MoS$_2$ NPs + NIR (100 μg/mL) and MoS$_2$-Van-FITC@CS hydrogel served as control groups. PBS + NIR served as a blank group. The cells underwent NIR irradiation at 808 nm (1.5 W/cm^2, 6 min). Scale bar = 1 cm. Results shown are mean ± SD, n = 5–7.

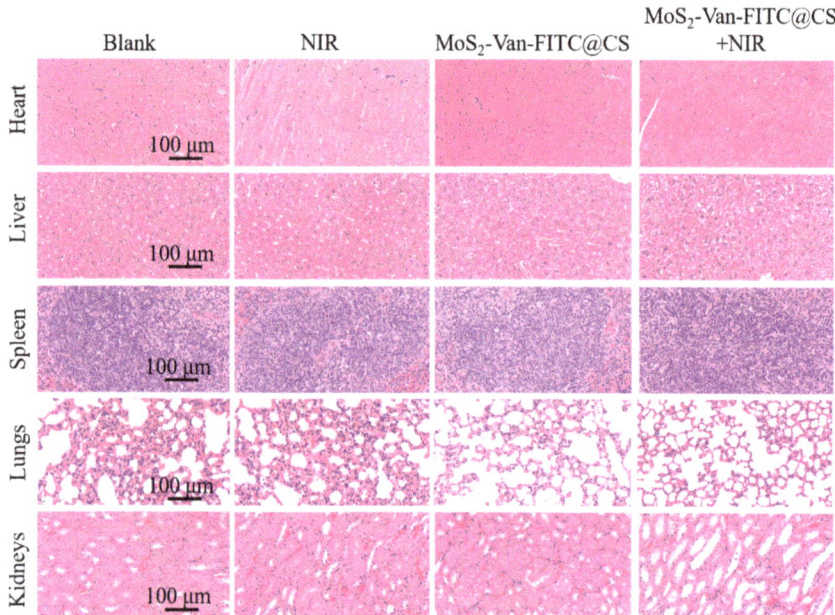

Figure 9. In vivo toxicity evaluation. The hematoxylin–eosin-stained images of major organs following different treatments of normal mice. Group 1 was the no treatment group, group 2 was the NIR irradiation (1.5 W/cm^2, 6 min) group, group 3 was the MoS$_2$-Van-FITC@CS group and group 4 was the MoS$_2$-Van-FITC@CS + NIR (1.5 W/cm^2, 6 min) group.

4. Conclusions

In this study, the Van-modified MoS$_2$-loaded nanosystem, which was encapsulated in a chitosan hydrogel, was established to examine its antibacterial activity and wound healing ability. The results showed that the thickness of MoS$_2$ NPs was <100 nm, whereas other experiments revealed that the surface of MoS$_2$ was successfully modified by Van. The antimicrobial activity was enhanced when Van was labeled. The photothermal characterization experiments confirmed that MoS$_2$ had good photothermal conversion efficiency, and cellular uptake assays verified the active capture of *S. aureus* by Van, which significantly improved the photothermal inhibition of bacterial growth. The results of in vitro experiments indicated that NIR could significantly increase the antibacterial activity of MoS$_2$ NPs, whereas those of flow cytometry showed that NPs could increase the apoptotic rate of bacterial cells. The morphological features of bacterial cells treated with MoS$_2$-Van-FITC NPs and NIR irradiation were examined, and NPs were observed to disrupt the integrity of the bacterial cell wall. Furthermore, MoS$_2$-Van-FITC combined with NIR irradiation could disrupt the cell morphology, induce apoptosis, and affect cell proliferation in vitro. In a wound healing assay, the MoS$_2$-Van-FITC@CS hydrogel could accelerate wound healing. In summary, MoS$_2$ in combination with NIR irradiation shows good applicability in the inhibition of bacterial growth, and the CS hydrogel in combination with a photothermal agent that actively traps *S. aureus* can disinfect the wound and maintain a moist environment to accelerate wound healing.

Supplementary Materials: The following supporting information can be downloaded at: https://www.mdpi.com/article/10.3390/nano12111865/s1, Figure S1: The thermal imaging images of MoS$_2$-Van-FITC@CS at different concentrations (50, 100 and 200 μg/mL) and NIR irradiation (808 nm; 2 W/cm^2); Figure S2: Thermal images of the test in MIC.

Author Contributions: Conceptualization, W.Z.; Data curation, W.Z., Z.K. and L.Z.; Formal analysis, W.Z. and Z.K.; Funding acquisition, P.S., W.L., L.G., Y.T. and F.G; Investigation, P.S., W.L. and C.T.; Methodology, P.S., W.L., C.T., F.G. and L.Z.; Project administration, L.G.; Resources, Y.T.; Writing—original draft, W.Z. and Z.K.; Writing—review & editing, Z.K. and L.Z. All authors have read and agreed to the published version of the manuscript.

Funding: The work was supported by the National natural science foundation of China (31671797), Anhui Nature Science Foundation (2008085QH397), Anhui Provincial Higher Education Institutes (KJ2020a0375 and KJ2021A0511), Anhui Polytechnic University (xjky2020064), Natural Science Foundation of Anhui University (KJ2020A118), Youth Key Talents Program of Wannan Medical College(wyqnyx202005).

Institutional Review Board Statement: The animal study protocol was approved by Suzhou University (Suzhou, China) (approval number: 202010A415).

Informed Consent Statement: Not applicable.

Data Availability Statement: Not applicable.

Conflicts of Interest: The authors have declared that no competing interests exist.

References

1. Vlazaki, M.; Huber, J.; Restif, O. Integrating mathematical models with experimental data to investigate the within-host dynamics of bacterial infections. *Pathog. Dis.* **2019**, *77*, 14. [CrossRef] [PubMed]
2. Li, W.; Song, P.; Xin, Y.; Kuang, Z.; Liu, Q.; Ge, F.; Zhu, L.; Zhang, X.; Tao, Y.; Zhang, W. The Effects of Luminescent CdSe Quantum Dot-Functionalized Antimicrobial Peptides Nanoparticles on Antibacterial Activity and Molecular Mechanism. *Int. J. Nanomed.* **2021**, *16*, 1849–1867. [CrossRef] [PubMed]
3. Tang, M.; Zhang, J.; Yang, C.; Zheng, Y.; Jiang, H. Gold Nanoclusters for Bacterial Detection and Infection Therapy. *Front. Chem.* **2020**, *8*, 181. [CrossRef] [PubMed]
4. Sun, D.; Zhang, W.; Mou, Z.; Chen, Y.; Guo, F.; Yang, E.; Wang, W. Transcriptome Analysis Reveals Silver Nanoparticle-Decorated Quercetin Antibacterial Molecular Mechanism. *ACS Appl. Mater. Interfaces* **2017**, *9*, 10047–10060. [CrossRef] [PubMed]
5. Yang, X.; Zhang, W.; Zhao, Z.; Li, N.; Mou, Z.; Sun, D.; Cai, Y.; Wang, W.; Lin, Y. Quercetin loading CdSe/ZnS nanoparticles as efficient antibacterial and anticancer materials. *J. Inorg. Biochem.* **2017**, *167*, 36–48. [CrossRef] [PubMed]
6. Carlie, S.; Boucher, C.E.; Bragg, R.R. Molecular basis of bacterial disinfectant resistance. *Drug Resist. Updat.* **2020**, *48*, 100672. [CrossRef]
7. Fernández, J.; Bert, F.; Nicolas-Chanoine, M.-H. The challenges of multi-drug-resistance in hepatology. *J. Hepatol.* **2016**, *65*, 1043–1054. [CrossRef]
8. Huang, S.; Liu, H.; Liao, K.; Hu, Q.; Guo, R.; Deng, K. Functionalized GO Nanovehicles with Nitric Oxide Release and Photothermal Activity-Based Hydrogels for Bacteria-Infected Wound Healing. *ACS Appl. Mater. Interfaces* **2020**, *12*, 28952–28964. [CrossRef]
9. Wang, W.; Li, Y.; Wang, W.; Gao, B.; Wang, Z. Palygorskite/silver nanoparticles incorporated polyamide thin film nanocomposite membranes with enhanced water permeating, antifouling and antimicrobial performance. *Chemosphere* **2019**, *236*, 124396. [CrossRef]
10. Sun, D.; Pang, X.; Cheng, Y.; Ming, J.; Xiang, S.; Zhang, C.; Lv, P.; Chu, C.; Chen, X.; Liu, G.; et al. Ultrasound-Switchable Nanozyme Augments Sonodynamic Therapy against Multidrug-Resistant Bacterial Infection. *ACS Nano* **2020**, *14*, 2063–2076. [CrossRef]
11. Tao, B.; Lin, C.; Deng, Y.; Yuan, Z.; Shen, X.; Chen, M.; He, Y.; Peng, Z.; Hu, Y.; Cai, K. Copper-nanoparticle-embedded hydrogel for killing bacteria and promoting wound healing with photothermal therapy. *J. Mater. Chem. B* **2019**, *7*, 2534–2548. [CrossRef] [PubMed]
12. Fan, X.; Yang, F.; Nie, C.; Yang, Y.; Ji, H.; He, C.; Cheng, C.; Zhao, C. Mussel-Inspired Synthesis of NIR-Responsive and Biocompatible Ag-Graphene 2D Nanoagents for Versatile Bacterial Disinfections. *ACS Appl. Mater. Interfaces* **2018**, *10*, 296–307. [CrossRef] [PubMed]
13. Cao, W.; Yue, L.; Wang, Z. High antibacterial activity of chitosan—Molybdenum disulfide nanocomposite. *Carbohydr. Polym.* **2019**, *215*, 226–234. [CrossRef] [PubMed]
14. Yin, W.; Yu, J.; Lv, F.; Yan, L.; Zheng, L.R.; Gu, Z.; Zhao, Y. Functionalized Nano-MoS2 with Peroxidase Catalytic and Near-Infrared Photothermal Activities for Safe and Synergetic Wound Antibacterial Applications. *ACS Nano* **2016**, *10*, 11000–11011. [CrossRef] [PubMed]
15. Liu, Y.; Guo, Z.; Li, F.; Xiao, Y.; Zhang, Y.; Bu, T.; Jia, P.; Zhe, T.; Wang, L. Multifunctional Magnetic Copper Ferrite Nanoparticles as Fenton-like Reaction and Near-Infrared Photothermal Agents for Synergetic Antibacterial Therapy. *ACS Appl. Mater. Interfaces* **2019**, *11*, 31649–31660. [CrossRef]

16. Zhao, X.; Chen, M.; Wang, H.; Xia, L.; Guo, M.; Jiang, S.; Wang, Q.; Li, X.; Yang, X. Synergistic antibacterial activity of streptomycin sulfate loaded PEG-MoS2/rGO nanoflakes assisted with near-infrared. *Mater. Sci. Eng. C Mater. Biol. Appl.* **2020**, *116*, 111221. [CrossRef]
17. McMullen, A.R.; Lainhart, W.; Wallace, M.A.; Shupe, A.; Burnham, C.D. Evaluation of telavancin susceptibility in isolates of Staphylococcus aureus with reduced susceptibility to vancomycin. *Eur. J. Clin. Microbiol. Infect. Dis.* **2019**, *38*, 2323–2330. [CrossRef]
18. Zhao, Z.; Yan, R.; Yi, X.; Li, J.; Rao, J.; Guo, Z.; Yang, Y.; Li, W.; Li, Y.Q.; Chen, C. Bacteria-Activated Theranostic Nanoprobes against Methicillin-Resistant Staphylococcus aureus Infection. *ACS Nano* **2017**, *11*, 4428–4438. [CrossRef]
19. Han, D.; Yan, Y.; Wang, J.; Zhao, M.; Duan, X.; Kong, L.; Wu, H.; Cheng, W.; Min, X.; Ding, S. An enzyme-free electrochemiluminesce aptasensor for the rapid detection of Staphylococcus aureus by the quenching effect of MoS2-PtNPs-vancomycin to S2O82−/O2 system. *Sens. Actuators B Chem.* **2019**, *288*, 586–593. [CrossRef]
20. Wu, Z.C.; Boger, D.L. Maxamycins: Durable Antibiotics Derived by Rational Redesign of Vancomycin. *Acc. Chem. Res.* **2020**, *53*, 2587–2599. [CrossRef]
21. Blaskovich, M.A.T.; Hansford, K.A.; Gong, Y.; Butler, M.S.; Muldoon, C.; Huang, J.X.; Ramu, S.; Silva, A.B.; Cheng, M.; Kavanagh, A.M.; et al. Protein-inspired antibiotics active against vancomycin- and daptomycin-resistant bacteria. *Nat. Commun.* **2018**, *9*, 22. [CrossRef] [PubMed]
22. Yang, C.; Ren, C.; Zhou, J.; Liu, J.; Zhang, Y.; Huang, F.; Ding, D.; Xu, B.; Liu, J. Dual Fluorescent- and Isotopic-Labelled Self-Assembling Vancomycin for in vivo Imaging of Bacterial Infections. *Angew. Chem. Int. Ed. Engl.* **2017**, *56*, 2356–2360. [CrossRef] [PubMed]
23. Li, J.; Yu, F.; Chen, G.; Liu, J.; Li, X.L.; Cheng, B.; Mo, X.M.; Chen, C.; Pan, J.F. Moist-Retaining, Self-Recoverable, Bioadhesive, and Transparent in Situ Forming Hydrogels To Accelerate Wound Healing. *ACS Appl. Mater. Interfaces* **2020**, *12*, 2023–2038. [CrossRef] [PubMed]
24. Nuutila, K.; Eriksson, E. Moist Wound Healing with Commonly Available Dressings. *Adv. Wound Care* **2021**, *10*, 685–698. [CrossRef] [PubMed]
25. Basha, S.I.; Ghosh, S.; Vinothkumar, K.; Ramesh, B.; Kumari, P.H.P.; Mohan, K.V.M.; Sukumar, E. Fumaric acid incorporated Ag/agar-agar hybrid hydrogel: A multifunctional avenue to tackle wound healing. *Mater. Sci. Eng. C Mater. Biol. Appl.* **2020**, *111*, 110743. [CrossRef]
26. Ganguly, S.; Das, P.; Itzhaki, E.; Hadad, E.; Gedanken, A.; Margel, S. Microwave-Synthesized Polysaccharide-Derived Carbon Dots as Therapeutic Cargoes and Toughening Agents for Elastomeric Gels. *ACS Appl. Mater. Interfaces* **2020**, *12*, 51940–51951. [CrossRef]
27. Zheng, Y.; Wang, W.; Zhao, J.; Wu, C.; Ye, C.; Huang, M.; Wang, S. Preparation of injectable temperature-sensitive chitosan-based hydrogel for combined hyperthermia and chemotherapy of colon cancer. *Carbohydr. Polym.* **2019**, *222*, 115039. [CrossRef]
28. Jiang, J.; Meng, X.; Wu, Z.; Qi, X. Modified chitosan thermosensitive hydrogel enables sustained and efficient anti-tumor therapy via intratumoral injection. *Carbohydr. Polym.* **2016**, *144*, 245–253. [CrossRef]
29. Zhao, H.; Wu, H.; Wu, J.; Li, J.; Wang, Y.; Zhang, Y.; Liu, H. Preparation of MoS2/WS2 nanosheets by liquid phase exfoliation with assistance of epigallocatechin gallate and study as an additive for high-performance lithium-sulfur batteries. *J. Colloid. Interface Sci.* **2019**, *552*, 554–562. [CrossRef]
30. Ye, J.; Li, X.; Zhao, J.; Mei, X.; Li, Q. A Facile Way to Fabricate High-Performance Solution-Processed n-MoS2/p-MoS2 Bilayer Photodetectors. *Nanoscale Res. Lett.* **2015**, *10*, 454. [CrossRef]
31. Yang, X.; Li, J.; Liang, T.; Ma, C.; Zhang, Y.; Chen, H.; Hanagata, N.; Su, H.; Xu, M. Antibacterial activity of two-dimensional MoS2 sheets. *Nanoscale* **2014**, *6*, 10126–10133. [CrossRef] [PubMed]
32. Zhang, W.; Mou, Z.; Wang, Y.; Chen, Y.; Yang, E.; Guo, F.; Sun, D.; Wang, W. Molybdenum disulfide nanosheets loaded with chitosan and silver nanoparticles effective antifungal activities: In vitro and in vivo. *Mater. Sci. Eng. C Mater. Biol. Appl.* **2019**, *97*, 486–497. [CrossRef] [PubMed]
33. Takai, H.; Kato, A.; Nakamura, T.; Tachibana, T.; Sakurai, T.; Nanami, M.; Suzuki, M. The importance of characterization of FITC-labeled antibodies used in tissue cross-reactivity studies. *Acta Histochem.* **2011**, *113*, 472–476. [CrossRef] [PubMed]
34. Liu, T.; Li, J.; Shao, Z.; Ma, K.; Zhang, Z.; Wang, B.; Zhang, Y. Encapsulation of mesenchymal stem cells in chitosan/beta-glycerophosphate hydrogel for seeding on a novel calcium phosphate cement scaffold. *Med. Eng. Phys.* **2018**, *56*, 9–15. [CrossRef]
35. Dang, Q.; Liu, K.; Zhang, Z.; Liu, C.; Liu, X.; Xin, Y.; Cheng, X.; Xu, T.; Cha, D.; Fan, B. Fabrication and evaluation of thermosensitive chitosan/collagen/alpha, beta-glycerophosphate hydrogels for tissue regeneration. *Carbohydr. Polym.* **2017**, *167*, 145–157. [CrossRef]
36. Liu, H.; Zhu, X.; Guo, H.; Huang, H.; Huang, S.; Huang, S.; Xue, W.; Zhu, P.; Guo, R. Nitric oxide released injectable hydrogel combined with synergistic photothermal therapy for antibacterial and accelerated wound healing. *Appl. Mater. Today* **2020**, *20*, 100781. [CrossRef]
37. Ganguly, S.; Das, P.; Das, T.K.; Ghosh, S.; Das, S.; Bose, M.; Mondal, M.; Das, A.K.; Das, N.C. Acoustic cavitation assisted destratified clay tactoid reinforced in situ elastomer-mimetic semi-IPN hydrogel for catalytic and bactericidal application. *Ultrason. Sonochem.* **2020**, *60*, 104797. [CrossRef]

38. Sun, D.; Zhang, W.; Li, N.; Zhao, Z.; Mou, Z.; Yang, E.; Wang, W. Silver nanoparticles-quercetin conjugation to siRNA against drug-resistant Bacillus subtilis for effective gene silencing: In vitro and in vivo. *Mater. Sci. Eng. C Mater. Biol. Appl.* **2016**, *63*, 522–534. [CrossRef]
39. Rekha, R.; Vaseeharan, B.; Vijayakumar, S.; Abinaya, M.; Govindarajan, M.; Alharbi, N.S.; Kadaikunnan, S.; Khaled, J.M.; Al-Anbr, M.N. Crustin-capped selenium nanowires against microbial pathogens and Japanese encephalitis mosquito vectors—Insights on their toxicity and internalization. *J. Trace Elem. Med. Biol.* **2019**, *51*, 191–203. [CrossRef]
40. Lu, B.-Y.; Zhu, G.-Y.; Yu, C.-H.; Chen, G.-Y.; Zhang, C.-L.; Zeng, X.; Chen, Q.-M.; Peng, Q. Functionalized graphene oxide nanosheets with unique three-in-one properties for efficient and tunable antibacterial applications. *Nano Res.* **2020**, *14*, 185–190. [CrossRef]
41. Sun, D.; Li, N.; Zhang, W.; Yang, E.; Mou, Z.; Zhao, Z.; Liu, H.; Wang, W. Quercetin-loaded PLGA nanoparticles: A highly effective antibacterial agent in vitro and anti-infection application in vivo. *J. Nanoparticle Res.* **2015**, *18*, 3. [CrossRef]
42. Li, W.; Wang, Y.; Qi, Y.; Zhong, D.; Xie, T.; Yao, K.; Yang, S.; Zhou, M. Cupriferous Silver Peroxysulfite Superpyramids as a Universal and Long-Lasting Agent to Eradicate Multidrug-Resistant Bacteria and Promote Wound Healing. *ACS Appl. Bio Mater.* **2020**, *4*, 3729–3738. [CrossRef] [PubMed]
43. Liang, Y.; Zhao, X.; Hu, T.; Chen, B.; Yin, Z.; Ma, P.X.; Guo, B. Adhesive Hemostatic Conducting Injectable Composite Hydrogels with Sustained Drug Release and Photothermal Antibacterial Activity to Promote Full-Thickness Skin Regeneration during Wound Healing. *Small* **2019**, *15*, e1900046. [CrossRef] [PubMed]
44. Zhang, W.; Ding, X.; Cheng, H.; Yin, C.; Yan, J.; Mou, Z.; Wang, W.; Cui, D.; Fan, C.; Sun, D. Dual-Targeted Gold Nanoprism for Recognition of Early Apoptosis, Dual-Model Imaging and Precise Cancer Photothermal Therapy. *Theranostics* **2019**, *9*, 5610–5625. [CrossRef]
45. Zhu, M.; Liu, X.; Tan, L.; Cui, Z.; Liang, Y.; Li, Z.; Yeung, K.W.K.; Wu, S. Photo-responsive chitosan/Ag/MoS2 for rapid bacteria-killing. *J. Hazard. Mater.* **2020**, *383*, 121122. [CrossRef]
46. Xu, M.; Zhang, K.; Liu, Y.; Wang, J.; Wang, K.; Zhang, Y. Multifunctional MoS2 nanosheets with Au NPs grown in situ for synergistic chemo-photothermal therapy. *Colloids Surf. B Biointerfaces* **2019**, *184*, 110551. [CrossRef]
47. Wang, C.; Gu, B.; Liu, Q.; Pang, Y.; Xiao, R.; Wang, S. Combined use of vancomycin-modified Ag-coated magnetic nanoparticles and secondary enhanced nanoparticles for rapid surface-enhanced Raman scattering detection of bacteria. *Int. J. Nanomed.* **2018**, *13*, 1159–1178. [CrossRef]
48. Vimberg, V.; Gazak, R.; Szucs, Z.; Borbas, A.; Herczegh, P.; Cavanagh, J.P.; Zieglerova, L.; Zavora, J.; Adamkova, V.; Novotna, G.B. Fluorescence assay to predict activity of the glycopeptide antibiotics. *J. Antibiot.* **2019**, *72*, 114–117. [CrossRef]

Article

Thermoresponsive Zinc TetraPhenylPorphyrin Photosensitizer/Dextran Graft Poly(N-IsoPropylAcrylAmide) Copolymer/Au Nanoparticles Hybrid Nanosystem: Potential for Photodynamic Therapy Applications

Oleg A. Yeshchenko [1,*], Nataliya V. Kutsevol [2,3], Anastasiya V. Tomchuk [1], Pavlo S. Khort [1], Pavlo A. Virych [2], Vasyl A. Chumachenko [2], Yulia I. Kuziv [2,3], Andrey I. Marinin [4], Lili Cheng [5] and Guochao Nie [5,*]

[1] Physics Department, Taras Shevchenko National University of Kyiv, 60 Volodymyrska Str., 01601 Kyiv, Ukraine; nastiona30@gmail.com (A.V.T.); mirason111@gmail.com (P.S.K.)
[2] Chemistry Department, Taras Shevchenko National University of Kyiv, 60 Volodymyrska Str., 01601 Kyiv, Ukraine; kutsevol@ukr.net (N.V.K.); sphaenodon@ukr.net (P.A.V.); chumachenko_va@ukr.net (V.A.C.); garaguts.yulia.fox@gmail.com (Y.I.K.)
[3] Institute Charles Sadron, 23 Rue du Loess, 67200 Strasbourg, France
[4] Problem Research Laboratory, National University of Food Technology, 68 Volodymyrska Str., 01601 Kyiv, Ukraine; andrii_marynin@ukr.net
[5] Guangxi Universities Key Lab of Complex System Optimization and Big Data Processing, Yulin Normal University, Yulin 537000, China; chenglili0428@163.com
* Correspondence: oleg.yeshchenko@knu.ua (O.A.Y.); bccu518@163.com (G.N.)

Abstract: The thermoresponsive Zinc TetraPhenylPorphyrin photosensitizer/Dextran poly (N-isopropylacrylamide) graft copolymer/Au Nanoparticles (ZnTPP/D-g-PNIPAM/AuNPs) triple hybrid nanosystem was synthesized in aqueous solution as a nanodrug for potential use in thermally driven and controlled photodynamic therapy applications. The aqueous solution of the nanosystem has demonstrated excellent stability in terms of aggregation and sedimentation several days after preparation. Optimal concentrations of the components of hybrid nanosystem providing the lowest level of aggregation and the highest plasmonic enhancement of electronic processes in the photosensitizer molecules have been determined. It has been revealed that the shrinking of D-g-PNIPAM macromolecule during a thermally induced phase transition leads to the release of both ZnTPP molecules and Au NPs from the ZnTPP/D-g-PNIPAM/AuNPs macromolecule and the strengthening of plasmonic enhancement of the electronic processes in ZnTPP molecules bound with the polymer macromolecule. The 2.7-fold enhancement of singlet oxygen photogeneration under resonant with surface plasmon resonance has been observed for ZnTPP/D-g-PNIPAM/AuNPs proving the plasmon nature of such effect. The data obtained in vitro on wild strains of *Staphylococcus aureus* have proved the high potential of such nanosystem for rapid photodynamic inactivation of microorganisms particular in wounds or ulcers on the body surface.

Keywords: thermoresponsive polymer; gold nanoparticles; photosensitizer; hybrid nanosystem; plasmon enhancement; singlet oxygen photogeneration; photodynamic therapy

1. Introduction

Photodynamic therapy (PDT) is a promising method of treatment that can be used to solve a number of problems, from the destruction of bacteria and viruses to the treatment of cancer. This method is based on the use of three factors: light, a special light-sensitive substance—photosensitizer (PS) and oxygen from the environment to generate cytotoxic oxygen or free radicals [1,2]. Although this technique is now most widely used to treat dermatological diseases and various skin malignancies, its prospects in the destruction of various pathogens cannot be ignored, especially given the permanent emergence of new, antibiotic-resistant bacterial strains [3,4]. The main advantage of antibacterial PDT (APDT) over conventional antibiotics is that the bacteria do not show resistance to PDT even after repeated sessions of treatment.

For advances in nanobiology and nanomedicine, there is an urgent need for new hybrid functional materials based on biocompatible polymers [5–10]. A variety of biocompatible water-soluble polymers can be used to increase the bioavailability of a variety of drugs, as well as to improve drug pharmacokinetics, such as controlled delivery of drugs directly into cells and their controlled release [11]. Polymers can be used as an effective matrix for in situ synthesis of metal nanoparticles (NPs) with narrow size distribution [12–15], preventing the aggregation of NPs. An encapsulation of photosensitizers in a polymer matrix avoids aggregation of hydrophobic photosensitizer molecules, which significantly increases their photodynamic efficiency in biological media [16].

Recently, thermoresponsive polymers have attracted much attention from researchers, since such polymers are the basis for the creation of locally temperature-controlled nanoactuators [17–19]. In particular, poly (N-isopropylacrylamide) (PNIPAM) were used as a potential basis for the fabrication of hybrid nanosystems for applications in biology and medicine [20]. In aqueous phase, PNIPAM undergoes a phase LCST transition (lower critical solution temperature) at 32 °C from hydrophilic to hydrophobic phase, which leads to a sharp shrinking of the polymer molecule [21–23]. Such temperature induced shrinking can be used for the controlled release (at a given temperature) of certain molecules (e.g., drugs) initially bound to the PNIPAM macromolecule [24]. Additionally, a temperature-induced change (decrease) in the distance between molecules and plasmonic nanoparticles (NPs) leads to a change (increase) in the strength of molecule–NP coupling. This, in principle, may change the magnitude of the plasmon enhancement of electronic processes in molecules, such as absorption, emission and scattering of light, photocatalysis, generation of singlet oxygen, etc. [25–28]. The use of PNIPAM with a star-like structure allows for an increase the phase transition point by approximately 2–4 °C compared to the corresponding temperature of PNIPAM with a linear structure [26]. In addition, the Au and Ag NPs grown in situ in a branched polymer in aqueous solution have significantly higher stability than NPs grown in a linear polymer [29]. The temperature and related laser-induced phase transitions and the possibility of changing its parameters for hybrid nanosystems based on PNIPAM with a star structure (D-g-PNIPAM) and Au (Ag) NPs was shown in recent works [30–33].

Porphyrin and its derivatives are the promising photosensitizers used in particular for antibacterial and antitumor therapy [34,35]. Their advantages include high stability, efficient absorption of visible light, low dark toxicity, long life in the triplet state, highly efficient photogeneration of singlet oxygen and ease of their modification. However, there is a serious problem—the vast majority of porphyrins are hydrophobic, and therefore in living organisms they have a tendency to aggregation, which significantly reduces their photodynamic efficiency [35]. In recent years, metal porphyrins have attracted considerable attention. The PDT activity of porphyrin metal complexes depends on the type of metal due to the paramagnetic effect [36]. Zinc is added to the porphyrin ring to provide ring stability and to maintain the pronounced photodynamic efficiency of the porphyrin based PS. Such porphyrin metal complexes, e.g., zinc tetraphenylporphyrin (ZnTPP), are similar to natural porphyrin and are widely used in biology and medicine. The presence of Zn atom in the porphyrin ring has been reported to reduce mitochondrial binding and promote cell membrane binding due to complexation with phospholipid phosphate groups, which enhances the PDT efficiency [37].

Gold nanoparticles (Au NPs) are widely used in many biological applications. They are less toxic than Ag NPs [38,39]. Recently, the high antitumor efficiency has been demonstrated for some gold-based compounds [40,41]. A well-known approach to enhance the various electronic processes (light absorption, fluorescence, Raman scattering, photocatalysis, etc.) in the molecules is to place the molecules near the plasmonic metal nanoparticles or nanostructures [25–28]. In such nanoparticles, the localized surface plasmon resonance

(LSPR) is excited by an external light, as a result, there is a significant increase in the electromagnetic field strength in the vicinity of plasmonic NPs, causing an increased optical response of the molecules located in the enhanced field area. Accordingly, the plasmon-enhanced light absorption by PS molecule located near the metal NP in a hybrid nanosystem would result in a more efficient photogeneration of singlet oxygen, therefore, an increase in the efficiency of PDT of the nanosystem [18,42–45]. In addition, small Au NPs have antibacterial activity against pathological bacteria due to high ability to penetrate the cell [46]. Another phenomenon with a plasmonic nature that can be used for PDT purposes is the photo-induced heating of metal NPs, plasmon heating. Due to an extremely low quantum yield of fluorescence of metal NPs, almost all of the light energy absorbed by the nanoparticle is converted to thermal energy. As a result, metal NPs act as highly efficient local nanoheaters. The effect of plasmon heating has a resonant character, i.e., it becomes strongest under the resonance of the frequencies of exciting light and LSPR in metal NP [47–49]. Thus, plasmon heating of metal NPs can be efficiently used for the photothermal therapy of cancer and bacterial diseases [50–53].

Here, we present the results of the chemical synthesis, the size and morphological characteristics, spectroscopic properties and APDT activity of the aqueous solution of ZnTPP/D-g-PNIPAM/AuNPs triple hybrid nanosystem. This hybrid nanosystem contains zinc tetraphenylporphyrin (ZnTPP) photosensitizer, thermoresponsive dextran poly (N-isopropylacrylamide) graft copolymer (D-g-PNIPAM) and Au NPs. Our work demonstrates that ZnTPP/D-g-PNIPAM/AuNPs nanosystem exhibits a significant plasmon enhancement of the singlet oxygen photogeneration, see scheme in Figure 1. ZnTPP/D-g-PNIPAM/AuNPs shows a high potential for thermally driven and controlled photodynamic rapid inactivation of microorganisms.

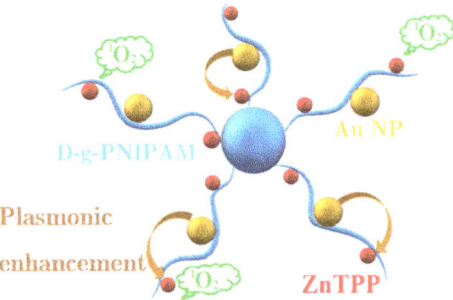

Figure 1. Scheme representing the structure of ZnTPP/D-g-PNIPAM/AuNPs hybrid macromolecule as well as the plasmon enhancement of singlet oxygen generation.

2. Experimental

2.1. Fabrication of ZnTPP/D-g-PNIPAM and ZnTPP/D-g-PNIPAM/AuNPs Hybrid Nanosystems

The procedures of synthesis and structural peculiarities of D-g-PNIPAM copolymers were reported previously. [14,30,31]. Initially, the copolymer with star-like structure containing dextran core with molecular weight $M_w = 70 \times 10^4$ g/mol and 15 PNIPAM grafts was synthesized. The molecular parameters of D-g-PNIPAM are as follows: the molecular weight parameters $M_v = 1.03 \times 10^6$ g/mol and $M_n = 0.67 \times 10^6$ g/mol; $M_v/M_n = 1.52$. The reduction of Au ions took place in a mixture of aqueous solutions of the D-g-PNIPAM polymer as a matrix and $HAuCl_4$ as a source of gold. The detailed description of synthesis and characterization of D-g-PNIPAM/Au NPs hybrid macromolecules in aqueous solution has been reported [30,31].

The ZnTPP powder was dissolved in ethanol at 40 °C. The obtained stock ZnTPP ethanol solution had concentration of 0.1 g/L. The aqueous stock solutions of D-g-PNIPAM ($c_{D\text{-}g\text{-}PNIPAM} = 1$ g/L) and D-g-PNIPAM/AuNPs ($c_{D\text{-}g\text{-}PNIPAM} = 0.8696$ g/L and $c_{Au} = 0.8529$ g/L) were diluted by water at 20 °C to obtain solutions with different concentrations. Finally,

0.03 mL of ZnTPP ethanol solution was mixed with 2.97 mL of diluted aqueous solutions of D-g-PNIPAM or D-g-PNIPAM/AuNPs. As a result, we obtained the aqueous solutions of ZnTPP/D-g-PNIPAM or ZnTPP/D-g-PNIPAM/AuNPs nanocomposite contained 0.001 g/L of ZnTPP and various concentrations of D-g-PNIPAM and/or Au.

2.2. Morphology and Size Characterization by TEM and DLS

The transmitted electron microscopy (TEM) measurements were performed by CM12 (FEI) microscope, and images were acquired using a Megaview SIS camera. Due to the much lower contrast of D-g-PNIPAM macromolecules compared to Au NPs, polymer macromolecules and ZnTPP molecules and aggregates are not visible on the TEM images. Therefore, the size of the studied nanosystems in solution were determined by the dynamic light scattering (DLS) method. The hydrodynamic particle size distribution (PSD) of the nanosystems in water was measured by DLS method on a Zetasizer Nano-ZS90 (Malvern Panalytical) which was equipped with a 5 mW He-Ne laser operating at 633 nm. The scattered light was detected at the angle of 173°. PSD were obtained from autocorrelation functions calculated by nonnegative truncated singular value decomposition method [54]. The DLS measurements were made in a temperature range including the LCST point. The PSD were measured at various time points (1 min—7 days) after mixing of ZnTPP ethanol solution with water, D-g-PNIPAM and D-g- PNIPAM/AuNPs aqueous solutions.

2.3. Absorption and Fluorescence Spectroscopy

The absorption spectra were measured with a Cary 60 UV-VIS spectrophotometer (Agilent). The fluorescence (FL) spectra were measured with a Shimadzu RF-6000 spectrofluorophotometer (Shimadzu) with an excitation wavelength at 421 nm. The solution samples were placed in 1 cm × 1 cm × 4 cm quartz cell. The spectra were measured in a temperature range including the LCST point. The absorption and FL measurements were carried out 1 day after mixing of ZnTPP ethanol solution with water, D-g-PNIPAM and D-g-PNIPAM/AuNPs aqueous solutions. The FL anisotropy were measured as $r = \frac{I_{VV} - GI_{VH}}{I_{VV} + 2GI_{VH}}$, where I_{ij} is the FL intensity, ij indices denote an orientation of polarizers before and after the sample respectively (V– vertical, H– horizontal), $G = I_{HV}/I_{HH}$ is the grating factor.

2.4. Biological Experiments

The antibacterial activity of the aqueous solutions of ZnTPP, ZnTPP/D-g-PNIPAM and ZnTPP/D-g-PNIPAM/AuNPs on wild strains of *Staphylococcus aureus* (*S. aureus*) was investigated. Bacteria were isolated on yolk-salt agar containing: meat–peptone agar—70%, sodium chloride—10%, yolk emulsion in 0.9% NaCl—20%, pH 7.3. The antibacterial activity of the nanocomposites was tested in a liquid medium of Mueller–Hinton №2 (g/L): casein hydrolysate—17.5, bull heart hydrolysate—2, water-soluble starch—1.5, pH—7.3. A suspension of *S. aureus* was prepared, then divided into 1 mL aliquots and incubated at 37 °C for 20 min, for the microorganisms to adapt to the changed conditions. The systems which were used in the investigations were also sensitive to temperature changes. Therefore, the study was performed at 28, 31, 33, 35, and 37 °C. The antibacterial activity of the nanocomposites was evaluated against control tubes that were under similar conditions. The LED-device Lika-LED (Photonics Plus) was used as a visible light source at 420 and 530 nm. The dose of irradiation was in the range of 3–18 J/mL with steps of 3 J/mL, the irradiation power was 0.1 J/s. The suspension was irradiated by light for 15 min after the addition of nanocomposites. The maximum duration of nanocomposites incubation in a suspension of *S. aureus* was not exceed 20 min.

The colony-forming units (CFU) number were counted in a Goryaev chamber. The microorganisms were stained for 1 min with acridine orange (concentration—$3.5 \cdot 10^{-3}$ M). The statistical analysis of the results was made by Shapiro–Wilk ($p > 0.05$) and Scheffe (ANOVA, $p < 0.05$) tests. The experiments were repeated three times.

3. Results and Discussion
3.1. Morphology and Size Study

Typical TEM images of D-g-PNIPAM/AuNPs nanohybrids obtained at 20 °C are shown in Figure 2. Au NPs have spherical shape and a mean radius of 3 nm. Since the TEM contrast of D-g-PNIPAM polymer is significantly lower compared to Au NPs, the polymer macromolecules cannot be visualized by TEM. Meanwhile, the mean radius of the D-g-PNIPAM/AuNPs hybrid macromolecules can be estimated as the radius of Au NPs clusters on the TEM image. This estimation gives the radius of D-g-PNIPAM/AuNPs hybrid macromolecule of about 25 nm, which agrees with the DLS data (see below).

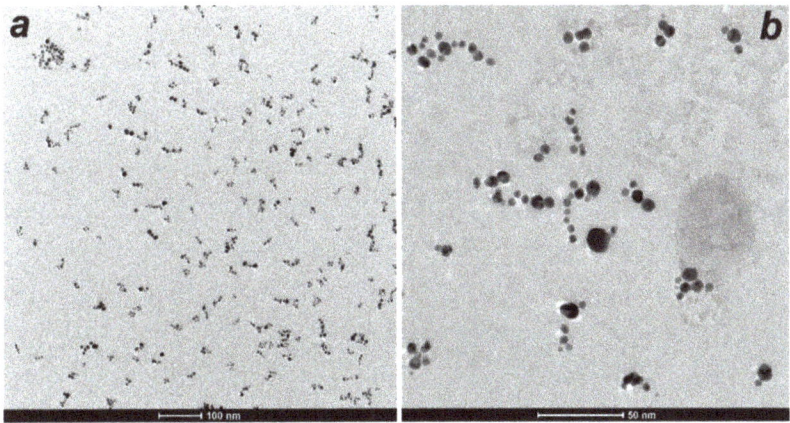

Figure 2. TEM images of D-g-PNIPAM/AuNPs nanohybrids at the temperature of 20 °C acquired at lower (**a**) and higher (**b**) magnification.

DLS measurements were performed to acquire the size of studied hybrid nanocomposites in solution. Figure 3a shows PSD change of D-g-PNIPAM during transition over LCST region at heating. The PNIPAM hydrodynamic radius (corresponding to PSD maximum) is $R_h^{max} = 25$ nm at 25 °C. At heating, D-g-PNIPAM macromolecules formed aggregates with $R_h^{max} = 160$ nm at LCST temperature of 34 °C. At subsequent heating to 40 °C higher than LCST, the D-g-PNIPAM aggregates decrease in size to 115 nm. Figure 3b demonstrates the PSD for ZnTPP in aqueous solution. In aqueous solution, the ZnTPP molecules form aggregates with a mean radius of 32 nm. It is due to hydrophobic nature of ZnTPP.

As we can see on Figure 4a, at room temperature (25 °C) the ZnTPP addition into aqueous solutions of D-g-PNIPAM and D-g-PNIPAM/AuNPs influences greatly on the average size of the respective macromolecules. Namely, for double-component ZnTPP/D-g-PNIPAM nanosystem, the average radius R_h^{max} shifts to 61 nm which is more than two times larger than for single-component one (D-g-PNIPAM). Size distribution of ZnTPP/D-g-PNIPAM/AuNPs has two-peaks. Those are high-intensity peak at 39 nm corresponding to ZnTPP/D-g-PNIPAM/AuNPs three-component nanosystem and low-intensity but still important peak at 3 nm corresponding to small AuNPs. Thus, hybrid macromolecules are expanded comparing to bare D-g-PNIPAM ones.

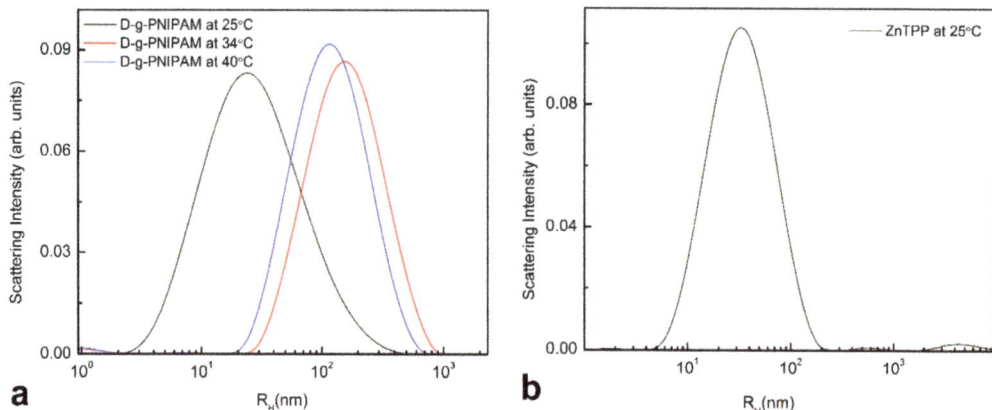

Figure 3. (**a**)—PSD of D-g-PNIPAM aqueous solution at temperatures lower (25 °C), equal (34 °C) and higher (40 °C) than LCST point. (**b**)—PSD of ZnTPP in ethanol mixed with water at 25 °C.

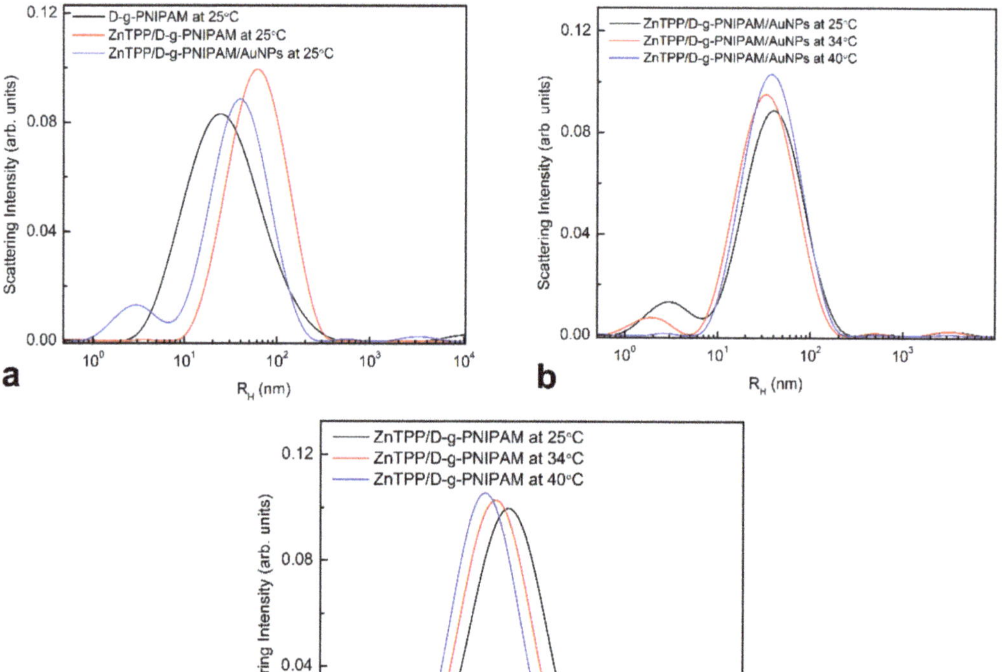

Figure 4. PSD of different solutions at 25 °C for comparison (**a**); triple system of ZnTPP/D-g-PNIPAM/AuNPs at 25, 34 and 40 °C (**b**); ZnTPP/D-g-PNIPAM at 25, 34 and 40 °C (**c**).

PSD of the three-component system at different temperatures (Figure 4a) shows that the main high intensity peak shifts towards smaller radius from 39 to 33 nm as the temperature rises from room 25 °C to LCST point at 34 °C. With further temperature increase, this peak shifts back to 39 nm. The observed size variation at temperature induced LCST transition is due to the shrinking of PNIPAM chains in ZnTPP/D-g-PNIPAM/AuNPs hybrid nanosystem. Such effect has been investigated for D-g-PNIPAM based multi-component hybrid nanosystems in our recent works [14,30–33]. It is important that the presence of additional components in D-g-PNIPAM based hybrid macromolecules prevents the system from aggregation at LCST transition which was observed for bare D-g-PNIPAM, see Figure 3a. This effect was explained previously [14]. Indeed, as we can see from Figure 4c, the same process takes place during heating of ZnTPP/D-g-PNIPAM system in temperature range of 25–40 °C, when the radius of ZnTPP/D-g-PNIPAM scatterers decreases gradually from 61 to 38 nm and no aggregation has been observed.

We also studied possible aging effect of the ZnTPP/D-g-PNIPAM/AuNPs nanosystem. No PSD changes were found during the period of observation up to 7 days and at the temperature variation in the temperature range of 25–40 °C, as shown in Figure S1a–c. Moreover, storage of the sample did not lead to any change of PSD at LCST transition. At 40 °C, we still can see well-defined and reproducible AuNPs peak at 2–3 nm despite low intensity of this mode, as shown in Figure S1d.

3.2. Absorption and Fluorescence Spectroscopy
3.2.1. Spectral Manifestations of ZnTPP Binding to D-g-PNIPAM and PNIPAM/AuNPs

Light absorption spectroscopy was performed on a solution of the photosensitizer ZnTPP in ethanol, a mixture of ZnTPP ethanol solution with water, aqueous solutions of polymer D-g-PNIPAM and hybrid nanosystem D-g-PNIPAM/AuNPs, Figure S2. Hereinafter, a mixture of ZnTPP ethanol solution with water and aqueous solutions of D-g-PNIPAM and D-g-PNIPAM/AuNPs will be referred to as aqueous solutions of ZnTPP, ZnTPP/D-g-PNIPAM and ZnTPP/D-g-PNIPAM/AuNPs, respectively. The absorption spectrum of ZnTPP in ethanol (Figure S2) has a structure typical for porphyrins in organic solvents. Namely, the spectrum contains low-intensity low-energy (530–620 nm) Q bands and intense high-energy (380–440 nm) B (Soret) band [55–58]. The aqueous solution of ZnTPP/D-g-PNIPAM/AuNPs has a LSPR absorption band of AuNPs with a maximum at 520 nm, Figure S2. The respective LSPR band is clearly seen in absorption spectrum of the reference aqueous solution of D-g-PNIPAM/AuNPs (Figure S2) that proves its plasmonic nature. The D-g-PNIPAM polymer absorption and FL spectra are in UV range at wavelengths shorter than 250 nm, which is outside the spectral range relevant to this work.

The FL spectra were also measured for ZnTPP ethanol solution, mixtures of ZnTPP ethanol solution with water, and aqueous solutions of D-g-PNIPAM and D-g-PNIPAM/AuNPs, Figure S3. The FL spectrum of ZnTPP ethanol solution (Figure S3) has a structure typical for porphyrins in organic solvents. There are the high-energy (602 nm) F_{00} and low-energy (655 nm) F_{01} bands [56,57] in the FL spectrum.

Mixing ethanol solution of ZnTPP with water causes the significant changes in the shape and intensity of absorption and FL spectra, Figures S2 and S3. Such changes originate from the aggregation of hydrophobic ZnTPP molecules in the water. However, the mixing ZnTPP ethanol solution with D-g-PNIPAM and especially with D-g-PNIPAM/AuNPs aqueous solutions leads to reverse changes in absorption and FL spectra. Most probably, this is due to the binding of ZnTPP molecules to D-g-PNIPAM polymer and especially to D-g-PNIPAM/AuNPs nanohybrids. The binding increases the solubility of ZnTPP in water. It reduces the size of ZnTPP aggregates that is in full agreement with the DLS data, Figure 4a. The absorption and FL spectra changes are discussed in detail in Supplementary Materials.

Accordingly, based on the fact of essential transformations of absorption and FL spectra occurring at mixing, we conclude that ZnTPP molecules bind to D-g-PNIPAM and D-g-PNIPAM/AuNPs macromolecules, and the binding is different in the presence and absence of AuNPs. Such conclusion is also proved by the FL anisotropy measurement data, which are given and discussed below. We also emphasize that, in the aqueous media, the aggregation of ZnTPP is the weakest in the solution of ZnTPP/D-g-PNIPAM/AuNPs. Therefore, it can be expected that the biological activity of the ZnTPP photosensitizer will be higher in the ZnTPP/D-g-PNIPAM/AuNPs aqueous solution.

3.2.2. Absorption and Fluorescence of ZnTPP/D-g-PNIPAM and ZnTPP/D-g-PNIPAM/AuNPs Nanosystems: Concentration Effects

The peculiarities of the interaction of ZnTPP molecules with D-g-PNIPAM and Au NPs in hybrid macromolecules have been revealed by the study of the impact of concentrations of photosensitizer, polymer and gold on the optical spectra of the nanohybrids. The relationship of the absorption and FL spectra of ZnTPP in ethanol and water with its concentration was studied in our recent works [59,60]. The dependence of ZnTPP FL intensity of ethanol solution on ZnTPP concentration was found to be non-monotonic, with a maximum FL intensity at ZnTPP concentration 0.005 g/L. Such non-monotonic dependence is caused by aggregation of ZnTPP molecules at high concentrations. Similar to ZnTPP in ethanol, the dependence of ZnTPP FL intensity in aqueous solution on ZnTPP concentration is also non-monotonic, with a maximum FL intensity at ZnTPP concentration 0.0025 g/L. The lower concentration for ZnTPP aqueous solution at the maximum FL intensity is due to the hydrophobicity of PS molecules, which promotes aggregation. Proceeding from this data, for further measurements the ZnTPP concentration of 0.001 g/L was chosen. This concentration corresponds to the middle of the growing linear region of dependence, i.e., fairly low concentrations at which aggregation is quite slight.

Next, the effect of polymer and Au concentration on magnitude of absorption (total optical density—integrated over the entire absorption spectrum) and FL (FL total intensity—integrated over the entire FL spectrum) of ZnTPP in aqueous solutions of ZnTPP/D-g-PNIPAM and ZnTPP/D-g-PNIPAM/AuNPs nanohybrids was investigated, Figure 5a,b respectively. The optical density and FL intensity at different concentrations were normalized against ZnTPP in aqueous solution. As shown in Figure 5, the concentration dependences of the optical density and FL intensity are the similar. Therefore, further we analyze only the concentration dependence of FL intensity. The dependences of ZnTPP FL intensity on D-g-PNIPAM concentration for ZnTPP/D-g-PNIPAM, and on Au concentration for ZnTPP/D-g-PNIPAM/AuNPs nanosystems are monotonically increasing in the whole studied range of concentrations, Figure 5. Proceeding from mentioned above, such dependences can be rationalized as a consequence of the fact that both the D-g-PNIPAM and D-g-PNIPAM/AuNPs inhibit the aggregation of ZnTPP molecules in aqueous medium. Herewith, this effect is significantly stronger for nanosystems containing Au NPs. In order to highlight the impact of Au NPs concentration on the ZnTPP FL intensity, the ratio of concentration dependences for ZnTPP/D-g-PNIPAM/AuNPs and ZnTPP/D-g-PNIPAM was calculated and plotted against ZnTPP concentration (Figure 5, triangles). It gives the dependence of the FL plasmon-enhancement factor on the gold concentration. It shows that the intensity of FL increases in 1.4 times in the concentration range of 0–0.008 g/L, reaches a maximum, and then decreases to 1.33 with a further increase of concentration to 0.2 g/L. The observed plasmonic enhancement of FL of photosensitizer molecules in the ZnTPP/D-g-PNIPAM/AuNPs nanosystem indicates that ZnTPP molecules and Au NPs are quite closely located inside the polymer macromolecule, which is due to binding of ZnTPP molecules to D-g-PNIPAM/AuNPs hybrid macromolecules.

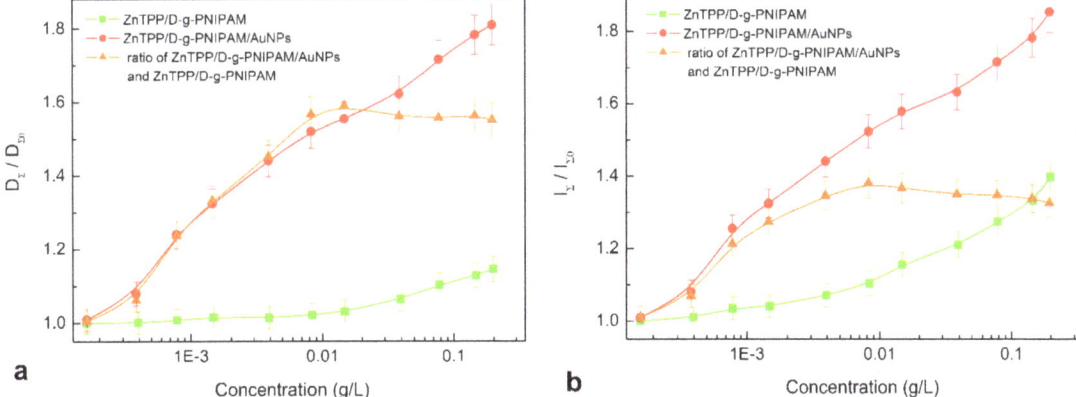

Figure 5. The dependence of normalized total optical density (**a**) and FL total intensity (**b**) of ZnTPP in ZnTPP/D-g-PNIPAM (squares) and ZnTPP/D-g-PNIPAM/AuNPs (circles) aqueous solutions on the concentration of polymer and gold respectively. Triangles—the dependence of ratio of optical densities and, respectively, FL intensities of ZnTPP/D-g-PNIPAM/AuNPs and ZnTPP/D-g-PNIPAM on the concentration of gold characterizing an impact of Au NPs on ZnTPP FL in the hybrid nanosystem. The optical density and FL intensity are normalized to the respective values for ZnTPP in aqueous solution. ZnTPP concentration is 0.001 g/L.

We analyze the physical mechanisms of the influence of Au NPs on FL of ZnTPP in hybrid macromolecules. By changing the Au NPs concentration, we thereby change the mean ZnTPP–AuNP distance inside the D-g-PNIPAM macromolecule. The distance change affects the strength of coupling of Au NPs and ZnTPP molecules [61–66]. The strength of coupling depends strongly on the overlap of LSPR in metal NP and the electronic energy spectrum of the fluorophore molecule. The coupling is stronger at shorter distances and higher spectral overlap. It is mentioned above that ZnTPP molecules and Au NPs are closely located inside the hybrid macromolecule network. In addition, there is a significant overlap of the LSPR absorption band of the Au NPs with the absorption and FL spectra of ZnTPP molecules, Figures S2 and S3. Thus, a strong coupling of Au NPs with ZnTPP molecules in ZnTPP/D-g-PNIPAM/AuNPs nanosystem is highly expected.

It is well known that there are two competing physical mechanisms for the donor (dye molecule-fluorophore)—acceptor (metal NP) pair that affect the FL intensity of the dye. The first mechanism is the plasmon enhancement. It strengthens with decreasing distance between the molecule and the metal NP [61–67]. The amplitude of plasmon field depends on the distance from the metal NP as $E_{sp} \propto R^{-3}$ [66,67]. The second mechanism is the non-radiative Förster resonant energy transfer (FRET) from the excited donor (fluorophore molecule) to the acceptor (metal NP) due to dipole-dipole interaction [61–63,67,68]. FRET leads to quenching of FL. The rate of FL quenching depends on the distance between the donor and acceptor as $\gamma_{FRET} \propto R^{-6}$ [68]. Such dependence limits the FRET to distances below 10 nm. The competition between plasmon enhancement and FRET quenching leads to an existence of the optimal NP-molecule distance (about 10 nm) providing the highest FL intensity [63]. At distances less than 10 nm, a small decrease in distance causes the strong FL quenching. Meanwhile, at distances larger than 10 nm, a decrease in distance causes the FL enhancement.

Thus, at lower concentrations of Au NPs in the range of 0–0.008 g/L, the ZnTPP–AuNP distance is too large for FRET. An increase in the concentration of Au NPs causes the shortening of ZnTPP–AuNP distance, which leads to the stronger plasmon enhancement of FL. At concentrations of Au NPs higher than 0.008 g/L, the ZnTPP–AuNP distance becomes short enough for FRET, which leads to FL quenching at the increase of Au NPs concentration. Therefore, the conclusion can be made that there is a certain optimal Au

NPs concentration, which provides the highest plasmon enhancement of optical processes involving the ZnTPP photosensitizer molecules, in particular light absorption, fluorescence and singlet oxygen generation.

Assumptions of the binding of ZnTPP molecules to the D-g-PNIPAM and D-g-PNIPAM/AuNPs can be checked directly by FL anisotropy r measurement. The FL anisotropy indicates how constrained the molecules is in its motion. For ZnTPP molecules in ethanol solution, r is 0.007 that indicates the almost free motion of ZnTPP molecules in ethanol. However, for ZnTPP molecules in water, r is 0.110 that indicates a higher constraint of the motion of ZnTPP molecules in water. Most probably, that is due to the formation of ZnTPP aggregates. Mixing ZnTPP with an aqueous solution of D-g-PNIPAM leads to a further increase of the FL anisotropy, which increases from 0.112 to 0.123 with increasing polymer concentration in the range of 0.00039–0.198 g/L, Figure 6. Meanwhile, in ZnTPP/D-g-PNIPAM/AuNPs aqueous solution, the r factor increases from 0.117 to 0.182 with increasing gold concentration in the range of 0.00039–0.195 g/L, Figure 6. Thus, the obtained results prove the fact of binding of ZnTPP molecules both with D-g-PNIPAM macromolecules and D-g-PNIPAM/AuNPs hybrid macromolecules. Thus, the FL anisotropy measurement data indicate that ZnTPP molecules bind better to hybrid macromolecules containing Au NPs that is in full agreement with data obtained from absorption and FL spectra of ZnTPP/D-g-PNIPAM and ZnTPP/D-g-PNIPAM/AuNPs hybrids, Figures S2 and S3 and above relevant discussions.

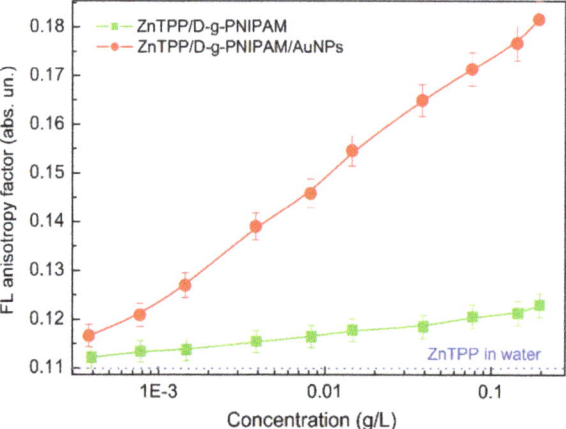

Figure 6. Concentration dependence of FL anisotropy for ZnTPP in aqueous solutions of ZnTPP/D-g-PNIPAM (squares) and ZnTPP/D-g-PNIPAM/AuNPs (circles). Squares and circles show the dependence on D-g-PNIPAM and AuNPs concentration respectively. Concentration of ZnTPP is 0.001 g/L. The blue dotted line corresponds to ZnTPP in water.

3.2.3. Thermally Induced Processes in ZnTPP/D-g-PNIPAM and ZnTPP/D-g-PNIPAM/AuNPs Nanohybrids

Since PNIPAM is a thermoresponsive polymer, it is reasonable to expect that the LCST phase transition should cause the thermally induced processes in ZnTPP/D-g-PNIPAM and ZnTPP/D-g-PNIPAM/AuNPs systems. The respective transformations were probed by light absorption (Figure 7) and FL (Figure 8) spectroscopy of ZnTPP molecules in aqueous solutions of ZnTPP/D-g-PNIPAM and ZnTPP/D-g-PNIPAM/AuNPs. It was revealed that, during heating, passing through the LCST point leads to 1.14 times decrease of light absorption by ZnTPP in ZnTPP/D-g-PNIPAM, while the temperature dependence of light absorption by ZnTPP in ZnTPP/D-g-PNIPAM/AuNPs is non-monotonic with slightly expressed maximum at LCST point, Figure 7a. Meanwhile, passing through the LCST point during heating leads to FL quenching both for ZnTPP/D-g-PNIPAM and

ZnTPP/D-g-PNIPAM/AuNPs, Figure 8a. Note that quenching is stronger for ZnTPP in ZnTPP/D-g-PNIPAM than in ZnTPP/D-g-PNIPAM/AuNPs. Indeed, the FL intensity decreased 2.02 and 1.85 times when the sample was heated from 20 to 48 °C for ZnTPP/D-g-PNIPAM and ZnTPP/D-g-PNIPAM/AuNPs respectively. The impact of temperature on the light absorption and FL is especially strong in the region of LCST point, i.e., at a temperature of about 35 °C.

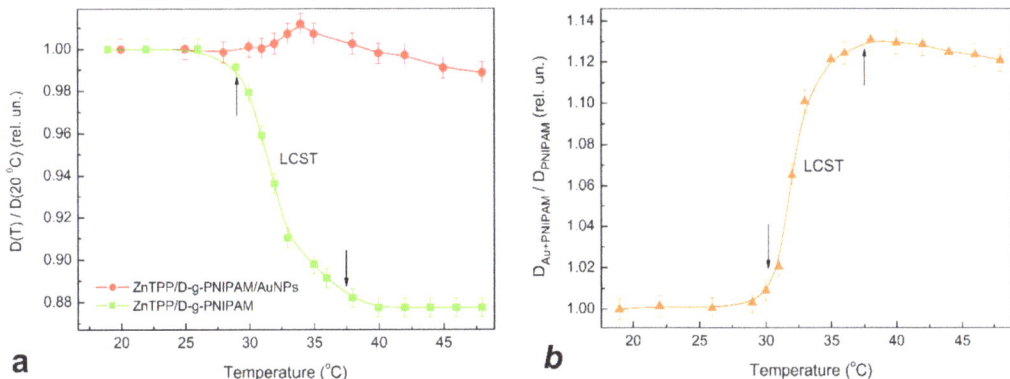

Figure 7. (**a**)—Dependence of total (B and Q bands) optical density on temperature for ZnTPP in aqueous solutions of ZnTPP/D-g-PNIPAM (squares) and ZnTPP/D-g-PNIPAM/AuNPs (circles) during heating. (**b**)—Dependence of the ratio of total optical densities for ZnTPP in ZnTPP/D-g-PNIPAM/AuNPs and ZnTPP/D-g-PNIPAM on temperature characterizing the impact of Au NPs on ZnTPP light absorption in ZnTPP/D-g-PNIPAM/AuNPs nanosystem. The optical density is normalized by value at 20 °C. Arrows show the LCST transition range. Concentrations: ZnTPP—0.001 g/L, D-g-PNIPAM—0.078 g/L, Au—0.077 g/L.

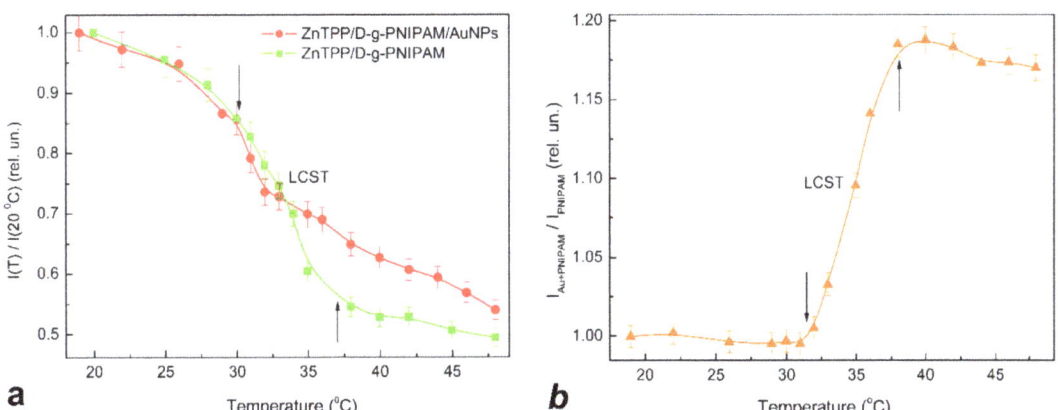

Figure 8. (**a**)—Temperature dependence of total FL intensity for aqueous solutions of ZnTPP/D-g-PNIPAM (squares) and ZnTPP/D-g-PNIPAM/AuNPs (circles) during heating. (**b**)—Temperature dependence of the ratio of total FL intensities for ZnTPP/D-g-PNIPAM/AuNPs and ZnTPP/D-g-PNIPAM characterizing the impact of Au NPs on FL of ZnTPP in ZnTPP/D-g-PNIPAM/AuNPs nanosystem. The FL intensity is normalized by value at 20 °C. Arrows show the LCST transition range. Concentrations: ZnTPP—0.001 g/L, D-g-PNIPAM—0.078 g/L, Au—0.077 g/L.

The observed behavior of the temperature dependences of absorption and FL in the temperature region of LCST transition can be caused by three physical mechanisms. First one is the release of the photosensitizer molecules out from the polymer macromolecule due to its shrinking at the LCST transition. Indeed, the released ZnTPP molecules are located far from the Au NPs in the spatial areas, where the plasmonic field is slight enough to enhance the FL. Thus, the release of ZnTPP should lead to a sharp decrease of plasmonic enhancement and, correspondingly, to a decrease of light absorption and FL quenching. Second one is the occurrence of FRET occurring due to the sharp decrease of the distance between ZnTPP molecules and Au NPs remaining bound to polymer macromolecule while it is shrinking. It leads to FL quenching at LCST transition. Meanwhile, the shortening of the distance between the ZnTPP molecules and Au NPs bound to D-g-PNIPAM can also lead to the opposite process (third mechanism), namely the intensification of the plasmonic enhancement of light absorption and FL of ZnTPP molecules bound to the polymer.

In order to find out which physical mechanisms are dominant, the temperature dependence of the FL anisotropy was measured, Figure 9. It clearly shows that LCST transition leads to sharp decrease of the FL anisotropy from 0.12 to 0.09 for ZnTPP/D-g-PNIPAM and from 0.17 to 0.11 for ZnTPP/D-g-PNIPAM/AuNPs. The observed sharp decrease of the FL anisotropy proves the fact of the release of the ZnTPP molecules when the polymer macromolecule is shrinking. It is reasonable to expect that the polymer shrinking also leads to the release of the Au NPs. Thus, the conclusion can be made that both ZnTPP/D-g-PNIPAM and ZnTPP/D-g-PNIPAM/AuNPs nanosystems are promising for thermally induced and controlled drug release.

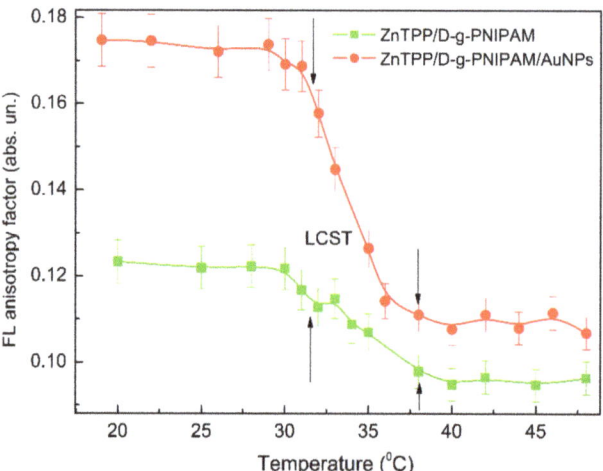

Figure 9. Temperature dependence of FL anisotropy factor for ZnTPP in aqueous solutions of ZnTPP/D-g-PNIPAM (squares) and ZnTPP/D-g-PNIPAM/AuNPs (circles) during heating. Arrows show the LCST transition range. Concentrations: ZnTPP—0.001 g/L, D-g-PNIPAM—0.078 g/L, Au—0.077 g/L.

In order to compare the impact of FRET and plasmonic enhancement on the light absorption, the temperature dependence of the ratio of total (integrated spectrally over B and Q bands) optical density for ZnTPP in ZnTPP/D-g-PNIPAM/AuNPs and one in ZnTPP/D-g-PNIPAM is calculated and presented in Figure 7b. Respectively, in order to compare the impact of the FRET and plasmonic enhancement on the FL intensity, the temperature dependence of the ratio of total FL intensities for ZnTPP in ZnTPP/D-g-PNIPAM/AuNPs and ZnTPP/D-g-PNIPAM is calculated and presented in Figure 8b. These ratio dependences characterize the influence of processes involved the Au NPs on light absorption and FL of the ZnTPP molecules which are bound to the polymer. The free ejected ZnTPP molecules do not contribute to this dependence. Figures 7b and 8b show that the polymer macromolecule shrinking at phase transition causes the sharp increase in light absorption and FL intensity for ZnTPP molecules bound with D-g-PNIPAM/AuNPs hybrid macromolecule, comparing to ZnTPP molecules bound with bare D-g-PNIPAM macromolecule. The observed increase in relative optical density and FL intensity is due to the strengthening of plasmonic enhancement, which was caused by the shortening of the mean distance between Au NPs and ZnTPP molecules bound to polymer macromolecule. Thus, one can conclude that the plasmonic enhancement prevails the FRET at the temperature induced phase transition at the determined optimal concentrations of ZnTPP and Au NPs. Most probably, the revealed temperature induced strengthening of plasmonic enhancement should also lead to the increase in efficiency of singlet oxygen generation in ZnTPP/D-g-PNIPAM/AuNPs at phase transition. Thus, summarizing the obtained data, we conclude that the temperature induced LCST transition in ZnTPP/D-g-PNIPAM/AuNPs nanosystem leads to the release of ZnTPP molecules and Au NPs from the macromolecule, as well as to the strengthening of plasmonic enhancement of the optical processes in ZnTPP molecules bound with the polymer macromolecules. Both these processes make the ZnTPP/D-g-PNIPAM/AuNPs nanosystem quite efficient for thermally driven and controlled PDT applications that is discussed in next Section 3.4.

3.3. Singlet Oxygen Photogeneration Enhancement in ZnTPP/D-g-PNIPAM/AuNPs System

An important characteristic of the efficiency of some molecular systems for use in PDT is the efficiency of photogeneration of singlet oxygen. Figure 10 shows the measured spectra of singlet oxygen emission from aqueous solutions of ZnTPP/D-g-PNIPAM and ZnTPP/D-g-PNIPAM/AuNPs systems, as well as the reference spectra of ZnTPP, D-g-PNIPAM and D-g-PNIPAM/Au NPs aqueous solutions. One can see that the aqueous solutions of D-g-PNIPAM and D-g-PNIPAM/AuNPs without ZnTPP demonstrate no singlet oxygen emission. Meanwhile, the spectra of ZnTPP containing systems show the emission peak at 1270 nm, a characteristic for singlet oxygen. Singlet oxygen emission spectra were detected for ZnTPP, ZnTPP/D-g-PNIPAM and ZnTPP/D-g-PNIPAM/AuNPs samples under excitation at 421 nm (Sore band), 553 nm and 595 nm (range of Q bands). Under excitations at 421 nm and 595 nm, the peak intensities are approximately the same for all ZnTPP containing systems. Additionally, under 553 nm excitation of aqueous solutions of ZnTPP and ZnTPP/D-g-PNIPAM, the peak intensities are close to the corresponding values obtained under 421 nm and 595 nm excitation. However, under 553 nm excitation of the ZnTPP/D-g-PNIPAM/AuNPs solution, a considerable 2.7-fold rise in intensity of the singlet oxygen peak is observed. Considering that the wavelength of 553 nm is resonant with LSPR in Au NPs (520 nm), it is reasonable to conclude that such enhancement of photogeneration of singlet oxygen has a plasmonic nature. Thus, the observed plasmonic enhancement of singlet oxygen photogeneration by ZnTPP/D-g-PNIPAM/AuNPs nanohybrid indicates its potential for PDT purposes that has proved by our biological studies discussed in Section 3.4.

3.4. Photodynamic Antibacterial Activity In Vitro

Thermoresponsive hydrogel systems are used as injectable gel-forming matrices. In the sol phase, therapeutic agents can be included in the system, and after injection to the target tissues, the solution is transformed into a gel and serves as a source of a drug. PNIPAM is one of the most common polymers of this type. Weak cytotoxicity of D-g-PNIPAM carrier against *S. aureus* at 37 °C was detected, Figure S4. The results are consistent with the literature data [69,70]. In Figures S4–S8, 11 and 12, we chose the same scale on the Y axis for the convenience of quantitative comparison of the bactericidal effect under different conditions of biological experiments.

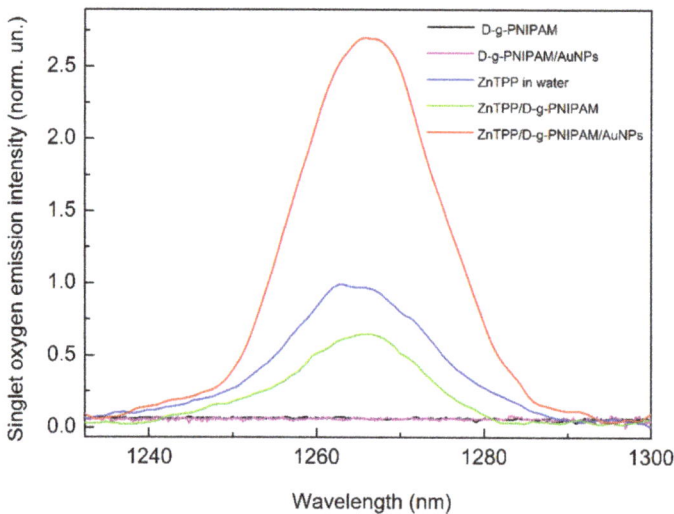

Figure 10. Emission spectra of singlet oxygen for ZnTPP, ZnTPP/D-g-PNIPAM and ZnTPP/D-g-PNIPAM/AuNPs aqueous solutions. The D-g-PNIPAM and D-g-PNIPAM/AuNPs aqueous solutions without ZnTPP demonstrate no emission from singlet oxygen. Excitation: 553 nm; concentrations: ZnTPP—0.001 g/L, D-g-PNIPAM—0.078 g/L, Au—0.077 g/L.

At 0.008 g/L concentration of D-g-PNIPAM polymer, there was no toxicity against *S. aureus*. Therefore, for further studies of the antibacterial activity of hybrid nanosystems, we chose a concentration of D-g-PNIPAM at 0.008 g/L to offset the toxic effects of the polymer and assess its response to temperature changes. LCST phase transition in D-g-PNIPAM occurs at 33–34 °C. Therefore, the copolymer is in the hydrophilic sol phase at 28 °C, and at human body temperature—in the hydrophobic gel phase. Comparison of antibacterial activity under such conditions indicates the retention of Au NPs in D-g-PNIPAM at temperatures below the LCST point, Figure S5. There was a 45–50% decrease of CFU amount at 37 °C for 0.08 and 0.008 g/L Au NPs concentrations after 25 min. Thus, one can assume that the polymer macromolecule shrinking at phase transition causes the release of Au NPs out to the solution.

Incubation of *S. aureus* suspension with 0.08 g/L D-g-PNIPAM/AuNPs and irradiation by 530 nm had a little effect on the antibacterial activity of nanocomposite at 37 °C, Figure S6. However, after reducing the concentration of D-g-PNIPAM/AuNPs to 0.008 g/L under similar conditions, the initial antibacterial activity of the nanocomposite increased by 40%.

The identified effects can be related to several factors. Probably, during shrinking of the Au NPs carrier, i.e., D-g-PNIPAM macromolecule, Au NPs are rapidly released into solution. The copolymer stabilizes AuNPs and prevents their aggregation. The high concentration of Au NPs increases the probability of their encounter and the formation of biologically inactive aggregates. Irradiation by light at 530 nm can accelerate the aggregation in the absence of a stabilizer. Some studies confirm the possibility of the aggregation under the light irradiation [31,33,71–73]. Reducing the D-g-PNIPAM/AuNPs concentration retains the necessary antibacterial properties when irradiated by light at 530 nm, Figure S6.

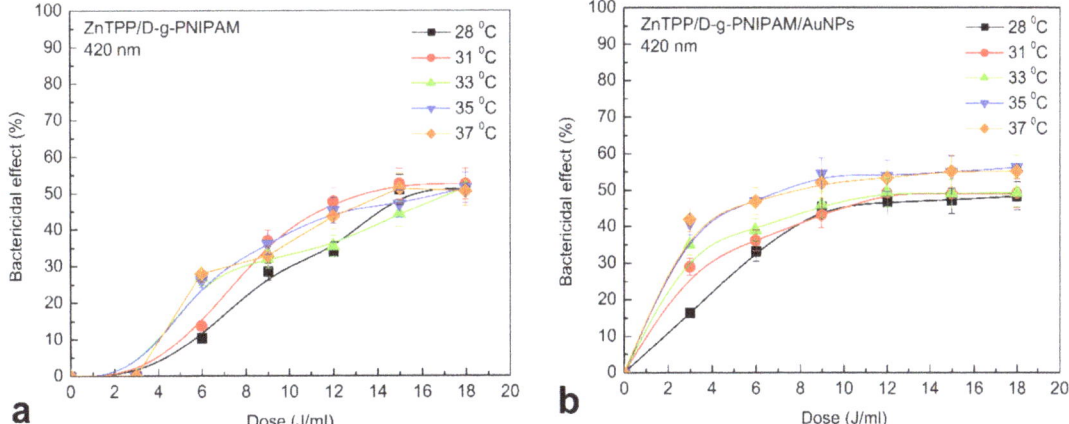

Figure 11. *S. aureus* inactivation at 28, 31, 33, 35, 37 °C after adding of ZnTPP/D-g-PNIPAM (**a**), ZnTPP/D-g-PNIPAM/AuNPs (**b**), and irradiation by light at 420 nm depending on the irradiation dose. Concentrations: AuNPs and D-g-PNIPAM—0.08 g/L, 0.008 g/L, 0.0008 g/L, ZnTPP—0.001 g/L. Power of light—0.1 J/s, dose of irradiation was in the range of 3–18 J/mL with 3 J/mL increment.

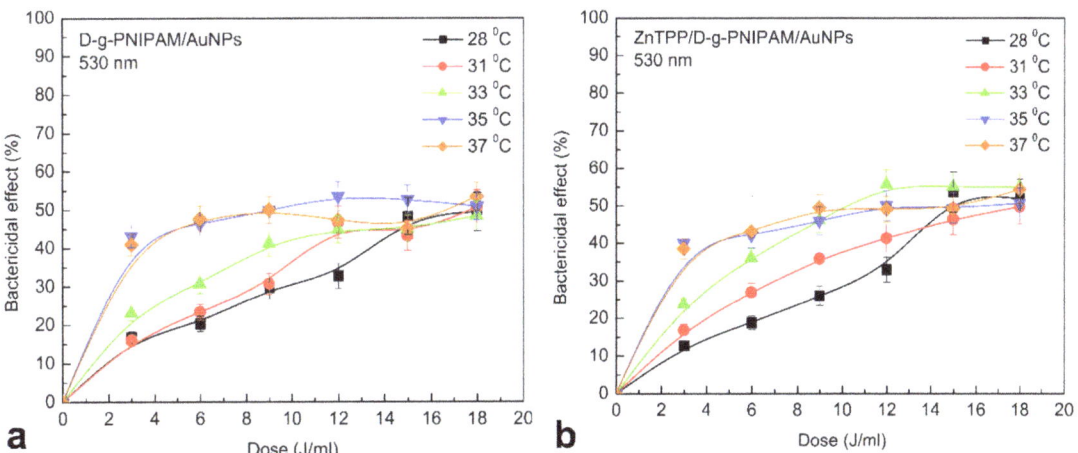

Figure 12. *S. aureus* inactivation at 28, 31, 33, 35, 37 °C after adding of D-g-PNIPAM/AuNPs (**a**), ZnTPP/D-g-PNIPAM/AuNPs (**b**), and irradiation by light at 530 nm depending on the irradiation dose. Concentrations: AuNPs and D-g-PNIPAM—0.08 g/L, 0.008 g/L, 0.0008 g/L, ZnTPP—0.001 g/L. Power of light—0.1 J/s, dose of irradiation was in the range of 3–18 J/mL with 3 J/mL increment.

Low sensitivity of the triple ZnTPP/D-g-PNIPAM/AuNPs system to the action of 420 nm light at 37 °C was revealed, Figure S7. Only irradiation of a bacterial suspension with 0.08 g/L nanocomposite concentration contributes to a linear increase in bactericidal activity from 33% to 75% depending on dose used. Additionally, for a 0.008 g/L Au NPs concentration and 3 J/mL irradiation dose, CFU amount further decrease by 25–30%. We assume that this is due to the features of the carrier. After the phase transition, it does not facilitate the delivery of active components to bacterial cells. In addition, to inactivate the bacterial cell, it is necessary to disrupt vital structures such as DNA, plasma membrane, enzymes etc. If the nanosystems do not penetrate the cells but are adsorbed on the cell wall or capsule instead, the damage of the cells is not critical. If such interactions are indeed present, nanocomposites can be used as the sensors for micro-organisms detection [74]. Irradiation by light at 530 nm of nanocomposite ZnTPP/D-g-PNIPAM/AuNPs contributes to the change of the number of CFU similar to the results for the system D-g-PNIPAM/AuNPs, Figure S8.

The results indicate that under such conditions, the Au NPs play a decisive role in the antibacterial activity, and a photosensitizer plays the secondary role at least at used concentrations. When irradiated by light, the antibacterial activity of nanocomposites depends on temperature. The change is clear at low dose (3–6 J/mL) of light irradiation (Figure 11). There are no detectable changes in the antibacterial properties under blue light irradiation at doses higher than 6 J/mL. The presence of Au NPs helps to increase the sensitivity of the nanosystems to 420 nm light (Figure 11b). The more pronounced temperature dependence was also found for ZnTPP/D-g-PNIPAM/AuNPs. This once again proves the decisive role of Au NPs in antibacterial activity. The results obtained at the irradiation of D-g-PNIPAM/AuNPs and ZnTPP/D-g-PNIPAM/AuNPs nanosystems by light at 530 nm resonant with LSPR in Au NPs prove this assumption (Figure 12). The rise of temperature closer to the point of LCST phase transition leads to a gradual increase in antibacterial activity. The antibacterial properties were the same at 35 and 37 °C.

The obtained temperature dependencies for irradiation of nanosystems at 420 and 530 nm indicate the determining role of Au NPs in the inactivation of microorganisms. The photosensitizer plays a secondary role in this process. However, it is noticeable that under the irradiation by 420 nm light resonant with ZnTPP Soret absorption band and irradiation doses lower than 10 J/mL, the antibacterial activity of ZnTPP/D-g-PNIPAM/AuNPs is essentially higher than one of D-g-PNIPAM/AuNPs, Figure 11. Most probably, this effect is caused by the plasmonic enhancement of the singlet oxygen generation by the ZnTPP molecules located in plasmonic field of Au NPs. As we discussed above, the local plasmon field enhances the optical processes in the photosensitizer molecules located in this field. Indeed, we revealed an enhancement of ZnTPP fluorescence in ZnTPP/D-g-PNIPAM/AuNPs nanosystem with the maximal FL enhancement at Au NPs concentration of 0.008 g/L (Figure 5). Another electronic process in ZnTPP molecules, that is probably plasmon enhanced, is the singlet oxygen generation. Such suggestion is proved by following observations. Figure 11 shows that at temperatures higher than LCST point, the antibacterial activity of ZnTPP/D-g-PNIPAM/AuNPs is elevated compared to ZnTPP/D-g-PNIPAM. Indeed, the macromolecule shrinking at the LCST transition leads to a decrease in ZnTPP–AuNP distance. As a result, the ZnTPP molecules are located in stronger plasmon field that should cause the enhancement of both singlet oxygen generation and FL by ZnTPP molecules. The singlet oxygen photogeneration enhancement causes an increase of antibacterial activity. In turn, as we discussed above, the LCST transition leads to enhancement of ZnTPP FL in ZnTPP/D-g-PNIPAM/AuNPs system comparing to ZnTPP/D-g-PNIPAM, Figure 11b.

Double and triple thermoresponsive nanosystems based on D-g-PNIPAM copolymer with Au NPs and ZnTPP photosensitizer, were irradiated by small doses of light of 420 and 530 nm. As the result, the number of CFU was decreased by 50% within 20 min. This bactericidal effect is not so strong when compared with antibiotics or other antibacterial chemicals [75,76]. Despite this, the investigated nanosystems have the significant potential to be used as broad-spectrum bactericidal agents with a reduced risk of developing resistance in bacteria. The LCST phase transition in D-g-PNIPAM provides a rapid increase in biological activity of D-g-PNIPAM based nanohybrids at temperatures lower than physiological. Given the results obtained on the properties of ZnTPP/D-g-PNIPAM/AuNPs nanohybrid, it can be applied on the body surface, wounds, or ulcers for rapid inactivation of microorganisms.

4. Conclusions

In conclusion, we have presented the results of the synthesis, the size and morphology characteristics, spectral properties and APDT activity of the aqueous solution of ZnTPP/D-g-PNIPAM/AuNPs triple hybrid nanosystem containing zinc tetraphenylporphyrin (ZnTPP) photosensitizer, thermoresponsive dextran poly (N-isopropylacrylamide) polymer (D-g-PNIPAM) and Au nanoparticles. Spectroscopic manifestations of binding of ZnTPP molecules to D-g-PNIPAM/AuNPs macromolecules have been obtained. The ZnTPP/D-g-PNIPAM/AuNPs nanohybrid has demonstrated high morphological stability (absence of the aggregation) up to 7 days after preparation. Optimal concentrations of the components of hybrid nanosystem providing the weak aggregation and the high plasmonic enhancement of electronic processes in photosensitizer molecules have been determined as following: ZnTPP—0.001 g/L, D-g-PNIPAM—0.078 g/L, Au—0.077 g/L. The shrinking of D-g-PNIPAM macromolecule at a thermally induced LCST phase transition has been revealed to lead to Au NPs and ZnTPP molecules release from the ZnTPP/D-g-PNIPAM/AuNPs macromolecule, as well as to strengthening of plasmon enhancement of the optical processes in ZnTPP molecules bound with the polymer macromolecule. The 2.7-fold plasmon enhancement of the photogeneration of singlet oxygen under excitation resonant with LSPR in Au NPs has been demonstrated for ZnTPP/D-g-PNIPAM/AuNPs system, which indicates its potential for PDT applications. The data obtained in vitro on *Staphylococcus aureus* wild strains have proved the potential of such nanosystem for rapid inactivation of microorganisms. In particular, the significant increase of PDT based bactericidal efficiency of ZnTPP/D-g-PNIPAM/AuNPs nanohybrid at the temperatures higher than LCST transition point has been observed at irradiation doses lower than 8 J/mL. This indicates that aqueous solution of ZnTPP/D-g-PNIPAM/AuNPs nanohybrid has a potential for thermally driven and controlled PDT applications at low light irradiation doses, in particular for rapid antibacterial PDT.

Supplementary Materials: The following supporting information can be downloaded at: https://www.mdpi.com/article/10.3390/nano12152655/s1, Aging effect on ZnTPP/D-g-PNIPAM/AuNPs nanosystem in aqueous solution from DLS measurements (Figure S1), Transformation of absorption spectrum of ZnTPP in water and aqueous solutions of hybrid nanosystems (Figure S2 and analysis), Transformation of FL spectrum of ZnTPP in water and aqueous solutions of hybrid nanosystems (Figure S3 and analysis), Remarks on the reabsorption effect on concentration dependence of FL, Additional information on photodynamic antibacterial activity of nanosystems in vitro (Figures S4–S8).

Author Contributions: Conceptualization, O.A.Y. and N.V.K.; methodology, O.A.Y., N.V.K., P.A.V. and V.A.C.; writing—original draft, O.A.Y.; writing—review & editing, O.A.Y., N.V.K., L.C. and G.N.; funding acquisition, O.A.Y., N.V.K. and G.N.; investigation, A.V.T., P.S.K., P.A.V., V.A.C., Y.I.K., A.I.M. and L.C.; formal analysis, A.V.T., P.S.K., P.A.V. and V.A.C. All authors have read and agreed to the published version of manuscript.

Funding: This work was financially supported by National Research Foundation of Ukraine Project (No. 2020.02/0022), Ministry of Education and Science of Ukraine Project (No. 0122U001818), Guangxi Innovation Driven Development Major Project (No. Guike AA20302013), Nanning Scientific Research and Technology Development Plan Project (No. RC20200001), "Yongjiang Plan" Project of Leading Talents of Innovation and Entrepreneurship in Nanning City (No. 2020024), Yulin City Science and Technology Transformation Project (No. 19040003), French PAUSE program for emergency welcome of Ukrainian scientists.

Data Availability Statement: Data are within the article and Supplementary Materials.

Conflicts of Interest: The authors declare no conflict of interest.

References

1. Abrahamse, H.; Hamblin, M.R. New photosensitizers for photodynamic therapy. *Biochem. J.* **2016**, *473*, 347–364. [CrossRef]
2. Jori, G.; Fabris, C.; Soncin, M.; Ferro, S.; Coppellotti, O.; Dei, D.; Fantetti, L.; Chiti, G.; Roncucci, G. Photodynamic therapy in the treatment of microbial infections: Basic principles and perspective applications. *Lasers Surg. Med.* **2006**, *38*, 889–905. [CrossRef] [PubMed]

3. Ghorbani, J.; Rahban, D.; Aghamiri, S.; Teymouri, A.; Bahador, A. Photosensitizers in antibacterial photodynamic therapy: An overview. *Laser Ther.* **2018**, *4*, 293–302. [CrossRef] [PubMed]
4. Huang, L.; Xuan, Y.; Koide, Y.; Zhiyentayev, T.; Tanaka, M.; Hamblin, M.R. Type I and type II mechanisms of antimicrobial photodynamic therapy: An in vitro study on gram-negative and gram-positive bacteria. *Lasers Surg. Med.* **2012**, *44*, 490. [CrossRef]
5. Bhatia, S. *Natural Polymer Drug Delivery Systems*; Springer: Cham, Switzerland, 2016.
6. Doberenz, F.; Zeng, K.; Willems, C.; Zhang, K.; Groth, T. Thermoresponsive polymers and their biomedical application in tissue engineering—A review. *J. Mater. Chem. B* **2020**, *8*, 607–628. [CrossRef]
7. Joglecar, M.; Trewyn, B.G. Polymer-based stimuli responsible nanosystems for biomedical applications. *Biotechnol. J.* **2013**, *8*, 931–945. [CrossRef]
8. Cabane, E.; Zhang, X.; Langowska, K.; Palivan, C.G.; Meier, W. Stimuli responsible polymers and their application in nanomedicine. *Biointerphases* **2012**, *7*, 9. [CrossRef]
9. Gong, C.; Qi, T.; Wei, X.; Qu, Y.; Wu, Q.; Luo, F.; Qian, Z. Thermosensitive polymeric hydrogeles as drug delivery systems. *Curr. Med. Chem.* **2016**, *20*, 79–94. [CrossRef]
10. Dal Lago, V.; França de Oliveira, L.; de Almeida Gonçalves, K.; Kobarg, J.; Cardoso, M.B. Size-selective silver nanoparticles: Future of biomedical devices with enhanced bactericidal properties. *J. Mater. Chem.* **2011**, *21*, 12267–12273. [CrossRef]
11. Karabasz, A.; Bzowska, M.; Szczepanowicz, K. Biomedical applications of multifunctional polymeric nanocarriers: A review of current literature. *Int. J. Nanomed.* **2020**, *15*, 8673–8696. [CrossRef]
12. Kutsevol, N.V.; Chumachenko, V.A.; Harahuts, I.I.; Marinin, A.I. Aging process of gold nanoparticles synthesized in situ in aqueous solutions of polyacrylamides. In *Chemical Engineering of Polymers: Production of Functional and Flexible Materials*; Mukbanianym, O.V., Abadie, M.J., Tatrishvili, T., Eds.; Apple Academic Press: Palm Bay, FL, USA, 2017; pp. 119–129.
13. Bulavin, L.; Kutsevol, N.; Chumachenko, V.; Soloviov, D.; Kuklin, A.; Marynin, A. SAXS combined with UV-Vis spectroscopy and QELS: Accurate characterization of silver sols synthesized in polymer matrices. *Nanoscale Res. Lett.* **2016**, *11*, 35. [CrossRef] [PubMed]
14. Chumachenko, V.; Kutsevol, N.; Harahuts, Y.; Rawiso, M.; Marinin, A.; Bulavin, L. Star-like dextran-graft-PNiPAM copolymers. Eff. Intern. Mol. Struct. Phase Transit. *J. Mol. Liq.* **2017**, *235*, 77–82.
15. Deng, S.; Gigliobianco, M.R.; Censi, R.; Di Martino, P. Polymeric nanocapsules as nanotechnological alternative for drug delivery system: Current status, challenges and opportunities. *Nanomaterials* **2020**, *10*, 847. [CrossRef] [PubMed]
16. Chumachenko, V.A.; Shton, I.O.; Shishko, E.D.; Kutsevol, N.V.; Marinin, A.I.; Gamaleia, N.F. Branched copolymers dextran-graft-polyacrylamide as nanocarriers for delivery of gold nanoparticles and photosensitizers to tumor cells. In *Nanophysics Nanophotonics Surface Studies and Applications, Springer Proceedings in Physics Series*; Fesenko, O., Yatsenko, L., Eds.; Springer: Berlin/Heidelberg, Germany, 2016; pp. 379–390.
17. Aguilar, M.R.; Elvira, C.; Gallardo, A.; Vázquez, B.; Román, J.S. Smart polymers and their applications as biomaterials. III Biomaterials. In *Topics in Tissue Engineering*; Ashammakhi, N., Reis, R., Chiellini, E., Eds.; Academic Press: Cambridge, MA, USA, 2007; Volume 3, pp. 1–27.
18. Gandhi, A.; Paul, A.; Sen, S.O.; Sen, K.K. Studies on thermoresponsive polymers: Phase behaviour, drug delivery and biomedical applications. *Asian J. Pharm. Sci.* **2015**, *10*, 99–107. [CrossRef]
19. Sedláček, O.; Černoch, P.; Kučka, J. Thermoresponsive polymers for nuclear medicine: Which polymer is the best? *Langmuir* **2016**, *32*, 6115–6122. [CrossRef]
20. Ma, Y.M.; Wei, D.X.; Yao, H.; Wu, L.P.; Chen, G.Q. Synthesis, characterization and application of thermoresponsive polyhydroxyalkanoate-graft-poly (N-isopropylacrylamide). *Biomacromolecules* **2016**, *17*, 2680–2690. [CrossRef]
21. Futscher, M.H.; Philipp, M.; Müller-Buschbaum, P.; Schulte, A. The role of backbone hydration of poly(N-isopropyl acrylamide) across the volume phase transition compared to its monomer. *Sci. Rep.* **2017**, *7*, 17012. [CrossRef]
22. Bischofberger, I.; Trappe, V. New aspects in the phase behaviour of poly-N-isopropyl acrylamide: Systematic temperature dependent shrinking of PNiPAM assemblies well beyond the LCST. *Sci. Rep.* **2015**, *5*, 15520. [CrossRef]
23. De Oliveira, T.E.; Marques, C.M.; Netz, P.A. Molecular dynamics study of the LCST transition in aqueous poly(N-n-propylacrylamide). *Phys. Chem. Chem. Phys.* **2018**, *20*, 10100. [CrossRef]
24. Lopez, V.C.; Hadgraft, J.; Snowden, M.J. The use of colloidal microgels as a (trans)dermal drug delivery system. *Int. J. Pharm.* **2005**, *292*, 137–147. [CrossRef] [PubMed]
25. Yu, H.; Peng, Y.; Yang, Y.; Li, Z.Y. Plasmon-enhanced light–matter interactions and applications. *NPJ Comput. Mater.* **2019**, *5*, 45. [CrossRef]
26. Baumberg, J.J.; Aizpurua, J.; Mikkelsen, M.H.; Smith, D.R. Extreme nanophotonics from ultrathin metallic gaps. *Nat. Mater.* **2019**, *18*, 668–678. [CrossRef] [PubMed]
27. Hou, W.; Cronin, S.B. A review of surface plasmon resonance-enhanced photocatalysis. *Adv. Funct. Mater.* **2013**, *23*, 1612–1619. [CrossRef]
28. Zhang, Y.; Aslan, K.; Previte, M.J.R.; Geddes, C.D. Metal-enhanced singlet oxygen generation: A consequence of plasmon enhanced triplet yields. *J. Fluoresc.* **2007**, *17*, 345–349. [CrossRef] [PubMed]
29. Kutsevol, N.V.; Chumachenko, V.A.; Rawiso, M.; Shkodich, V.F.; Stoyanov, O.V. Star-like polymers dextran-polyacrylamide: The prospects of application for nanotechnology. *J. Struct. Chem.* **2015**, *56*, 1016–1023. [CrossRef]

30. Yeshchenko, O.A.; Naumenko, A.P.; Kutsevol, N.V.; Maskova, D.O.; Harahuts, I.I.; Chumachenko, V.A.; Marinin, A.I. Anomalous inverse hysteresis of phase transition in thermosensitive dextran-graft-PNIPAM copolymer/Au nanoparticles hybrid nanosystem. *J. Phys. Chem. C* **2018**, *122*, 8003–8010. [CrossRef]
31. Yeshchenko, O.A.; Naumenko, A.P.; Kutsevol, N.V.; Harahuts, I.I. Laser-driven structural transformations in dextran-graft-PNIPAM copolymer/Au nanoparticles hybrid nanosystem: The role of plasmon heating and attractive optical forces. *RSC Adv.* **2018**, *8*, 38400–38409. [CrossRef]
32. Chumachenko, V.; Kutsevol, N.; Harahuts, I.; Soloviov, D.; Bulavin, L.; Yeshchenko, O.; Naumenko, A.; Nadtoka, O.; Marinin, A. Temperature driven transformation in dextran-graft-PNIPAM/embedded silver nanoparticle hybrid system. *Int. J. Polym. Sci.* **2019**, *2019*, 3765614. [CrossRef]
33. Yeshchenko, O.A.; Bartenev, A.O.; Naumneko, A.P.; Kutsevol, N.V.; Harahuts, I.I.; Marinin, A.I. Laser-driven aggregation in dextran-graft-PNIPAM/silver nanoparticles hybrid nanosystem: Plasmonic effects. *Ukr. J. Phys.* **2020**, *65*, 254–267. [CrossRef]
34. Martinez De Pinillos Bayona, A.; Mroz, P.; Thunshelle, C.; Hamblin, M.R. Design features for optimization of tetrapyrrole macrocycles as antimicrobial and anticancer photosensitizers. *Chem. Biol. Drug Des.* **2017**, *89*, 192–206. [CrossRef]
35. Lin, Y.; Zhou, T.; Bai, R.; Xie, Y. Chemical approaches for the enhancement of porphyrin skeleton-based photodynamic therapy. *J. Enzym. Inhib. Med. Chem.* **2020**, *35*, 1080–1099. [CrossRef] [PubMed]
36. Zhang, Z.; Yu, H.J.; Wu, S.; Huang, H.; Si, L.P.; Liu, H.Y.; Shi, L.; Zhang, H.T. Synthesis, characterization, and photodynamic therapy activity of 5,10,15,20-tetrakis(carboxyl)porphyrin. *Bioorganic Med. Chem.* **2019**, *27*, 2598–2608. [CrossRef] [PubMed]
37. Elms, J.; Beckett, P.N.; Griffin, P.; Curran, A.D. Mechanisms of isocyanate sensitisation. An in vitro approach. *Toxicol. Vitr.* **2001**, *15*, 631. [CrossRef]
38. Shamaila, S.; Zafar, N.; Riaz, S.; Sharif, R.; Nazir, J.; Naseem, S. Gold nanoparticles: An efficient antimicrobial agent against enteric bacterial human pathogen. *Nanomaterials* **2016**, *6*, 71. [CrossRef] [PubMed]
39. Kutsevol, N.; Harahuts, Y.; Shton, I.; Borikun, T.; Storchai, D.; Lukianova, N.; Chekhun, V. In vitro study of toxicity of hybrid gold-polymer composites. *Mol. Cryst. Liq. Cryst.* **2018**, *671*, 1–8. [CrossRef]
40. Milacic, V.; Dou, Q.P. The tumor proteasome as a novel target for gold(III) complexes: Implications for breast cancer therapy. *Coord. Chem. Rev.* **2009**, *19*, 398–403. [CrossRef]
41. Lammer, A.D.; Cook, M.E.; Sessler, J.L. Synthesis and anti-cancer activities of a water soluble gold(III) porphyrin. *J. Porphyr. Phthalocyanines* **2015**, *19*, 398–403. [CrossRef]
42. Macia, N.; Kabanov, V.; Côté-Cyr, M.; Heyne, B. Roles of near and far fields in plasmon-enhanced singlet oxygen production. *J. Phys. Chem. Lett.* **2019**, *10*, 3654–3660. [CrossRef]
43. Macia, N.; Kabanov, V.; Heyne, B. Rationalizing the plasmonic contributions to the enhancement of singlet oxygen production. *J. Phys. Chem. C* **2020**, *124*, 3768–3777. [CrossRef]
44. Planas, O.; Macia, N.; Agut, M.; Nonell, S.; Heyne, B. Distance-dependent plasmon-enhanced singlet oxygen production and emission for bacterial inactivation. *J. Am. Chem. Soc.* **2016**, *138*, 2762–2768. [CrossRef]
45. Tavakkoli Yaraki, M.; Daqiqeh Rezaei, S.; Tan, Y.N. Simulation guided design of silver nanostructures for plasmon-enhanced fluorescence singlet oxygen generation and SERS applications. *Phys. Chem. Chem. Phys.* **2020**, *22*, 5673–5687. [CrossRef] [PubMed]
46. Mohamed, M.M.; Fouad, S.A.; Elshoky, H.A.; Mohammed, G.M.; Salaheldin, T.A. Antibacterial effect of gold nanoparticles against corynebacterium pseudotuberculosis. *Int. J. Vet. Sci. Med.* **2017**, *5*, 23–29. [CrossRef] [PubMed]
47. Yeshchenko, O.A.; Kutsevol, N.V.; Naumenko, A.P. Light-induced heating of gold nanoparticles in colloidal solution: Dependence on detuning from surface plasmon resonance. *Plasmonics* **2016**, *11*, 345–350. [CrossRef]
48. Yeshchenko, O.A.; Kozachenko, V.V. Light-induced heating of dense 2D ensemble of gold nanoparticles: Dependence on detuning from surface plasmon resonance. *J. Nanoparticle Res.* **2015**, *17*, 296. [CrossRef]
49. Harahuts, Y.I.; Pavlov, V.A.; Mokrinskaya, E.V.; Chuprina, N.G.; Davidenko, N.A.; Naumenko, A.P.; Bezugla, T.M.; Kutsevol, N.V. The study of Au sol synthesized in uncharged and charged star-like copolymers under light irradiation. *Funct. Mater.* **2019**, *4*, 723–728.
50. Kutsevol, N.; Kuziv, Y.; Bezugla, T.; Virych, P.; Marynin, A.; Borikun, T.; Lukianova, N.; Virych, P.; Chekhun, V. Application of new multicomponent nanosystems for overcoming doxorubicin resistance in breast cancer therapy. *Appl. Nanosci.* **2022**, *12*, 427–437. [CrossRef]
51. Millenbaugh, N.J.; Baskin, J.B.; DeSilva, M.N.; Elliot, W.R.; Glickman, R.D. Photothermal killing of Staphylococcus aureus using antibody-targeted gold nanoparticles. *Int. J. Nanomed.* **2015**, *10*, 1953–1960. [CrossRef]
52. Lim, Z.Z.J.; Li, J.E.J.; Ng, C.T.; Yung, L.Y.L.; Bay, B.H. Gold nanoparticles in cancer therapy. *Acta Pharmacol. Sin.* **2011**, *32*, 983–990. [CrossRef]
53. Von Maltzahn, G.; Park, J.H.; Lin, K.Y.; Singh, N.; Schwöppe, C.; Mesters, R.; Berdel, W.E.; Ruoslahti, E.; Sailor, M.J.; Bhatia, S.N. Nanoparticles that communicate in vivo to amplify tumour targeting. *Nat. Mater.* **2011**, *10*, 545–552. [CrossRef]
54. Yuan, X.; Liu, Z.; Wang, Y.; Xu, Y.; Zhang, W.; Mu, T. The non-negative truncated singular value decomposition for adaptive sampling of particle size distribution in dynamic light scattering inversion. *J. Quant. Spectrosc. Radiat. Transf.* **2020**, *246*, 106917. [CrossRef]
55. Marsh, D.F.; Mink, L.M. Microscale synthesis and electronic absorption spectroscopy of tetraphenylporphyrin H2(TPP) and metalloporphyrins ZnII(TPP) and NiII(TPP). *J. Chem. Educ.* **1996**, *73*, 1181. [CrossRef]

56. Strachan, J.P.; Gentemann, S.; Seth, J.; Kalsbeck, W.A.; Lindsey, J.S.; Holten, D.; Bocian, D.F. Effects of orbital ordering on electronic communication in multiporphyrin arrays. *J. Am. Chem. Soc.* **1997**, *119*, 11191–11201. [CrossRef]
57. Harriman, A. Luminescence of porphyrins and metalloporphyrins. Part 1—Zinc(II) nickel(II) and manganese(II) porphyrins. *J. Chem. Soc. Faraday Trans. 1 Phys. Chem. Condens. Phases* **1980**, *6*, 1978–1985. [CrossRef]
58. Nguyen, K.A.; Day, P.N.; Pachter, R.; Tretiak, S.; Chernyak, V.; Mukamel, S. Analysis of absorption spectra of zinc porphyrin, zinc meso-tetraphenylporphyrin and halogenated derivatives. *J. Phys. Chem. A* **2002**, *106*, 10285–10293. [CrossRef]
59. Yeshchenko, O.A.; Kutsevol, N.V.; Tomchuk, A.V.; Khort, P.S.; Virych, P.A.; Chumachenko, V.A.; Kuziv, Y.I.; Naumenko, A.P.; Marinin, A.I. Plasmonic enhancement of the antibacterial photodynamic efficiency of a zinc tetraphenylporphyrin photosensitizer/dextran graft polyacrylamide anionic copolymer/Au nanoparticles hybrid nanosystem. *RSC Adv.* **2022**, *12*, 11–23. [CrossRef] [PubMed]
60. Yeshchenko, O.A.; Kutsevol, N.V.; Tomchuk, A.V.; Khort, P.S.; Virych, P.A.; Chumachenko, V.A.; Kuziv, Y.I.; Naumenko, A.P.; Marinin, A.I. Zinc tetraphenylporphyrin/dextran-graft-polyacrylamide/gold nanoparticles hybrid nanosystem for photodynamic therapy: Plasmonic enhancement effect. *Nanomed. Res. J.* **2022**, *7*, 173–188.
61. Törmö, P.; Barnes, W.L. Strong coupling between surface plasmon polaritons and emitters: A Review. *Rep. Prog. Phys.* **2015**, *78*, 013901. [CrossRef]
62. Rodarte, A.L.; Tao, A.R. Plasmon-exciton coupling between metallic nanoparticles and dye monomers. *J. Phys. Chem. C* **2017**, *121*, 3496–3502. [CrossRef]
63. Anger, P.; Bharadwaj, P.L. Novotny, Enhancement and quenching of single-molecule fluorescence. *Phys. Rev. Lett.* **2006**, *96*, 113002. [CrossRef] [PubMed]
64. Yeshchenko, O.A.; Khort, P.S.; Kutsevol, N.V.; Prokopets, V.M.; Kapush, O.; Dzhagan, V. Temperature driven plasmon-exciton coupling in thermoresponsive dextran-graft-PNIPAM/Au nanoparticle/CdTe quantum dots hybrid nanosystem. *Plasmonics* **2021**, *16*, 1137–1150. [CrossRef]
65. Roller, E.M.; Argyropoulos, C.; Högele, A.; Liedl, T.; Pilo-Pais, M. Plasmon-exciton coupling using DNA templates. *Nano Lett.* **2016**, *16*, 5962–5966. [CrossRef] [PubMed]
66. Dolinnyi, A.I. Nanometric rulers based on plasmon coupling in pairs of gold nanoparticles. *J. Phys. Chem. C* **2015**, *119*, 4990–5001. [CrossRef]
67. Su, Q.; Jiang, C.; Gou, D.; Long, Y. Surface plasmon-assisted fluorescence enhancing and quenching: From theory to application. *ACS Appl. Bio Mater.* **2021**, *4*, 4684–4705. [CrossRef] [PubMed]
68. Medintz, I.; Hildebrandt, N. *FRET—Förster Resonance Energy Transfer: From Theory to Applications*; Wiley: Hoboken, NJ, USA, 2013.
69. Ashraf, S.; Park, H.; Park, H.; Lee, S.-H. Snapshot of phase transition in thermoresponsive hydrogel PNIPAM: Role in drug delivery and tissue engineering. *Macromol. Res.* **2016**, *24*, 297–304. [CrossRef]
70. Garcia-Pinel, B.; Ortega-Rodríguez, A.; Porras-Alcalá, C.; Cabeza, L.; Contreras-Cáceres, R.; Ortiz, R.; Díaz, A.; Moscoso, A.; Sarabia, F.; Prados, J.; et al. Magnetically active pNIPAM nanosystems as temperature-sensitive biocompatible structures for controlled drug delivery. *Artif. Cells Nanomed. Biotechnol.* **2020**, *48*, 1022–1035. [CrossRef] [PubMed]
71. Zhou, J.; Sedev, R.; Beattie, D.; Ralston, J. Light-induced aggregation of colloidal gold nanoparticles capped by thymine derivatives. *Langmuir* **2008**, *24*, 4506–4511. [CrossRef] [PubMed]
72. Schmarsow, R.N.; dell'Erba, I.E.; Villaola, M.S.; Hoppe, C.E.; Zucchi, I.A.; Schroeder, W.F. Effect of light intensity on the aggregation behavior of primary particles during in situ photochemical synthesis of gold/polymer nanocomposites. *Langmuir* **2020**, *36*, 13759–13768. [CrossRef] [PubMed]
73. Bhattacharya, S.; Narasimha, S.; Roy, A.; Banerjee, S. Does shining light on gold colloids influence aggregation? *Sci. Rep.* **2014**, *4*, 5213. [CrossRef]
74. Verma, M.S.; Wei, S.C.; Rogowski, J.L.; Tsuji, J.M.; Chen, P.Z.; Lin, C.W.; Jones, L.; Gu, F.X. Interactions between bacterial surface and nanoparticles govern the performance of "chemical nose" biosensors. *Biosens. Bioelectron.* **2016**, *15*, 115–125. [CrossRef]
75. Gold, B.; Smith, R.; Nguyen, Q.; Roberts, J.; Ling, Y.; Lopez Quezada, L.; Somersan, S.; Warrier, T.; Little, D.; Pingle, M.; et al. Novel cephalosporins selectively active on nonreplicating mycobacterium tuberculosis. *J. Med. Chem.* **2016**, *59*, 6027–6044. [CrossRef]
76. Kavitha, S.; Harikrishnan, A.; Jeevaratnam, K. Characterization and evaluation of antibacterial efficacy of a novel antibiotic-type compound from a probiotic strain Lactobacillus plantarum KJB23 against food-borne pathogens. *LWT* **2020**, *118*, 108759. [CrossRef]

 *nanomaterials*

Article

Evidence of Au(II) and Au(0) States in Bovine Serum Albumin-Au Nanoclusters Revealed by CW-EPR/LEPR and Peculiarities in HR-TEM/STEM Imaging

Radek Ostruszka [1], Giorgio Zoppellaro [2,*], Ondřej Tomanec [2], Dominik Pinkas [3], Vlada Filimonenko [3] and Karolína Šišková [1,*]

[1] Department of Experimental Physics, Faculty of Science, Palacký University, tř. 17. Listopadu 12, 77900 Olomouc, Czech Republic; radek.ostruszka@upol.cz
[2] Regional Centre of Advanced Technologies and Materials, Faculty of Science, Palacký University, tř. 17. Listopadu 12, 77900 Olomouc, Czech Republic; ondrej.tomanec@upol.cz
[3] Institute of Molecular Genetics of the Czech Academy of Sciences, Microscopy Centre, Electron Microscopy Core Facility, Vídeňská 1083, 14220 Prague, Czech Republic; dominik.pinkas@img.cas.cz (D.P.); vlada.filimonenko@img.cas.cz (V.F.)
* Correspondence: giorgio.zoppellaro@upol.cz (G.Z.); karolina.siskova@upol.cz (K.Š.)

Abstract: Bovine serum albumin-embedded Au nanoclusters (BSA-AuNCs) are thoroughly probed by continuous wave electron paramagnetic resonance (CW-EPR), light-induced EPR (LEPR), and sequences of microscopic investigations performed via high-resolution transmission electron microscopy (HR-TEM), scanning transmission electron microscopy (STEM), and energy dispersive X-ray analysis (EDS). To the best of our knowledge, this is the first report analyzing the BSA-AuNCs by CW-EPR/LEPR technique. Besides the presence of Au(0) and Au(I) oxidation states in BSA-AuNCs, the authors observe a significant amount of Au(II), which may result from a disproportionation event occurring within NCs: 2Au(I) → Au(II) + Au(0). Based on the LEPR experiments, and by comparing the behavior of BSA versus BSA-AuNCs under UV light irradiation (at 325 nm) during light off-on-off cycles, any energy and/or charge transfer event occurring between BSA and AuNCs during photoexcitation can be excluded. According to CW-EPR results, the Au nano assemblies within BSA-AuNCs are estimated to contain 6–8 Au units per fluorescent cluster. Direct observation of BSA-AuNCs by STEM and HR-TEM techniques confirms the presence of such diameters of gold nanoclusters in BSA-AuNCs. Moreover, in situ formation and migration of Au nanostructures are observed and evidenced after application of either a focused electron beam from HR-TEM, or an X-ray from EDS experiments.

Keywords: gold nanostructures; fluorescent nanoprobe; noble metal nanocrystal; protein nanocomposite

Citation: Ostruszka, R.; Zoppellaro, G.; Tomanec, O.; Pinkas, D.; Filimonenko, V.; Šišková, K. Evidence of Au(II) and Au(0) States in Bovine Serum Albumin-Au Nanoclusters Revealed by CW-EPR/LEPR and Peculiarities in HR-TEM/STEM Imaging. *Nanomaterials* **2022**, *12*, 1425. https://doi.org/10.3390/nano12091425

Academic Editors: Wei Chen and Derong Cao

Received: 28 March 2022
Accepted: 19 April 2022
Published: 22 April 2022

Publisher's Note: MDPI stays neutral with regard to jurisdictional claims in published maps and institutional affiliations.

Copyright: © 2022 by the authors. Licensee MDPI, Basel, Switzerland. This article is an open access article distributed under the terms and conditions of the Creative Commons Attribution (CC BY) license (https://creativecommons.org/licenses/by/4.0/).

1. Introduction

Fluorescent gold nanoclusters (AuNCs) embedded in bovine serum albumin (BSA) have been extensively studied in the literature since 2009, following the publication of Xie et al. [1] However, many open questions remain unsolved, especially for the witnessed dependence of the fluorescent properties with the sizes of Au nanostructures. [2–42]. The discussion remains controversial because many reports describe fluorescent AuNCs where the origin of such phenomenon is linked to the small size of the nanoparticles (NPs) (e.g., Burt et al. [37], Zhang and Wang [38], Zheng et al. [39]), while other reports suggest that the key to the fluorescence properties is intimately associated to the oxidation state of the gold cluster, Au(III)- [3] and/or Au(I)-complexes [40]. In Table 1, examples of selected BSA-AuNCs preparations are given, including their synthetic parameters, as reported in literature, in conjunction with the AuNCs size, oxidation state of Au when known, and maximum (maxima) of their fluorescence emissions.

Table 1. Details of selected BSA-AuNCs preparations and properties as reported in literature.

Reference Number	Oxidation State of Au	Size of BSA-AuNCs [nm]	Emission Wavelength Maximum [nm]	Quantum Yield [%]	Synthesis Conditions (X = BSA + HAuCl4)
[1]	Au^0, Au$^+$ XPS	≈0.8 TEM	640	6	X –(2 min)> NaOH → incubation at 37 °C for 12 h
[2]	n.a.	≈1 TEM	n.a.	n.a.	X → NaBH$_4$ → incubation at RT for 1 h
[3]	n.a.	6.3 ± 2.9 (pH 12) * 3.3 ± 1.4 (pH 10) * 1.6 ± 0.7 (pH 9) *	640 (pH 12) n.a. (pH 10) 440 (pH 9)	n.a.	same as [1], NaBH$_4$ used in later steps
[4]	Au^0, Au$^+$ XPS	4.2 ± 0.5 TEM	676	4.14	X → NaOH → MW (incubation at 80 °C for 4 min)
[4]	Au^0 XPS + FQ	3.1 ± 0.4 TEM	436	1.94	X → NaOH → MW (incubation at 135 °C for 4 min)
[7]	n.a.	n.a.	640, 710	n.a.	X → incubation at 37 °C overnight
[15]	Au^0, Au$^+$ XPS	2.1 ± 0.3 TEM	650	1.9	X → NaOH → MW (120 W, 2 min)
[16]	Au^0, Au$^+$ XPS	n.a.	656	n.a.	same as [1]
[27]	n.a.	n.a.	705	n.a.	X –(1 h) > NaOH → incubation at 45 °C for 4 h
[30]	Au^0, Au$^+$ XPS	n.a.	635	n.a.	same as [1]
[32]	Au^0, Au$^+$ XPS	4–6 TEM	650	≈8	X –(30 min)> NaOH → incubation at 50 °C for 3–4 h
[34]	Au^0, Au$^+$ XPS	n.a.	685	≈5.5	X → ascorbic acid → NaOH → incubation at 37 °C for 5 h
[35]	Au^0, Au$^+$ XPS	1.6 HR-TEM	604	n.a.	X –(2 min)> NaOH → MW and then incubation at 37 °C for 12 h X –(10 min)> NaOH → incubation at 37 °C for 24 h
[36]	Au^0, Au$^+$ XPS	n.a.	620	n.a.	X → NaOH → incubation at 37 °C for 12 h + dialysis for 48 h
[37]	n.a.	average < 2 TEM 1.5–1.8 HAADF-STEM	n.a.	n.a.	X → NaBH$_4$ → incubation for 3 h
[41]	n.a.	3.6 DLS	650	6.2	X –(90 s)> NaOH → MW (150 W, 10 s)
[42]	n.a.	≈5 TEM, HR-TEM	645	n.a.	X → NaOH → MW (300 W, 6 min)

Notes: n.a. = not available, * = sizes obtained after the reduction performed by addition of NaBH$_4$ to Au(III)-BSA systems which were generated at given pH values, MW = microwave radiation heating, XPS = X-ray photoelectron spectroscopy, FQ = fluorescence quenching, TEM = transmission electron microscopy, HAADF-STEM = high-angle annular dark-field scanning transmission electron microscopy, HR-TEM = high resolution transmission electron microscopy.

From the size perspectives, AuNCs are considered intermediate systems, falling between dimensions of isolated Au atoms (0D, zero dimensional) and Au nanoparticles (3D dimensional). The fluorescent properties emerge and are tuned by the presence of discrete energy levels that mirror the nanoparticle size variance, which approach the Fermi wavelength of free electrons (e.g., [38,43]). More detailed descriptions of different types of luminescent Au nanostructures and the related emission mechanisms, including not only the free-electron model, but also surface ligand effects, are thoroughly discussed in recent reviews (e.g., [39,44–46]).

The oxidation state of Au is undoubtedly an important parameter to unveil for understanding the physical basis of the fluorescence phenomenon. [44] In this context, analysis of the fluorescent properties of Au clusters entrapped by proteins, such as bovine serum albumin (BSA), is particularly debated. Mostly, Au(0) and Au(I) oxidation states are observed and screened by XPS (X-ray photoelectron spectroscopy) in BSA-AuNCs [1,2,4,15,16,30,32,34–36]. The pseudo-polymeric structure, such as SR-(Au-SR)x- (x = 1 or 2) motif, is envisioned to be present, and from XPS results the authors [2] demonstrated that Au-Au and Au-S bonds form in the BSA-AuNCs system. It should be noted that the occurrence of the staple motif arrangement (-S-Au(I)-S-Au(I)-S-) determined in BSA-AuNCs (as in e.g., [7]) shares similarities with the arrangement observed in thiol-protected $Au_{25}(SR)_{18}$ clusters prepared by Zhu M. et al. [47,48]. Based on XRD analysis, a core-shell structure of the thiol-protected Au_{25} cluster was determined in [47]. The same group of researchers revealed the intrinsic magnetism of these thiolate-protected Au_{25} super atoms in 2009 [49]. On the contrary, the magnetic properties of BSA-AuNCs have not been investigated so far, to the best of our knowledge. Therefore, the presence of such a property represents one of the aims tackled in this study.

Importantly, the oxidation state of Au in the final AuNCs is closely related to the experimental conditions. There are obvious differences in the presence and/or absence of a strong reduction agent in the course of AuNCs formation as shown in Table 1. In synthetic procedures where $NaBH_4$ is used, BSA represents solely a template (matrix) that prevents the occurrence of coalescence processes in AuNCs at room temperature. On the contrary, when the use of $NaBH_4$ is avoided (as in our previous work [41]), BSA acts as a reducing agent (probably due to tyrosine residues [1] at increased pH above its pK_a value) and, simultaneously, as a capping agent for AuNCs. Interestingly, even syntheses exploiting solely BSA as the reducing agent can lead to different fluorescent properties of resulting AuNCs, as shown in Table 1. The reason for such differences in fluorescent characteristics may be either the type of heating (incubator at 37 °C vs. MW heating), or the initial concentration of BSA—see Table 1.

Several literature reports that describe organometallic complexes containing Au(I) sites show the existence of strong Au–Au bonds and the term "aurophilicity" was therefore introduced in 1989 to describe the unusual bond properties [50]. Aurophilic interaction is thought to be as strong as a hydrogen bond and shows the presence of bond length as shorter than the sum of two van der Waals radii of gold (3.80 Å). This bond characteristic is suggested to be one of the relevant factors promoting the fluorescence effect of Au(I) complexes in the UV-vis region, when investigated in the solid state [50]. The fluorescent property may then vanish in the solution, but there are scenarios where the luminescence is restored, at least for high concentrations, when a suitable solvent is used [51,52] and/or increases in the solution when AuNCs undergo self-assembly into nanoribbon structures [53]. Importantly, many Au(I) ions, being complexed with thiols and/or compounds containing other functional groups, such as carboxylates and amides, show fluorescent effects when they are partially ordered (from aurophillic interactions) [51,52]. This might also be the case of Au(I) in BSA-AuNCs.

Concerning another oxidation state of Au (besides Au(I) and Au(0)) being evidenced in BSA-AuNCs, however implicitly (i.e., based on additional chemical reduction using $NaBH_4$ and not directly by XPS), Dixon and Egusa [3] reported the presence of a significant amount of Au(III) in 2018. Terminologically, they label the systems as BSA-Au(III) complexes, but, simultaneously, they admit that there is no disproval of the existence of neutral Au(0) NCs in BSA-Au samples [3]. They [3] summarized that the red fluorescence that emerged due to Au(III) cations complexed with BSA at or above the pH value (9.7 +/− 0.2), which was responsible for the conformational change of BSA structure from normal (N) to aged (A) conformation. Therefore, they [3] concluded that upon BSA conformational change, some Cys–Cys bonds were expected to be sufficiently exposed and solvent-accessible and might interact with Au(III). They also experimentally proved multiple specific Au(III) binding sites in BSA and postulated that the UV-excitable red fluorescence of the BSA-Au(III) complex is due to the internal and potentially cascaded energy transfers among chromophores [3]. Alternatively, they admit strong electron delocalization as a source of the red fluorescence [3].

Furthermore, based on Table 1, it is obvious that sizes of AuNCs, the oxidation state of Au, and consequently their luminescent properties are strongly dependent, among others, on the exact conditions of their synthesis. Moreover, purification (if applied), storage conditions, and time elapsed before the characterization of AuNCs is completed, may represent other factors influencing the fluorescence of the final NCs as demonstrated in our previous work [41]. Obviously (Table 1), visualization of tiny AuNCs is usually performed by TEM and HR-TEM (high-resolution transmission electron microscopy). Particularly, STEM (scanning transmission electron microscopy) using High-Angle Annular Dark Field (HAADF) mode is necessary to be employed in the case of BSA-AuNCs [37]; otherwise, the TEM resolution is obscured by the enveloping protein (BSA) and the accurate size distribution cannot be obtained as stated in [2]. Interestingly, in many research papers (e.g., [3,4,42,54]), the TEM images of AuNCs exceed the size characteristic for their fluorescent properties (maximum of 2 nm in diameter). Besides HR-TEM/STEM, the determination of BSA-AuNCs sizes is often based on mass spectrometry measurements (MS) [55,56]. This is an indirect method where the MS size of BSA is subtracted from the MS size of the BSA-AuNCs system. However, this leads to the determination of the number of Au atoms per BSA. Such estimation does not provide any information about the real size of AuNCs because Au atoms are not localized on a single place in BSA; indeed, they can be spread on different cysteine and other residues of the protein [3,41]. Therefore, it seems that direct visualization of AuNCs as well as determination of individual AuNCs sizes within BSA represent challenging tasks.

As above mentioned, the magnetic properties of BSA-AuNCs have not been investigated so far and, therefore, it is one of the aims of the present study. From the analysis of literature results [57–66], many NPs (mostly prepared using thiolate residues) comprising engineered Au(0) metal clusters give a wide range of magnetic behaviors (diamagnetic, Pauli paramagnetism, antiferro-, and ferromagnetic responses) that depend on the system size (1.8–4.4 nm), the metal's local symmetry, metal coordination environments, and unbalanced/charged states (neutral, positive, and negative metal–organic clusters). For example, the 25 gold (Au(0)) atoms NC stabilized by 18 thiolate ligands prepared by Zhu and co-authors ($Au_{25}(SR)_{18}$, with R = phenylethyl) shows, in frozen CH_2Cl_2/toluene solution at T = 8 K an EPR (Electron Paramagnetic Resonance) signal characteristic for $S = \frac{1}{2}$ systems with g-values consistent with an orthorhombic symmetry at 2.56(x), 2.36(y), and 1.82(z) [49]. Similarly, Agrachev and co-authors reported the phenylethanethiolate Au complex (neutral), $Au_{25}(SC_2Ph)_{18}$, encoding a radius of 13.2 Å, which also exhibits EPR signatures associated to spin $S = \frac{1}{2}$. The EPR spectrum of the complex at T = 5 K is rather broad (~200 mT) for $Au_{25}(SR)_{18}$ and anisotropic, with g-tensor components at 2.53 (x), 2.36 (y), and 1.82 (z) [67].

In this report, the authors used CW-EPR (Continuous Wave EPR), light-induced EPR (LEPR), HR-TEM, and STEM to thoroughly investigate the formation of BSA-AuNCs in a solution, which were intentionally prepared by Xie's type of synthesis [1]. So far, EPR has been employed for the investigation of the emerging magnetism of atomically precise AuNCs of different sizes (e.g., $Au_{25}(SR)_{18}$, $Au_{133}(TBBT)_{52}$), which were prepared by a multistep synthesis involving $NaBH_4$ as a reduction agent [5,6,49]. The EPR/LEPR study of BSA-AuNCs (further labelled as AuBSA in the present study to be directly distinguished from the other published results for BSA-AuNCs) is more complicated than the previous EPR studies of atomically precise AuNCs [5,6,49] because BSA alone reveals photochemically induced radicals' generation, which results in damages of the BSA structure and oxidation of cysteine and tryptophan residues [68,69]. Several years ago, Lassmann and coworkers [68] analyzed, using the LEPR technique, the formation of C and S radical centers in BSA by UV-photolysis, especially thiyl (R-CH2-S•) and perthiyl radicals (R-CH2-S-S•), revealing that radiation damages are produced in the amino acid chain in a dose-dependent manner, i.e., the number of radicals increases by increasing the UV light irradiation time. The authors found that in bovine serum albumin (BSA) the thiyl radical (R-CH2-S•) exhibits axial anisotropy, with $g_{//}$ = 2.17, $g_\perp$ = 2.008, ΔB_{pp} = 1.8 mT, pH = 7, T = 80 K) and the cysteine × HCl radical (phosphate buffer, pH = 3) shows similar features, with $g_{//}$ = 2.11, $g_\perp$ = 2.011, ΔB_{pp} = 3.2 mT, T = 80 K). The $g_{//}$ component was always found to be very broad and weak, to the point that it becomes barely detectable, while the $g_\perp$ component gives the dominating, most intense signal. Together with the thiyl radicals, other types of radical species were demonstrated to form as well, which include perthiyl centres (R-CH2-S-S•) that exhibits rhombic character, and likewise, C-radical centers. In BSA (pH = 7, T = 80 K), EPR signals for the perthiyl specie show resonances at g_1 = 2.057, g_2 = 2.027, and g_3 = 2.002, while the C-radical species have g_{avg} ~ 2.000 [68].

Therefore, the main aim of this study is to thoroughly investigate the properties of the AuBSA system with the aid of selected techniques (EPR/LEPR, HR-TEM/STEM). The following key findings are presented: (i) redox processes take place upon entrapment of Au(III) in BSA, in which the protein backbone, in absence of other reducing equivalents, provide the electrons needed for Au(III) reduction; (ii) the Au uptake process by BSA in the solution occurs on two types of sites, i.e., cysteine (Cys) and oxygen/carboxylate donor residues (e.g., tyrosine, Tyr); (iii) the magnetic interaction among cationic forms of Au arises from an admixture of Au(0), Au(I), and Au(II); and (iv) UV irradiation of the AuBSA system induces irreversible damages, and radicals are generated in BSA upon UV light irradiation. It is disclosed that no energy transfer between entrapped Au clusters and the BSA protein occurs under photoexcitation, thus the photoluminescence phenomenon is governed by the intrinsic properties of the Au system (its size and the presence of various spin active redox states) and both variables equally matter. On the other hand, in view of the extended application of these materials in biomedical scenarios, correlation between their experimental parameters (e.g., concentration in solution) and cell's toxicity (with and upon photoexcitation) should be analyzed with great caution, because observation of an enhancement of toxic effects on cells by light irradiation might be hampered by light-induced radical's formation on the protein itself and not solely to the photoexcited states of the gold nanoassembly.

2. Materials and Methods

2.1. Chemicals

Bovine serum albumin (>98%, BSA), gold(III) chloride trihydrate (HAuCl$_4$·3H$_2$O), and sodium hydroxide (NaOH) were purchased from Sigma-Aldrich (Saint Louis, MO, USA) and used as received (without any further purification) for all experiments. Deionized (DI) water prepared by purging Milli-Q purified water (Millipore Corp., Bedford, MA, USA) was used in all experiments.

2.2. Synthesis of AuBSA System

The procedure of our AuBSA system preparation was a slightly modified version of a method published by Xie et al. [1]. In a typical experiment, BSA solution (1 mL, 1 mM) was mixed with aqueous HAuCl$_4$ solution (1 mL, 10 mM) under vigorous stirring (600 rpm). After 90 s, NaOH solution (1 M) was added to obtain a basic environment (pH ≈ 12), which induced the reduction capability of BSA. Ninety seconds later, the mixed solution was heated up in a microwave oven for 10 s (power was set to 150 W). After 2 h of ageing at room temperature, the samples were dialyzed with a 12 kDa cut-off dialysis membrane against DI water. Dialysis was performed at room temperature for 24 h, with DI water being changed twice: once after one hour and then again after the second hour. Dialyzed samples were stored in the dark at room temperature.

2.3. Characterization of AuBSA System

Fluorescence measurements of AuNCs were performed on a JASCO F8500 (Jasco, Tokyo, Japan) spectrofluorometer using a 1 cm quartz cuvette and 2.5 nm slits. Emission spectra were measured in the range of 500–850 nm with the data interval of 1 nm and a scan speed of 100 nm/min. The excitation wavelength was set to 480 nm. Excitation-emission 3D maps were measured in the excitation range of 250–500 nm with the data interval of 2 nm and in the emission range of 250–850 nm with the data interval of 1 nm and a scan speed of 5000 nm/min (Figure 1). All spectra were corrected to avoid any deviations of instrumental components. Samples were diluted with DI water so that the protein concentration was 2 mg/mL for emission spectra and 0.25 mg/mL for excitation-emission 3D maps.

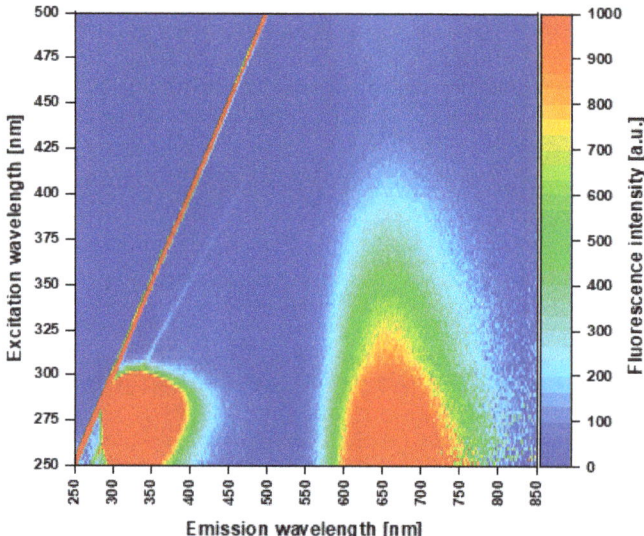

Figure 1. Fluorescence 3D excitation-emission map of AuBSA. Note: Strong Rayleigh (red line) and weak Raman (blue line) scattering first-order maxima are visible as straight lines in the map and represent artifacts that do not belong to AuBSA.

The quantum yield of fluorescence (QY, Φ) was calculated by the following equation:

$$\Phi = \Phi_s \cdot \frac{F \cdot (1 - 10^{-A_s}) \cdot n^2}{F_s \cdot (1 - 10^{-A}) \cdot n_s^2} \quad (1)$$

where F is the integrated fluorescence intensity, A is the absorbance, n is the index of refraction, and subscript s indicates the standard. DCM, 4-(dicyanomethylene)-2-methyl-6-(4-dimethylaminostyryl)-4H-pyran, dissolved in ethanol (99,8%, Lach-Ner, Neratovice, Czech Republic) was used as a standard (Φ_s = 0.437 ± 0.024) [70].

Absorbance was measured on a Specord 250 Plus—223G1032 (Analytik Jena, Jena, Germany) using a 1 cm quartz cuvette and double beam arrangement. As a reference, a 1 cm quartz cuvette filled with DI water was used.

CW-EPR spectra were recorded on a JEOL JES-X-320 spectrometer (JEOL, Tokyo, Japan) operating at the X-band frequency (~9.0–9.1 GHz) equipped with a variable-temperature controller (He, N2) ES-CT470 apparatus. The cavity quality factor (Q) was kept above 6000 in all measurements. Highly pure quartz tubes were employed (Suprasil, Wilmad, ≤0.5 OD), and accuracy on g-values was obtained against the Mn(II)/MgO standard (JEOL standard). The spectra were acquired by carefully monitoring that signal saturation from the applied microwave power did not occur during signal's acquisition. In situ light excitation EPR experiments (LEPR) were performed using a HeCd laser source operating @325 nm (max cw power of 200 mW) from Kimmon Koha Co. Ltd. (Tokyo, Japan). The UV-light was shined directly onto the sample, kept frozen inside the cavity EPR resonator, through its dedicated optical window. The light-off to light-on process was operated by an on-off light-shutter mechanism. Filling factors were kept the same (200 µL) in all experiments. Experimental parameters used for all the EPR traces shown in Figure 2A:

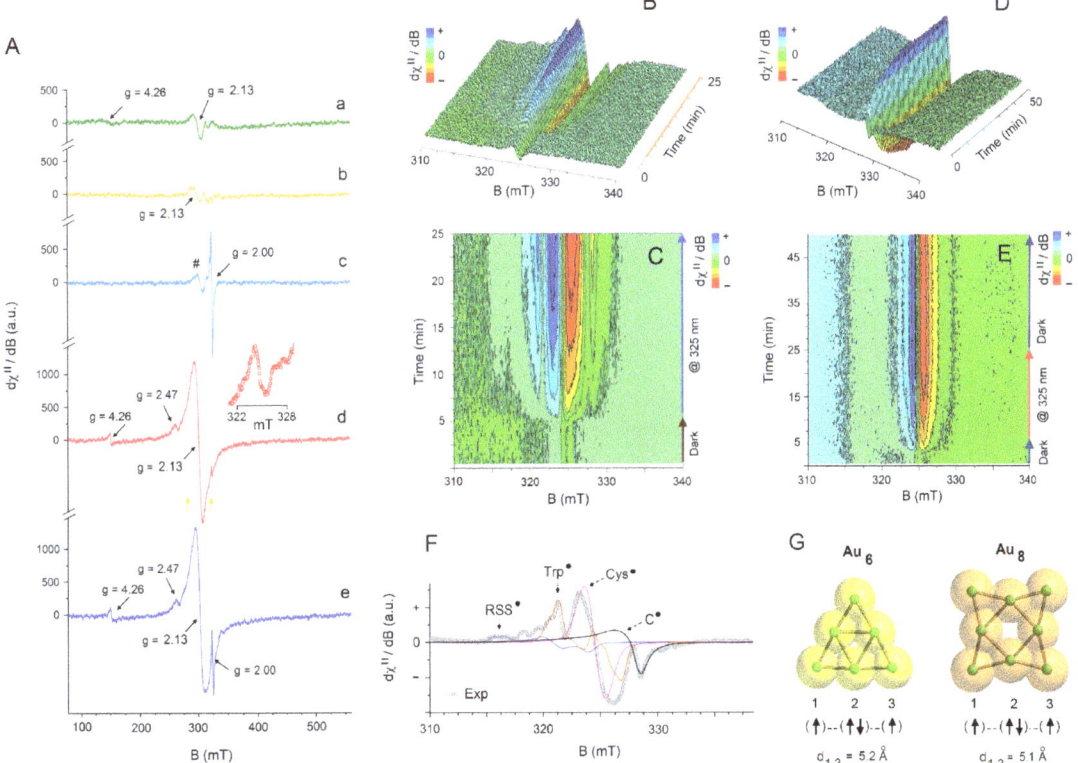

Figure 2. Panel (**A**) CW X-band Electron Paramagnetic Resonance (EPR) spectra recorded in water at T = 90 K under dark conditions of (a) HAuCl$_4$, (b) BSA, and (d) AuBSA. The LEPR spectra obtained under in situ UV irradiation (@325 nm, for 10 min) of the frozen matrix solutions for the various samples (T = 90 K) are given as trace (c) for BSA and trace (e) for AuBSA. Note that in trace (c), the symbol (#) indicates the region in which the g_x component of the thyil radical (Cys•) in BSA is expected to be observed. The in situ 3D/2D LEPR experiment (T = 90 K, water) for BSA is shown in panel (**B**) and panel (**C**); panel (**D**) and panel (**E**) show the 3D/2D LEPR experiment for AuBSA. Panel (**F**) gives the experimental EPR spectrum (T = 90 K) together with computer simulations of the individual spin components obtained for neat BSA in water after 20 min of continuous UV irradiation. Panel (**G**) shows the structural organization of gold nanoclusters (Au$_6$ and Au$_8$) derived from DFT theory (neutral forms, coordinate files taken from ref. [67]) encoding the minimum dimensions for the Au nano-assemblies in BSA that are compatible with the distances calculated by EPR for a pair of spin–spin interacting Au(II)/Au(0) centers (labelled as 1,3).

9.0821–9.0878 GHz frequency, 100 kHz modulation frequency, 0.03 s time constant, 2.00 mW applied power, 1.0 mT modulation width, 4 min sweep time, and 3 scans accumulated and averaged. Cavity background was subtracted. Experimental parameters for the dark to light sequential EPR spectra shown in Figure 2B,C: T = 90 K, 9.0802 GHz frequency, 100 kHz modulation frequency, 0.03 s time constant, 1.60 mW applied power, 0.5 mT modulation width, 30 s sweep time, and 5 min under dark followed by 20 min under light irradiation. Experimental parameters for the dark to light sequential EPR spectra shown in Figure 2D,E: T = 90 K, 9.0808 GHz frequency, 100 kHz modulation frequency, 0.03 s time constant, 1.60 mW applied power, 0.5 mT modulation width, 30 s sweep time, 5 min under dark conditions followed by 20 min under light irradiation, and then another 25 min under dark conditions. Experimental parameters for the EPR spectrum shown in Figure 2F: T = 90 K, 9.0802 GHz frequency, 100 kHz modulation frequency, 0.03 s time constant, 1.60 mW applied power, 0.5 mT modulation width, 1 min sweep time, and 5 scans accumulated and averaged. Cavity background subtracted.

HR-TEM measurements were independently performed by two different operators on two working places: (i) in the Regional Centre of Advanced Technologies and Materials (RCPTM) in Olomouc (Czech Republic) and (ii) in the Institute of Molecular Genetics (IMG) in Prague (Czech Republic). In both cases, AuBSA (the final protein concentration of 1 mg/mL) was drop-casted (2–4 µL) on glow discharged TEM copper grids covered with either a continuous 4-nm-thick carbon foil or lacey carbon and allowed them to dry spontaneously at room temperature. It is worth noting that for successful measurements, lacey carbon-copper grids (300 mesh) were used similarly as in ref. [37], and an electron beam was focused within the holes of the carbon film to obtain images of the protein-conjugated Au nanostructures without the interference of the carbon support (i.e., the measurements of samples deposited on 4-nm-thick carbon films were not successful). Then, the samples were measured by HR-TEM Titan G2 60–300 (FEI, Hillsboro, OR, USA) with an image corrector with an accelerating voltage of 300 kV in RCPTM. Images were taken with a BM UltraScan CCD camera (Gatan, Pleasanton, CA, USA). Energy Dispersive Spectrometry (EDS) was performed in STEM mode by a Super-X system with four silicon drift detectors (Bruker, Billerica, MA, USA). STEM images were taken with an HAADF detector 3000 (Fishione, Export, PA, USA). HAADF mode of STEM is intentionally used to better visualize NCs because heavier elements appear bright, while lighter elements appear dark. Imaging and data acquisition in IMG were performed with JEM-F200 TEM (JEOL, Tokyo, Japan) operated at 200 kV. TEM Images were acquired using an XF 416 CMOS camera (TVIPS, Gauting, Germany) and STEM images were acquired using the HAADF detector with a detecting angle of 24.4–89.4 mrad at a camera length set to 250 mm. EDS data were acquired using a JED 2300 X-ray spectrometer (JEOL, Tokyo, Japan) with a single 100 mm^2 (0.98 sr) windowless SDD detector. Presented elemental maps are calculated from the raw spectra using a standardless Zeta factor method embedded in the Jeol Analysis Station software.

3. Results and Discussion

3.1. Fluorescent Properties of AuBSA

Figure 1 displays characteristic fluorescent properties of AuBSA samples. According to the 3D excitation-emission fluorescence map (Figure 1), two distinct asymmetric excitation-emission maxima (EX/EM) with bathochromic tails appeared in the ultraviolet-visible (UV-vis) spectral region: 275/345 nm (EX/EM) and 275/655 nm (EX/EM). This is in accordance with our previous work [41] as well as with the fluorescent features of NCs prepared in a similar manner (see Table 1), i.e., using BSA simultaneously as a reductant of Au(III) precursors and a template for AuNCs formation. While the former EX/EM maximum and its bathochromic tail of a low intensity (being located at around 325/400 nm EX/EM) could be attributed to Trp and oxidized Tyr residues of BSA and/or blue-emissive small AuNCs, respectively; the latter intensive EX/EM maximum can stem from Au(x)-BSA, where x represents the oxidation state of gold reaching the values of 0, I, II, (proved in this study for the first time), and III. Obviously, AuBSA can be excited by employing any wavelength in the range of 250–500 nm, as clearly demonstrated in Figure 1.

The characteristic fluorescent spectrum of AuBSA (EX 480 nm) where the center of emission (positioned at 655 nm) is marked off is shown in Figure S1 in Supporting Information for the sake of a direct comparison with fluorescence spectra of AuNCs being prepared by other authors using similar synthetic procedures (see Table 1). Namely, the visible excitations of fluorescent species serving as bio-imaging probes are preferred in practical applications because of the tissue optical window [71]. Fluorescence quantum yield (QY) of AuBSA was determined according to equation (1) and revealed the mean value of 6.7 ± 0.1% (details presented in Table S1 in Supporting Information) with respect to DCM dissolved in ethanol.

3.2. CW-EPR and LEPR Experiments and Analysis

To address in more details the electronic/magnetic characteristics of AuBSA, without and under in situ UV-light irradiation (@325 nm), CW-EPR and LEPR spectroscopic techniques were employed. In addition, as reference systems, the resonance fingerprints of $HAuCl_4$ (Au(III) precursor) dissolved in DI water, as well as the neat BSA protein (in DI water), were recorded under identical experimental conditions to unveil the electronic changes that might occur in the BSA protein after decoration/entrapment of Au; see Figure 2.

3.2.1. CW-EPR and LEPR Analysis of Precursors ($HAuCl_4$ and BSA)

Figure 2A (trace a) shows the X-band resonance signal acquired at T = 90 K in dark conditions of the aqueous solution of $HAuCl_4$ (8.1 mg/0.2 mL). As expected, the Au^{3+} cation does not show any strong EPR resonance signals, which agrees with the diamagnetic nature of its ground state electronic configuration ($[Xe]4f^{14}5d^8$). Only a very weak derivative signal emerges at g_{iso} = 2.13, which is superimposed to a broad dispersion signal extending over 100 mT that becomes visible above the background noise at a magnetic field higher than 300 mT. These weak resonances suggest that a small fraction of Au is present in the reduced forms, Au^0 ($[Xe]4f^{14}5d^{10}6s^1$) and/or Au^{2+} ($[Xe]4f^{14}5d^9$), in the hydrated Au^{3+} cations [72]. The additional resonance signal that appears in the low field region (Figure 2A, trace a), at g = 4.26, indicates that the reduced Au centers, Au^0 or Au^{2+}, are magnetically interacting, either via a direct or super-exchange pathway (e.g., O or OH- bridged $Au^{0/2+}$ dimers), forming aggregates with dimensions <1 nm, and giving an effective integer spin system in which the resonance signal at g = 4.26 represents the half-field transition (Δm_s = 2). Upon in situ irradiation by UV-light of the frozen $HAuCl_4$ aqueous solution at T = 90 K, no changes in the overall resonance envelope were observed (spectrum not shown).

Different behavior is recorded for BSA in water. The EPR resonance signal observed under dark conditions is shown in Figure 2A (trace b) (T = 90 K, neat BSA, 7.0 mg/0.2 mL in DI water). A cluster of very weak resonances develop just above the background noise at g_{avg} = 2.13. No other detectable high or low field resonance components are observed. These resonance signal originate from either not fully compensated cavity background noise, even after its subtraction from the sample resonance signal, or from the presence of the very minute inclusion of metal impurities in BSA, such as Cu^{2+} or low spin Fe^{3+}. During UV-light irradiation (Figure 2A, trace c) a strong signal immediately appears at g_{iso} = 2.00, which arises from radiation damages induced in the BSA protein. These spin containing sites are based on cysteine (Cys•), tryptophan (Trp•) moieties and other C-centred radicals (C•) [68,69].

In full agreement with previous literature reports [68], the authors observed the dynamic formation of the radical species in BSA during an UV light-off (5 min) to UV light-on (20 min) EPR signal acquisition sequence, as shown in the 3D reconstructed CW-EPR/LEPR resonance plot given in Figure 2B and in the 2D CW-EPR/LEPR plot in Figure 2C. Individual CW-EPR spectra were acquired under fast-scan detection mode (30 s acquisition time for each sequential resonance spectrum). The authors noticed that formation of thiyl, tryptophan and C-centered radicals represent the early oxidation events in BSA and start to form as soon as the UV light is applied to the frozen sample kept inside the cavity resonator. The perthiyl radicals (RSS•), on the other hand, develop much later in a slower process, after application of a much longer UV irradiation time (e.g., detected after 8 min of UV irradiation in the frozen water matrix at T = 90 K). Figure 2F shows a well-resolved EPR resonance signal recorded for BSA in water irradiated in situ at 325 nm for 20 min at 90 K, which unveils the resonance features of the perthiyl radical (RSS•, g_x = 2.052, g_y = 2.018 g_z = 2.002) overlapped to tryptophan-based radicals (Trp•, $g_{y,z}$ = 2.003, g_x = 2.002, $A_N(xx,yy,zz)$ = 2.0 G, 2.0 G, 10.0 G; $A_{H\beta1}(xx,yy,zz)$ = 28.3 G, 28.3 G, 28.3 G; $A_{H\beta2}(xx,yy,zz)$ = 13.0 G, 13.0 G, 13.0 G). Note that the region around g = 2.000 clearly contains the overlapped resonance contributions of various spin species, but it is dominated by the $g_{y,z}$ components of the thiyl radical (Cys•, g_y = 2.003, g_z = 2.002) [68]. Unfortunately, the g_x component of this radical specie is expected to fall around 2.16–2.17 and it is known to be rather weak and broad; thus, its resonance feature remains obscured by the presence of clustered signals seen already in EPR spectrums recorded under dark conditions (see Figure 2A, trace b). Additional resonances are observed in the spectrum and originate from the hyperfine interactions of the electron spin moments of the radical species with the nuclear spin moments of hydrogen (^{1}H, I = 1/2, natural abundance 99.98%), nitrogen (^{14}N, I = 1, natural abundance 99.60%) and sulfur nuclei (^{33}S, I = 3/2, natural abundance 0.76%) present in cysteine (Cys) and tryptophan (Trp) amino acid residues. For simplicity, the detailed EPR simulation parameters used for plotting the various resonance envelopes of these spin species in BSA, shown in Figure 2F, are given in the Supporting Information (Figure S2). Therefore, a threshold of 10 min under UV-light irradiation time was used in the following analysis of AuBSA during photoexcitation, in agreement with the suggested irradiation time employed by Lassman and co-authors [68].

3.2.2. CW-EPR and LEPR Analysis of AuBSA

The resonance features of the AuBSA system are much different from those observed in neat BSA. Figure 2A (trace d) shows the EPR spectrum of the AuBSA recorded under dark conditions in DI water at T = 90 K (7.2 mg/0.2 mL). Unlike the EPR signatures shown earlier by neat BSA recorded under dark conditions (Figure 2A, trace b), AuBSA exhibits a strong derivative signal at g_{avg} = 2.13 (ΔB_{pp} = 13 mT) with clear shoulders developing rather symmetrically at low and high magnetic fields: one absorption component at a g-value of 2.47 and a derivative component at low field, with a g-value of 4.26. All these resonance signals arise from spin containing Au centers. The intensity of the EPR resonances coming from Au centers in AuBSA (Figure 2A, trace d) significantly increased in comparison to those seen in the EPR spectrum of HAuCl$_4$ (Figure 2A, trace a); thus, a reduction process of Au(III) promoted by BSA takes place during protein incubation with the gold precursor, either to Au(I), Au(II), or to Au(0), in agreement with literature reports [1,10,16,41,73,74]. However, Au(I) is a diamagnetic cation and cannot produce the resonance features seen in the EPR spectrum. Moreover, the EPR spectrum shown in Figure 2A, trace (d), differs substantially from those reported for thiolate Au^0 nanoclusters [5,6,49,57–66].

According to literature data, the reduction process of Au(III) by BSA leads to the formation of gold nanoparticles (AuNCs) in which the BSA protein is directly interacting with metal centers through cysteine (Cys) and methionine (Met) residues. [3,74]. Furthermore, it is also expected that Tyr residues of BSA are the moieties that provide the electron equivalents needed for driving the Au(III) reduction process [1,41], and thus, these are the amino acids that undergo oxidation. Although very weak, at $g \sim 2.00$, the presence of organic radical signals in AuBSA emerges in the EPR spectrum, but its resolution could not be improved (Figure 2A, trace d, inset). In this framework, the authors performed a set of test-experiments where either Tyr and/or Cys molecules were incubated with Au(III) salt under the same experimental conditions used in AuBSA synthesis and the resulting systems (denoted as Au-Tyr, Au-Cys) were characterized by UV-vis extinction measurements and TEM imaging. The appearance of surface plasmon resonance (SPR) peak (around 525 nm), which is characteristic for AuNPs—see Figure S3A in Supporting Information—in the Au-Tyr system confirms the following results: (i) direct reduction of Au(III) to Au(0) by Tyr and (ii) key function of the BSA protein as a matrix/template in limiting the growth of the gold nanocluster during AuBSA synthesis. No SPR was detected in the UV-vis spectrum of Au-Cys. TEM images (Figure S3B,C in Supporting Information) give further evidence that AuNPs, of various sizes, are formed in the Au-Tyr system.

Based on the EPR results, it is suggested that while small, nanosized Au aggregates are indeed formed during the incubation and redox process between Au and BSA (size of Au nano-assembly estimated ~ 1 nm), an admixture of oxidation states of entrapped and clustered Au cations is indeed present. The authors speculate that the presence of both Au^{2+} and Au^0, together with the dominant Au^+ specie, is generated during the entrapment process of Au in BSA, in which a disproportionation event might occur in the nanocluster (NC), $2Au^+ \rightarrow Au^{2+} + Au^0$, following the initial electron delivery by BSA to Au^{3+} and reorganization of the electron distribution within the Au cluster in response to its anisotropic interactions with the BSA protein backbone. Thus, the weak signal at $g = 2.47$ and the more intense signal at $g = 2.13$ can be associated with the co-existence of two gold-based spin active species, Au^{2+} and Au^0. Moreover, the observed EPR features cannot simply be described, in our opinion, in terms of isolated $S = 1/2$ states, but rather to an admixture of doublets and triplet states. Simulated CW-EPR spectra of AuBSA in dark conditions are shown in Figure S4 in the Supporting Information. The transition observed at $g = 4.26$ may thus represent the half-field component ($\Delta m_s = 2$) of the $S = 1$ $Au^{0/2+}$ interacting species, having for the $\Delta m_s = 1$ transition a g_{eff} value of 2.13, as observed here. Assuming an axial zero-field-splitting for the high spin component of ~20.5 mT (see Figure 2A, trace d, the orange arrows indicate the estimated 2D), from a simple point dipole approximation [75], $|D|$ (MHz) = 77,924 $(g_{obs}/g_e)/r^3$, the through-space distance (r) between interacting $S = 1/2$ centers should fall at ~5.1 Å. The through-space distance derived from EPR analysis translates into the smallest dimension for the entrapped Au cluster in BSA of Au_6-Au_8 gold units. Figure 2G shows the density functional theory (DFT) calculated structures (neutral forms) [67] for an Au_6 (triangular) and Au_8 (4-fold edge-capped square) assemblies. From the EPR results, it is suggested that the Au(II)/Au(0) centers that are magnetically interacting to form the $S = 1$ system adopt the spin exchange coupling pathway Au(II)-Au(I)-Au(0), thus the spin containing sites Au(II)/Au(0) are located at the corners of the Au_6 or Au_8 nano assemblies. Such an assumption corresponds well with the early observations of Dixon and Egusa [3] who discovered that the maximum number of Au being incorporated into BSA is less than approximately 30 per BSA, but the minimum number of Au per BSA to yield red fluorescence is less than 7.

Upon in situ irradiation of AuBSA (10 min of UV-light irradiation) at $T = 90$ K, the overall EPR signatures addressable to the Au system did not change, both in intensity and overall signal-shape. However, the clear appearance of a resonance signal around $g = 2.00$ (Figure 2A, trace e) points towards the formation of radical species, as described above, in BSA. Most importantly, these spin containing centers that formed during UV irradiation do not appear to magnetically interact with the Au NCs; in fact, the intensity of the half-field component did not increase at all, and no other additional signals arising from a new admixture of high spin species become evident in the EPR spectrum. Therefore, these BSA radical centers (Cys•, Trp•, C•) formed during photoexcitation must involve sites that are far from the amino acid (Cys) residues that directly interact with the Au nano assemblies.

Further validation of the present findings was obtained upon following (via LEPR measurements) the in situ variation of the entire resonance line of AuBSA using an irradiation cycle of light-off to light-on and back to light-off cycle, as shown in the 3D reconstructed CW-EPR/LEPR resonance plot given in Figure 2D and in the 2D CW-EPR/LEPR plot in Figure 2E. Individual CW-EPR spectra were acquired similarly to those employed for neat BSA under fast-field scan detection mode (30 s acquisition time for each sequential resonance spectrum). The authors observed that after application of UV light, the radiation damages induced in the protein backbone in AuBSA are quickly triggered, likewise in the neat BSA system, but there is no clear impact on the EPR signals associated with the EPR signal associated to the gold species. Then, when UV light is turned off, no significant changes are observed in the radical canters of BSA (hence the protein damages are not reversible), nor changes in the EPR resonance features of the gold centers are seen. Therefore, there is no charge or energy transfer processes involved between AuNCs and the BSA protein (at 90 K), although it is a process that, sometimes, becomes observable in BSA-AuNCs, as stated in literature reports [3,9,10]. Our finding supports the results of Yamamoto et al. [40], who discussed a high photo-stability of their BSA-Au$_{25}$NCs, albeit the authors are dealing with much smaller AuNCs (Au$_{6-8}$) in AuBSA as derived from EPR spectra analysis.

3.3. HR-TEM, STEM, and EDS Analysis of AuBSA

To directly visualize our NCs in AuBSA samples, HR-TEM and STEM analysis were performed. These techniques exploit electrons that represent reduction species par excellence. Therefore, by using electrons in studies of cationic metal-organic compounds (obviously our case), irreversible transformation (i.e., reduction) may occur. Consequently, the results of visualization cannot report on the initial nanomaterial, but on the transformed one, as already known from the literature [76].

EDS was performed on the places of holes in the lacey-carbon Cu grid covered by an AuBSA sample, similarly as in ref. [37]. It turned out that Au is well distributed along with S elements being attributed to the protein (Figure 3a). Figure 3b,c demonstrates STEM images of the same area on the AuBSA sample prior and after the scanning of EDS for several occasions, respectively. Red circles pinpoint the areas where the changes are the most apparent. Similarly, Figure 3d,e shows the same area of AuBSA prior and after the HR-TEM measurement being performed by using a higher magnification (i.e., higher density of electron beam). White ellipses emphasize different spots on the AuBSA sample in these two images—see Figure 3d,e. The authors can sum up that Au nanoparticles (NPs, sizes exceeding 2 nm) are formed by the focused electron beam if working with a higher magnification of HR-TEM and/or if using EDS mapping. Our observation can be a clue and indirectly confirms why there are discrepancies in sizes of fluorescent NCs presented in the literature (e.g., [3,4,42,54]) and summarized in Table 1. Another set of STEM images taken on the same AuBSA sample using a different HR-TEM/STEM machine (located at IMG in Prague) is shown in Figure S5 in Supporting Information. Independently, these additional measurements confirmed that it is possible to visualize NCs of sizes well below 2 nm without inducing their size modification. Moreover, based on TEM, HR-TEM, and STEM images, particle size distributions (PSD) were determined, and appropriate histograms are shown in Figure S6 in Supporting Information. The average AuNC sizes of 1.01 ± 0.24 nm and 1.18 ± 0.21 nm were derived from TEM and STEM images of AuBSA samples, respectively. On the other hand, the average AuNC size in the same AuBSA sample increased to 2.28 ± 0.63 nm after HR-TEM and/or EDS measurements. This corroborates the above-discussed observation about NPs in situ formation. Lattice planes of Au(111) and Au(200) were detected in NCs of AuBSA sample in a particular HR-TEM image as provided in Figure S7 in Supporting Information.

Furthermore, HR-TEM and STEM images of the same spot on the AuBSA sample are compared in Figure 3f,g. Indeed, the HR-TEM image (Figure 3f) was taken first, followed by STEM imaging (being recorded for a better contrast, Figure 3g). Red circles demonstrate the shifts of Au nanostructures, which migrated right after HR-TEM imaging. The observed Au NPs formation and their further migration upon HR-TEM imaging can be explained by considering the cationic state of Au in AuBSA (as determined by EPR measurements here) and reduction ability of electrons employed during TEM imaging. Indeed, Au(I) as well as Au(II) being attached to BSA, most probably through sulfur atoms (as evidenced by EPR and EDS in this study), can be reduced to Au(0). Since this Au(0) may be less strongly connected with the protein backbone than cationic Au in AuBSA, it can migrate on the carbon support of the TEM grid (which is more pronounced in cases of lacey-carbon rather than on 4-nm carbon coating of the Cu grid). By meeting other Au(0) atoms, they can coalesce and form larger Au nanostructures than NCs, i.e., NPs as evidenced in Figure 3c,e,g, and confirmed also by an increased value of average AuNC size (2.28 ± 0.63 nm). The above-mentioned observation of Au NPs formation and their migration during HR-TEM imaging could also explain why the difference in size of red and blue emitting systems in ref. [1] was not as clear as expected. Logically, this supports the idea that AuNCs samples might be influenced by the interaction with the focused electron beam and the resulting images do not show the real structure of AuNCs. In other words, the presence of any cationic Au center (Au^+ and/or Au^{2+} and/or Au^{3+}) in BSA-AuNCs render such systems as being much more sensitive to techniques that use high energy electrons and/or X-ray irradiation and can explain the size and oxidation state discrepancies of Au nanoassembly in BSA-AuNCs found within the scientific literature (compared in Table 1).

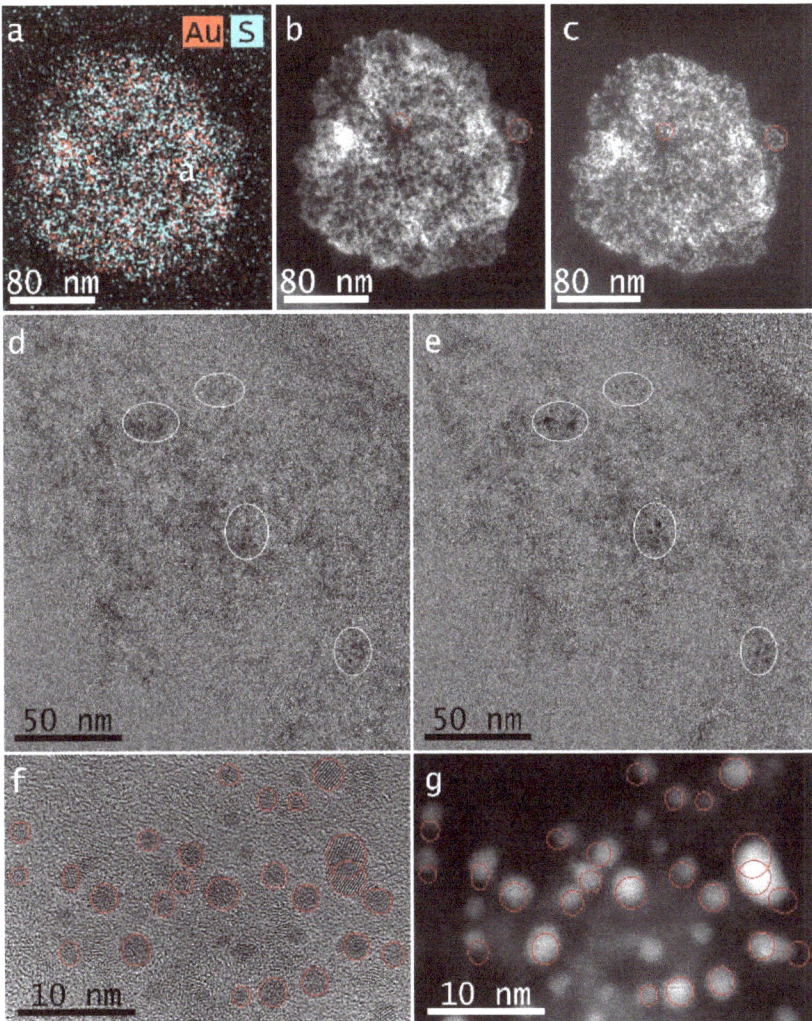

Figure 3. (**a**) Elemental map of Au and S distributions in AuBSA determined by EDS. STEM images of AuBSA prior (**b**) and after (**c**) the performed EDS analysis. HR-TEM images of the same area of AuBSA prior (**d**) and after (**e**) the exploitation of a higher magnification. HR-TEM (**f**) and STEM (**g**) images of the same spot on AuBSA. Colored and/or white circles indicate appearance of new NPs in (**b**–**e**) and/or migration of NPs in (**f**,**g**).

Based on our results, the authors would like to emphasize that results strongly depend on the type of experimental techniques employed for AuNCs characterization and subsequent data interpretation. Many powerful experimental techniques frequently used for nanomaterials' characterization exploit electrons and/or are working with strong irradiation (X ray), thus affecting the electronic structure of the studied material, e.g., TEM, XAS, EDS, etc., respectively. On the contrary, EPR represents a mild technique which uses the magnetic field and microwave irradiation and, thus, can be envisaged as a very powerful technique suitable for probing AuBSA systems in particular, as well as BSA-AuNCs in general, because it does not physically alter the oxidation state of the nanomaterial in the course of data acquisition. Moreover, by using LEPR, any changes in the electronic structure of a particular nanomaterial exposed to electromagnetic irradiation (UV in our case here) can be investigated and evidenced in situ.

4. Conclusions

The present work demonstrates that a redox process that leads to an admixture of Au(0), Au(I), and Au(II) is taking place upon entrapment of Au(III) in BSA and several Au atoms of AuNCs are directly interacting through Cys and Tyr residues with BSA. The authors observed that Tyr molecules are able to reduce Au(III) to Au(0) in the form of AuNPs, while Cys did not provide the electron equivalents needed for the Au(III) reduction. It is also demonstrated that UV irradiation induces irreversible damages in AuBSA and radical species are generated during photoexcitation in the protein backbone. Since there is no evidence of electronic communication between AuNCs and BSA according to the LEPR experiments (in situ off-on-off cycles under UV light, at 90 K), any energy and/or charge transfer can be excluded as a key mechanism for expressing the bright fluorescence phenomenon observed, with maximum emission at 655 nm, in AuBSA. Furthermore, evidence was given for the in situ growth of AuNCs to AuNPs and migration of Au nanostructures during HR-TEM and EDS data acquisition. This work further reinforces the knowledge that results obtained by using sophisticated techniques exploiting electrons and/or X-ray irradiation in the investigation of gold nanostructures embedded in protein matrices must be considered carefully, because the experimental evidence might be severely hampered by induced radiation damages on the protein structure.

Supplementary Materials: The following supporting information can be downloaded at: https://www.mdpi.com/article/10.3390/nano12091425/s1, Figure S1: Corrected emission spectrum of AuBSA in deionized water; Table S1: Quantum yield and wavelength of maximum fluorescence of AuBSA; Figure S2. CW-EPR spectrum obtained after 20 min of UV irradiation of BSA in water. Experimental parameters: T = 90 K, 9.0802 GHz frequency, 100 kHz modulation frequency, 0.03 s time constant, 1.60 mW applied power, 0.5 mT modulation width, 6 × 100 Gain, 1 min sweep time, and 5 scans accumulated and averaged. The simulation of the various spin components is shown together with their sum (red-line); Figure S3. Extinction spectra of Au-Tyr in deionized water 1 min and 45 s after synthesis (grey line) and 61 min and 45 s after synthesis (wine line). (B) and (C) TEM images of Au-Tyr recorded at different magnifications; Figure S4. CW-EPR spectrum of AuBSA in water recorded at T = 90 K in dark conditions with the EPR simulation of the diverse spin-components associated to the Au nanoclusters; Figure S5: Further STEM images of AuBSA; Figure S6: Particle size distributions derived from TEM, STEM and HR-TEM images; Figure S7: Lattice spaces determined for several AuNCs of AuBSA in a particular HR-TEM image.

Author Contributions: Conceptualization: K.Š. and G.Z.; data curation: G.Z., R.O., O.T. and D.P.; formal analysis: G.Z., R.O., O.T. and D.P.; funding acquisition: K.Š. and V.F.; investigation: G.Z., R.O., O.T. and D.P.; methodology: G.Z. and K.Š.; project administration: K.Š.; supervision: K.Š.; validation: G.Z., R.O., O.T. and D.P.; visualization: G.Z., R.O., O.T. and D.P.; writing—original draft: K.Š.; writing—review and editing: G.Z. and K.Š. All authors have read and agreed to the published version of the manuscript.

Funding: K.Š. thanks, for financial support, the Grant Agency of the Czech Republic (grant no. 19-03207S). K.Š. and R.O. thank the Internal Grant Agency of Palacký University (grant numbers IGA_PrF_2021_003 and IGA_PrF_2022_003). D.P. and V.F. acknowledge the MEYS CR (project no. LM2018129 Czech BioImaging) and ERDF projects (CZ.02.1.01/0.0/0.0/18_046/0016045, CZ.02.1.01/0.0/0.0/16_013/0001775) for funding. G.Z. and O.T. thank the ERDF/ESF project "Nano4Future" (project no. CZ.02.1.01/0.0/0.0/16_019/0000754).

Institutional Review Board Statement: Not applicable.

Informed Consent Statement: Not applicable.

Data Availability Statement: Not applicable.

Acknowledgments: Monika Opatíková is acknowledged for TEM imaging of Au-Tyr system and Erik Dostál is thanked for this and Au-Cys system preparation and UV-vis extinction measurements.

Conflicts of Interest: The authors declare no conflict of interest.

Abbreviations

AuNCs: gold nanoclusters; AuBSA: abbreviation used in this study for the particular AuNCs embedded in BSA being prepared by the described synthetic procedure; BSA: bovine serum albumin; BSA-AuNCs: general abbreviation for AuNCs embedded in BSA being published within the scientific literature, i.e., regardless the consideration of the exact experimental conditions; CW-EPR: continuous wave electron paramagnetic resonance; DFT: density functional theory; EDS: energy dispersive X-ray analysis; EX/EM: excitation/emission; EXAFS: extended X-ray absorption fine structure; HAADF: high-angle annular dark field; HR-TEM: high resolution transmission electron microscopy; l-DOS: local density of states; LEPR: light-induced electron paramagnetic resonance; MS: mass spectrometry; MW: microwaves; NCs: nanoclusters; SPR: surface plasmon resonance; STEM: scanning transmission electron microscopy; UV: ultraviolet; vis: visible; XANES: X-ray absorption near-edge structure; XAS: X-ray absorption spectroscopy; XPS: X-ray photoelectron spectroscopy; XRD: X-ray diffraction.

References

1. Xie, J.; Zheng, Y.; Ying, J.Y. Protein-Directed Synthesis of Highly Fluorescent Gold Nanoclusters. *J. Am. Chem. Soc.* **2009**, *131*, 888–889. [CrossRef] [PubMed]
2. Simms, G.A.; Padmos, J.D.; Zhang, P. Structural and Electronic Properties of Protein/Thiolate-Protected Gold Nanocluster with "Staple" Motif: A XAS, L-DOS, and XPS Study. *J. Chem. Phys.* **2009**, *131*, 214703. [CrossRef]
3. Dixon, J.M.; Egusa, S. Conformational Change-Induced Fluorescence of Bovine Serum Albumin-Gold Complexes. *J. Am. Chem. Soc.* **2018**, *140*, 2265–2271. [CrossRef] [PubMed]
4. Chuang, K.T.; Lin, Y.W. Microwave-Assisted Formation of Gold Nanoclusters Capped in Bovine Serum Albumin and Exhibiting Red or Blue Emission. *J. Phys. Chem. C* **2017**, *121*, 26997–27003. [CrossRef]
5. Antonello, S.; Perera, N.V.; Ruzzi, M.; Gascón, J.A.; Maran, F. Interplay of Charge State, Lability, and Magnetism in the Molecule-like $Au_{25}(SR)_{18}$ Cluster. *J. Am. Chem. Soc.* **2013**, *135*, 15585–15594. [CrossRef] [PubMed]
6. Zeng, C.; Weitz, A.; Withers, G.; Higaki, T.; Zhao, S.; Chen, Y.; Gil, R.R.; Hendrich, M.; Jin, R. Controlling Magnetism of Au133(TBBT)52 Nanoclusters at Single Electron Level and Implication for Nonmetal to Metal Transition. *Chem. Sci.* **2019**, *10*, 9684–9691. [CrossRef] [PubMed]
7. Wen, X.; Yu, P.; Toh, Y.R.; Tang, J. Structure-Correlated Dual Fluorescent Bands in BSA-Protected Au_{25} Nanoclusters. *J. Phys. Chem. C* **2012**, *116*, 11830–11836. [CrossRef]
8. Wen, X.; Yu, P.; Toh, Y.R.; Hsu, A.C.; Lee, Y.C.; Tang, J. Fluorescence Dynamics in BSA-Protected Au_{25} Nanoclusters. *J. Phys. Chem. C* **2012**, *116*, 19032–19038. [CrossRef]
9. Raut, S.; Chib, R.; Butler, S.; Borejdo, J.; Gryczynski, Z.; Gryczynski, I. Evidence of Energy Transfer from Tryptophan to BSA/HSA Protected Gold Nanoclusters. *Methods Appl. Fluoresc.* **2014**, *2*, 035004. [CrossRef]
10. Russell, B.A.; Kubiak-Ossowska, K.; Mulheran, P.A.; Birch, D.J.S.; Chen, Y. Locating the Nucleation Sites for Protein Encapsulated Gold Nanoclusters: A Molecular Dynamics and Fluorescence Study. *Phys. Chem. Chem. Phys.* **2015**, *17*, 21935–21941. [CrossRef]
11. Chang, H.; Karan, N.S.; Shin, K.; Bootharaju, M.S.; Nah, S.; Chae, S.I.; Baek, W.; Lee, S.; Kim, J.; Son, Y.J.; et al. Highly Fluorescent Gold Cluster Assembly. *J. Am. Chem. Soc.* **2021**, *143*, 326–334. [CrossRef] [PubMed]
12. Cao, X.L.; Li, H.W.; Yue, Y.; Wu, Y. PH-Induced Conformational Changes of BSA in Fluorescent AuNCs@BSA and Its Effects on NCs Emission. *Vib. Spectrosc.* **2013**, *65*, 186–192. [CrossRef]
13. Cui, M.; Zhao, Y.; Song, Q. Synthesis, Optical Properties and Applications of Ultra-Small Luminescent Gold Nanoclusters. *TrAC—Trends Anal. Chem.* **2014**, *57*, 73–82. [CrossRef]
14. Bhowal, A.C.; Pandit, S.; Kundu, S. Fluorescence Emission and Interaction Mechanism of Protein-Coated Gold and Copper Nanoclusters as Ion Sensors in Different Ionic Environments. *J. Phys. D. Appl. Phys.* **2019**, *52*, 015302. [CrossRef]
15. Hsu, N.Y.; Lin, Y.W. Microwave-Assisted Synthesis of Bovine Serum Albumin-Gold Nanoclusters and Their Fluorescence-Quenched Sensing of Hg2+ Ions. *New J. Chem.* **2016**, *40*, 1155–1161. [CrossRef]

16. Lin, H.; Imakita, K.; Fujii, M.; Sun, C.; Chen, B.; Kanno, T.; Sugimoto, H. New Insights into the Red Luminescent Bovine Serum Albumin Conjugated Gold Nanospecies. *J. Alloys Compd.* **2017**, *691*, 860–865. [CrossRef]
17. Chib, R.; Butler, S.; Raut, S.; Shah, S.; Borejdo, J.; Gryczynski, Z.; Gryczynski, I. Effect of Quencher, Denaturants, Temperature and PH on the Fluorescent Properties of BSA Protected Gold Nanoclusters. *J. Lumin.* **2015**, *168*, 62–68. [CrossRef]
18. Kawasaki, H.; Hamaguchi, K.; Osaka, I.; Arakawa, R. Ph-Dependent Synthesis of Pepsin-Mediated Gold Nanoclusters with Blue Green and Red Fluorescent Emission. *Adv. Funct. Mater.* **2011**, *21*, 3508–3515. [CrossRef]
19. Liu, J.; Duchesne, P.N.; Yu, M.; Jiang, X.; Ning, X.; Vinluan, R.D., III; Zhang, P.; Zheng, J. Luminescent Gold Nanoparticles with Size-Independent Emission. *Angew. Chem.* **2016**, *128*, 9040–9044. [CrossRef]
20. Liu, C.; Zhang, X.; Han, X.; Fang, Y.; Liu, X.; Wang, X.; Waterhouse, G.I.N.; Xu, C.; Yin, H.; Gao, X. Polypeptide-Templated Au Nanoclusters with Red and Blue Fluorescence Emissions for Multimodal Imaging of Cell Nuclei. *ACS Appl. Bio Mater.* **2020**, *3*, 1934–1943. [CrossRef]
21. Shang, L.; Stockmar, F.; Azadfar, N.; Nienhaus, G.U. Intracellular Thermometry by Using Fluorescent Gold Nanoclusters. *Angew. Chem. Int. Ed.* **2013**, *52*, 11154–11157. [CrossRef] [PubMed]
22. Wei, Z.; Pan, Y.; Hou, G.; Ran, X.; Chi, Z.; He, Y.; Kuang, Y.; Wang, X.; Liu, R.; Guo, L. Excellent Multiphoton Excitation Fluorescence with Large Multiphoton Absorption Cross Sections of Arginine-Modified Gold Nanoclusters for Bioimaging. *ACS Appl. Mater. Interfaces* **2022**, *14*, 2452–2463. [CrossRef] [PubMed]
23. Wen, F.; Dong, Y.; Feng, L.; Wang, S.; Zhang, S.; Zhang, X. Horseradish Peroxidase Functionalized Fluorescent Gold Nanoclusters for Hydrogen Peroxide Sensing. *Anal. Chem.* **2011**, *83*, 1193–1196. [CrossRef] [PubMed]
24. Wu, Z. Anti-Galvanic Reduction of Thiolate-Protected Gold and Silver Nanoparticles. *Angew. Chem.* **2012**, *51*, 2934–2938. [CrossRef] [PubMed]
25. Wu, Z.; Jin, R. On the Ligand's Role in the Fluorescence of Gold Nanoclusters. *Nano Lett.* **2010**, *10*, 2568–2573. [CrossRef]
26. Wu, Y.T.; Shanmugam, C.; Tseng, W.B.; Hiseh, M.M.; Tseng, W.L. A Gold Nanocluster-Based Fluorescent Probe for Simultaneous PH and Temperature Sensing and Its Application to Cellular Imaging and Logic Gates. *Nanoscale* **2016**, *8*, 11210–11216. [CrossRef]
27. Xu, Y.; Sherwood, J.; Qin, Y.; Crowley, D.; Bonizzoni, M.; Bao, Y. The Role of Protein Characteristics in the Formation and Fluorescence of Au Nanoclusters. *Nanoscale* **2014**, *6*, 1515–1524. [CrossRef]
28. Yue, Y.; Liu, T.Y.; Li, H.W.; Liu, Z.; Wu, Y. Microwave-Assisted Synthesis of BSA-Protected Small Gold Nanoclusters and Their Fluorescence-Enhanced Sensing of Silver(i) Ions. *Nanoscale* **2012**, *4*, 2251–2254. [CrossRef]
29. Yue, Y.; Li, H.W.; Liu, T.Y.; Wu, Y. Exploring the Role of Ligand-BSA in the Response of BSA-Protected Gold-Nanoclusters to Silver (I) Ions by FT-IR and Circular Dichroism Spectra. *Vib. Spectrosc.* **2014**, *74*, 137–141. [CrossRef]
30. Zhang, M.; Dang, Y.Q.; Liu, T.Y.; Li, H.W.; Wu, Y.; Li, Q.; Wang, K.; Zou, B. Pressure-Induced Fluorescence Enhancement of the BSA-Protected Gold Nanoclusters and the Corresponding Conformational Changes of Protein. *J. Phys. Chem. C* **2013**, *117*, 639–647. [CrossRef]
31. Zhou, M.; Du, X.; Wang, H.; Jin, R. The Critical Number of Gold Atoms for a Metallic State Nanocluster: Resolving a Decades-Long Question. *ACS Nano* **2021**, *15*, 13980–13992. [CrossRef] [PubMed]
32. Govindaraju, S.; Ankireddy, S.R.; Viswanath, B.; Kim, J.; Yun, K. Fluorescent Gold Nanoclusters for Selective Detection of Dopamine in Cerebrospinal Fluid. *Sci. Rep.* **2017**, *7*, 1–12. [CrossRef] [PubMed]
33. Chang, T.K.; Cheng, T.M.; Chu, H.L.; Tan, S.H.; Kuo, J.C.; Hsu, P.H.; Su, C.Y.; Chen, H.M.; Lee, C.M.; Kuo, T.R. Metabolic Mechanism Investigation of Antibacterial Active Cysteine-Conjugated Gold Nanoclusters in Escherichia Coli. *ACS Sustain. Chem. Eng.* **2019**, *7*, 15479–15486. [CrossRef]
34. Le Guével, X.; Hötzer, B.; Jung, G.; Hollemeyer, K.; Trouillet, V.; Schneider, M. Formation of Fluorescent Metal (Au, Ag) Nanoclusters Capped in Bovine Serum Albumin Followed by Fluorescence and Spectroscopy. *J. Phys. Chem. C* **2011**, *115*, 10955–10963. [CrossRef]
35. Li, H.W.; Yue, Y.; Liu, T.Y.; Li, D.; Wu, Y. Fluorescence-Enhanced Sensing Mechanism of BSA-Protected Small Gold-Nanoclusters to Silver(I) Ions in Aqueous Solutions. *J. Phys. Chem. C* **2013**, *117*, 16159–16165. [CrossRef]
36. Wang, X.; Wu, P.; Hou, X.; Lv, Y. An Ascorbic Acid Sensor Based on Protein-Modified Au Nanoclusters. *Analyst* **2013**, *138*, 229–233. [CrossRef]
37. Burt, J.L.; Gutiérrez-Wing, C.; Miki-Yoshida, M.; José-Yacamán, M. Noble-Metal Nanoparticles Directly Conjugated to Globular Proteins. *Langmuir* **2004**, *20*, 11778–11783. [CrossRef]
38. Zhang, L.; Wang, E. Metal Nanoclusters: New Fluorescent Probes for Sensors and Bioimaging. *Nano Today* **2014**, *9*, 132–157. [CrossRef]
39. Zheng, J.; Zhou, C.; Yu, M.; Liu, J. Different Sized Luminescent Gold Nanoparticles. *Nanoscale* **2012**, *4*, 4073–4083. [CrossRef]
40. Yamamoto, M.; Osaka, I.; Yamashita, K.; Hasegawa, H.; Arakawa, R.; Kawasaki, H. Effects of Ligand Species and Cluster Size of Biomolecule-Protected Au Nanoclusters on Eff Iciency of Singlet-Oxygen Generation. *J. Lumin.* **2016**, *180*, 315–320. [CrossRef]
41. Andrýsková, P.; Šišková, K.M.; Michetschlägerová, Š.; Jiráková, K.; Kubala, M.; Jirák, D. The Effect of Fatty Acids and Bsa Purity on Synthesis and Properties of Fluorescent Gold Nanoclusters. *Nanomaterials* **2020**, *10*, 343. [CrossRef] [PubMed]
42. Yan, L.; Cai, Y.; Zheng, B.; Yuan, H.; Guo, Y.; Xiao, D.; Choi, M.M.F. Microwave-Assisted Synthesis of BSA-Stabilized and HSA-Protected Gold Nanoclusters with Red Emission. *J. Mater. Chem.* **2012**, *22*, 1000–1005. [CrossRef]
43. Shang, L.; Dong, S.; Nienhaus, G.U. Ultra-Small Fluorescent Metal Nanoclusters: Synthesis and Biological Applications. *Nano Today* **2011**, *6*, 401–418. [CrossRef]

44. Sonia; Komal; Kukreti, S.; Kaushik, M. Gold Nanoclusters: An Ultrasmall Platform for Multifaceted Applications. *Talanta* **2021**, *234*, 122623. [CrossRef]
45. Nienhaus, K.; Wang, H.; Nienhaus, G.U. Nanoparticles for Biomedical Applications: Exploring and Exploiting Molecular Interactions at the Nano-Bio Interface. *Mater. Today Adv.* **2020**, *5*, 100036. [CrossRef]
46. Zheng, Y.; Lai, L.; Liu, W.; Jiang, H.; Wang, X. Recent Advances in Biomedical Applications of Fluorescent Gold Nanoclusters. *Adv. Colloid Interface Sci.* **2017**, *242*, 1–16. [CrossRef] [PubMed]
47. Zhu, M.; Aikens, C.M.; Hollander, F.J.; Schatz, G.C.; Jin, R. Correlating the Crystal Structure of A Thiol-Protected Au_{25} Cluster and Optical Properties. *J. Am. Chem. Soc.* **2008**, *130*, 5883–5885. [CrossRef]
48. Zhu, M.; Lanni, E.; Garg, N.; Bier, M.E.; Jin, R. Kinetically Controlled, High-Yield Synthesis of Au_{25} Clusters. *J. Am. Chem. Soc.* **2008**, *130*, 1138–1139. [CrossRef]
49. Zhu, M.; Aikens, C.M.; Hendrich, M.P.; Gupta, R.; Qian, H.; Schatz, G.C.; Jin, R. Reversible Switching of Magnetism in Thiolate-Protected Au_{25} Superatoms. *J. Am. Chem. Soc.* **2009**, *131*, 2490–2492. [CrossRef]
50. Schmidbaur, H. The Aurophilicity Phenomenon: A Decade of Experimental Findings, Theoretical Concepts and Emerging Applications. *Gold Bull.* **2000**, *33*, 3–10. [CrossRef]
51. Schmidbaur, H.; Schier, A. A Briefing on Aurophilicity. *Chem. Soc. Rev.* **2008**, *37*, 1931–1951. [CrossRef] [PubMed]
52. Schmidbaur, H.; Schier, A. Aurophilic Interactions as a Subject of Current Research: An up-Date. *Chem. Soc. Rev.* **2012**, *41*, 370–412. [CrossRef] [PubMed]
53. Wu, Z.; Du, Y.; Liu, J.; Yao, Q.; Chen, T.; Cao, Y.; Zhang, H.; Xie, J. Aurophilic Interactions in the Self-Assembly of Gold Nanoclusters into Nanoribbons with Enhanced Luminescence. *Angew. Chem. Int. Ed.* **2019**, *58*, 8139–8144. [CrossRef] [PubMed]
54. Rehman, F.U.; Du, T.; Shaikh, S.; Jiang, X.; Chen, Y.; Li, X.; Yi, H.; Hui, J.; Chen, B.; Selke, M.; et al. Nano in Nano: Biosynthesized Gold and Iron Nanoclusters Cargo Neoplastic Exosomes for Cancer Status Biomarking. *Nanomed. Nanotechnol. Biol. Med.* **2018**, *14*, 2619–2631. [CrossRef]
55. Fernández-Iglesias, N.; Bettmer, J. Synthesis, Purification and Mass Spectrometric Characterisation of a Fluorescent Au9@BSA Nanocluster and Its Enzymatic Digestion by Trypsin. *Nanoscale* **2014**, *6*, 716–721. [CrossRef]
56. Mathew, M.S.; Baksi, A.; Pradeep, T.; Joseph, K. Choline-Induced Selective Fluorescence Quenching of Acetylcholinesterase Conjugated Au@BSA Clusters. *Biosens. Bioelectron.* **2016**, *81*, 68–74. [CrossRef]
57. Hori, H.; Teranishi, T.; Nakae, Y.; Seino, Y.; Miyake, M.; Yamada, S. Anomalous Magnetic Polarization Effect of Pd and Au Nano-Particles. *Phys. Lett. Sect. A Gen. At. Solid State Phys.* **1999**, *263*, 406–410. [CrossRef]
58. Brust, M.; Walker, M.; Bethell, D.; Schiffrin, D.J.; Whyman, R. Synthesis of Thiol-Derivatised Gold Nanoparticles in a Two-Phase Liquid-Liquid System. *J. Chem. Soc. Chem. Commun.* **1994**, *7*, 801–802. [CrossRef]
59. Gréget, R.; Nealon, G.; Vileno, B.; Turek, P.; Mény, C.; Ott, F.; Derory, A.; Voirin, E.; Rivière, E.; Rogalev, A.; et al. Magnetic Properties of Gold Nanoparticles A Room-Temperature Quantum Effect. *ChemPhysChem* **2012**, *13*, 3092–3097. [CrossRef]
60. Muñoz-Márquez, M.A.; Guerrero, E.; Fernández, A.; Crespo, P.; Hernando, A.; Lucena, R.; Conesa, J.C. Permanent Magnetism in Phosphine- and Chlorine-Capped Gold: From Clusters to Nanoparticles. *J. Nanoparticle Res.* **2010**, *12*, 1307–1318. [CrossRef]
61. Crespo, P.; García, M.A.; Fernández Pinel, E.; Multigner, M.; Alcántara, D.; De La Fuente, J.M.; Penadés, S.; Hernando, A. Fe Impurities Weaken the Ferromagnetic Behavior in Au Nanoparticles. *Phys. Rev. Lett.* **2006**, *97*, 1–4. [CrossRef] [PubMed]
62. Cirri, A.; Silakov, A.; Jensen, L.; Lear, B.J. Probing Ligand-Induced Modulation of Metallic States in Small Gold Nanoparticles Using Conduction Electron Spin Resonance. *Phys. Chem. Chem. Phys.* **2016**, *18*, 25443–25451. [CrossRef] [PubMed]
63. Donnio, B.; García-Vázquez, P.; Gallani, J.-L.; Guillon, D.; Terazzi, E. Dendronized Ferromagnetic Gold Nanoparticles Self-Organized in a Thermotropic Cubic Phase. *Adv. Mater.* **2007**, *19*, 3534–3539. [CrossRef]
64. Cirri, A.; Silakov, A.; Lear, B.J. Ligand Control over the Electronic Properties within the Metallic Core of Gold Nanoparticles. *Angew. Chem. Int. Ed.* **2015**, *54*, 11750–11753. [CrossRef] [PubMed]
65. Yamamoto, Y.; Miura, T.; Suzuki, M.; Kawamura, N.; Miyagawa, H.; Nakamura, T.; Kobayashi, K.; Teranishi, T.; Hori, H. Direct Observation of Ferromagnetic Spin Polarization in Gold Nanoparticles. *Phys. Rev. Lett.* **2004**, *93*, 1–4. [CrossRef] [PubMed]
66. Cirri, A.; Silakov, A.; Jensen, L.; Lear, B.J. Chain Length and Solvent Control over the Electronic Properties of Alkanethiolate-Protected Gold Nanoparticles at the Molecule-to-Metal Transition. *J. Am. Chem. Soc.* **2016**, *138*, 15987–15993. [CrossRef]
67. Agrachev, M.; Antonello, S.; Dainese, T.; Ruzzi, M.; Zoleo, A.; Aprà, E.; Govind, N.; Fortunelli, A.; Sementa, L.; Maran, F. Magnetic Ordering in Gold Nanoclusters. *ACS Omega* **2017**, *2*, 2607–2617. [CrossRef]
68. Lassmann, G.; Kolberg, M.; Bleifuss, G.; Gräslund, A.; Sjöberg, B.M.; Lubitz, W. Protein Thiyl Radicals in Disordered Systems: A Comparative Epr Study at Low Temperature. *Phys. Chem. Chem. Phys.* **2003**, *5*, 2442–2453. [CrossRef]
69. Silvester Julie, A.; Timmins Graham, S.D. Photodynamically Generated Bovine Serum Albumin Radicals: Evidence for Damage Transfer and Oxidation at Cysteine and Tryptophan Residues. *Free Radic. Biol. Med.* **1998**, *24*, 754–766. [CrossRef]
70. Rurack, K.; Spieles, M. Fluorescence Quantum Yields of a Series of Red and Near-Infrared Dyes Emitting at 600–1000 Nm. *Anal. Chem.* **2011**, *83*, 1232–1242. [CrossRef]
71. Lyons, S.K.; Patrick, P.S.; Brindle, K.M. Imaging Mouse Cancer Models in Vivo Using Reporter Transgenes. *Cold Spring Harb. Protoc.* **2013**, *2013*, 685–699. [CrossRef] [PubMed]
72. Nealon, G.L.; Donnio, B.; Greget, R.; Kappler, J.P.; Terazzi, E.; Gallani, J.L. Magnetism in Gold Nanoparticles. *Nanoscale* **2012**, *4*, 5244–5258. [CrossRef] [PubMed]

73. Chevrier, D.M.; Thanthirige, V.D.; Luo, Z.; Driscoll, S.; Cho, P.; Macdonald, M.A.; Yao, Q.; Guda, R.; Xie, J.; Johnson, E.R.; et al. Structure and Formation of Highly Luminescent Protein-Stabilized Gold Clusters. *Chem. Sci.* **2018**, *9*, 2782–2790. [CrossRef] [PubMed]
74. Hsu, Y.C.; Hung, M.J.; Chen, Y.A.; Wang, T.F.; Ou, Y.R.; Chen, S.H. Identifying Reducing and Capping Sites of Protein-Encapsulated Gold Nanoclusters. *Molecules* **2019**, *24*, 1630. [CrossRef] [PubMed]
75. Anderson, P.W. New Approach to the Theory of Superexchange Interactions. *Career Theor. Phys. A 2nd Ed.* **2005**, *115*, 100–111. [CrossRef]
76. Andres, J.; Longo, E.; Gouveia, A.F.; Gracia, L.; Oliveira, M.C. In Situ Formation of Metal Nanoparticles through Electron Beam Irradiation: Modeling Real Materials from First-Principles Calculations. *J. Mater. Sci. Eng.* **2018**, *7*, 3. [CrossRef]

Article

Plasmonic Ag Nanoparticle-Loaded n-p Bi₂O₂CO₃/α-Bi₂O₃ Heterojunction Microtubes with Enhanced Visible-Light-Driven Photocatalytic Activity

Haibin Li [1,2], Xiang Luo [1], Ziwen Long [1], Guoyou Huang [1] and Ligang Zhu [2,*]

[1] College of Materials Science and Engineering, Changsha University of Science and Technology, Changsha 410114, China; coastllee@hotmail.com (H.L.); xiangl0926@163.com (X.L.); lzw19961020@163.com (Z.L.); hgyviny123viny@163.com (G.H.)

[2] Guangxi Key Laboratory of Agricultural Resources Chemistry and Biotechnology, College of Chemistry and Food Science, Yulin Normal University, Yulin 537000, China

* Correspondence: zhuligang@ylu.edu.cn

Citation: Li, H.; Luo, X.; Long, Z.; Huang, G.; Zhu, L. Plasmonic Ag Nanoparticle-Loaded n-p Bi₂O₂CO₃/α-Bi₂O₃ Heterojunction Microtubes with Enhanced Visible-Light-Driven Photocatalytic Activity. *Nanomaterials* **2022**, *12*, 1608. https://doi.org/10.3390/nano12091608

Academic Editor: Diego Cazorla-Amorós

Received: 13 April 2022
Accepted: 7 May 2022
Published: 9 May 2022

Publisher's Note: MDPI stays neutral with regard to jurisdictional claims in published maps and institutional affiliations.

Copyright: © 2022 by the authors. Licensee MDPI, Basel, Switzerland. This article is an open access article distributed under the terms and conditions of the Creative Commons Attribution (CC BY) license (https://creativecommons.org/licenses/by/4.0/).

Abstract: In this study, n-p Bi₂O₂CO₃/α-Bi₂O₃ heterojunction microtubes were prepared via a one-step solvothermal route in an H₂O-ethylenediamine mixed solvent for the first time. Then, Ag nanoparticles were loaded onto the microtubes using a photo-deposition process. It was found that a Bi₂O₂CO₃/α-Bi₂O₃ heterostructure was formed as a result of the in situ carbonatization of α-Bi₂O₃microtubes on the surface. The photocatalytic activities of α-Bi₂O₃ microtubes, Bi₂O₂CO₃/α-Bi₂O₃ microtubes, and Ag nanoparticle-loaded Bi₂O₂CO₃/α-Bi₂O₃ microtubes were evaluated based on their degradation of methyl orange under visible-light irradiation ($\lambda > 420$ nm). The results indicated that Bi₂O₂CO₃/α-Bi₂O₃ with a Bi₂O₂CO₃ mass fraction of 6.1% exhibited higher photocatalytic activity than α-Bi₂O₃. Loading the microtubes with Ag nanoparticles significantly improved the photocatalytic activity of Bi₂O₂CO₃/α-Bi₂O₃. This should be ascribed to the internal static electric field built at the heterojunction interface of Bi₂O₂CO₃ and α-Bi₂O₃ resulting in superior electron conductivity due to the Ag nanoparticles; additionally, the heterojunction at the interfaces between two semiconductors and Ag nanoparticles and the local electromagnetic field induced by the surface plasmon resonance effect of Ag nanoparticles effectively facilitate the photoinduced charge carrier transfer and separation of α-Bi₂O₃. Furthermore, loading of Ag nanoparticles leads to the formation of new reactive sites, and a new reactive species $\cdot O^{2-}$ for photocatalysis, compared with Bi₂O₂CO₃/α-Bi₂O₃.

Keywords: α-Bi₂O₃; Bi₂O₂CO₃; silver; heterojunction; microtube; photocatalysis

1. Introduction

In the past decades, photocatalytic technology through semiconductor oxides for the purification and treatment of polluted water and air has been extensively studied. Recent research activity in the field of heterogeneous photocatalysis is focused on exploiting novel and more efficient photocatalysts capable of using visible light for the degradation of organic contaminants. Many Bi-based semiconductors, such as BiVO₄ [1], Bi₂O₃ [2], Bi₂WO₆ [3], Bi₂O₂CO₃ [4], Bi₂MoO₆ [5], and BiPO₄ [6] have been developed as visible-light-driven photocatalysts. Among them, Bi₂O₃ has received significant attention in recent years. It is well known that Bi₂O₃ is a p-type semiconductor with five crystallographic polymorphs denoted as monoclinic α-Bi₂O₃, tetragonal β-Bi₂O₃, cubic (BCC) γ-Bi₂O₃, cubic (FCC) δ-Bi₂O₃, and triclinic ω-Bi₂O₃ [2]. Monoclinic α-Bi₂O₃, which is nontoxic and chemically stable in aqueous solution under irradiation, has been proved to be a visible-light-driven photocatalyst, owing to its narrow band-gap energy (band gap around 2.6–2.8 eV). However, as a photocatalyst, α-Bi₂O₃ suffered severe problems in practical applications due to its low quantum yield, which is normally caused by the rapid recombination of its charge carriers [2]. Thus, novel photocatalysts based on α-Bi$_2$O$_3$ are required to be further explored in order to achieve increases in quantum efficiency and successes in practical applications.

Coupling a p-type α-Bi$_2$O$_3$ with another n-type semiconductor with matching band potentials to form a p-n heterojunction has been demonstrated to be an effective strategy to enhance the quantum yield. Driven by the internal static electric field built at the heterojunction interface, the photogenerated charges can transport from one semiconductor to another, thus improving the electron–hole pairs separation and interfacial charge transfer efficiency [7]. Bi$_2$O$_2$CO$_3$ is an n-type semiconductor with a band gap of 3.55 eV. Growing attention has been paid to it, since Zhang et al. reported for the first time the application of Bi$_2$O$_2$CO$_3$ as a photocatalyst in the degradation of methyl orange in aqueous solution under UV light irradiation [8]. Since α-Bi$_2$O$_3$ and Bi$_2$O$_2$CO$_3$ are intrinsic p-type and n-type semiconductors, respectively; thus theoretically, an n-p Bi$_2$O$_2$CO$_3$/α-Bi$_2$O$_3$ heterojunction is formed when the two dissimilar crystalline semiconductors combine. The reason for this is that the conduction band edge for α-Bi$_2$O$_3$ is much higher than that for Bi$_2$O$_2$CO$_3$. As a well-defined interface is the key to improving the catalytic activities of heterojunction photocatalysts by facilitating charge transfer and separation, it is of great significance to develop a facile route to fabricate Bi$_2$O$_2$CO$_3$/α-Bi$_2$O$_3$ heterostructures with effective contacts between Bi$_2$O$_2$CO$_3$ and α-Bi$_2$O$_3$.

Noble metal nanoparticles (NPs), such as Au NPs [9,10], Pt NPs [11,12], Ru NPs [13,14], Ag NPs [15,16], and so on, have been used as co-catalysts to work with photocatalysts for enhanced photocatalytic performance, not only because they play the crucial roles of being photoinduced electron trappers due to their superior electron conductivities, but also because of the surface plasmon resonance (SPR) effect caused by the mutual oscillation between incident light and the electrons on the surface of noble metal NPs. Ag nanoparticles are a good choice for constructing noble metal NPs/semiconductor heterostructures, due to their facile preparation and relatively low cost. So far, several Ag NP-hybridized heterostructures have been reported, including Ag-Cu$_2$O/PANI [17], Ag/ZnO@CF [18], Ag/AgCl/Ag$_2$MoO$_4$ [19], Ag/ZnO/3Dgraphene [20], Ag/GO/TiO$_2$ [21], Bi$_2$WO$_6$/Ag$_3$PO$_4$-Ag [22], and g-C$_3$N$_4$/Ag/TiO$_2$ [23], with enhanced photocatalytic activity. To the best of our knowledge, no study has been performed on synthesis and photocatalytic application of Ag NP-loaded Bi$_2$O$_2$CO$_3$/α-Bi$_2$O$_3$ heterostructure composite systems.

In the present study, novel n-p Bi$_2$O$_2$CO$_3$/α-Bi$_2$O$_3$ heterojunction microtubes with hexagonal cross sections were prepared via a facile one-step template- and surfactant-free solvothermal method for the first time. As Bi$_2$O$_2$CO$_3$ was formed via in situ carbonatization of α-Bi$_2$O$_3$ microtubes on the surface, this method is more conducive to generate well-defined Bi$_2$O$_2$CO$_3$/α-Bi$_2$O$_3$ heterojunction interfaces than two-step strategies. Co-catalyst Ag nanoparticles were evenly loaded on the surface of Bi$_2$O$_2$CO$_3$/α-Bi$_2$O$_3$ heterojunction microtubes, using a photo-deposition process to construct a novel Ag/Bi$_2$O$_2$CO$_3$/α-Bi$_2$O$_3$ microtube ternary system to further enhance the photocatalytic activity. The photocatalytic performances of the as-prepared samples were evaluated by examining the degradation of methyl orange (MO) under visible light (λ > 420 nm) irradiation.

2. Materials and Methods

2.1. Synthesis of Bi$_2$O$_2$CO$_3$/α-Bi$_2$O$_3$ Heterostructure Microtubes

Bismuth nitrate pentahydrate and ethylenediamine were purchased from Xilong Scientific Co., Ltd (Shantou, China) and Taicang Hushi Reagent Co., Ltd (Taicang, China), respectively. All reagents were of AR grade, and used without further purification. Distilled water was used in all experiments. As illustrated in Figure 1, in a typical synthesis, 0.00175 mol of Bi(NO$_3$)$_3$·5H$_2$O was added into the ethylenediamine (en)–water mixture (80 mL), with a certain volume ratio of ethylenediamine and water (V$_{en}$:V$_{water}$). After being stirred for 30 min, the resulting faint yellow suspension (donated as precursor) was transferred into a 100-milliliter Teflon-lined stainless steel autoclave. The autoclave was sealed and maintained at 140 °C for 10 h and then cooled down to room temperature. The resulting precipitate was centrifuged, rinsed repeatedly with distilled water and ethanol, then dried at 80 °C in air to obtain the $Bi_2O_2CO_3/\alpha\text{-}Bi_2O_3$ heterostructure microtubes.

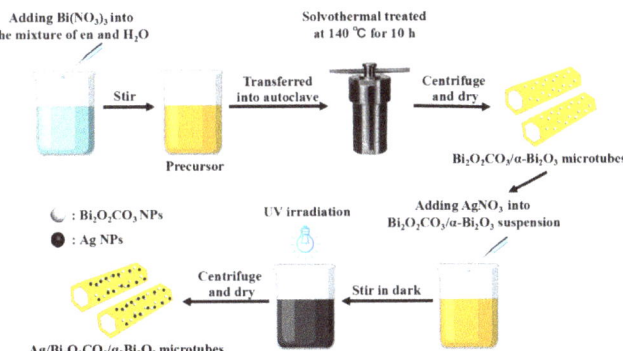

Figure 1. Schematic illustration for the synthesis of Ag NP-loaded $Bi_2O_2CO_3/\alpha\text{-}Bi_2O_3$ heterostructure microtubes.

2.2. Synthesis of Ag NP-Loaded $Bi_2O_2CO_3/\alpha\text{-}Bi_2O_3$ Heterostructure Microtubes

The fabrication of Ag NP-loaded $Bi_2O_2CO_3/\alpha\text{-}Bi_2O_3$ heterostructure microtubes was conducted as follows. First, 0.5 g of $Bi_2O_2CO_3/\alpha\text{-}Bi_2O_3$ heterostructure microtubes was dispersed into the $AgNO_3$ ((AR grade, Sinopharm Chemical Reagent Co., Ltd, Shanghai, China) aqueous solution under stirring. The theoretical loading amount of silver was set at 3 wt% in the $Ag/Bi_2O_2CO_3/\alpha\text{-}Bi_2O_3$ sample. After being ultrasonically treated for 10 min, the suspension was further magnetically stirred for 10 h in the dark, followed by UV illumination for 2 h under stirring. The black powder was centrifuged, rinsed with distilled water repeatedly to purify the product, and finally dried at 80 °C in air.

2.3. Characterization

The crystalline structure of the samples was analyzed by a Rigaku D/Max 2500 powder diffractometer (XRD) (Tokyo, Japan) with Cu Kα radiation (λ = 1.5406 Å). The morphology of the as-prepared samples was characterized by field-emission scanning electron microscopy (FESEM, FEI SIRION 200, Hillsboro, OR, USA), and transmission electron microscopy (TEM, Philips Tecnai 20 G2 S-TWIN, Hillsboro, OR, USA). X-ray photoelectron spectroscopy (XPS) data of the samples were determined with a K-Alpha 1063 electron spectrometer from Thermo Fisher Scientific (East Grinstead, West Sussex, UK) using 72W Al Kα radiation. Infrared spectroscopy analysis (IR) of the samples was performed on an AVATAR360 IR analyzer (Madison, WI, USA). UV-vis diffuse reflectance spectra (UV-vis) were measured with a Specord 200 UV spectrophotometer (Schönwalde-Glien, Germany).

2.4. Photocatalytic Experiments

The photocatalytic properties of the as-prepared samples were assessed by degradation of MO under the irradiation of visible light (λ > 420 nm). First, 0.5 g of photocatalyst was added to 100 mL of 10 mg/L MO aqueous solution. Then, the suspension was magnetically stirred in the dark for 1h before commencing the photocatalytic reactions, to allow the system to reach an adsorption/desorption equilibrium. All photocatalytic reactions were carried out in a laboratory constructed photo-reactor under visible light irradiation from a 500W Xe lamp equipped with a 420-nanometer cutoff filter. The photocatalytic system was magnetically stirred simultaneously during the course of illumination. At given time intervals, 3.5-milliliter aliquots of the aqueous solution were collected and centrifuged. The concentrations of MO solution were evaluated by measuring its absorption on a UNICO

UV-2100 spectrophotometer (Palo Alto, CA, USA) at 463 nm, from which the photocatalytic activity was calculated.

3. Results and Discussion

XRD was used to analyze the phase composition and crystal structure of the samples. Figure 2 shows the XRD patterns of the samples produced at 140 °C for 10 h in the ethylenediamine–water mixture with various ratios of $V_{en}:V_{water}$. For all the samples, the diffraction peaks are sharp, and the intensity of the diffraction is high, indicating that the products are well-crystallized. In addition, the diffraction peaks assigned to α-Bi_2O_3 (JCPDS Card No. 71-2274) are accompanied by three characteristic peaks of $Bi_2O_2CO_3$ (JCPDS Card No. 41-1488) at 12.9°, 23.8°, and 30.2°. No peaks of any additional phases were detected, indicating that the products exhibit a coexistence of both α-Bi_2O_3 and $Bi_2O_2CO_3$ phases. Furthermore, when increasing the ratio of $V_{en}:V_{water}$, the intensity of the characteristic peaks attributed to $Bi_2O_2CO_3$ gradually increases, whereas the intensity of the diffraction peaks assigned to α-Bi_2O_3 decreases. The mass fractions of the $Bi_2O_2CO_3$ in the samples are 0%, 6.1%, 15.5%, 36.7%, 47.9%, and 51.3% for the samples prepared at $V_{en}:V_{water}$ ratios of 1:7, 2:6, 3:5, 4:4, 5:3, and 6:2, respectively, which were estimated from XRD intensity data by using the formula as expressed by Equation (1):

$$R_C = \frac{I_C}{I_C + I_O} \quad (1)$$

where I_C and I_O are the integrated intensities of $Bi_2O_2CO_3$ (013) and α-Bi_2O_3 (113) diffraction peaks, respectively. It can be inferred that the ratio of $V_{en}:V_{water}$ plays a key role in the phase composition of the products, and that a larger proportion of en favors the generation of $Bi_2O_2CO_3$.

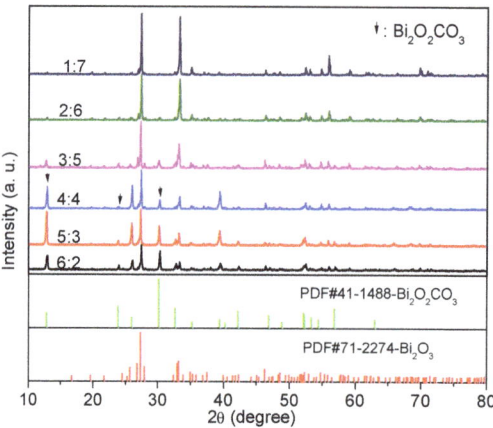

Figure 2. XRD patterns of the samples prepared at 140 °C for 10 h in the ethylenediamine–water mixture with various ratios of $V_{en}:V_{water}$.

How are the α-Bi_2O_3 and $Bi_2O_2CO_3$ generated? Why does the proportion of en in the mixed solvent have such a significant effect on the generation of $Bi_2O_2CO_3$? In order to answer these questions, XRD investigations on the precursor and the products obtained at 140 °C for 1, 3, 5, 7.5, 10, and 12.5 h in the en–water mixture with a $V_{en}:V_{water}$ ratio of 2:6 were carried out. The results are presented in Figure 3. For the precursor and the products obtained after solvothermal treatment for 1 h, 3 h, 5 h, and 7.5 h, all the diffraction peaks can be readily indexed to a pure α-Bi_2O_3 (JCPDS Card No. 71-2274) phase, revealing that α-Bi_2O_3 was formed before solvothermal treatment, and that a pure α-Bi_2O_3 phase could be maintained via controlling the reaction time using this technique. Moreover, the diffraction peaks of the solvothermal-treated products are much narrower than that of the precursor, and the peak intensities of the solvothermal-treated products are much higher, indicating that solvothermal treatment improved the crystallinity of the products. As the time increased to 10 h, the diffraction pattern of the sample indexed to the mixture of α-Bi_2O_3 and $Bi_2O_2CO_3$ (JCPDS Card No. 41-1488). Three weak peaks at 12.9°, 23.8°, and 30.2° can be attributed to $Bi_2O_2CO_3$. Further prolonging the time to 12.5 h, the intensity of the peaks indexed to $Bi_2O_2CO_3$ increases, suggesting an increase in the amount of $Bi_2O_2CO_3$. From the XRD results, it can be seen that the $Bi_2O_2CO_3$/α-Bi_2O_3 composite is derived from α-Bi_2O_3, but not formed at the precursor stage.

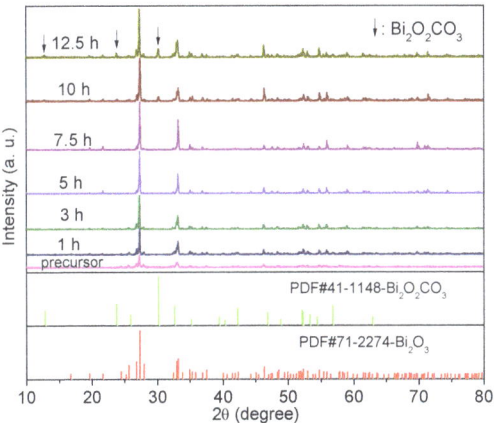

Figure 3. XRD patterns of the precursor and the samples obtained at 140 °C for 1, 3, 5, 7.5, 10 h, and 12.5 h in the ethylenediamine–water mixture with a V_{en}:V_{water} ratio of 2:6.

This is also supported by FT-IR spectra of the precursor and the products obtained after solvothermal treatment for 7.5 h and 10 h (Figure 4). For all the samples, the weak adsorptions at 1460, 1384, and 1315 cm^{-1} may be attributed to the carbonated species formed by the reactions between the surface hydroxyl groups and atmospheric CO_2. The peaks at around 545, 505, and 430 cm^{-1} are due to the vibration of Bi-O bonds in BiO_6 octahedral units [24,25]. It is necessary to mention that only the product obtained after solvothermal treatment for 10 h shows an extra band at 850 cm^{-1}, which is ascribed to the CO_3^{2-}, indicating the formation of $Bi_2O_2CO_3$ at this stage [24,25].

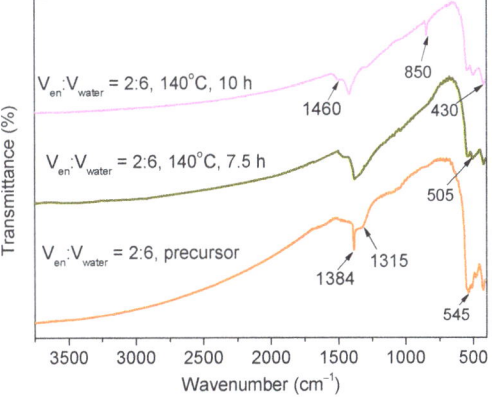

Figure 4. FT-IR spectra of the precursor and the samples obtained at 140 °C for 7.5 h and 10 h in the ethylenediamine–water mixture with a V_{en}:V_{water} ratio of 2:6.

Based on the XRD and FT-IR analyses, formation of the $Bi_2O_2CO_3/\alpha\text{-}Bi_2O_3$ composite in the present solvothermal process could be described by following reactions:

$$H_2NCH_2CH_2NH_2 + 2H_2O \rightarrow H_3\overset{+}{N}CH_2CH_2\overset{+}{N}H_3 + 2OH^- \quad (2)$$

$$Bi^{3+} + 3OH^- \rightarrow Bi(OH)_3 \downarrow \quad (3)$$

$$2Bi(OH)_3 \rightarrow Bi_2O_3 + 3H_2O \quad (4)$$

$$CO_2 + 2OH^- \rightarrow CO_3^{2-} + H_2O \quad (5)$$

$$Bi_2O_3 + CO_3^{2-} + H_2O \rightarrow Bi_2O_2CO_3 + 2OH^- \quad (6)$$

When $Bi(NO_3)_3 \cdot 5H_2O$ was added to the en–water mixture with a $V_{en}:V_{water}$ ratio of 2:6, the reaction was performed in a strong alkali condition, as indicated in Equation (2). Abundant hydroxide ions firstly reacted with Bi^{3+} to produce $Bi(OH)_3$, which then dehydrated to form $\alpha\text{-}Bi_2O_3$ under vigorous stirring, as illustrated in Equations (3) and (4). Due to the presence of en, the mixed solvent easily captured CO_2 from the air to generate CO_3^{2-} before being transferred into the autoclave. In prolonging the solvothermal treatment time to 10 h, a small amount of obtained $\alpha\text{-}Bi_2O_3$ reacted with CO_3^{2-} in the solvent to give rise to $Bi_2O_2CO_3$, as summarized in Equations (5) and (6) [26]. It can be concluded that $Bi_2O_2CO_3$ was formed by in situ carbonatization of $\alpha\text{-}Bi_2O_3$. A larger proportion of en in the solvent captures more CO_2 to generate more CO_3^{2-}, resulting in a higher ratio of $Bi_2O_2CO_3$ in the product.

Figure 5a,b show the SEM images of the products obtained by solvothermal treatment at 140 °C for 10 h in the ethylenediamine–water mixture with $V_{en}:V_{water}$ ratios of 1:7 and 2:6, respectively. It can be seen that both samples consist of microtubes. The magnified image of the microtubes presented in the left insert of Figure 5b clearly demonstrates that the microtubes have well-defined hexagonal cross sections. The SEM image with low magnification (Figure 5c) reveals that the products obtained in the ethylenediamine–water mixture with a $V_{en}:V_{water}$ ratio of 2:6 are almost entirely microtubes with lengths of 5–30 μm, and side lengths of 0.2–1 μm, indicating the high yield of microtubes in this condition. However, when the $V_{en}:V_{water}$ ratio was controlled at 4:4, 5:3, and 6:2, the as-prepared products contain microtubes and a lot of irregular particles, as presented in Figure 5d–f, respectively. This indicates that the proportion of en in the mixed solvent also has a significant effect on the morphology of the products. More en in the solvent captures more CO_2 to generate more CO_3^{2-}, which makes more $\alpha\text{-}Bi_2O_3$ carbonatized, resulting in the destruction of microtubes.

Figure 6a presents the TEM image of the obtained $\alpha\text{-}Bi_2O_3$ microtube prepared at a $V_{en}:V_{water}$ ratio of 2:6 for 7.5 h. There is a contrast between the inner and outside parts of the sample, confirming its tubular structure. The lattice spacing of about 0.34 nm between adjacent lattice planes in the insert corresponds to the interplanar spacing of the (002) plane of $\alpha\text{-}Bi_2O_3$. Figure 6b shows the TEM image of $Bi_2O_2CO_3/\alpha\text{-}Bi_2O_3$ heterojunction microtubes prepared at a $V_{en}:V_{water}$ ratio of 2:6 for 10 h. It can be clearly seen that a lot of nanoparticles highly disperse on the surface of $\alpha\text{-}Bi_2O_3$ microtubes, which are considered to be $Bi_2O_2CO_3$ particles. No "support-free" $Bi_2O_2CO_3$ nanoparticles are found, indicating that those nanoparticles are strongly anchored to the $\alpha\text{-}Bi_2O_3$ microtubes. From the HRTEM image of the sample shown in Figure 6c, it can be seen that the lattice structure of $\alpha\text{-}Bi_2O_3$ is very orderly and different from that of $Bi_2O_2CO_3$ nanoparticles. The measured lattice fringes of 0.34 nm well match the (002) crystallographic planes of $\alpha\text{-}Bi_2O_3$. In particular, it can be well confirmed that the $Bi_2O_2CO_3$ nanoparticles are anchored on the surface of the $\alpha\text{-}Bi_2O_3$ substrate, forming a good attachment. The obvious interface between the $Bi_2O_2CO_3$ nanoparticles and the $\alpha\text{-}Bi_2O_3$ microtubes shown in HRTEM images implies the formation of a well-defined heterojunction structure. Because $\alpha\text{-}Bi_2O_3$ and $Bi_2O_2CO_3$ are p-type and n-type semiconductors, respectively, the heterojunction can be considered to be a well-defined and well-formed p–n junction.

Figure 5. SEM images of the samples prepared at 140 °C for 10 h in the en–water mixture with various ratios of $V_{en}:V_{water}$: (**a**) 1:7, (**b**,**c**) 2:6, (**d**) 4:4, (**e**) 5:3, and (**f**) 6:2.

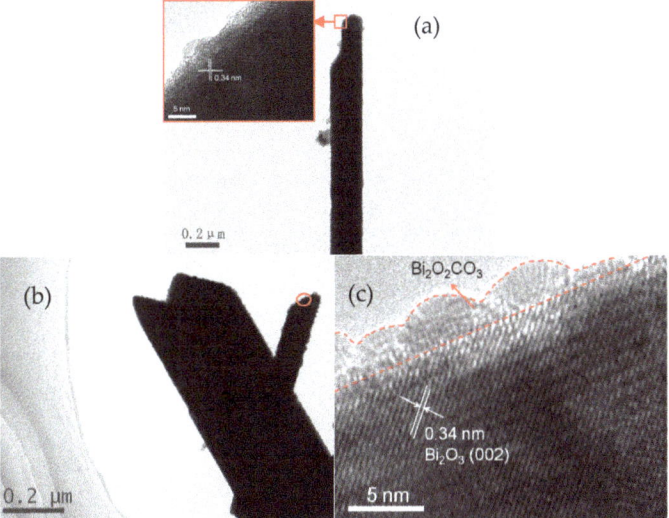

Figure 6. TEM images of (**a**) α-Bi_2O_3 microtubes (insert: HRTEM) and (**b**) $Bi_2O_2CO_3$/α-Bi_2O_3 microtubes; an HRTEM image of (**c**) $Bi_2O_2CO_3$/α-Bi_2O_3 microtubes.

Figure 7 shows the high-resolution XPS spectra of Bi, O, and Ag in Ag NP-loaded $Bi_2O_2CO_3/\alpha\text{-}Bi_2O_3$ heterojunction microtubes with R_c of 6.1%. As observed in the XPS spectrum of Bi 4f (Figure 7a), two strong peaks at 163.8 and 158.5 eV are assigned to Bi $4f_{5/2}$ and Bi $4f_{7/2}$, respectively, confirming that the bismuth species in the sample are Bi^{3+} cations [27]. In the O 1s XPS spectrum (Figure 7b), the O 1s region is fitted by two peaks at 529.6 and 531.3 eV, which are attributed to the oxygen in the Bi–O bond and carbonate species, respectively [27]. Figure 7c presents the Ag 3d XPS spectrum, with two peaks at 368.3 and 374.3 eV, which correspond to Ag $3d_{5/2}$ and Ag $3d_{3/2}$, respectively, suggesting that the silver species in the sample is metallic silver, as the bonding energy corresponding to Ag $3d_{5/2}$ of metallic Ag and Ag_2O are 368.25 eV and 367.70 eV, respectively, according to the previous report [28].

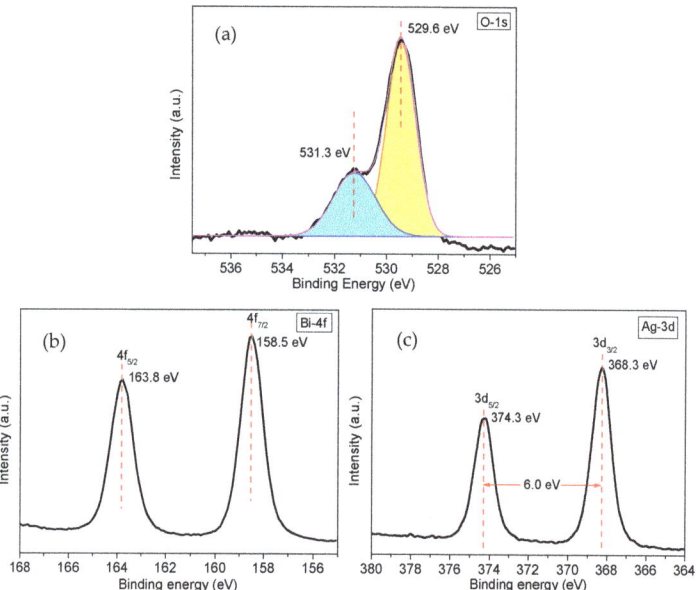

Figure 7. High-resolution XPS spectra of (**a**) O 1s, (**b**) Bi 4f, and (**c**) Ag 3d.

The TEM image of Ag NP-loaded $Bi_2O_2CO_3/\alpha\text{-}Bi_2O_3$ heterojunction microtubes with R_c of 6.1% is shown in Figure 8a. As seen from the image, many nanoparticles are evenly dispersed on the surface of microtubes, and strongly anchored. HRTEM was carried out to verify the nanoparticles, as shown in Figure 8b. The lattice structure of nanoparticles anchored on the surface of microtubes is very orderly, and obviously different from that of the microtubes. The measured lattice fringes of 0.245 nm well match the (200) crystallographic planes of metallic Ag, suggesting that Ag NP-loaded $Bi_2O_2CO_3/\alpha\text{-}Bi_2O_3$ heterojunction microtubes are achieved by this strategy.

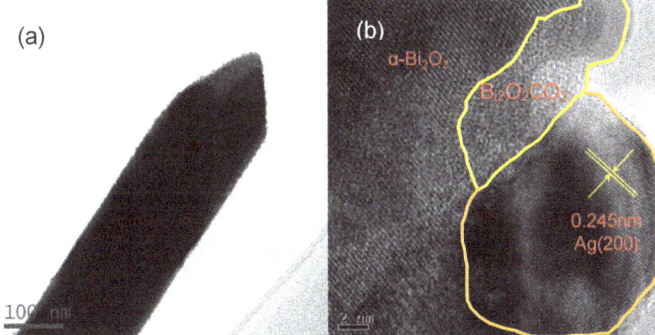

Figure 8. TEM (**a**) and HRTEM (**b**) images of Ag-loaded $Bi_2O_2CO_3/\alpha$-Bi_2O_3 heterojunction microtube.

Figure 9 shows the UV–vis diffuse reflectance spectra of α-Bi_2O_3 microtubes, $Bi_2O_2CO_3/\alpha$-Bi_2O_3 heterojunction microtubes, and Ag NP-loaded $Bi_2O_2CO_3/\alpha$-Bi_2O_3 heterojunction microtubes. The α-Bi_2O_3 microtubes prepared at V_{en}:V_{water} = 1:7 exhibit strong absorption in the visible range in addition to the UV range. The absorption edge occurs at about 450 nm. The spectrum is steep, indicating that the absorption of visible light is not due to the transition from impurity levels, but to the band-gap transition. The $Bi_2O_2CO_3/\alpha$-Bi_2O_3 heterojunction microtubes with R_c of 6.1% and 51.3% show dual absorption edges at 365 and 450 nm, which are related to their mixed-phase structure. Moreover, the absorbance in the 360–450 nm range of $Bi_2O_2CO_3/\alpha$-Bi_2O_3 is much weaker compared with that of α-Bi_2O_3 due to the its substantial $Bi_2O_2CO_3$ phase content. The band-gap energies were estimated to be 2.75 and 3.4 eV for α-Bi_2O_3 and $Bi_2O_2CO_3$, respectively, and were calculated from the formula $\lambda_g = 1239.8/E_g$, where λ_g is the band-gap wavelength, and E_g is the bandgap energy [29]. Ag NP-loaded $Bi_2O_2CO_3/\alpha$-Bi_2O_3 heterojunction microtubes with R_c of 6.1% show an extended absorption in the visible region, which is due to the typical surface plasmon band exhibited by the Ag nanoparticles [30].

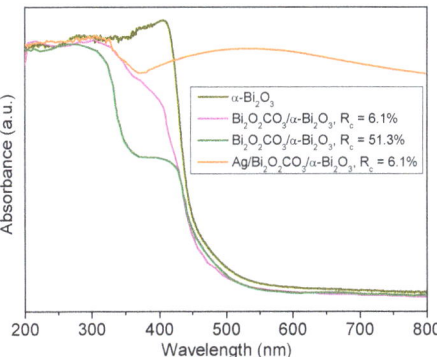

Figure 9. UV–vis diffuse reflectance spectra of α-Bi_2O_3, $Bi_2O_2CO_3/\alpha$-Bi_2O_3, and Ag/$Bi_2O_2CO_3/\alpha$-Bi_2O_3.

Photodegradation of MO under visible light irradiation was carried out to estimate the photocatalytic performance of the as-prepared samples. The photodegradation efficiencies of MO as a function of irradiation time by α-Bi_2O_3, $Bi_2O_2CO_3/\alpha$-Bi_2O_3 with R_c of 6.7%, $Bi_2O_2CO_3/\alpha$-Bi_2O_3 with R_c of 15.5%, Ag/$Bi_2O_2CO_3/\alpha$-Bi_2O_3 with R_c of 6.7%, as well as in the absence of photocatalysts, are presented in Figure 10. It can be seen that all the samples show visible light photocatalytic activities. After 140 min of irradiation, the photodegradation efficiencies of MO by α-Bi$_2$O$_3$, Bi$_2$O$_2$CO$_3$/α-Bi$_2$O$_3$ with R$_c$ of 6.7%, and Bi$_2$O$_2$CO$_3$/α-Bi$_2$O$_3$ with R$_c$ of 15.5%, reach 69%, 100%, and 65%, respectively. For Ag/Bi$_2$O$_2$CO$_3$/α-Bi$_2$O$_3$ with R$_c$ of 6.7%, it reaches 100% after 60 min. Generally, the overall photocatalytic activity of a semiconductor is primarily dictated by surface area, photoabsorption ability, and the separation and transporting rates of photoinduced electron/hole pairs in the catalysts [31]. Since α-Bi$_2$O$_3$, Bi$_2$O$_2$CO$_3$/α-Bi$_2$O$_3$ with R$_c$ of 6.7%, and Ag/Bi$_2$O$_2$CO$_3$/α-Bi$_2$O$_3$ possess similar size and morphology, the enhanced photocatalytic activities of Ag/Bi$_2$O$_2$CO$_3$/α-Bi$_2$O$_3$ and Bi$_2$O$_2$CO$_3$/α-Bi$_2$O$_3$ with R$_c$ of 6.7% should be ascribed to the improved separation and transporting rates of photoinduced electron/hole pairs.

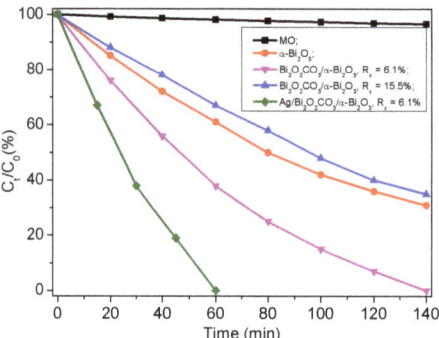

Figure 10. The residual MO at different irradiation time for the as-prepared samples.

Photogenerated electrons, holes, ·O$_2^-$, and ·OH are considered to be major reactive species in organics photodegradation [32]. MO can be degraded into CO$_2$, H$_2$O, and other products by those reactive species [33]. In order to clarify the reaction mechanism further, 1 mmol of various scavengers was introduced to explore the specific reactive species that might play important roles in MO degradation by Ag/Bi$_2$O$_2$CO$_3$/α-Bi$_2$O$_3$. Benzoquinone (BQ), ethylene diaminetetraacetic acid (EDTA), and tertiary butanol (TBA) were used as the scavengers for ·O$_2^-$, holes, and ·OH, respectively [34]. Figure 11 shows the photodegradation efficiencies of MO by Ag/Bi$_2$O$_2$CO$_3$/α-Bi$_2$O$_3$ in the presence of these scavengers under visible light irradiation for 60 min. Both BQ and TBA show suppression of the degradation rate of MO, with TBA exhibiting a stronger suppressing effect. Meanwhile, EDTA shows a much weaker suppressing effect than BQ and TBA, suggesting that ·OH and ·O$_2^-$ are the major reactive species responsible for the photodegradation of MO by Ag/Bi$_2$O$_2$CO$_3$/α-Bi$_2$O$_3$.

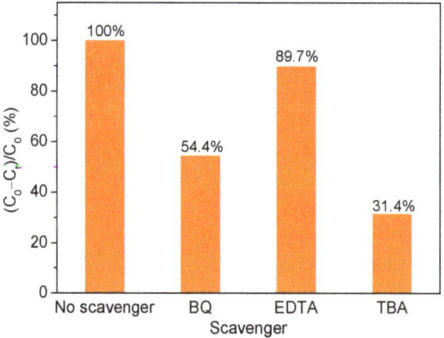

Figure 11. The photodegradation rates of MO by Ag/Bi$_2$O$_2$CO$_3$/α-Bi$_2$O$_3$ after 60 min in the presence of various scavengers.

The effects of $Bi_2O_2CO_3/\alpha\text{-}Bi_2O_3$ and Ag NPs on the efficiency of photoinduced electrons and holes separation were investigated by the photocurrent tests, as shown in Figure 12. The photocurrent intensities of the samples follow the order of $Ag/Bi_2O_2CO_3/\alpha\text{-}Bi_2O_3 > Bi_2O_2CO_3/\alpha\text{-}Bi_2O_3 > \alpha\text{-}Bi_2O_3$. As demonstrated in the previous research, higher photocurrent intensity means higher separation efficiency of the photoinduced electron/hole pairs. The photocurrent measurement results suggest that the formation of $Bi_2O_2CO_3/\alpha\text{-}Bi_2O_3$ heterostructures improves charge carrier transfer and separation of $\alpha\text{-}Bi_2O_3$, while loading of Ag NPs on the heterostructures further enhances this effect. It is consistent with the photocatalytic performance.

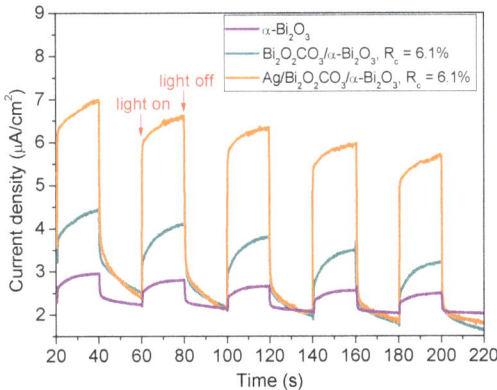

Figure 12. Photocurrent responses of different samples under visible light.

According to the experimental results, we believe that there are four major reasons responsible for the enhanced photodegradation of MO by Ag NP-loaded $Bi_2O_2CO_3/\alpha\text{-}Bi_2O_3$ heterojunction microtubes, as illustrated in Figure 13. Firstly, $Bi_2O_2CO_3/\alpha\text{-}Bi_2O_3$ heterojunction facilitates the charge separation. As reported in the previous work, $\alpha\text{-}Bi_2O_3$ is a p-type semiconductor, while $Bi_2O_2CO_3$ is determined as an n-type material. The conduction band edge of $\alpha\text{-}Bi_2O_3$ and $Bi_2O_2CO_3$ at the point of zero charge (pH_{zpc}) can be theoretically predicted from the formula $E_{CB}^0 = X - E_c - 0.5E_g$, where X is the absolute electronegativity of the semiconductor, and E_c is the energy of free electrons on the hydrogen scale (4.5 eV) [35]. The values of X are 5.95 eV for $\alpha\text{-}Bi_2O_3$ and 6.35 eV for $Bi_2O_2CO_3$, while the estimated E_g is 2.75 eV for $\alpha\text{-}Bi_2O_3$ and 3.4 eV for $Bi_2O_2CO_3$. Given the formula above, the calculated E_{CB} and E_{VB} values are 0.075 eV and 2.825 eV for $\alpha\text{-}Bi_2O_3$, respectively, and 0.15 eV and 3.55 eV for $Bi_2O_2CO_3$, respectively. Therefore, both the conduction band (CB) and valence band (VB) of $Bi_2O_2CO_3$ are considered to be at lower levels than those of $\alpha\text{-}Bi_2O_3$. Thus, a Type II p-n heterojunction is formed at the interfaces as Bi_2O_3 and $Bi_2O_2CO_3$ are closely joined together. When $Bi_2O_2CO_3/\alpha\text{-}Bi_2O_3$ heterojunction microtubes are exposed to visible light irradiation, the electrons in the VB of $\alpha\text{-}Bi_2O_3$ are excited to its CB, leaving holes in the VB. However, for $Bi_2O_2CO_3$, the electrons in the VB cannot be excited because of the wide bandgap of 3.4 eV. Due to the internal field resulting from the potential of band energy difference between $\alpha\text{-}Bi_2O_3$ and $Bi_2O_2CO_3$, there is a great tendency for $\alpha\text{-}Bi_2O_3$ to transfer its photoexcited electrons into the CB of $Bi_2O_2CO_3$, facilitating electron-hole separation in $\alpha\text{-}Bi_2O_3$, and providing more holes for photocatalytic reactions. Secondly, as the Ag NPs loaded on the surface of $Bi_2O_2CO_3/\alpha\text{-}Bi_2O_3$ heterojunction microtubes are in close contact with $\alpha\text{-}Bi_2O_3$ or $Bi_2O_2CO_3$, the electrons in the CB of $\alpha\text{-}Bi_2O_3$ and $Bi_2O_2CO_3$ will transfer to the Ag NPs because of the superior electron conductivity of Ag NPs, along with the formation of heterojunctions at the interface between two semiconductors and the Ag NPs as a result of their work function differences, further suppressing charge carrier recombination [30]. Thirdly, as mentioned above, the valence bands of $\alpha\text{-}Bi_2O_3$ are located at a deep position of about 2.825 eV versus NHE, which is more positive than that of ·OH/OH⁻ (1.9 eV vs. NHE), indicating that the photogenerated holes in the VB of α-Bi$_2$O$_3$ can react with OH⁻ to produce ·OH for oxidation of MO [35,36]. Meanwhile, the conduction band potentials of α-Bi$_2$O$_3$ and Bi$_2$O$_2$CO$_3$ are close to +0.075 eV and +0.15 eV versus NHE, respectively, which are more positive than that of O$_2$/·O$_2$⁻ (−0.33 eV vs. NHE). Thus, it is impossible for the adsorption oxygen to capture an electron from the conduction bands of α-Bi$_2$O$_3$ and Bi$_2$O$_2$CO$_3$ to form active oxygen species (·O$_2$⁻) [35,36]. However, the electrons transferred to Ag NPs from the CBs of α-Bi$_2$O$_3$ and Bi$_2$O$_2$CO$_3$ in Ag/Bi$_2$O$_2$CO$_3$/α-Bi$_2$O$_3$ might be trapped by oxygen molecules in the solutions to form ·O$_2$⁻ for reaction [30,35,36]. This means that loading Ag NPs onto the surface of Bi$_2$O$_2$CO$_3$/α-Bi$_2$O$_3$ can bring another benefit that leads to the formation of new reaction active sites, and a new reactive species ·O$_2$⁻, enhancing the photocatalytic activity of Bi$_2$O$_2$CO$_3$/α-Bi$_2$O$_3$. The possible reactions in the Ag/Bi$_2$O$_2$CO$_3$/α-Bi$_2$O$_3$ ternary photocatalytic system are illustrated by the following equations:

$$Bi_2O_3 + hv \rightarrow Bi_2O_3(h^+ + e^-) \quad (7)$$

$$Bi_2O_3(e^-) + Bi_2O_2CO_3 \rightarrow Bi_2O_3 + Bi_2O_2CO_3(e^-) \quad (8)$$

$$Bi_2O_3(e^-) + Ag \rightarrow Bi_2O_3 + Ag(e^-) \quad (9)$$

$$Bi_2O_2CO_3(e^-) + Ag \rightarrow Bi_2O_2CO_3 + Ag(e^-) \quad (10)$$

$$Bi_2O_3(h^+) + OH^- \rightarrow Bi_2O_3 + \cdot OH \quad (11)$$

$$Ag(e^-) + O_2 \rightarrow Ag + \cdot O_2^- \quad (12)$$

$$\cdot OH/ \cdot O_2^- + MO \rightarrow Product \quad (13)$$

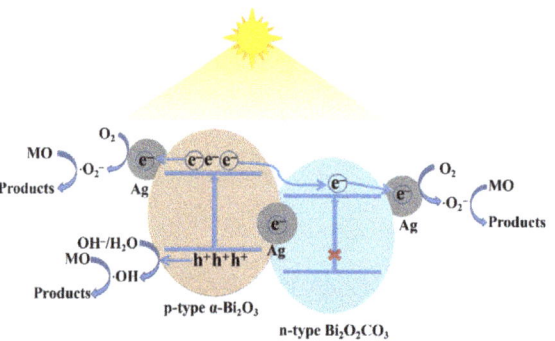

Figure 13. Schematic illustration of the proposed possible mechanism for photodegradation of MO by Ag/Bi$_2$O$_2$CO$_3$/α-Bi$_2$O$_3$ under visible light irradiation.

Lastly, the surface plasmon resonance effect caused by the mutual oscillation between incident light and the electrons on the surface of metallic Ag NPs causes the rise of a local electromagnetic field [35]. Under the influence of this local electromagnetic field, the photogenerated electron/hole pairs on the α-Bi$_2$O$_3$ surface are effectively separated, which also enhances photocatalytic activity.

Figure 14 presents the results of repeated experiments on photodegradation of MO by Ag/Bi$_2$O$_2$CO$_3$/α-Bi$_2$O$_3$ under visible light irradiation. After each run, the photocatalysts were collected by centrifugation, followed by ultrasonic cleaning with distilled water. As shown in the image, no significant loss is found after four successive cycles; 89.8% of MO was degraded in the fifth run after 60 min of visible light irradiation, suggesting that the sample is stale and not photo-corroded in the photocatalytic reactions.

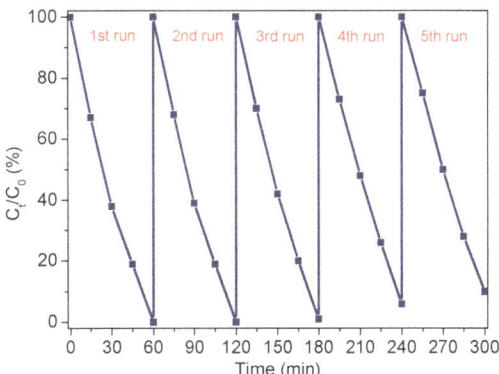

Figure 14. Cyclic photodegradation curve for Ag/Bi$_2$O$_2$CO$_3$/α-Bi$_2$O$_3$.

4. Conclusions

In summary, we have developed a facile solvothermal approach to prepare n-p Bi$_2$O$_2$CO$_3$/α-Bi$_2$O$_3$ heterojunction microtubes. Plasmonic Ag NPs were loaded onto the Bi$_2$O$_2$CO$_3$/α-Bi$_2$O$_3$ microtubes using a simple photo-deposition process, to construct an Ag/Bi$_2$O$_2$CO$_3$/α-Bi$_2$O$_3$ ternary photocatalytic system. This Ag/Bi$_2$O$_2$CO$_3$/α-Bi$_2$O$_3$ ternary system showed much higher photocatalytic activity than α-Bi$_2$O$_3$ and Bi$_2$O$_2$CO$_3$/α-Bi$_2$O$_3$. Under visible light irradiation, the well-defined interfaces between Bi$_2$O$_2$CO$_3$ and α-Bi$_2$O$_3$ in the heterojunctions due to the in situ carbonation of α-Bi$_2$O$_3$ on the surface into Bi$_2$O$_2$CO$_3$, facilitate the transfer of photoinduced electrons from the CB of α-Bi$_2$O$_3$ to that of Bi$_2$O$_2$CO$_3$. Meanwhile, the superior electron conductivity of Ag NPs, the heterojunction at the interface between two semiconductors and Ag NPs, and the local electromagnetic field induced by the surface plasmon resonance effect of Ag NPs, further promote the transfer of photoinduced electrons and suppress the recombination of hole/electron pairs, leaving more holes in the VB of α-Bi$_2$O$_3$ to produce more ·OH for photodegradation of MO. After the photoinduced electrons in the CB of α-Bi$_2$O$_3$ and Bi$_2$O$_2$CO$_3$ that cannot form ·O$_2^-$ are transferred to Ag NPs, they combine with O$_2$ to form ·O$_2^-$, which means that loading of Ag NPs onto Bi$_2$O$_2$CO$_3$/α-Bi$_2$O$_3$ creates new reaction active sites and a new reactive species ·O$_2^-$ for photocatalysis, compared with Bi$_2$O$_2$CO$_3$/α-Bi$_2$O$_3$.

Author Contributions: Conceptualization, H.L. and L.Z.; formal analysis and data curation, X.L., Z.L., and G.H.; writing—original draft preparation, H.L., X.L., and G.H.; writing—review and editing, all authors; funding acquisition, L.Z. and H.L. All authors have read and agreed to the published version of the manuscript.

Funding: The Talent Project of Yulin Normal University (No. G2020ZK14).

Institutional Review Board Statement: Not applicable.

Informed Consent Statement: Not applicable.

Data Availability Statement: Data are contained within the article.

Acknowledgments: Authors acknowledge the financial support from the Talent Project of Yulin Normal University (No. G2020ZK14).

Conflicts of Interest: The authors declare no conflict of interest.

References

1. Rather, R.A.; Mehta, A.; Lu, Y.; Valant, M.; Fang, M.; Liu, W. Influence of exposed facets, morphology and hetero-interfaces of BiVO$_4$ on photocatalytic water oxidation: A review. *Int. J. Hydrog. Energy* **2021**, *46*, 21866–21888. [CrossRef]
2. Zahid, A.H.; Han, Q. A review on the preparation, microstructure, and photocatalytic performance of Bi$_2$O$_3$ in polymorphs. *Nanoscale* **2021**, *13*, 17687–17724. [CrossRef]

3. Zhu, Z.; Wan, S.; Zhao, Y.; Qin, Y.; Ge, X.; Zhong, Q.; Bu, Y. Recent progress in Bi_2WO_6-based photocatalysts for clean energy and environmental remediation: Competitiveness, challenges, and future perspectives. *Nano Sel.* **2021**, *2*, 187–215. [CrossRef]
4. Yu, L.; Zhang, X.; Li, G.; Cao, Y.; Shao, Y.; Li, D. Highly efficient $Bi_2O_2CO_3$/BiOCl photocatalyst based on heterojunction with enhanced dye-sensitization under visible light. *Appl. Catal. B Environ.* **2016**, *187*, 301–309. [CrossRef]
5. Yin, G.; Jia, Y.; Lin, Y.; Zhang, C.C.; Zhu, Z.; Ma, Y. A review on the hierarchical Bi_2MoO_6 nanostructures for photocatalysis application. *New J. Chem.* **2021**, *46*, 906–918. [CrossRef]
6. Kumar, R.; Raizada, P.; Khan, A.A.P.; Nguyen, V.H.; Van Le, Q.; Ghotekar, S.; Selvasembian, R.; Gandhi, V.; Singh, A.; Singh, P. Recent progress in emerging $BiPO_4$-based photocatalysts: Synthesis, properties, modification strategies, and photocatalytic applications. *J. Mater. Sci. Technol.* **2022**, *108*, 208–225. [CrossRef]
7. Theerthagiri, J.; Chandrasekaran, S.; Salla, S.; Elakkiya, V.; Senthil, R.A.; Nithyadharseni, P.; Maiyalagan, T.; Micheal, K.; Ayeshamariam, A.; Arasu, M.V.; et al. Recent developments of metal oxide based heterostructures for photocatalytic applications towards environmental remediation. *J. Solid State Chem.* **2018**, *267*, 35–52. [CrossRef]
8. Zheng, Y.; Duan, F.; Chen, M.; Xie, Y. Synthetic $Bi_2O_2CO_3$ nanostructures: Novel photocatalyst with controlled special surface exposed. *J. Mol. Catal. A Chem.* **2010**, *317*, 34–40. [CrossRef]
9. Orooji, Y.; Tanhaei, B.; Ayati, A.; Tabrizi, S.H.; Alizadeh, M.; Bamoharram, F.F.; Karimi, F.; Salmanpour, S.; Rouhi, J.; Afshar, S.; et al. Heterogeneous UV-Switchable Au nanoparticles decorated tungstophosphoric acid/TiO_2 for efficient photocatalytic degradation process. *Chemosphere* **2021**, *281*, 130795. [CrossRef]
10. Li, L.; Zhang, Q.; Wang, X.; Zhang, J.; Gu, H.; Dai, W.L. Au Nanoparticles Embedded in Carbon Self-Doping g-C_3N_4: Facile Photodeposition Method for Superior Photocatalytic H_2 Evolution. *J. Phys. Chem. C* **2021**, *125*, 10964–10973. [CrossRef]
11. Zhang, X.; Yang, P. Pt nanoparticles embedded spine-like g-C_3N_4 nanostructures with superior photocatalytic activity for H_2 generation and CO_2 reduction. *Nanotechnology* **2021**, *32*, 175401. [CrossRef] [PubMed]
12. Guo, Z.; Zhao, Y.; Shi, H.; Yuan, X.; Zhen, W.; He, L.; Che, H.; Xue, C.; Mu, J. $MoSe_2$/g-C_3N_4 heterojunction coupled with Pt nanoparticles for enhanced photocatalytic hydrogen evolution. *J. Phys. Chem. Solids* **2021**, *156*, 110137. [CrossRef]
13. Álvarez-Prada, I.; Peral, D.; Song, M.; Muñoz, J.; Romero, N.; Escriche, L.; Acharjya, A.; Thomas, A.; Schomäcker, R.; Schwarze, M.; et al. Ruthenium nanoparticles supported on carbon-based nanoallotropes as co-catalyst to enhance the photocatalytic hydrogen evolution activity of carbon nitride. *Renew. Energy* **2021**, *168*, 668–675. [CrossRef]
14. Xu, W.; Li, X.; Peng, C.; Yang, G.; Cao, Y.; Wang, H.; Peng, F.; Yu, H. One-pot synthesis of Ru/Nb_2O_5@Nb_2C ternary photocatalysts for water splitting by harnessing hydrothermal redox reactions. *Appl. Catal. B Environ.* **2022**, *303*, 120910. [CrossRef]
15. Ren, T.; Dang, Y.; Xiao, Y.; Hu, Q.; Deng, D.; Chen, J.; He, P. Depositing Ag nanoparticles on g-C_3N_4 by facile silver mirror reaction for enhanced photocatalytic hydrogen production. *Inorg. Chem. Commun.* **2021**, *123*, 108367. [CrossRef]
16. Li, Y.; Wang, H.; Xie, J.; Hou, J.; Song, X.; Dionysiou, D.D. Bi_2WO_6-TiO_2/starch composite films with Ag nanoparticle irradiated by γ-ray used for the visible light photocatalytic degradation of ethylene. *Chem. Eng. J.* **2021**, *421*, 129986. [CrossRef]
17. Ma, C.; Yang, Z.; Wang, W.; Zhang, M.; Hao, X.; Zhu, S.; Chen, S. Fabrication of Ag-Cu_2O/PANI nanocomposites for visible-light photocatalysis triggering super antibacterial activity. *J. Mater. Chem. C* **2020**, *8*, 2888–2898. [CrossRef]
18. Liang, H.; Li, T.; Zhang, J.; Zhou, D.; Hu, C.; An, X.; Liu, R.; Liu, H. 3-D hierarchical Ag/ZnO@CF for synergistically removing phenol and Cr (VI): Heterogeneous vs. homogeneous photocatalysis. *J. Colloid Interface Sci.* **2020**, *558*, 85–94. [CrossRef]
19. Jiao, Z.; Zhang, J.; Liu, Z.; Ma, Z. Ag/AgCl/Ag_2MoO_4 composites for visible-light-driven photocatalysis. *J. Photochem. Photobiol. A Chem.* **2019**, *371*, 67–75. [CrossRef]
20. Kheirabadi, M.; Samadi, M.; Asadian, E.; Zhou, Y.; Dong, C.; Zhang, J.; Moshfegh, A.Z. Well-designed Ag/ZnO/3D graphene structure for dye removal: Adsorption, photocatalysis and physical separation capabilities. *J. Colloid Interface Sci.* **2019**, *537*, 66–78. [CrossRef]
21. de Almeida, G.C.; Mohallem ND, S.; Viana, M.M. Ag/GO/TiO_2 nanocomposites: The role of the interfacial charge transfer for application in photocatalysis. *Nanotechnology* **2021**, *33*, 035710. [CrossRef] [PubMed]
22. Amiri, M.; Dashtian, K.; Ghaedi, M.; Mosleh, S.; Jannesar, R. Bi_2WO_6/Ag_3PO_4-Ag Z-scheme heterojunction as a new plasmonic visible-light-driven photocatalyst: Performance evaluation and mechanism study. *New J. Chem.* **2019**, *43*, 1275–1284. [CrossRef]
23. Chen, Y.; Huang, W.; He, D.; Situ, Y.; Huang, H. Construction of heterostructured g-C_3N_4/Ag/TiO_2 microspheres with enhanced photocatalysis performance under visible-light irradiation. *ACS Appl. Mater. Interfaces* **2014**, *6*, 14405–14414. [CrossRef] [PubMed]
24. Wu, Z.; Zeng, D.; Liu, X.; Yu, C.; Yang, K.; Liu, M. Hierarchical δ-Bi_2O_3/$Bi_2O_2CO_3$ composite microspheres: Phase transformation fabrication, characterization and high photocatalytic performance. *Res. Chem. Intermed.* **2018**, *44*, 5995–6010. [CrossRef]
25. Guo, G.; Yan, H. Zn-doped $Bi_2O_2CO_3$: Synthesis, characterization and photocatalytic properties. *Chem. Phys.* **2020**, *538*, 110920. [CrossRef]
26. Taylor, P.; Sunder, S.; Lopata, V.J. Structure, spectra, and stability of solid bismuth carbonates. *Can. J. Chem.* **1984**, *62*, 2863–2873. [CrossRef]
27. Yu, C.; Zhou, W.; Zhu, L.; Li, G.; Yang, K.; Jin, R. Integrating plasmonic Au nanorods with dendritic like α-Bi_2O_3/$Bi_2O_2CO_3$ heterostructures for superior visible-light-driven photocatalysis. *Appl. Catal. B Environ.* **2016**, *184*, 1–11. [CrossRef]
28. Ge, L.; Han, C.; Liu, J.; Li, Y. Enhanced visible light photocatalytic activity of novel polymeric g-C_3N_4 loaded with Ag nanoparticles. *Appl. Catal. A Gen.* **2011**, *409*, 215–222. [CrossRef]
29. Liu, Y.; Ouyang, S.; Guo, W.; Zong, H.; Cui, X.; Jin, Z.; Yang, G. Ultrafast one-step synthesis of N and Ti^{3+} codoped TiO_2 nanosheets via energetic material deflagration. *Nano Res.* **2018**, *11*, 4735–4743. [CrossRef]

30. Ren, J.; Wang, W.; Sun, S.; Zhang, L.; Chang, J. Enhanced photocatalytic activity of Bi_2WO_6 loaded with Ag nanoparticles under visible light irradiation. *Appl. Catal. B: Environ.* **2009**, *92*, 50–55. [CrossRef]
31. Guan, M.L.; Ma, D.K.; Hu, S.W.; Chen, Y.J.; Huang, S.M. From hollow olive-shaped $BiVO_4$ to n-p core-shell $BiVO_4@Bi_2O_3$ microspheres: Controlled synthesis and enhanced visible-light-responsive photocatalytic properties. *Inorg. Chem.* **2011**, *50*, 800–805. [CrossRef] [PubMed]
32. Shi, J.; Li, J.; Huang, X.; Tan, Y. Synthesis and enhanced photocatalytic activity of regularly shaped Cu_2O nanowire polyhedra. *Nano Res.* **2011**, *4*, 448–459. [CrossRef]
33. Ong, S.A.; Min, O.M.; Ho, L.N.; Wong, Y.S. Comparative study on photocatalytic degradation of mono azo dye acid orange 7 and methyl orange under solar light irradiation. *Water Air Soil Pollut.* **2012**, *223*, 5483–5493. [CrossRef]
34. Dong, G.; Ho, W.; Zhang, L. Photocatalytic NO removal on BiOI surface: The change from nonselective oxidation to selective oxidation. *Appl. Catal. B Environ.* **2015**, *168*, 490–496. [CrossRef]
35. Tun, P.P.; Wang, J.; Khaing, T.T.; Wu, X.; Zhang, G. Fabrication of functionalized plasmonic Ag loaded Bi_2O_3/montmorillonite nanocomposites for efficient photocatalytic removal of antibiotics and organic dyes. *J. Alloy. Compd.* **2020**, *818*, 152836. [CrossRef]
36. Majhi, D.; Mishra, A.K.; Das, K.; Bariki, R.; Mishra, B.G. Plasmonic Ag nanoparticle decorated $Bi_2O_3/CuBi_2O_4$ photocatalyst for expeditious degradation of 17α-ethinylestradiol and Cr(VI)reduction: Insight into electron transfer mechanism and enhanced photocatalytic activity. *Chem. Eng. J.* **2021**, *413*, 127506. [CrossRef]

Article

Luminescence Reduced Graphene Oxide Based Photothermal Purification of Seawater for Drinkable Purpose

Jin Huang [1,2], Zhen Chu [1,2], Christina Xing [3], Wenting Li [4], Zhongxin Liu [1,2,*] and Wei Chen [4,*]

1. Key Laboratory of Advanced Materials of Tropical Island Resources of Ministry of Education, School of Materials and Chemical Engineering, Hainan University, Haikou 570228, China; 68828813@hainanu.edu.cn (J.H.); 19085216210007@hainanu.edu.cn (Z.C.)
2. Hainan Provincial Key Lab of Fine Chem, School of Materials and Chemical Engineering, Hainan University, Haikou 570228, China
3. Department of Physics, The University of Texas at Arlington, Arlington, TX 76019-0059, USA; christina.xing@uta.edu
4. Key Laboratory of Functional Molecular Engineering of Guangdong Province, School of Chemistry and Chemical Engineering, South China University of Technology, Guangzhou 510641, China; wenting@scut.edu.cn
* Correspondence: liuzhongxin@hainanu.edu.cn (Z.L.); weichen@uta.edu (W.C.)

Abstract: Getting drinking water from seawater is a hope and long-term goal that has long been explored. Here, we report graphene-loaded nonwoven fabric membranes for seawater purification based on photothermal heating. The photothermal membrane of non-woven fabric loaded with graphene oxide has high light absorption and strong heating effect, and its evaporation rate about 5 times higher than that of non-woven fabric. Under the condition of light intensity of 1 kW m^{-2}, the evaporation rate can reach 1.33 kg m^{-2} h^{-1}. The results of cell activity test showed that the concentration of bacteria after photothermal membrane treatment decreased significantly. The photothermal membrane can be used for many times without greatly reducing the evaporation efficiency, which means that it is suitable for regional water purification and seawater desalination.

Keywords: graphene oxide nonwoven membrane; solar evaporation; air–water interface; water purification; photothermal conversion; cell viability; cytotoxicity

1. Introduction

The shortage of safe drinking water is one of the most severe global challenges facing human society. Desalination is one of the most cost-effective ways to increase drinking water supply. In recent years, the development of seawater desalination industry has made great progress. For example, as one of the most widely used desalination technologies, reverse osmosis (RO) has a recovery rate of about 50% and a consumption of only 2 kw·h·m^{-3}. However, due to the need for complex infrastructure and high energy consumption, this technology is still unavailable in most underdeveloped countries. In addition, with the increase of feed water salinity, the working pressure, scaling probability and fuel consumption of the reverse osmosis system increase significantly [1–4]. Therefore, the development of new green seawater desalination technology is particularly important. At present, solar powered desalination is becoming one of the most promising technologies to increase the supply of clean water, because of the abundance of solar energy and seawater and their negligible carbon footprint, especially in remote areas lacking electricity or infrastructure [5–8].

To promote the further development of direct solar desalination, great efforts have been made to improve the efficiency of solar thermal conversion, prolong the service life of equipment and reduce the cost of solar desalination [9–13]. Some studies show that the evaporator with optimized structure design can produce fresh water and collect the by-product salt in salt water at the same time, to achieve zero liquid discharge. To strengthen the relationship between water and energy and alleviate the energy shortage, various solar powered hydropower hydrogen production systems have also attracted people's attention. This system converts waste heat into electric energy, and then drives electrochemical decomposition of water, which not only improves the overall energy conversion efficiency, but also promotes the production of green fuel [14,15]. According to the mechanism of photothermal conversion, photothermal materials are mainly divided into three categories, such as plasma metal nanoparticles, semiconductors and carbon-based materials. Among them, carbon-based materials are ideal for solar evaporative power generation because of their excellent thermal properties, low cost and abundant source materials [16,17]. As an important carbon-based material, graphene oxide (Go) or reduced graphene oxide (rGO) has thin nanostructures, large surface area, low molar specific heat, stable mechanical strength and excellent light absorption covering the whole solar spectral range (250–2500 nm). Therefore, they are directly used as light absorbers or photothermal layers for various substrates (such as wood, sponge, polymer membrane and natural fiber) [18–20].

To better transfer heat to the water body that needs evaporation, good heat management performance and excellent light absorbing materials are two key factors. For the heat management strategy, the traditional self-floating evaporation configuration generally has large heat loss due to the direct contact of the light absorbing material with the water body. In contrast, the use of thermal insulator assisted evaporation device can minimize the heat and conduction loss of photothermal materials [21,22]. At the same time, interface heating can limit the absorbed heat energy to a small amount of interface water in the upper layer, to shorten the start-up time of steam generation and increase the water evaporation rate. Although some transport auxiliary evaporators have been reported, their evaporation performance is still hindered due to the lack of good design structure. Therefore, the evaporation material must have a highly porous structure and hydrophilic channels to promote the transmission of water and the escape of water vapor. Non-woven fabric is made of polyester fiber and polyester fiber (PET for short). It has the characteristics of antibacterial, alkali corrosion and strong hydrophilicity [23,24]. In addition, Yu and others think that the hydrogel polymer network can restrict the water mass in the molecular grid, effectively reduce the enthalpy of vaporization, which is beneficial to improve the evaporation performance of the self-floating gel evaporator [25–27]. However, the combination of nanofiber hydrogel and rGO as a feasible strategy to improve the evaporative performance of the transport assisted evaporator and the mechanism of controlling the interfacial evaporation need further study.

Here, we report an approach based on reduced graphene oxide (rGO)-loaded non-woven membrane for localizing evaporation, which is cost-effective and environmentally friendly. The hydrophilic non-woven transports water to the hot region by capillary forces. The rGO non-woven composite membrane has a superior mechanical stability. This method can be used repeatedly for many times. Under 1 sun intensity (1 kW m^{-2}), the evaporation rate of photothermal membrane reached 1.33 kg·m^{-2}·h^{-1}. In addition, we tested the cell viability on the original seawater and the purified water. The results show the purified water has no toxicity, while the seawater is toxic to cells at a certain concentration. This indicates this is an excellent method for purification of drinkable water from seawater. The simple design of the system, along with the low costs, makes it possible for practical applications as an outstanding solution to the long-term challenge of drinking water issues.

2. Experiment Section

2.1. Materials

Graphite powder is purchased from Tianjin Dengke Chemical Reagent Co, Ltd. (Tianjin, China), H_2SO_4, $KMnO_4$ and H_2O_2 are obtained from Aldrich, Deionized water, self-made in the laboratory, non-woven fabric is provided by Hainan Xinglong Nonwoven Company (Hainan, China).

2.2. Preparation of Graphene Oxide (Go)

GO is synthesized by the modified Hummers methods [28–30]. Graphite (1 g) are mixed in 23 mL H_2SO_4, and the mixture is stirred for 2 min in an ice bath at a temperature less than 20 °C. Then, add 3 g $KMnO_4$ and stir for 3 h. The ice bath is then removed, and the mixture is stirred at 35 °C for 8 h. Next, 100 mL water is dropped gently. It is then diluted with 50 mL water and 5 mL H_2O_2. The mixture is washed with HCl (5%) and deionized (DI) water and centrifugation, freeze-drying to obtain samples.

2.3. Preparation of rGO-Based Nonwoven Fabric Membrane (RGM)

First, 0.05 g graphite oxide be dispersed in 10 mL DI water by sonication for 1 h to make an aqueous dispersion. Cut a round non-woven fabric into graphene oxide solution, taking it out, and by centrifugation, removing the excess graphene oxide, and cycling it several times. Then, the non-woven membrane loaded with GO was placed in an oven at 180 °C and reacted for 5 h to obtain RGM photothermal membrane [31].

2.4. Characterization

The structural analysis of graphene oxide was characterized 31 by power X-ray diffraction using a Bruker D8-advance. The Raman spectra were obtained by a confocal laser Raman spectrometer. The size and morphology were obtained by scanning electron microscopy (SEM, Hitachi S-4800). The transmission electron microscope (TEM) was characterized by a JEOL JEM 2100. The real-time temperatures of the samples were measured by an IR thermography (HT-02, Xin Site China). The 980 nm IR laser with tunable powers was generated by a diode laser system (BWT, Beijing, China).

2.5. Cell Toxicity Studies by MTT (3-(4,5-Dimethylthiazol-2-yl)-2,5-diphenyltetrazolium Bromide) Assay

The cytotoxicity of the seawater and the purified water were evaluated by means of MTT assay on Hela Cells. Hela Cells-CCL-2 were purchased from ATCC American Type Culture Collection. Cells were cultured in high-glucose Dulbecco's Modified Eagle's Medium (H-DMEM) containing 10% FBS and 1% penicillin-streptomycin at 37 °C in a humidified environment containing 5% CO_2. Before the experiment, the cells were pre-cultured until confluence was reached. MTT assay is a colorimetric assay which is in widespread use for assaying cell proliferation and cytotoxicity [32]. First, 10,000 cells/well were seeded in 96 well plates and incubated for 24 h for the completion of cell attachment. Cells were divided into three groups: seawater, purified water and control (untreated) groups. Then, the plates were incubated for another 24 h to allow cells to uptake water. MTT assay was prepared by diluting 5 mg/mL stock solution with media by a factor of 10. Then, 100 µL of MTT assay was added to each well, replacing old media. The 96 well plates were then incubated for 3 h at 37 °C under humidified atmosphere. After incubation for 3 h, MTT solution was removed and 100 µL DMSO (dimethyl sulfoxide) was added to solubilize formazan crystal. The formazan crystal becomes purple-colored with DMSO dissolution. The viability of cells was directly dependent on the absorption of formazan solution.

After adding water, 96 well plates be incubated for 3 h, then 0.5 mg/mL MTT solution was added to each well and incubated for 3 h. Yellow colored MTT converted into purple colored formazan crystal. Formazan is insoluble in aqueous solution. Formazan is solubilized by adding 100 Ul DMSO to each well. The absorbance of formazan is directly proportional live cell counts and can be employed to present a relative cell viability as compared to control. The absorbance of Formazan solution was measured using a multiskan FC microplate photometer (Fisher Scientific, Hampton, NH, USA) at 540 nm.

Cell viability was calculated as follows:

$$\text{Cell viability} = \frac{\text{The absorbance of the treatment group}}{\text{The absorbance of the control group}} \times 100\%$$

3. Results and Discussion

3.1. Preparation of the Graphene Oxide-Loaded Hydrophilic Nonwoven

Figure 1 shows the transmission electron microscope (TEM) images of the prepared GO with average size of ~100 nm, in which the GO exhibits lamellar structure with some laminations slightly stacked. There is a slight sliding between the graphene oxide layered structures. The magnified TEM image on the right shows that the microscopic size of the entire sheet is nanoscale.

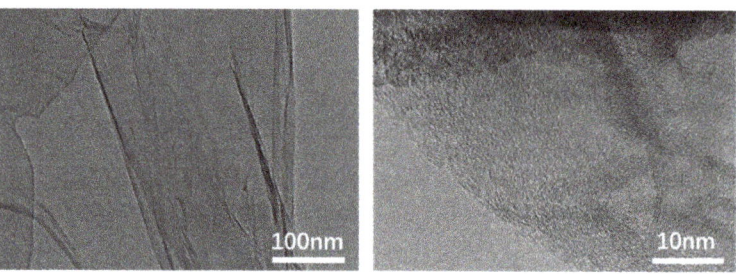

Figure 1. TEM image of GO, On the right is a HRTEM image.

Figure 2 is the Raman spectrum of GO. The D peak and the G peak are Raman characteristic peaks of carbon atom crystals, respectively, and their diffraction peaks are centered at 1300 cm^{-1} and 1580 cm^{-1}, respectively. D peak indicates some defects of the carbon atom lattice, and the G peak indicates the in-plane stretching vibration state of the sp^2 hybridization of carbon atoms. The D peak and G peak of GO independently developed by the laboratory are at 1359 cm^{-1} and 1615 cm^{-1}, respectively. I_D/I_G is the intensity ratio of D peak to G peak. This ratio can refer to the intensity relationship between the two peaks. In the Figure 2, I_D/I_G is about 0.8, which is less than the generally reported literature value [33,34], indicating that the properties of the synthesized GO are closer to the characteristics of graphene oxide, the fewer the defects on its surface and the higher the chemical stability, which will help to improve the stability and service life of the photothermal membrane.

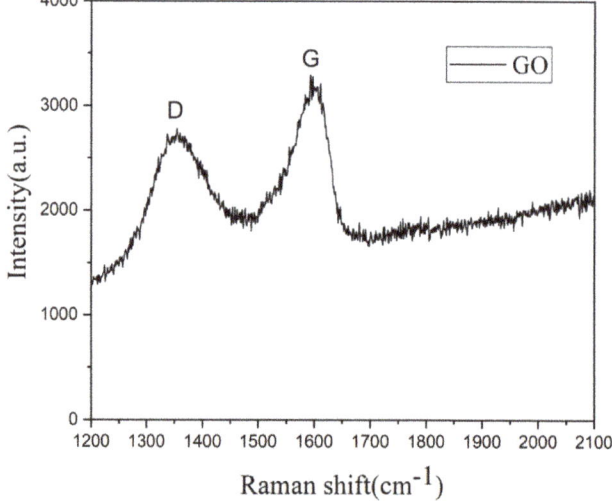

Figure 2. Raman spectra of the prepared GO.

As shown in Figure 3, there is a sharp diffraction peak at 9.96°, slightly lower than the characteristic peak of standard GO centered at about 11° [5,35]. This may be due to the intercalation of H$^+$ in the go layer during the stripping preparation of GO, which increases the distance between the crystal planes of GO. This expansion layer can aggravate the exposure rate of the edge and improve the overall light absorption effect. We have found through experiments that after the ice bath is removed, the normal temperature stage is about 30 °C. The longer the reaction time, the better the oxidation.

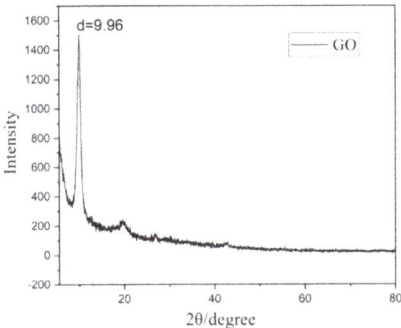

Figure 3. XRD fluorescence spectra of GO.

In the subsequent EDS test (Figure 4), the oxygen element on the surface of GO decreased significantly after thermal reduction. The molar ratio of carbon to oxygen decreased from 0.58:1 to 0.41:1, which indicates that thermal reduction removes some oxygen-containing functional groups of GO, which also makes RGM show weak hydrophilicity in subsequent tests.

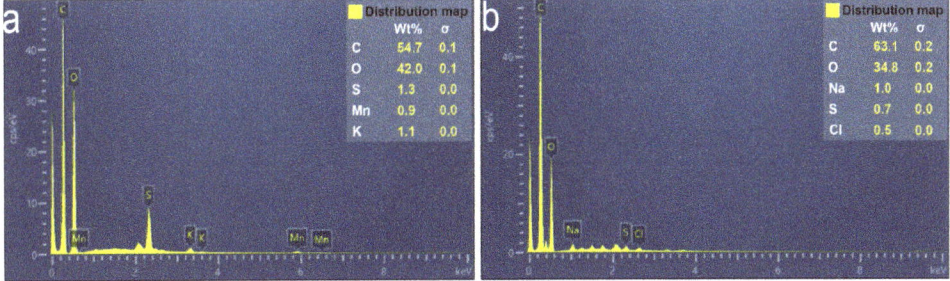

Figure 4. EDS spectra of GO (**a**) and rGO (**b**).

Figure 5a,b shows that the luminescence spectrum of rGO is a wide peak, that is, the luminescence wavelength is in a large range, and its luminescence depends on the excitation energy. Its emission spectrum and excitation spectrum are symmetrical, which is in line with the luminescence characteristics of organic fluorescent molecules. Figure 5c is the absorption spectrum of rGO dispersion sample tested. In the Figure 5c, the peak of 227 nm comes from the π–π* energy level transition in the C=C bond of sp^2 hybrid, which reflects the increase of sp^2 structure order in go, because part of sp^2 hybrid structure has been restored during the reaction. In addition, there is a weak shoulder peak near 320 nm, corresponding to the electron n–π* transition in C=O, which proves that the oxidation degree of rGO is high [36]. FTIR can show the functional groups on the rGO surface in more detail. It can be seen from the Figure 5d that the rGO surface includes many oxygen-containing groups, including C=O, C-O-C, COOH and O-H, and -C=C-, also has a strong vibration peak, which means that there are many sp^2 clusters inside the rGO.

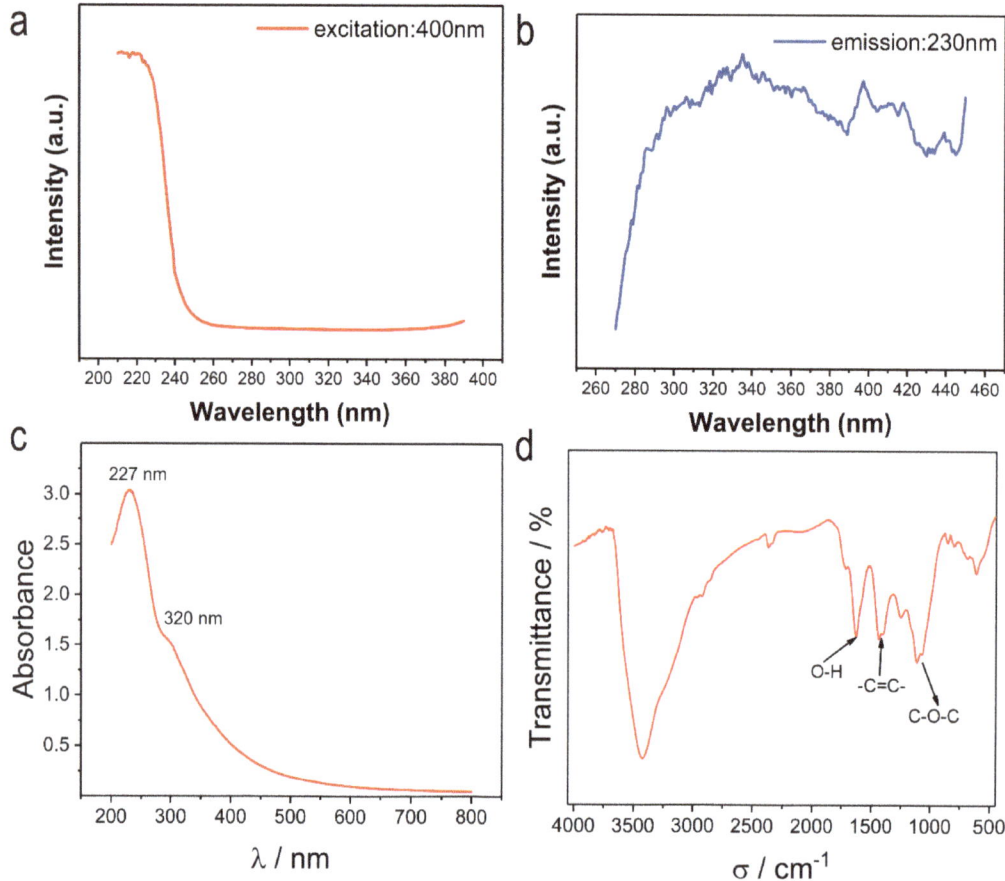

Figure 5. (**a**,**b**) Excitation (Ex: 400 nm) and emission (Em: 230 nm) spectra of rGO. (**c**) UV-Vis absorption spectrum of rGO. (**d**) FTIR spectra of rGO.

The above results show that when the emission wavelength is 340 nm, there is an obvious excitation peak at 230 nm, which corresponds to the results of π–π* energy level transition in C=C bond and electron transition in C=O of sp^2 hybrid and corresponds to the energy level shown in the absorption spectrum. In addition, the non-radiative transition of electron hole pair in a small amount of oxygen-containing functional group defects that may be contained in rGO can also excite rGO luminescence.

Figure 6 shows the actual picture of the RGM material in two states. The left picture is the actual photo in the dry state. It can be seen that the rGO is loaded on the non-woven composite membrane in the whole state is black. However, white inclusions can be observed due to the influence of the white non-woven fabric on the color of the membrane. When the RGM is in a wet state, the color becomes darker. The main reason for this phenomenon is that there is a very thin water layer on the surface of the photothermal membrane in the wet state. When the light shines on the surface of the wet photothermal membrane, the light path will be refracted due to the existence of the water layer, that is, the light will shift inward. The light that should have escaped is absorbed, and, finally, the color of the wet photothermal Membrane is deepened [37].

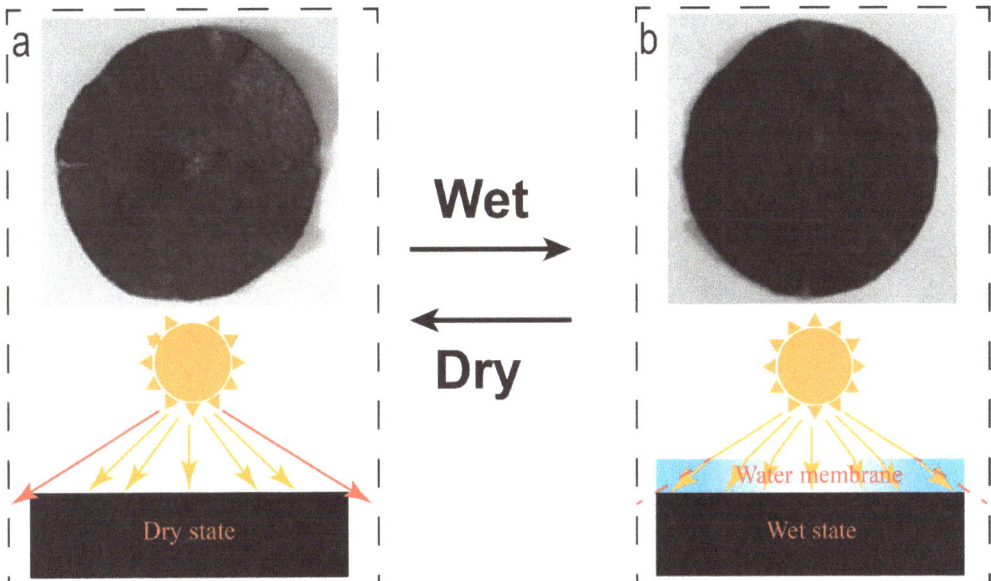

Figure 6. (**a**,**b**) Physical picture of color difference of photothermal membrane under dry and wet conditions (upper part) and corresponding schematic diagram (lower part, the yellow lines indicate absorbed light and red lines indicate escaping light).

According to the ultraviolet spectrum (Figure 7), the absorption of RGM has a significant change, and greatly enhances its ability to absorb light. The absorbance of RGM composite is about 93% in the wavelength range of 500–2500 nm, which is much higher than the light absorption of non-woven fabrics (8%). The introduction of rGO greatly improved the light absorption performance of the composite material.

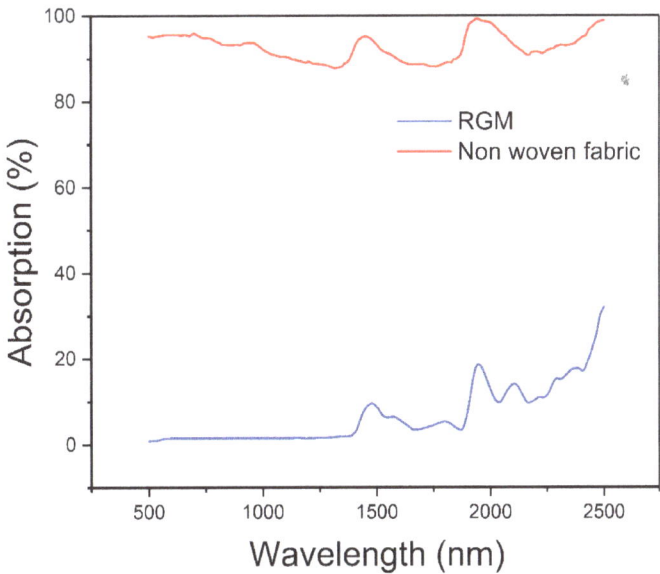

Figure 7. Absorption of RGM and non-woven fabric in the range of 500–2500 nm.

The SEM image of the pristine non-woven fabric in Figure 8a shows that the fibers were interspersed in a random network structure. The non-woven fibers are smooth and free of impurities, and the fiber bundle can transport water through capillary action. Loading rGO improves the photothermal conversion efficiency. Magnification shows the rGO sheets are well coated with the fabric. (Figure 8b). Compared with the initial non-woven fabric, the surface of the photothermal membrane after rGO loading and heat treatment is rougher. This rough surface can make the process of multiple reflection of light on its surface, prolong the optical path and, finally, improve the overall light absorption. However, it should be noted that the rGO layer does not block the pores between non-woven fabrics, which are conducive to the escape of steam, which is conducive to the evaporation process of photothermal water.

Figure 8. SEM images of the graphene oxide-loaded nonwoven fabric show in different amplified times: (**a1,a2**) pristine non-woven fabric of different resolutions; (**b1,b2**) modified non-woven fabric with a 0.5% graphene oxide of different resolutions.

3.2. Light-Induced Evaporation Enhancement via the rGO-Loaded Hydrophilic Nonwoven

To demonstrate the high light absorption of the RGM, water evaporation was performed in four different forms under different light intensities. The water body naturally evaporates, the black non-woven fabric is at the bottom, white, black non-woven fabric on the surface. Through the underlying non-woven core, it looks like the pump to absorb the upper layer of water evaporation.

We can observe that when irradiation with the 1 sun light source for 30 min, the black non-woven fabric is fixed at the bottom of the water container, the evaporation rate is not significantly increased. When the black non-woven fabric is on the top, and the amount of water after evaporation reaches about 0.84 g, four times the black non-woven fabric at the bottom, and two times the white non-woven fabric on the surface (Figure 9).

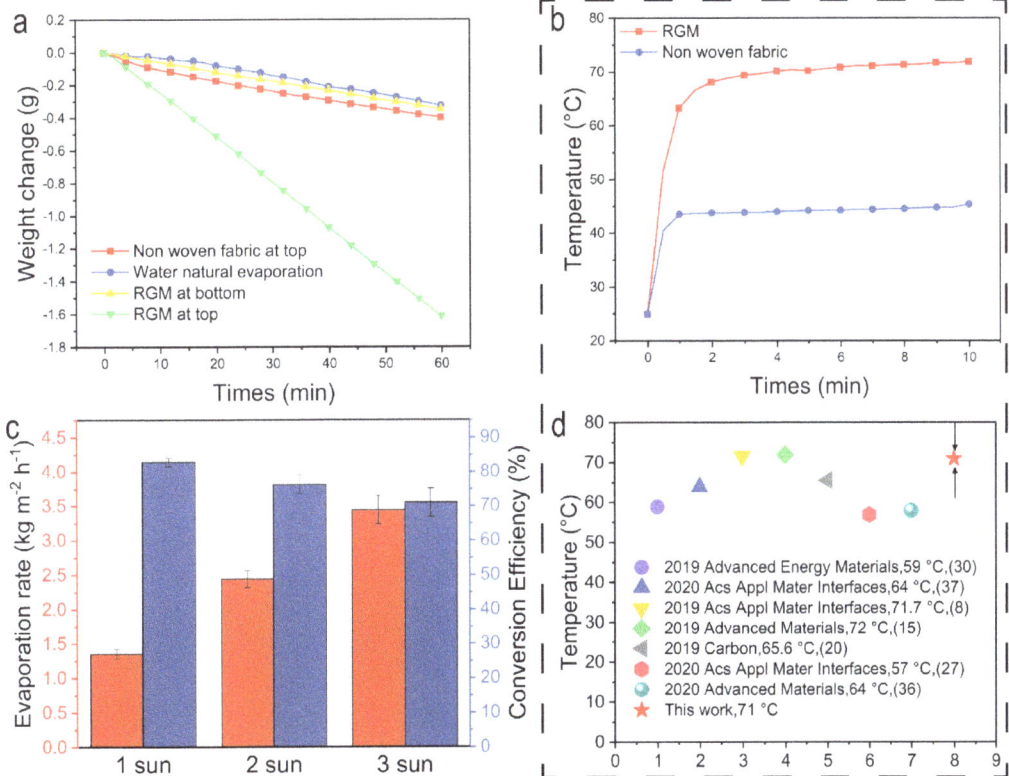

Figure 9. (a) Weight change diagram under four evaporation modes (natural evaporation of pure water, evaporation of non-woven fabric at the top, evaporation of RGM at the bottom and evaporation of RGM at the top). (b) Temperature rises curve of non-woven fabric and RGM under 1 sun condition. (c) Evaporation rate and photothermal conversion efficiency under different light intensities. (d) Comparison of heating efficiency in evaporator previously reported (reference numbers are in brackets).

The results in Figure 9a show that there is little difference between the evaporation rates of non-woven fabric at the top, pure water natural evaporation and RGM at the bottom under one light intensity (1 kW·m^{-2}). Natural evaporation of pure water and bottom evaporation of RGM because light acts directly on many water bodies, although the RGM photothermal membrane at the bottom can provide a certain amount of heat for the local water body at the bottom, it cannot heat the overall water volume, so the water evaporation rate under these two evaporation modes is relatively low. Due to its low light absorption (~8%), the non-woven membrane will have obvious reflection and transmission effects on its surface, which makes the membrane surface very limited to convert light into heat. Therefore, even if the non-woven fabric evaporates at the interface, its evaporation flux will not be significantly improved compared with the first two. This is also illustrated by the temperature rise curves of non-woven fabrics and RGM (Figure 9b). The above results show that RGM has better heating performance than previous studies (Figure 9d). Among them, the surface of RGM can be heated up to 71 °C under one light intensity, which is much higher than the heating effect of non-woven fabric under the same conditions (43 °C). In the experiment, the evaporation rate can be estimated simply by the mass loss. The evaporation rate of the above RGM increased with the increase of the radiant light intensity, and the evaporation rate increased from 1.33 kg·m^{-2}·h^{-1} to

3.42 kg·m^{-2}·h^{-1} (Figure 9c), while the evaporation rate of the other three types is only about 0.28 kg·m^{-2}·h^{-1}.

The following equation calculated the solar energy conversion efficiency:

$$\eta_{eva} = \frac{\dot{m}h_{LV}}{q}$$

where η_{eva} is the evaporation efficiency, $\dot{m}$ represents the stable evaporation rates (kg·m^{-2}·h^{-1}), h_{LV} is the total enthalpy of sensible heat and the liquid-vapor phase change (kJ·kg^{-1}) and q represents the power density of laser irradiation (kW·m^{-2}). In our results, the evaporation efficiency is 50.4%. From Figure 9c, it can be shown that the relationship between evaporation efficiency and evaporation rate under different light density of laser indifferent form [38].

It should be noted that when the light intensity increases from one to three, the evaporation rate and photothermal conversion efficiency will not increase in equal proportion. On the contrary, with the increase of light intensity, the increase proportion of evaporation rate decreases. This may be because RGM shows a weak hydrophilic effect (Figure 10), so RGM will not show strong capillarity and water transport capacity like other super hydrophilic materials [39–41]. It is this effect that will make the water content on the membrane surface relatively less. Therefore, with the increase of light intensity, the energy obtained by the evaporation system will increase, but the water content on the membrane surface is not enough to support the high-energy state of the system, which will reduce the photothermal conversion efficiency.

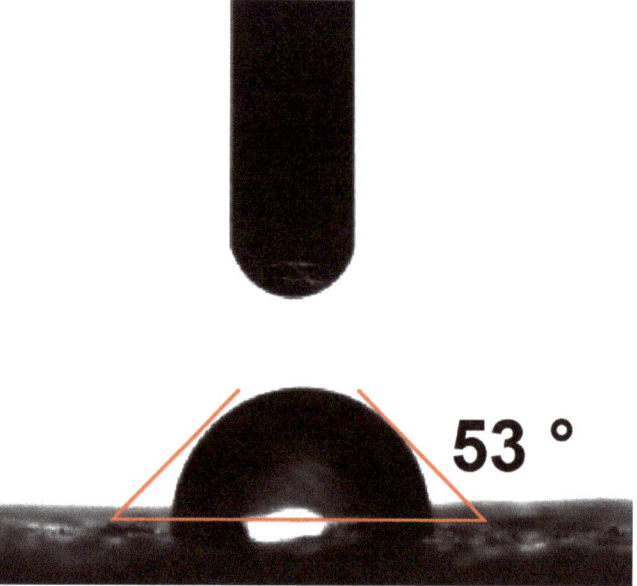

Figure 10. Contact angle test of RGM.

To further verify the stability of the rGO loaded on the nonwoven fabric, we sonicated the rGO nonwoven composite membrane for 30 min. From Figure 11, we can see that there is a slight loss of rGO on the beaker after ultrasound, but the change is not significant, which indicates that the stability of rGO loaded on the nonwoven fabric is relatively excellent.

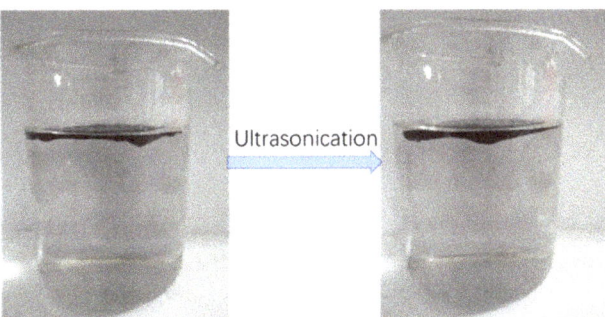

Figure 11. The stability comparison of RGM which are sonicated for 30 min.

To verify the stability of rGO-loaded nonwoven fabric, we tested the evaporation performance of RGM 10 times under the irradiation of 1 sun light intensity. As shown in Figure 12, we can see that the points in the Figure 12 are unevenly distributed, not in a linear trend. By linear fitting, we can see that the evaporation performance of the entire process material is like that. Near the closing line, it indicates the evaporation mechanical stability of the entire composite membrane is better. The evaporation performance does not increase with amounts of evaporations but has a wide range.

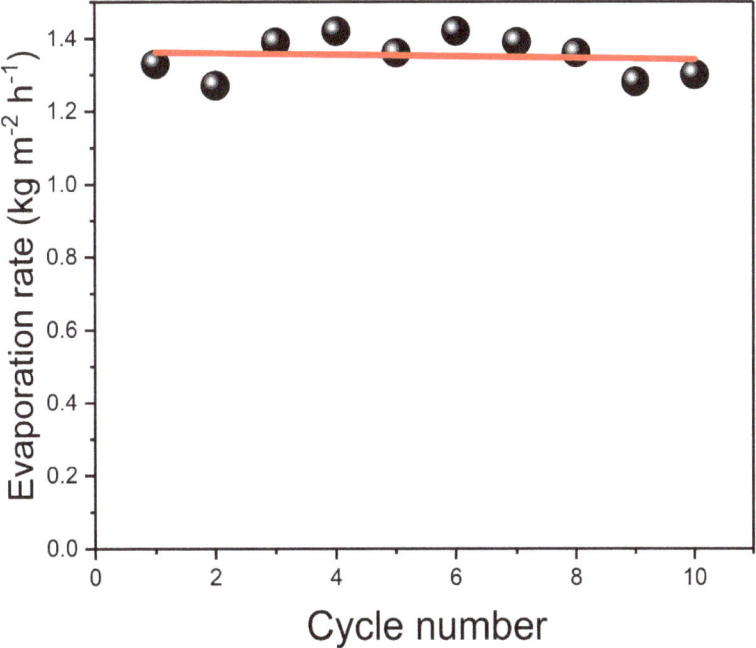

Figure 12. The evaporation rate of RGM on the surface as the function of cycle number.

3.3. Outdoor Simplified Device Evaporation Experiment

The solar water purification model was simplified, and the experiment was placed on the roof of the Li Yunqiang Experimental building of Hainan University. The rGO-loaded nonwoven fabric has a length of 15 cm, a width of 12 cm and a thickness of 2 mm. It was placed on a simple treated heat-insulating device, which is a foam strip connected by a wire. It was connected by a water-absorbent non-woven fabric in the middle of each foam strip, so that the water at the bottom can be transported to the top through the hydrophilicity of the nonwoven fabric for solar evaporation. The evaporated water condensed into water on a transparent round plastic lid, and then fell by gravity into our collection container. The advantage of this type of evaporation device is simple, and such a set of water evaporation and purification devices can be built at any time in remote areas we recorded the optical density of sunlight from 12:00 to 18:00 outdoors, and the average heat flux is 0.7 kW·m^{-2}. In the case of solar radiation, we recorded the surface temperature, air temperature and water volume of the samples separately. From the results in Figure 13, the surface temperature of the sample was obviously changed. When the sunlight irradiates it for about 30 min, the surface temperature of the sample reached 53 °C. Subsequent temperature is somewhat reduced, which may cause a portion of the heat to be removed from the closed system convection, resulting in a decrease in the surface temperature of the sample. Simultaneously, we can also observe from our chart that our water purification is increasing with time. After 6 h, the water volume reached 59 mL. This simple and efficient water purification material combined with the device is suitable for applications in remote areas. It is also worth noting that the bacteria content in the treated seawater is significantly lower than in the original seawater, which indicates the go photothermal membrane has excellent sterilization ability. This is critical for practical applications.

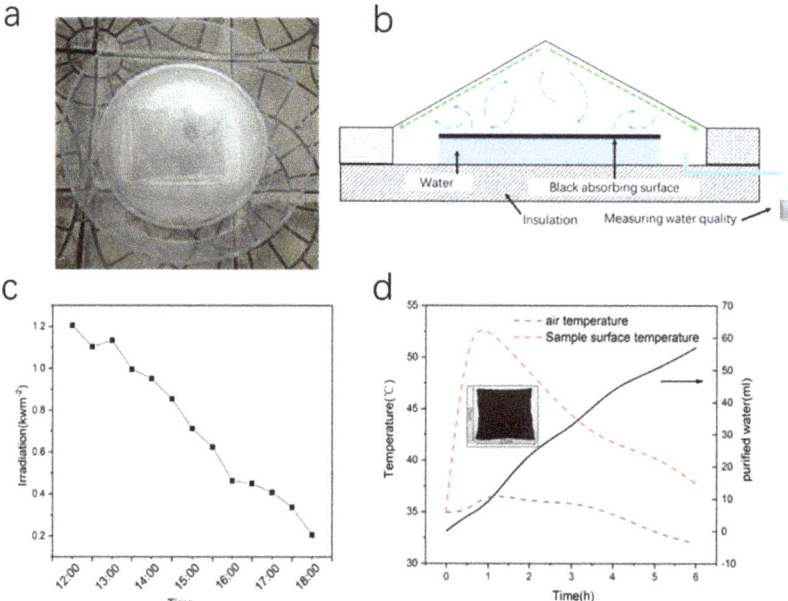

Figure 13. (a) A photograph of a large-scale rGO-loaded nonwoven fabric membrane device. (b) A schematic diagram of the device collecting water by evaporation. (c) Solar radiation recorded over time on a sunny day from 12:00 to 18:00. (d) Temperature change with time. The red dotted line represents the surface temperature of the sample, the purple dotted line represents air temperature, the black dotted line represents water evaporation quality changes with time, the illustration shows the size of the sample.

To further prove that the purified water is drinkable, the cytotoxicity of the seawater and the purified water were evaluatedby means of MTT assay on Hela Cells. The results in Figure 14 indicate the purified water have no cell toxicities, while the seawater is highly toxic at high concentrations. This is strong evidence to prove that the method designed is viable for drinking water purification from seawater.

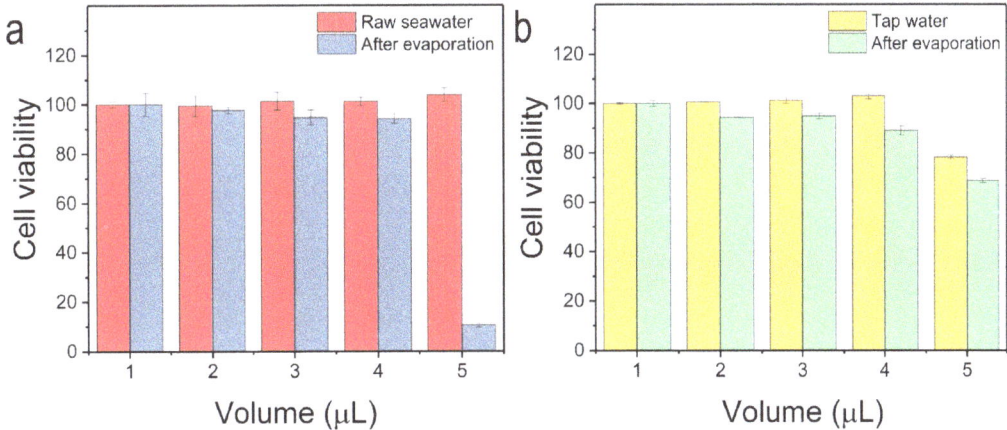

Figure 14. (**a**,**b**) Cell activity before and after seawater and tap water treatment.

To verify the stability of the materials for seawater desalination, we prepared a 20% sodium chloride solution as a simulating seawater. We conducted a five-day evaporation experiment outdoors. From Figure 15, we can see that the desalination amount of seawater is relatively stable. Figure 15, after evaporation, the surface of the sample leaves a large area of sodium chloride crystals, and the crystals can be washed off with water, which means it can be used repeatedly.

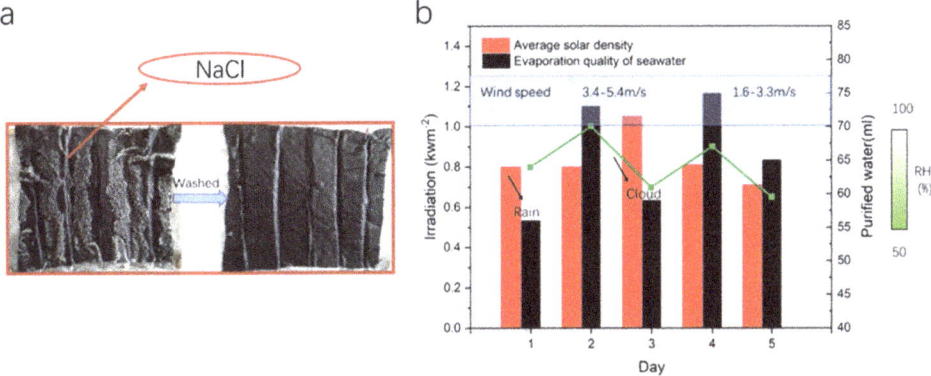

Figure 15. (**a**) A photograph of the surface of the sample after simulating seawater evaporation. (**b**) Cycle days of seawater evaporation. The red column represents the average optical density of sunlight. The black column represents the evaporation quality of water in the afternoon. Green dotted line diagram represents daily air humidity. Blue wireframe represents wind speed.

4. Conclusions

In summary, the high-photothermal conversion composite material can be produced in a large scale with a low-cost by a physical evaporation phase inversion method in which rGO was loaded into the nonwoven fabric as a main photothermal conversion material. The non-woven fabric is rGO composite membrane, and its mechanical properties are relatively proficient. It can be used repeatedly for seawater purification after sonication based on photothermal heating. The rGO-loaded nonwoven fabric has a high permeability and an evaporation rate of five–six times higher than conventional evaporation. For the solar density of about 726 kw/m^2, our method can purify 2.35 L of seawater to be evaporated into water vapor for purification. Cell viability studies indicate the purified water has no cell toxicities, while the seawater is highly toxic at high concentrations. Our method designed is viable for drinking water purification from seawater.

Author Contributions: Conceptualization, J.H. and Z.C.; methodology, Z.L.; software, Z.C.; validation, J.H., Z.C. and Z.L.; formal analysis, Z.C.; investigation, J.H. and W.L.; resources, Z.L.; data curation, Z.C.; writing—original draft preparation, Z.L.; writing—review and editing, Z.L. and W.C.; visualization, Z.L.; supervision, Z.L. and W.C.; project administration, Z.L. and W.C.; funding acquisition, Z.L. and W.C.; Editing—C.X. All authors have read and agreed to the published version of the manuscript.

Funding: This research was funded by High-level talents project of Hainan Basic and Applied Basic Research Program grant number 2019RC141 and partially by UT Arlington.

Institutional Review Board Statement: This study was approved by the institutional review committee of our Universities.

Informed Consent Statement: Informed consent was obtained from all subjects involved in the study.

Acknowledgments: W.C. would like to thank the support from Solgro Inc., Guangxi Jialouyuan Medical Inc. and UT Arlington distinguished award.

Conflicts of Interest: The authors declare no conflict of interest.

References

1. Elimelech, M.; Phillip, W.A. The Future of Seawater Desalination: Energy, Technology, and the Environment. *Science* **2011**, *333*, 712–717. [CrossRef] [PubMed]
2. Zodrow, K.R.; Li, Q.; Buono, R.M.; Chen, W.; Daigger, G.; Dueñas-Osorio, L.; Elimelech, M.; Huang, X.; Jiang, G.; Kim, J.H.; et al. Advanced Materials, Technologies, and Complex Systems Analyses: Emerging Opportunities to Enhance Urban Water Security. *Environ. Sci. Technol.* **2017**, *51*, 10274–10281. [CrossRef] [PubMed]
3. Finnerty, C.; Zhang, L.; Sedlak, D.L.; Nelson, K.L.; Mi, B. Synthetic Graphene Oxide Leaf for Solar Desalination with Zero Liquid Discharge. *Environ. Sci. Technol.* **2017**, *51*, 11701–11709. [CrossRef] [PubMed]
4. Tong, T.; Elimelech, M. The Global Rise of Zero Liquid Discharge for Wastewater Management: Drivers, Technologies, and Future Directions. *Environ. Sci. Technol.* **2016**, *50*, 6846–6855. [CrossRef]
5. Li, X.; Xu, W.; Tang, M.; Zhou, L.; Zhu, B.; Zhu, S.; Zhu, J. Graphene Oxide-Based Efficient and Scalable Solar Desalination under One Sun with a Confined 2d Water Path. *Proc. Natl. Acad. Sci. USA* **2016**, *113*, 13953–13958. [CrossRef]
6. Liu, Y.; Lou, J.; Ni, M.; Song, C.; Wu, J.; Dasgupta, N.P.; Tao, P.; Shang, W.; Deng, T. Bioinspired Bifunctional Membrane for Efficient Clean Water Generation. *ACS Appl. Mater. Interfaces* **2015**, *8*, 772–779. [CrossRef]
7. Tao, P.; Ni, G.; Song, C.; Shang, W.; Wu, J.; Zhu, J.; Chen, G.; Deng, T. Solar-driven interfacial evaporation. *Nat. Energy* **2018**, *3*, 1031–1041. [CrossRef]
8. Fang, Q.; Li, T.; Chen, Z.; Lin, H.; Wang, P.; Liu, F. Full Biomass-Derived Solar Stills for Robust and Stable Evaporation To Collect Clean Water from Various Water-Bearing Media. *ACS Appl. Mater. Interfaces* **2019**, *11*, 10672–10679. [CrossRef]
9. Xu, J.; Wang, Z.; Chang, C.; Fu, B.; Tao, P.; Song, C.; Shang, W.; Deng, T. Solar-driven interfacial desalination for simultaneous freshwater and salt generation. *Desalination* **2020**, *484*, 114423. [CrossRef]
10. Zhou, Y.; Ding, T.; Gao, M.; Chan, K.H.; Cheng, Y.; He, J.; Ho, G.W. Controlled heterogeneous water distribution and evaporation towards enhanced photothermal water-electricity-hydrogen production. *Nano Energy* **2020**, *77*, 105102. [CrossRef]
11. Gao, M.; Peh, C.K.; Zhu, L.; Yilmaz, G.; Ho, G.W. Photothermal Catalytic Gel Featuring Spectral and Thermal Management for Parallel Freshwater and Hydrogen Production. *Adv. Energy Mater.* **2020**, *10*, 2000925. [CrossRef]
12. Zhu, L.; Gao, M.; Peh, C.K.N.; Ho, G.W. Recent progress in solar-driven interfacial water evaporation: Advanced designs and applications. *Nano Energy* **2018**, *57*, 507–518. [CrossRef]
13. Gao, M.; Connor, P.K.N.; Ho, G.W. Plasmonic photothermic directed broadband sunlight harnessing for seawater catalysis and desalination. *Energy Environ. Sci.* **2016**, *9*, 3151–3160. [CrossRef]
14. Zhu, M.; Li, Y.; Chen, F.; Zhu, X.; Dai, J.; Li, Y.; Yang, Z.; Yan, X.; Song, J.; Wang, Y.; et al. Plasmonic Wood for High-Efficiency Solar Steam Generation. *Adv. Energy Mater.* **2017**, *8*, 1701028. [CrossRef]
15. Wang, X.; Liu, Q.; Wu, S.; Xu, B.; Xu, H. Multilayer Polypyrrole Nanosheets with Self-Organized Surface Structures for Flexible and Efficient Solar–Thermal Energy Conversion. *Adv. Mater.* **2019**, *31*, e1807716. [CrossRef]
16. Liu, H.; Chen, C.; Wen, H.; Guo, R.; Williams, N.A.; Wang, B.; Chen, F.; Hu, L. Narrow bandgap semiconductor decorated wood membrane for high-efficiency solar-assisted water purification. *J. Mater. Chem. A* **2018**, *6*, 18839–18846. [CrossRef]
17. Li, Y.; Gao, T.; Yang, Z.; Chen, C.; Luo, W.; Song, J.; Hitz, E.; Jia, C.; Zhou, Y.; Liu, B.; et al. 3D-Printed, All-in-One Evaporator for High-Efficiency Solar Steam Generation under 1 Sun Illumination. *Adv. Mater.* **2017**, *29*, 1700981. [CrossRef]
18. Wu, X.; George, Y.; Chen, W.Z.; Liu, X.; Xu, H. A Plant-Transpiration-Process-Inspired Strategy for Highly Efficient Solar Evaporation. *Adv. Sustain. Syst.* **2017**, *1*, 1700046. [CrossRef]

19. Chen, C.; Li, Y.; Song, J.; Yang, Z.; Kuang, Y.; Hitz, E.; Jia, C.; Gong, A.; Jiang, F.; Zhu, J.Y.; et al. Highly Flexible and Efficient Solar Steam Generation Device. *Adv. Mater.* **2017**, *29*, 1701756. [CrossRef]
20. Huo, B.; Jiang, D.; Cao, X.; Liang, H.; Liu, Z.; Li, C.; Liu, J. N-Doped Graphene/Carbon Hybrid Aerogels for Efficient Solar Steam Generation. *Carbon* **2019**, *142*, 13–19. [CrossRef]
21. Zhang, Q.; Xiao, X.; Wang, G.; Ming, X.; Liu, X.; Wang, H.; Yang, H.; Xu, W.; Wang, X. Silk-Based Systems for Highly Efficient Photothermal Conversion under One Sun: Portability, Flexibility, and Durability. *J. Mater. Chem. A* **2018**, *6*, 17212–17219. [CrossRef]
22. Chen, T.; Wang, S.; Wu, Z.; Wang, X.; Peng, J.; Wu, B.; Cui, J.; Fang, X.; Xie, Y.; Zheng, N. A cake making strategy to prepare reduced graphene oxide wrapped plant fiber sponges for high-efficiency solar steam generation. *J. Mater. Chem. A* **2018**, *6*, 14571–14576. [CrossRef]
23. Shi, L.; Wang, Y.; Zhang, L.; Wang, P. Rational design of a bi-layered reduced graphene oxide film on polystyrene foam for solar-driven interfacial water evaporation. *J. Mater. Chem. A* **2016**, *5*, 16212–16219. [CrossRef]
24. Liu, K.-K.; Jiang, Q.; Tadepalli, S.; Raliya, R.; Biswas, P.; Naik, R.R.; Singamaneni, S. Wood–Graphene Oxide Composite for Highly Efficient Solar Steam Generation and Desalination. *ACS Appl. Mater. Interfaces* **2017**, *9*, 7675–7681. [CrossRef] [PubMed]
25. Wang, Y.; Wang, C.; Song, X.; Megarajan, S.K.; Jiang, H. A facile nanocomposite strategy to fabricate a rGO–MWCNT photothermal layer for efficient water evaporation. *J. Mater. Chem. A* **2017**, *6*, 963–971. [CrossRef]
26. Zhang, Y.; Zhao, D.; Yu, F.; Yang, C.; Lou, J.; Liu, Y.; Chen, Y.; Wang, Z.; Tao, P.; Shang, W.; et al. Floating Rgo-Based Black Membranes for Solar Driven Sterilization. *Nanoscale* **2017**, *9*, 19384–19389. [CrossRef] [PubMed]
27. Ma, N.; Fu, Q.; Hong, Y.; Hao, X.; Wang, X.; Ju, J.; Sun, J. Processing Natural Wood into an Efficient and Durable Solar Steam Generation Device. *ACS Appl. Mater Interfaces* **2020**, *12*, 18165–18173. [CrossRef] [PubMed]
28. Hummers, W.S.; Offeman, R.E. Preparation of Graphitic Oxide. *J. Am. Chem. Soc.* **1958**, *80*, 1339. [CrossRef]
29. Gao, J.; Liu, F.; Liu, Y.; Ma, N.; Wang, Z.; Zhang, X. Environment-Friendly Method To Produce Graphene That Employs Vitamin C and Amino Acid. *Chem. Mater.* **2010**, *22*, 2213–2218. [CrossRef]
30. Li, K.; Chang, T.; Li, Z.; Yang, H.; Fu, F.; Li, T.; Ho, J.S.; Chen, P.Y. Biomimetic Mxene Textures with Enhanced Light-to-Heat Conversion for Solar Steam Generation and Wearable Thermal Management. *Adv. Energy Mater.* **2019**, *9*, 1901687. [CrossRef]
31. Zhou, J.; Sun, Z.; Chen, M.; Wang, J.; Qiao, W.; Long, D.; Ling, L. Macroscopic and Mechanically Robust Hollow Carbon Spheres with Superior Oil Adsorption and Light-to-Heat Evaporation Properties. *Adv. Funct. Mater.* **2016**, *26*, 5368–5375. [CrossRef]
32. Li, N.; Qiao, L.; He, J.; Wang, S.; Yu, L.; Murto, P.; Li, X.; Xu, X. Solar-Driven Interfacial Evaporation and Self-Powered Water Wave Detection Based on an All-Cellulose Monolithic Design. *Adv. Funct. Mater.* **2020**, *31*, 2008681. [CrossRef]
33. Shi, M.-M.; Bao, D.; Li, S.-J.; Wulan, B.-R.; Yan, J.-M.; Jiang, Q. Anchoring PdCu Amorphous Nanocluster on Graphene for Electrochemical Reduction of N_2 to NH_3 under Ambient Conditions in Aqueous Solution. *Adv. Energy Mater.* **2018**, *8*, 1800124. [CrossRef]
34. Wang, Z.; Xue, Z.; Zhang, M.; Wang, Y.; Xie, X.; Chu, P.; Zhou, P.; Di, Z.; Wang, X. Germanium-Assisted Direct Growth of Graphene on Arbitrary Dielectric Substrates for Heating Devices. *Small* **2017**, *13*, 1700929. [CrossRef]
35. Hu, X.; Xu, W.; Zhou, L.; Tan, Y.; Wang, Y.; Zhu, S.; Zhu, J. Tailoring Graphene Oxide-Based Aerogels for Efficient Solar Steam Generation under One Sun. *Adv. Mater.* **2016**, *29*, 1604031. [CrossRef]
36. Eda, G.; Lin, Y.-Y.; Mattevi, C.; Yamaguchi, H.; Chen, H.-A.; Chen, I.-S.; Chen, C.-W.; Chhowalla, M. Blue Photoluminescence from Chemically Derived Graphene Oxide. *Adv. Mater.* **2010**, *22*, 505–509. [CrossRef]
37. Yuan, Y.; Dong, C.; Gu, J.; Liu, Q.; Xu, J.; Zhou, C.; Song, G.; Chen, W.; Yao, L.; Zhang, D. A Scalable Nickel–Cellulose Hybrid Metamaterial with Broadband Light Absorption for Efficient Solar Distillation. *Adv. Mater.* **2020**, *32*, e1907975. [CrossRef]
38. Wang, Q.; Guo, Q.; Jia, F.; Li, Y.; Song, S. Facile Preparation of Three-Dimensional MoS_2 Aerogels for Highly Efficient Solar Desalination. *ACS Appl. Mater. Interfaces* **2020**, *12*, 32673–32680. [CrossRef]
39. Song, G.; Yuan, Y.; Liu, J.; Liu, Q.; Zhang, W.; Fang, J.; Gu, J.; Ma, D.; Zhang, D. Biomimetic Superstructures Assembled from Au Nanostars and Nanospheres for Efficient Solar Evaporation. *Adv. Sustain. Syst.* **2019**, *3*, 1900003. [CrossRef]
40. Zhou, L.; Tan, Y.; Wang, J.; Xu, W.; Yuan, Y.; Cai, W.; Zhu, S.; Zhu, J. 3D self-assembly of aluminium nanoparticles for plasmon-enhanced solar desalination. *Nat. Photonics* **2016**, *10*, 393–398. [CrossRef]
41. Zhang, H.; Li, L.; Jiang, B.; Zhang, Q.; Ma, J.; Tang, D.; Song, Y. Highly Thermal Insulated and Super-Hydrophilic Corn Straw for Efficient Solar Vapor Generation. *ACS Appl. Mater. Interfaces* **2020**, *12*, 16503–16511. [CrossRef] [PubMed]

Article

Spray-Assisted Interfacial Polymerization to Form Cu$^{II/I}$@CMC-PANI Film: An Efficient Dip Catalyst for A^3 Reaction

Zhian Xu [1], Liang Xiao [1], Xuetao Fan [1], Dongtao Lin [1], Liting Ma [2], Guochao Nie [2,*] and Yiqun Li [1,*]

[1] Department of Chemistry, College of Chemistry and Materials Science, Jinan University, Guangzhou 511443, China; zhian_xu@outlook.com (Z.X.); liang_xi728@163.com (L.X.); lovefandou1018@gmail.com (X.F.); Dongtao_Lin@163.com (D.L.)
[2] Photoelectric Information Center, School of Physics and Telecom, Yulin Normal University, Yulin 537000, China; mikt13f@163.com
* Correspondence: bccu518@163.com (G.N.); tlyq@jnu.edu.cn (Y.L.)

Abstract: A novel and interesting method for the preparation of carboxymethylcellulose–polyaniline film-supported copper catalyst (Cu$^{II/I}$@CMC-PANI) has been developed via spray-assisted interfacial polymerization. Using copper sulfate as an initiator, spraying technology was introduced to form a unique interface that is perfectly beneficial to the polymerization of aniline monomers onto carboxymethylcellulose macromolecule chains. To further confirm the composition and structure of the as-prepared hybrid film, it was systematically characterized by inductively coupled plasma (ICP), Fourier transform infrared spectroscopy (FTIR), X-ray photoelectron spectroscopy (XPS), X-ray diffraction (XRD), scanning electron microscopy (SEM), energy-dispersive X-ray spectroscopy (EDS), and thermogravimetric analysis (TGA) techniques. The Cu content in the fresh Cu$^{II/I}$@CMC-PANI film was determined to be 1.805 mmol/g, and spherical nanoparticles with an average size of ca. 10.04 nm could be observed in the hybrid film. The Cu$^{II/I}$@CMC-PANI hybrid film was exerted as a dip catalyst to catalyze the aldehyde–alkyne–amine (A^3) coupling reactions. High yields of the products (up to 97%) were obtained in this catalytic system, and the catalyst could be easily picked up from the reaction mixture by tweezers and reused for at least six consecutive runs, without any discernible losses in its activity in the model reaction. The dip catalyst of Cu$^{II/I}$@CMC-PANI, with easy fabrication, convenient deployment, superior catalytic activity, and great reusability, is expected to be very useful in organic synthesis.

Keywords: interfacial polymerization; spraying method; carboxymethylcellulose; polyaniline; copper catalyst; dip catalyst; A^3 reaction

Citation: Xu, Z.; Xiao, L.; Fan, X.; Lin, D.; Ma, L.; Nie, G.; Li, Y. Spray-Assisted Interfacial Polymerization to Form Cu$^{II/I}$@CMC-PANI Film: An Efficient Dip Catalyst for A^3 Reaction. *Nanomaterials* **2022**, *12*, 1641. https://doi.org/10.3390/nano12101641

Academic Editor: Francisco Alonso

Received: 31 March 2022
Accepted: 2 May 2022
Published: 11 May 2022

Publisher's Note: MDPI stays neutral with regard to jurisdictional claims in published maps and institutional affiliations.

Copyright: © 2022 by the authors. Licensee MDPI, Basel, Switzerland. This article is an open access article distributed under the terms and conditions of the Creative Commons Attribution (CC BY) license (https://creativecommons.org/licenses/by/4.0/).

1. Introduction

In 2010, a pioneering researcher Radhakrishnan developed the concept of "dip catalyst", and successively presented an Ag nanoparticle-embedded PVA thin film for the reduction of 4-nitrophenol by sodium borohydride [1]. Since then, the dip catalyst, which can switch the reaction on and off instantaneously, by merely dipping in and out of the reaction vessels, has drawn increasing attention for its ease of fabrication, excellent catalytic performance, convenient separation and reusability, and environmental friendliness [2]. To date, the dip catalyst has emerged as one of the most powerful tools in catalysis, synthetic methodology, materials science, and environmental science. A considerable number of dip catalysts have been well documented in the last dozen years, and the range of reported dip catalysts includes Pt@GS (dip coating) [3], CMC-Ni-BC (coating) [4], gold nanoparticle-loaded filter paper (impregnated into the filter paper) [5], palladium nanoparticle-loaded cellulose paper (dip coating) [6], Pd@filter paper (dip coating) [7], Pt-PVA thin film (spin coating) [8], Pd-PVA (spin coating) [9], AgNPs@NH$_2$-CP (chemical modification) [10],

Cu@CS-FP (layer coating) [11], Ag/CH-FP (layer coating) [12], and so on. Common strategies for the fabrication of a dip catalyst previously included dip-coating [3–7] and spin-coating technology [8,9], chemical modification of thin films with a variety of functional units [10], and layer coating over thin films with functional material [11,12]. However, these preparation procedures suffered from one or more drawbacks, such as tedious operation, being time consuming, involving multiple step reactions, and possessing low catalytic performance. Therefore, searching for novel and convenient protocols for the fabrication of a dip catalyst with excellent catalytic activity and reusability is still in great demand.

Copper is a low-toxicity, economical, sustainable, and readily available metal elemental, belonging to the $[Ar]^3d^9$ transition metal, with various oxidation states, such as Cu(0), Cu(I), Cu(II), and Cu(III). Depending on its unique properties and characteristics, copper can effectively catalyze various organic reactions, such as Suzuki–Miyaura cross-coupling reactions [13,14], azide–alkyne cycloadditions [15], and Ullmann-type coupling reactions [16,17]. Performing the transition metal-catalyzed three-component reactions of aldehydes, alkynes, and amines (commonly called A^3 reactions) in an atom-economic way is of importance, as it produces valuable N-containing heteroatom propargylamine products [18]. A great number of transition metal catalysts, including Cu, Ag, Au, and so on, have been exploited for the A^3 coupling process. Several reviews have outlined the latest progress in A^3 couplings well, and a large variety of copper catalysts have been criticized in these surveys [18–20]. However, there are no pertinent reports of a copper-based dip catalyst being applied to a one-pot A^3 coupling reaction for the synthesis of versatile propargylamines.

We envisaged that polymer thin films generated in situ by interfacial polymerization, initiated by metal salts and embedded metal catalysts simultaneously, would be a class of easily fabricated, efficient, and reusable dip catalysts. Interfacial polymerization is a process that utilizes the interfacial layer (liquid–liquid layer or liquid–gas layer, etc.) to provide a unique space to constrain the polymerization for the preparation of polymer films or membranes [21]. Polymer materials generated in situ by interfacial polymerization have multiple advantages, including mild production conditions, high molecular weight, excellent uniformity [21], and high permeation rates [22]. The interfacial polymerization method also permits the creation of films that are highly resistant to destruction by exposure to harsh environments [22]. These merits of the films prepared by interfacial polymerization make them beneficial to be a dip catalyst.

Polyaniline exhibits good affinity to metal ions, through the unique coordination with its nitrogen atoms [23] and delocalized π-π conjugate system [24]; thus, it is a useful metal supportive material to form PANI-based metal catalysts (Metal@PANI), such as Pd@PANI [25,26], Fe@PANI [27], and Ag@PANI [28]. However, as most developed catalysts were reported in the preparation of powder-based PANI hybrids, using PANI to form a film is difficult, due to its strong rigidity, even when using an advanced method, such as the interfacial polymerization method mentioned above. The traditional method for the preparation of PANI films is the solution casting technique, using various solvents, such as *m*-cresol solution [29], to dissolve PANIs. To develop a new approach, a template method was developed by using synthetic polymer film as a soft template for the polymerization of aniline monomer on its surface, and, thus, assisted the formation of PANI film [30]. Carboxymethylcellulose (CMC) is non-toxic, biodegrade, economic, and eco-friendly, and it bears a great number of carboxymethyl (–CH$_2$COO$^-$) and free hydroxyl (–OH) groups on its glucose-unit chain, providing it with a splendid capacity to coordinate with various metal cations [31], and to form H-bonds with molecules containing active groups, such as hydroxy-, amino-, and carboxylic groups.

Based on the properties of PANI and CMC, in our previous work, we fabricated CMC-PANI hybrids via the one-pot and one-step oxidative polymerization of aniline, with CMC as the soft template and CuSO$_4$ as the initiator [32]. However, the CuSO$_4$NPs@CMC/PANI hybrids were obtained as dark green powders, which are not appropriate for application in dip catalysis. Interfacial polymerization may be an alternative approach to address the

PANI shaping issue. Unfortunately, many attempts made on interfacial polymerization by various conventional methods, including pouring, dropping, immersing, and so on, failed to obtain the desired CMC-PANI hybrid film. It was reported that the occurrence of interfacial polymerization needs an interface between two phases to provide a unique reaction space [33,34]. However, the interfacial polymerization of aniline monomer onto the surface of the CMC-Na macromolecular chain is hard to perform using $CuSO_4$ solution as an initiator, by conventional approaches, such as pouring, dropping, or immersing methods, as these two phases are both water soluble. Creatively, the spraying method was introduced into the process to solve this problem. By spraying copper sulfate solution onto the surface of the CMC and aniline mixture, aniline monomers polymerized immediately to rapidly form an extremely thin layer of CMC-PANI film. This thin layer served as an interface to provide a suitable space for further interfacial polymerization. Concerning the above facts, innovatively, we have attempted to fabricate carboxymethylcellulose–polyaniline film-supported copper catalyst ($Cu^{II/I}$@CMC-PANI) in situ via spray-assisted interfacial polymerization, and further explored it as a dip catalyst for A^3 reactions.

2. Experimental

2.1. Materials and Instrumentations

All chemicals were purchased from commercial sources and used as received, without further purification.

Gas chromatography (GC) was performed on a Shimadzu GCMS-QP2020 (Kyoto, Japan). The copper content of the catalyst was determined by inductively coupled plasma atomic emission spectrometry (ICP-AES), using a Thermo Fisher Scientific X Series 2 instrument (Waltham, MA, USA). Fourier transform infrared spectra (FT-IR) were collected on a PerkinElmer FT-IR Spectrometer Spectrum Two (Waltham, MA, America) within the spectral range of 4000–400 cm^{-1}. X-ray photoelectron spectroscopy (XPS) data were obtained on a Thermo Fisher Scientific K-Alpha instrument (Waltham, MA, America). X-ray powder diffraction (XRD) data were collected on a Rigaku MiniFlex600 diffractometer (Tokyo, Japan), using Cu Kα radiation in a range of Bragg's angles (5°–80°). Scanning electron microscopy (SEM) and energy-dispersive spectroscopy (EDS) were conducted with a Zeiss Sigma 300 instrument (Oberkochen, Germany). Thermogravimetric analysis (TGA) and derivative thermogravimetric analysis (DTG) were performed on a Mettler TGA/DSC3+ (Zurich, Switzerland), under a nitrogen atmosphere from 30 to 800 °C in a 50 mL·min^{-1} N_2 flow and at a ramp rate of 10 °C·min^{-1}. ^1H NMR (300 MHz) and ^{13}C NMR (75 MHz) spectra were obtained with a Bruker 300 Avance instrument (Karlsruhe, Germany), with $CDCl_3$ as the solvent and TMS as the internal standard. HRMS was determined by using Agilent 6545 Q-TOF MS (Santa Clara, CA, USA).

2.2. Preparation of $Cu^{II/I}$@CMC-PANI Film

CMC-Na (0.242 g, 1 mmol) was added to 30 mL aqueous methanol solution (MeOH/H_2O, v/v = 1/2) at room temperature, with continuous stirring until it completely dissolved. Then, aniline monomer (0.0911 g, 1 mmol) was dropped in above the solution to form a homogeneous mixture. This mixture was poured into a Petri dish, and $CuSO_4$ solution (5 wt.%) was finely and evenly misted onto the surface of the mixture using a handheld sprayer. Immediately, a light green interface (extremely thin film) was formed and isolated the CMC-Na/aniline mixture from the $CuSO_4$ solution. The resultant system was stood for 48 h to form a dark green film. The as-formed film was washed thoroughly with methanol and water to remove unreacted aniline and CMC-Na, and dried to afford the $Cu^{II/I}$@CMC-PANI film. The schematic illustration of the preparation process of the $Cu^{II/I}$@CMC-PANI film is shown in Scheme 1.

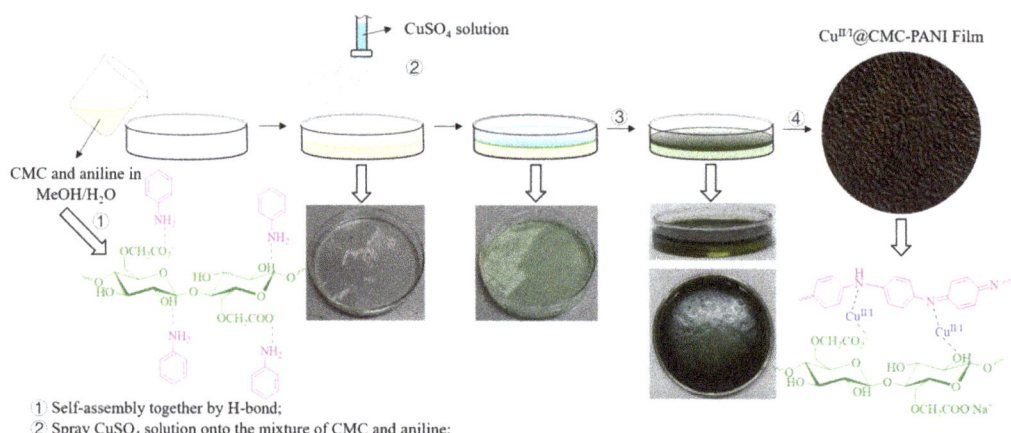

Scheme 1. Schematic illustration of the preparation of Cu$^{II/I}$@CMC-PANI film.

2.3. General Procedure for A^3 Coupling Reactions Catalysed by Cu$^{II/I}$@CMC-PANI Dip Catalyst

Aldehyde (1.0 mmol), amine (1.2 mmol), terminal alkyne (1.5 mmol) and a catalytic amount of Cu$^{II/I}$@CMC-PANI film (15 mg, 2.7 mol% of Cu) were added to 2 mL toluene in a sealed vessel, and the mixture was rigorously stirred at 110 °C for the specific time and monitored by TLC. As the reaction was completed, the dip catalyst was picked up with tweezers and washed several times with ethyl acetate and dried at 80 °C for the next run. The remaining mixture was extracted with ethyl acetate (3 × 4 mL). Then, the combined organic phase was washed with brine and dried over anhydrous Na$_2$SO$_4$. After removal of the organic solvent by a vacuum rotary evaporator, the crude product was purified by column chromatography on silica gel to afford the corresponding propargylamine. All the products, except product (i), are known, and their ^{1}H NMR and ^{13}C NMR data were found to be identical to those reported in previous literature. The new compound (i) was fully characterized by FT-IR, ^{1}H NMR, ^{13}C NMR, and HRMS. All these data and spectra have been concluded in Supplementary Materials.

3. Results and Discussion

3.1. Synthesis and Characterization of Catalyst

The synthetic route of the Cu$^{II/I}$@CMC-PANI film is outlined in Scheme 1. Before the addition of CuSO$_4$ solution, the CMC-Na and aniline monomers were self-assembled together by H-bond interactions [35]. Then, the CuSO$_4$ solution was finely and evenly misted onto the surface of the CMC-Na/aniline mixture, and a light green interface (an extremely thin film) was generated immediately, which separated the CMC-Na/aniline mixture and CuSO$_4$ solution to provide an interfacial layer for aniline to polymerize (Scheme 2). In this process, the CuSO$_4$ solution not only acted as the initiator, triggering the polymerization reaction, but also as the coupling reagent, combining the CMC molecular chain and PANI chain together. In the meantime, Cu(II)/Cu(I) species were deposited into the as-formed CMC-PANI film and stabilized by complexation with carboxylic groups (–COO$^-$), hydroxyl groups (–OH), and nitrogen atoms (–NH– and –N=) to form the target dip catalyst. The Cu content in fresh Cu$^{II/I}$@CMC-PANI film was determined by ICP-AES to be 1.805 mmol/g.

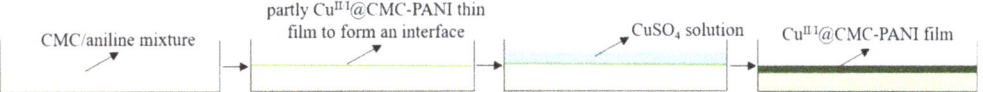

Scheme 2. Detailed illustration of spray-assisted interfacial polymerization process.

To verify the composition and structures of the $Cu^{II/I}$@CMC-PANI film, analytical techniques, including FT-IR, XPS, XRD, SEM, EDS, TGA, and DGT, were performed.

The FT-IR spectra of CMC-Na (curve a), PANI (curve b), and the obverse side (curve c) and back side (curve d) of the $Cu^{II/I}$@CMC-PANI film are presented in Figure 1. As shown in the spectrum of CMC-Na (Figure 1, curve a), the characteristic peaks appeared at 1606 cm^{-1} and 1424 cm^{-1}, related to the asymmetric and symmetric stretching vibration of carboxylic (–COO$^-$) groups, respectively [36,37]. In the spectra of PANI (Figure 1, curve b), the peaks at 1558 cm^{-1}, coupled with 1141 cm^{-1}, and 1482 cm^{-1}, along with 805 cm^{-1}, are assigned to the stretching vibration of quinoid and benzenoid rings, correspondingly [32,38]. Due to the interfacial polymerization, the $Cu^{II/I}$@CMC-PANI film owns obverse and back sides, and they are composed of different components. The characteristic peaks of CMC (carboxylic groups: 1579 cm^{-1} and 1413 cm^{-1}) and PANI (quinoid: 1107 cm^{-1} and benzenoid: 1501 cm^{-1}) were both observed in the spectra of the catalyst obverse side (Figure 1, curve c), which confirmed that aniline monomers succeeded to the polymerization of PANI onto the CMC molecular chain. However, on the back side, only peaks of carboxylate groups (1582 cm^{-1} and 1414 cm^{-1}) appeared (Figure 1, curve d), which means that PANI mostly exists on the obverse side of the catalyst.

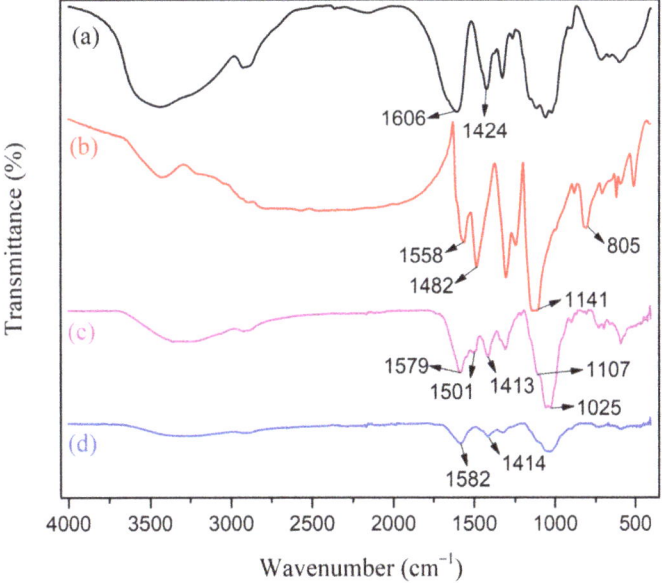

Figure 1. The FT-IR spectra of CMC-Na (a), PANI (b), the obverse side of $Cu^{II/I}$@CMC-PANI film (c), and the back side of $Cu^{II/I}$@CMC-PANI film (d).

XPS was carried out to further demonstrate the existence of all elements and the oxidation state of copper. Presented in the survey scan spectrum (Figure 2a), C, N, O, and Cu elements are found to be 39.33%, 4.57%, 48.20%, and 7.90%, respectively, in fresh $Cu^{II/I}$@CMC-PANI film. The peaks at 932.78 eV ($Cu_{2p3/2}$) and 952.48 eV ($Cu_{2p1/2}$) in Figure 2e, and the peak at 571.12 eV (Cu_{LM2}) in Figure 2f, were assigned to the presence of Cu(I) [39]. Meanwhile, the shoulder peaks (934.83 eV and 954.63 eV) and satellites peaks (940.18 eV, 944.07 eV, and 962.56 eV) illustrate that Cu (II) also existed in the catalyst [40]. Notably, nearly all the Cu (II) in the recovered catalyst was turned into Cu(I), evidenced by the disappearance of shoulder peaks and satellites, as well as the presence of peaks at 932.37 eV ($Cu_{2p3/2}$) and 952.57 eV ($Cu_{2p1/2}$) in Figure 2g, and the peak at 571.74 eV (Cu_{LM2}) [40] in Figure 2h, which means that Cu(I) species may be the true catalyst for A^3 reactions.

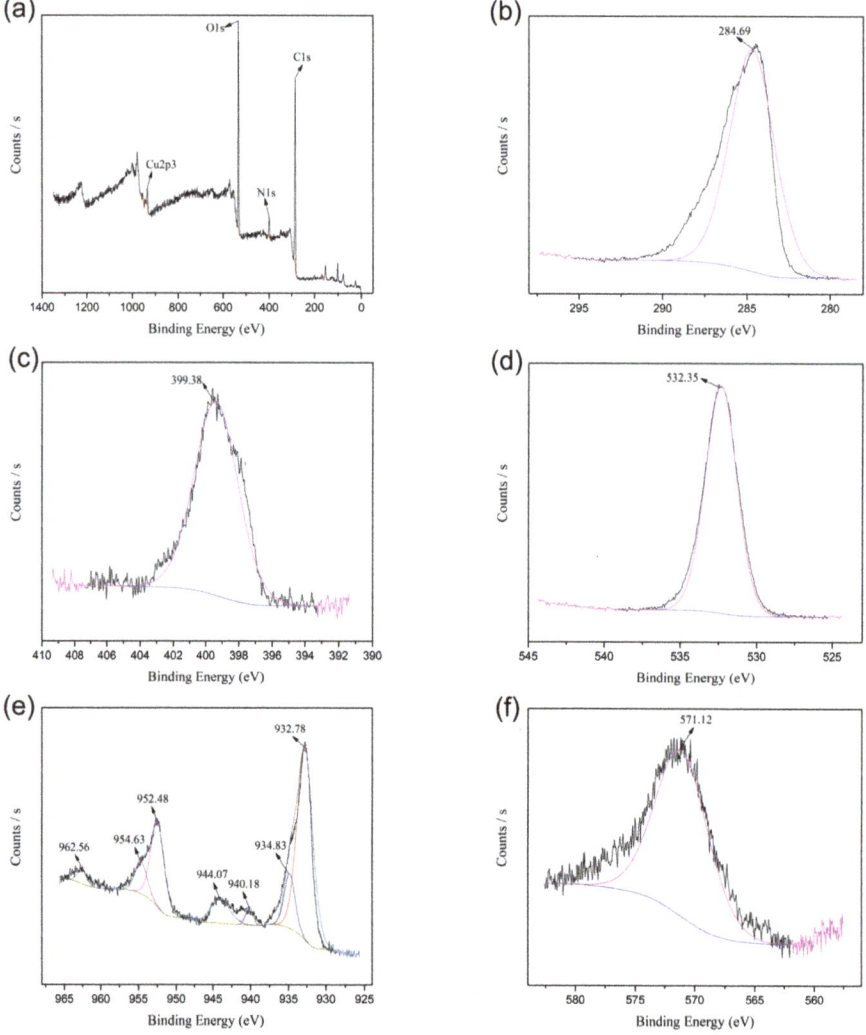

Figure 2. *Cont.*

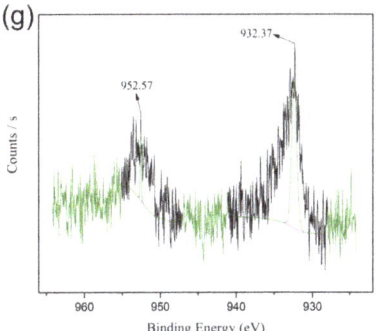
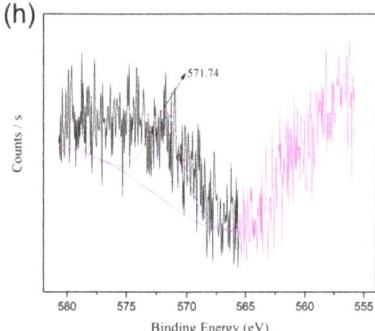

Figure 2. The XPS survey scan (**a**) of $Cu^{II/I}$@CMC-PANI film and its high-resolution spectra of C_{1s} (**b**), N_{1s} (**c**), O_{1s} (**d**), Cu_{2p} (**e**) and Cu_{LM2} (**f**); XPS high-resolution spectra of Cu_{2p} (**g**) and Cu_{LM2} (**h**) of catalyst recovered from A^3 reaction.

To further confirm the composition of the $Cu^{II/I}$@CMC-PANI film, XRD was performed. The amorphous peak at $2\theta = 20.6°$ in curve (a) and the broad peak at $2\theta = 25.2°$ in curve (b) (Figure 3) were attributed to the pristine CMC-Na [41] and PANI powders [42], respectively, showing an extremely low degree of crystallinity. The peak at $2\theta = 20.6°$ can also be observed in curve (c) of the catalyst, but the peak of PANI was shifted to $2\theta = 21.8°$, due to the interaction between PANI and CMC-Na [43]. The appearance of CMC and PANI peaks in curve (c) (Figure 3) proves that CMC and PANI were merged successfully into the film. Moreover, as can be found in curve (c) of the catalyst, the films consist of mixed phases of $6CuO \cdot Cu_2O$ (JCPDS 03-0879), CuO (JCPDS 44-0706), Cu_2O (JCPDS 35-1091), and Cu_4O_3 (JCPDS 49-1830). Generally, the $6CuO \cdot Cu_2O$ phase is observed to be an intermediate phase between Cu_2O and CuO [44], and Cu_4O_3 can be written as $Cu(I)_2Cu(II)_2O_3$ [45]. These observations conclude that Cu (II) and Cu(I) species exist in the catalyst. The size of the particles can be calculated using the Scherrer equation, as follows: $D = K\lambda/(\beta\cos\theta) = \frac{0.89 \times 0.15405}{\frac{0.534}{180} \times 3.14 \times \cos\frac{28.5}{2}} = 15.19$ (nm). These data are approximate, with the value of 10.04 nm obtained from the SEM images (Figure 4e,f), due to the Scherrer equation being applicable to nanocrystals with perfect crystallinity, and there may be a certain number of errors to the particles without high crystallinity.

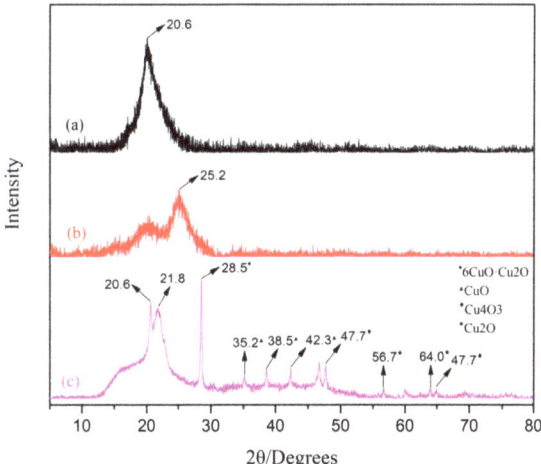

Figure 3. The XRD patterns of the CMC-Na (a), PANI (b), and $Cu^{II/I}$@CMC-PANI film (c).

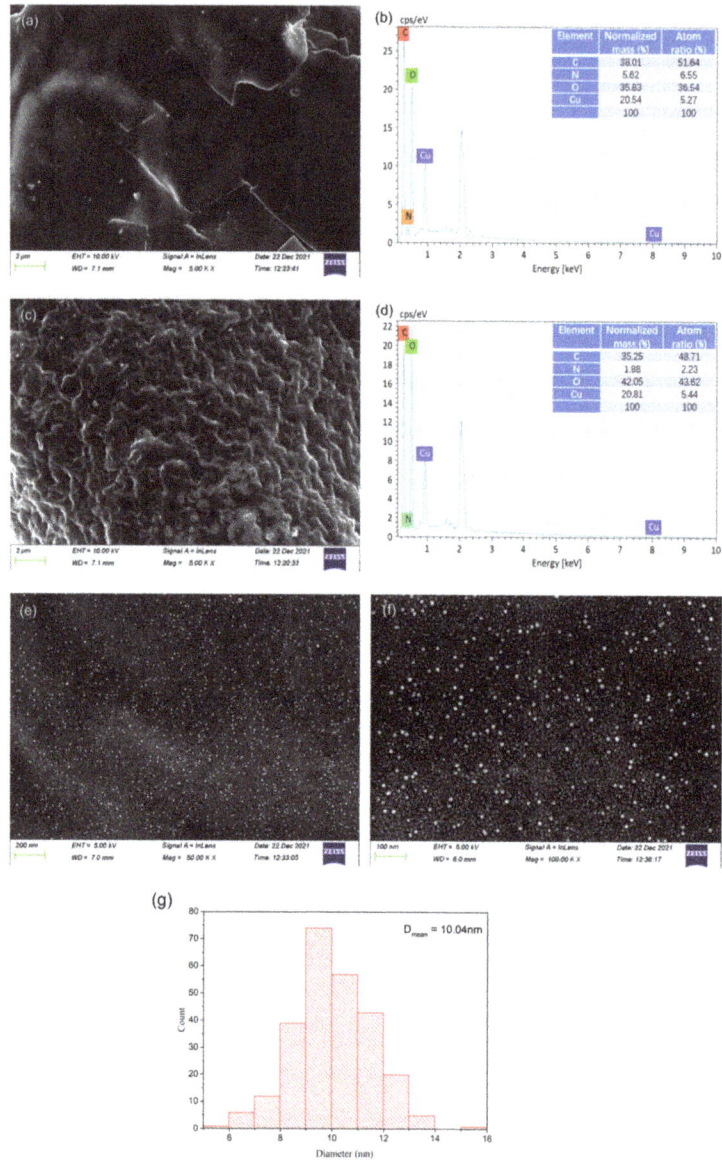

Figure 4. SEM images (**a,e,f**) and EDS data (**b**) of the obverse side of $Cu^{II/I}$@CMC-PANI film; SEM images (**c**) and EDS data (**d**) of its back side; and histogram of particle size (**g**).

SEM and EDS were carried out in order to study the morphology and components of the $Cu^{II/I}$@CMC-PANI film. The SEM images show a smooth surface for the obverse side (Figure 4a) and a rough surface for the back side (Figure 4c). It is worth noting that by increasing the magnification, spherical nanoparticles can be observed (Figure 4e,f), which may be attributed to the uniformly distributed copper oxides loaded in the film. The particle size histogram (Figure 4g) shows that the average particle size is approximately 10.04 nm in diameter. The EDS images of the obverse and back sides of $Cu^{II/I}$@CMC-PANI are shown in Figure 4b,d. EDS clearly showed the presence of the nonmetallic elements C,

N, and O, and the metallic element Cu in the dip catalyst. Significantly, the N content of the obverse side of the film (Figure 4b) is much higher than that of the back side (Figure 4d). This observation matched very well with the finding obtained from FT-IR (Figure 1), which illustrated that PANI mostly exists on the obverse side of the dip catalyst. The elemental mapping images (Figure 5), coupled with SEM, evidenced that the C, N, O, and Cu elements distributed throughout the catalyst in a homogeneous manner. The uniform distribution of Cu makes the catalyst work steadily.

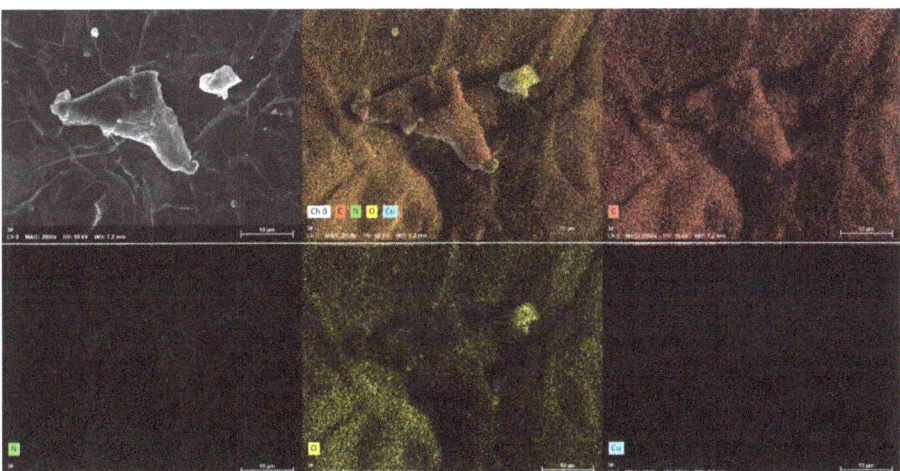

Figure 5. SEM corresponding elemental mapping images of $Cu^{II/I}$@CMC-PANI film.

To investigate the thermal behavior of the $Cu^{II/I}$@CMC-PANI film, TGA and a corresponding DTG analysis of CMC-Na, PANI, and the $Cu^{II/I}$@CMC-PANI film were performed, and the results are displayed in Figure 6. Below 100 °C, all the samples (Figure 6a–c) showed a slight mass loss, which can be attributed to the release of adsorbed moisture and volatile impurities. After that, CMC-Na (Figure 6a) presented a sharp mass loss at around 290 °C, which corresponds to the decomposition of its glucose-unit chain and side carboxylic groups [46]. PANI (Figure 6b) exhibited two sharp decreases in mass at around 253 °C and 522 °C, which may be caused by the deprotonation of PANI and decomposition of the backbone units of PANI, respectively [30,47]. In the curve of the dip catalyst (Figure 6c), a mass loss in the temperature range of 198–400 °C appeared. This may account for the combined influence of the decomposition of CMC and PANI chains. This result indicated that the dip catalyst is stable up to nearly 200 °C, and that it is applicable for further A^3 reactions.

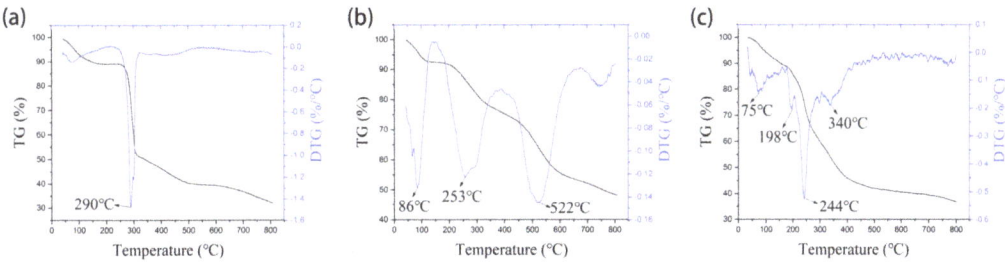

Figure 6. The TGA and DTG curves of CMC (**a**), PANI (**b**) and $Cu^{II/I}$@CMC-PANI film (**c**).

3.2. Application of $Cu^{II/I}$@CMC-PANI Dip Catalyst in Three-Component A^3 Coupling Reactions

First, the optimal reaction conditions for A^3 reactions were explored in the presence of the $Cu^{II/I}$@CMC-PANI dip catalyst (Table 1). Morpholine (1.2 mmol), *p*-chlorobenzaldehyde (1.0 mmol), and phenylacetylene (1.5 mmol) were selected as the model substrates. Initially, different solvents, including H_2O, DMSO, DMF, CH_3CN, EtOH, *n*-BuOH, and toluene, were screened to assess efficiency. It was found that the product yields increased with the decrease in solvent polarity. The highest yield was obtained when toluene was used as the solvent (Table 1, entry 7). Subsequently, to establish the fact that copper in the dip catalyst plays a key role in the A^3 reaction, the model reaction was performed in toluene at 110 °C in the absence of a catalyst, and in the presence of 1.8 mol% and 2.7 mol% of Cu in the catalyst, the product yields obtained were 0%, 93%, and 97% (Table 1, entries 7–9), respectively. When further increasing the dose of the catalyst to 3.6 mol%, no significant elevation in the product yield was observed (Table 1, entry 10). These results suggested that copper is essential for this A^3 model reaction, and 2.7 mol% of Cu is the optimal amount for the model reaction. Finally, the reaction temperature was explored, and it was found that decreasing the temperature from 110 °C to 90 °C, and even to 70 °C, had a considerable negative impact on the product yields (Table 1, entries 11–12). Therefore, it was concluded that the optimal condition involved morpholine (1.2 mmol), *p*-chlorobenzaldehyde (1.0 mmol), phenylacetylene (1.5 mmol), and the $Cu^{II/I}$@CMC-PANI dip catalyst (2.7 mol% of Cu) in toluene at 110 °C.

Table 1. Optimization of model A^3 reactions catalyzed by $Cu^{II/I}$@CMC-PANI dip catalyst [a].

Entry	Solvent	Dose of Catalyst (Mol% of Cu)	Temp. (°C)	Time (h)	Yield [b] (%)
1	H_2O	2.7	100	12	N.R. [c]
2	EtOH	2.7	78	12	6
3	*n*-BuOH	2.7	110	8	55
4	DMSO	2.7	110	8	38
5	DMF	2.7	110	8	51
6	CH_3CN	2.7	80	12	trace [d]
7	toluene	2.7	110	6	97
8	toluene	0	110	8	trace [d]
9	toluene	1.8	110	8	93
10	toluene	3.6	110	8	97
11	toluene	2.7	90	8	12
12	toluene	2.7	70	24	trace [d]

[a] Reaction conditions: *p*-chlorobenzaldehyde (1.0 mmol), morpholine (1.2 mmol), phenylacetylene (1.5 mmol) and solvent (2 mL). [b] Isolated yields. [c] No reaction. [d] Observed by TLC.

Next, with the optimal reaction condition in hand, the substrate scope of the reaction was examined (Table 2). Aryl aldehydes, bearing both electron-withdrawing and electron-donating groups, proceeded well to afford excellent yields (Table 2, entries 1, 3, 4, and 6), while salicylaldehyde was difficult to conduct the reaction with (Table 2, entry 5), due to the intramolecular H-bond, which improves the stability of its aldehyde group. Phenylacetylene, with electron-withdrawing groups on its aromatic ring, afforded lower yields, due to the fact that this group reduces its nucleophilic activity (Table 2, entries 7–10). Because it is hard to form a Cu-alkyne intermediate with aliphatic alkyne, a trace product was observed when using aliphatic alkyne as a substrate (Table 2, entry 11). As the imine intermediate formed, aniline afforded no desired product (Table 2, entry 15). Unfortunately, aliphatic aldehydes afforded dissatisfactory yields (Table 2, entries 16–17), due to their low reactivity.

Table 2. A^3 coupling reactions catalyzed by $Cu^{II/I}$@CMC-PANI film to synthesize a variety of propargylamines [a].

Entry	R	R'	HNR"$_2$	Time (h)	Yield [b] (%)
1	4-Cl-Ph	Ph	morpholine	6	97 (a)
2	Ph	Ph	morpholine	8	91 (b)
3	4-Me-Ph	Ph	morpholine	8	83 (c)
4	2-MeO-Ph	Ph	morpholine	8	97 (d)
5	2-OH-Ph	Ph	morpholine	48	39 (e)
6	3-NO$_2$-Ph	Ph	morpholine	48	86 (f)
7	Ph	3-Me-Ph	morpholine	16	87 (g)
8	Ph	4-MeO-Ph	morpholine	32	97 (h)
9	Ph	4-CF$_3$-Ph	morpholine	48	83 (i)
10	Ph	4-Cl-Ph	morpholine	24	88 (j)
11	Ph	n-hexyl	morpholine	48	trace (k) [c]
12	Ph	Ph	piperidine	48	84 (l)
13	4-Cl-Ph	Ph	piperidine	48	86 (m)
14	4-Cl-Ph	Ph	diethylamine	48	58 (n)
15	Ph	Ph	aniline	48	N.P. (o) [d]
16	Et	Ph	morpholine	8	54 (p)
17	i-Pr	Ph	morpholine	8	30 (q)

[a] Reaction conditions: aldehyde (1.0 mmol), amine (1.2 mmol), alkyne (1.5 mmol), $Cu^{II/I}$@CMC-PANI film (2.7 mol% of Cu), toluene (2 mL), at 110 °C. [b] All yields are isolated. [c] Observed by TLC. [d] No product, and generated imine was observed by GC-MS.

Furthermore, the recyclability and reusability of the $Cu^{II/I}$@CMC-PANI film for the A^3 model reaction was assessed. As shown in Scheme 3, the film could be recycled and reused successfully for up to six consecutive cycles, without significant losses in its activity. The Cu content, as measured by ICP-AES, in the recovered catalyst after two cycles was 1.705 mmol/g, which is slightly lower than that of 1.805 mmol/g in the fresh catalyst. Thus, to further investigate the Cu leaching of the catalyst in the A^3 reaction, a hot filtration test was performed. The model A^3 reaction was carried out at 110 °C for 1 h (36% conversion), and then the dip catalyst was picked up. The remaining mixture was further stirred for 5 h without the catalyst and a slight increase in conversation (36% to 50%) was observed, indeed suggesting that there was a small amount of Cu species leached out from the catalyst into the reaction mixture, which is consistent with the result of the ICP-AES analysis.

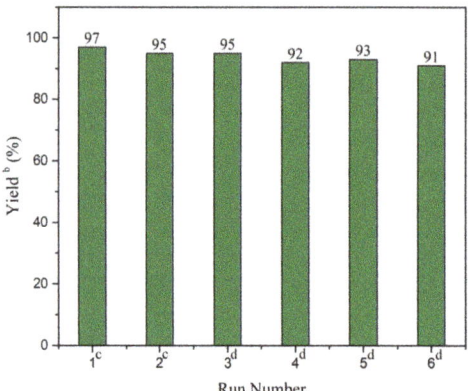

Scheme 3. Reusability of $Cu^{II/I}$@CMC-PANI dip catalyst in model A^3 reaction [a]. [a] Reaction conditions: *p*-chlorobenzaldehyde (1.0 mmol), morpholine (1.2 mmol), phenylacetylene (1.5 mmol), $Cu^{II/I}$@CMC-PANI dip catalyst (2.7 mol% of Cu), toluene (2 mL), at 110 °C. [b] Isolated yields. [c] Reacted for 6 h. [d] Reacted for 10 h.

4. Conclusions

A novel and practical method for the fabrication of a $Cu^{II/I}$@CMC-PANI film dip catalyst was developed via spray-assisted interfacial polymerization. In the preparation process, carboxymethylcellulose and copper sulfate served as the soft template and initiator, respectively, for the polymerization of aniline monomers, and the spraying method provided a unique interfacial layer to facilitate polymerization. This as-prepared film catalyst was well characterized by ICP, FT-IR, XPS, XRD, SEM, EDS, and TGA techniques. These characterized results proved that aniline had successfully polymerized to PANI onto the CMC molecular chain, and that Cu(II)/Cu(I) oxides, at the nano scale, were homogeneously loaded into the CMC-PANI film at the same time. The obverse and back side of the resultant film had different morphologies, due to their different PANI contents. In terms of application, the as-synthesized $Cu^{II/I}$@CMC-PANI film dip catalyst possessed good thermal stability up to 200 °C, and could be efficiently applied in the A^3 coupling reaction, due to its excellent recyclability and tolerance of a broad scope of substrates.

Supplementary Materials: The following supporting information can be downloaded at: https://www.mdpi.com/article/10.3390/nano12101641/s1, The spectra data, ^{1}H and ^{13}C NMR for all products (**a–q**), FT-IR ^{19}F NMR and HRMS for new product (**i**). References [48–61] are cited in the Supplementary Materials.

Author Contributions: Investigation, data curation, formal analysis, methodology, visualization, writing—original draft, Z.X.; investigation, methodology, L.X., X.F., D.L. and L.M.; conceptualization, supervision, methodology, resources, funding acquisition, project administration, writing—review and editing, G.N. and Y.L. All authors have read and agreed to the published version of the manuscript.

Funding: This work was financially supported by The Natural Science Foundation of Guangdong Province (No. 2020A1515010399), Guangxi Innovation Driven Development Major Project (No. Guike AA20302013), Nanning Scientific Research and Technology Development Plan Project (No. RC20200001), "Yongjiang Plan" Project of Leading Talents of Innovation and Entrepreneurship in Nanning City (No. 2020024), and Yulin City Science and Technology Transformation Project (No. 19040003).

Data Availability Statement: Data are within the article and Supplementary Materials.

Conflicts of Interest: The authors declare no conflict of interest.

References

1. Hariprasad, E.; Radhakrishnan, T.P. A highly efficient and extensively reusable "dip catalyst" based on a silver-nanoparticle-embedded polymer thin film. *Chem.-Eur. J.* **2010**, *16*, 14378–14384. [CrossRef] [PubMed]
2. Madhuri, U.D.; Saha, J.; Radhakrishnan, T.P. 'Dip Catalysts' based on polymer-metal nanocomposite thin films: Combining soft-chemical fabrication with efficient application and monitoring. *ChemNanoMat* **2018**, *4*, 1191–1201. [CrossRef]
3. Shaikh, M.N.; Aziz, M.A.; Yamani, Z.H. Facile hydrogenation of cinnamaldehyde to cinnamyl ether by employing a highly re-usable "dip-catalyst" containing Pt nanoparticles on a green support. *Catal. Sci. Technol.* **2020**, *10*, 6544–6551. [CrossRef]
4. Kamal, T.; Ahmad, I.; Khan, S.B.; Asiri, A.M. Anionic polysaccharide stabilized nickel nanoparticles-coated bacterial cellulose as a highly efficient dip-catalyst for pollutants reduction. *React. Funct. Polym.* **2019**, *145*, 104395. [CrossRef]
5. Zheng, G.; Polavarapu, L.; Liz-Marzan, L.M.; Pastoriza-Santos, I.; Perez-Juste, J. Gold nanoparticle-loaded filter paper: A recyclable dip-catalyst for real-time reaction monitoring by surface enhanced Raman scattering. *Chem. Commun.* **2015**, *51*, 4572–4575. [CrossRef]
6. Zheng, G.; Kaefer, K.; Mourdikoudis, S.; Polavarapu, L.; Vaz, B.; Cartmell, S.E.; Bouleghlimat, A.; Buurma, N.J.; Yate, L.; de Lera, A.R.; et al. Palladium nanoparticle-loaded cellulose paper: A highly efficient, robust, and recyclable self-assembled composite catalytic system. *J. Phys. Chem. Lett.* **2015**, *6*, 230–238. [CrossRef]
7. Nishikata, T.; Tsutsumi, H.; Gao, L.; Kojima, K.; Chikama, K.; Nagashima, H. Adhesive catalyst immobilization of palladium nanoparticles on cotton and filter paper: Applications to reusable catalysts for sequential catalytic reactions. *Adv. Synth. Catal.* **2014**, *356*, 951–960. [CrossRef]
8. Madhuri, U.D.; Rao, V.K.; Hariprasad, E.; Radhakrishnan, T.P. In situ fabricated platinum—Poly(vinyl alcohol) nanocomposite thin film: A highly reusable 'dip catalyst' for hydrogenation. *Mater. Res. Express* **2016**, *3*, 045018. [CrossRef]
9. Hariprasad, E.; Radhakrishnan, T.P. Palladium nanoparticle-embedded polymer thin film "dip catalyst" for Suzuki–Miyaura reaction. *ACS Catal.* **2012**, *2*, 1179–1186. [CrossRef]
10. Liang, M.; Zhang, G.; Feng, Y.; Li, R.; Hou, P.; Zhang, J.; Wang, J. Facile synthesis of silver nanoparticles on amino-modified cellulose paper and their catalytic properties. *J. Mater. Sci.* **2017**, *53*, 1568–1579. [CrossRef]
11. Feiz, E.; Mahyari, M.; Ghaieni, H.R.; Tavangar, S. Copper on chitosan-modified cellulose filter paper as an efficient dip catalyst for ATRP of MMA. *Sci. Rep.* **2021**, *11*, 8257. [CrossRef] [PubMed]
12. Ahmad, I.; Kamal, T.; Khan, S.B.; Asiri, A.M. An efficient and easily retrievable dip catalyst based on silver nanoparticles/chitosan-coated cellulose filter paper. *Cellulose* **2016**, *23*, 3577–3588. [CrossRef]
13. Hemming, D.; Fritzemeier, R.; Westcott, S.A.; Santos, W.L.; Steel, P.G. Copper-boryl mediated organic synthesis. *Chem. Soc. Rev.* **2018**, *47*, 7477–7494. [CrossRef] [PubMed]
14. Lakshmidevi, J.; Naidu, B.R.; Venkateswarlu, K. CuI in biorenewable basic medium: Three novel and low E-factor Suzuki-Miyaura cross-coupling reactions. *Mol. Catal.* **2022**, *522*, 112237. [CrossRef]
15. Venkateswarlu, K. Ashes from organic waste as reagents in synthetic chemistry: A review. *Environ. Chem. Lett.* **2021**, *19*, 3887–3950. [CrossRef]
16. Gan, Z.; Yan, Q.; Li, G.; Li, Q.; Dou, X.; Li, G.Y.; Yang, D. Copper-catalyzed domino synthesis of sulfur-containing heterocycles using carbon disulfide as a building block. *Adv. Synth. Catal.* **2019**, *361*, 4558–4567. [CrossRef]
17. Wu, X.; Ma, P.; Wang, J. Copper-catalyzed direct synthesis of arylated 8-aminoquinolines through chelation assistance. *Appl. Organomet. Chem.* **2022**, *36*, e6578. [CrossRef]
18. Volkova, Y.; Baranin, S.; Zavarzin, I. A^3 coupling reaction in the synthesis of heterocyclic compounds. *Adv. Synth. Catal.* **2020**, *363*, 40–61. [CrossRef]
19. Nasrollahzadeh, M.; Sajjadi, M.; Ghorbannezhad, F.; Sajadi, S.M. A review on recent advances in the application of nanocatalysts in A^3 coupling reactions. *Chem. Rec.* **2018**, *18*, 1409–1473. [CrossRef]
20. Peshkov, V.A.; Pereshivko, O.P.; Van der Eycken, E.V. A walk around the A^3-coupling. *Chem. Soc. Rev.* **2012**, *41*, 3790–3807. [CrossRef]
21. Zhang, F.; Fan, J.B.; Wang, S. Interfacial polymerization: From chemistry to functional materials. *Angew. Chem. Int. Ed. Engl.* **2020**, *59*, 21840–21856. [CrossRef] [PubMed]
22. Chai, G.Y.; Krantz, W.B. Formation and characterization of polyamide membranes via interfacial polymerization. *J. Membrane Sci.* **1994**, *93*, 175–192. [CrossRef]
23. Eskandari, E.; Kosari, M.; Farahani, M.H.D.A.; Khiavi, N.D.; Saeedikhani, M.; Katal, R.; Zarinejad, M. A review on polyaniline-based materials applications in heavy metals removal and catalytic processes. *Sep. Purif. Technol.* **2020**, *231*, 115901. [CrossRef]
24. Song, E.; Choi, J.-W. Conducting polyaniline nanowire and its applications in chemiresistive sensing. *Nanomaterials* **2013**, *3*, 498–523. [CrossRef]
25. Yu, L.; Han, Z.; Ding, Y. Gram-scale preparation of Pd@PANI: A practical catalyst reagent for copper-free and ligand-free sonogashira couplings. *Org. Process. Res. Dev.* **2016**, *20*, 2124–2129. [CrossRef]
26. Yu, L.; Han, Z. Palladium nanoparticles on polyaniline (Pd@PANI): A practical catalyst for Suzuki cross-couplings. *Mater. Lett.* **2016**, *184*, 312–314. [CrossRef]
27. Shi, B.; Zhao, C.; Ji, Y.; Shi, J.; Yang, H. Promotion effect of PANI on Fe-PANI/Zeolite as an active and recyclable Fenton-like catalyst under near-neutral condition. *Appl. Surf. Sci.* **2020**, *508*, 145298. [CrossRef]

28. Wang, X.; Shen, Y.; Xie, A.; Chen, S. One-step synthesis of Ag@PANI nanocomposites and their application to detection of mercury. *Mater. Chem. Phys.* **2013**, *140*, 487–492. [CrossRef]
29. Ahmed, F.; Kumar, S.; Arshi, N.; Anwar, M.S.; Su-Yeon, L.; Kil, G.-S.; Park, D.W.; Koo, B.H.; Lee, C.G. Preparation and characterizations of polyaniline (PANI)/ZnO nanocomposites film using solution casting method. *Thin Solid Films* **2011**, *519*, 8375–8378. [CrossRef]
30. Lin, Z.; Cao, N.; Sun, Z.; Li, W.; Sun, Y.; Zhang, H.; Pang, J.; Jiang, Z. Based on confined polymerization: In situ synthesis of PANI/PEEK composite film in one-step. *Adv. Sci.* **2022**, *9*, e2103706. [CrossRef]
31. Heinze, T. New ionic polymers by cellulose functionalization. *Macromol. Chem. Phys.* **1998**, *199*, 2341–2364. [CrossRef]
32. Xu, Z.; Xu, J.; Li, Y. $CuSO_4$ nanoparticles loaded on carboxymethylcellulose/polyaniline composites: A highly efficient catalyst with enhanced catalytic activity in the synthesis of propargylamines, benzofurans, and 1,2,3-triazoles. *Appl. Organomet. Chem.* **2021**, *35*, e6349. [CrossRef]
33. Raaijmakers, M.J.T.; Benes, N.E. Current trends in interfacial polymerization chemistry. *Prog. Polym. Sci.* **2016**, *63*, 86–142. [CrossRef]
34. Song, Y.; Fan, J.-B.; Wang, S. Recent progress in interfacial polymerization. *Mater. Chem. Front.* **2017**, *1*, 1028–1040. [CrossRef]
35. Megha, R.; Ravikiran, Y.T.; Kotresh, S.; Vijaya Kumari, S.C.; Raj Prakash, H.G.; Thomas, S. Carboxymethyl cellulose: An efficient material in enhancing alternating current conductivity of HCl doped polyaniline. *Cellulose* **2017**, *25*, 1147–1158. [CrossRef]
36. Fathi Achachlouei, B.; Zahedi, Y. Fabrication and characterization of CMC-based nanocomposites reinforced with sodium montmorillonite and TiO_2 nanomaterials. *Carbohydr. Polym.* **2018**, *199*, 415–425. [CrossRef]
37. Xiao, J.; Lu, Z.; Li, Y. Carboxymethylcellulose-supported palladium nanoparticles generated in situ from palladium(II) carboxymethylcellulose: An efficient and reusable catalyst for Suzuki–Miyaura and Mizoroki–Heck reactions. *Ind. Eng. Chem. Res.* **2015**, *54*, 790–797. [CrossRef]
38. Zhang, L.; Liu, P.; Su, Z. Preparation of PANI–TiO_2 nanocomposites and their solid-phase photocatalytic degradation. *Polym. Degrad. Stabil.* **2006**, *91*, 2213–2219. [CrossRef]
39. Poulston, S.; Parlett, P.M.; Stone, P.; Bowker, M. Surface oxidation and reduction of CuO and Cu_2O studied using XPS and XAES. *Surf. Interface Anal.* **1996**, *24*, 811–820. [CrossRef]
40. Liu, S.; Zhong, H.; Liu, G.; Xu, Z. Cu(I)/Cu(II) mixed-valence surface complexes of S-[(2-hydroxyamino)-2-oxoethyl]-N,N-dibutyldithiocarbamate: Hydrophobic mechanism to malachite flotation. *J. Colloid Interface Sci.* **2018**, *512*, 701–712. [CrossRef]
41. Liang, T.; Sun, G.; Cao, L.; Li, J.; Wang, L. A pH and NH_3 sensing intelligent film based on artemisia sphaerocephala krasch. gum and red cabbage anthocyanins anchored by carboxymethyl cellulose sodium added as a host complex. *Food Hydrocolloid.* **2019**, *87*, 858–868. [CrossRef]
42. Kai, W. Electrodeposition synthesis of PANI/MnO_2/graphene composite materials and its electrochemical performance. *Int. J. Electrochem. Sc.* **2017**, *12*, 8306–8314. [CrossRef]
43. Buron, C.C.; Lakard, B.; Monnin, A.F.; Moutarlier, V.; Lakard, S. Elaboration and characterization of polyaniline films electrodeposited on tin oxides. *Synth. Met.* **2011**, *161*, 2162–2169. [CrossRef]
44. Lee, S.H. The characteristics of Cu_2O thin films deposited using RF-magnetron sputtering method with nitrogen-ambient. *ETRI J.* **2013**, *35*, 1156–1159. [CrossRef]
45. Morgan, P.; Partin, D.; Chamberland, B.; O'Keeffe, M. Synthesis of paramelaconite: Cu_4O_3. *J. Solid State Chem.* **1996**, *121*, 33–37. [CrossRef]
46. Calegari, F.; da Silva, B.C.; Tedim, J.; Ferreira, M.G.S.; Berton, M.A.C.; Marino, C.E.B. Benzotriazole encapsulation in spray-dried carboxymethylcellulose microspheres for active corrosion protection of carbon steel. *Prog. Org. Coat.* **2020**, *138*, 105329. [CrossRef]
47. Kotal, M.; Thakur, A.K.; Bhowmick, A.K. Polyaniline-carbon nanofiber composite by a chemical grafting approach and its supercapacitor application. *ACS Appl. Mater. Interfaces* **2013**, *5*, 8374–8386. [CrossRef]
48. Kidwai, M.; Jahan, A. Nafion® NR50 catalyzed A^3-coupling for the synthesis of propargylamines via C-H activation. *J. Iran. Chem. Soc.* **2011**, *8*, 462–469. [CrossRef]
49. Samai, S.; Nandi, G.C.; Singh, M.S. An efficient and facile one-pot synthesis of propargylamines by three-component coupling of aldehydes, amines, and alkynes via C–H activation catalyzed by $NiCl_2$. *Tetrahedron Lett.* **2010**, *51*, 5555–5558. [CrossRef]
50. Zhu, W.; Qian, W.; Zhang, Y. Synthesis of 1, 3-diaryl-3-aminopropynes via the dethiolation of thioamides promoted by the samarium/samarium diiodide mixed reagent. *J. Chem. Res.* **2005**, *2005*, 410–412. [CrossRef]
51. Ren, G.; Zhang, J.; Duan, Z.; Cui, M.; Wu, Y. A simple and economic synthesis of propargylamines by CuI-catalyzed three-component coupling reaction with succinic acid as additive. *Aust. J. Chem.* **2009**, *62*, 75–81. [CrossRef]
52. Namitharan, K.; Pitchumani, K. Nickel-catalyzed solvent-free three-component coupling of aldehyde, alkyne and amine. *Eur. J. Org. Chem.* **2010**, *2010*, 411–415. [CrossRef]
53. Fodor, A.; Kiss, A.; Debreczeni, N.; Hell, Z.; Gresits, I. A simple method for the preparation of propargylamines using molecular sieve modified with copper(II). *Org. Biomol. Chem.* **2010**, *8*, 4575–4581. [CrossRef] [PubMed]
54. Samanta, S.; Hajra, A. Divergent synthesis of allenylsulfonamide and enaminonesulfonamide via In(III)-catalyzed couplings of propargylamine and N-fluorobenzenesulfonimide. *J. Org. Chem.* **2018**, *83*, 13157–13265. [CrossRef] [PubMed]
55. Zhang, Q.; Chen, J.-X.; Gao, W.-X.; Ding, J.-C.; Wu, H.-Y. Copper-catalyzed one-pot synthesis of propargylamines via C-H activation in PEG. *Appl. Organomet. Chem.* **2010**, *24*, 809–812. [CrossRef]

56. Sun, R.; Liu, J.; Yang, S.; Chen, M.; Sun, N.; Chen, H.; Xie, X.; You, X.; Li, S.; Liu, Y. Cp_2TiCl_2-catalyzed cis-hydroalumination of propargylic amines with Red-Al: Stereoselective synthesis of Z-configured allylic amines. *Chem. Commun.* **2015**, *51*, 6426–6429. [CrossRef]
57. Munshi, A.M.; Agarwal, V.; Ho, D.; Raston, C.L.; Saunders, M.; Smith, N.M.; Iyer, K.S. Magnetically directed assembly of nanocrystals for catalytic control of a three-component coupling reaction. *Cryst. Growth Des.* **2016**, *16*, 4773–4776. [CrossRef]
58. Wang, L.; Cai, C. Reusable polymer-anchored amino acid copper complex for the synthesis of propargylamines. *J. Chem. Res.* **2008**, *2008*, 538–541. [CrossRef]
59. Zhou, Y.; He, T.; Wang, Z. Nanoparticles of silver oxide immobilized on different templates: Highly efficient catalysts for three-component coupling of aldehyde-amine-alkyne. *Arkivoc* **2008**, *8*, 80–90. [CrossRef]
60. Han, L.; Li, S.J.; Zhang, X.T.; Tian, S.K. Aromatic aza-claisen rearrangement of arylpropargylammonium salts generated in situ from arynes and tertiary propargylamines. *Chem.-Eur. J.* **2021**, *27*, 3091–3097. [CrossRef]
61. Idzik, K.; Cabaj, J.; Sooducho, J.; Abdel-Fattah, A.A. Classical benzotriazole-mediated α-aminoalkylations of alkynes: Synthesis and characterization of alk-2-yn-1-amines as amphiphilic materials. *Helv. Chim. Acta* **2010**, *90*, 1672–1680. [CrossRef]

MDPI
St. Alban-Anlage 66
4052 Basel
Switzerland
Tel. +41 61 683 77 34
Fax +41 61 302 89 18
www.mdpi.com

Nanomaterials Editorial Office
E-mail: nanomaterials@mdpi.com
www.mdpi.com/journal/nanomaterials

www.ingramcontent.com/pod-product-compliance
Lightning Source LLC
LaVergne TN
LVHW070446100526
838202LV00014B/1678